Lecture Notes in Computer Science 15277

The series Lecture Notes in Computer Science (LNCS), including its subseries Lecture Notes in Artificial Intelligence (LNAI) and Lecture Notes in Bioinformatics (LNBI), has established itself as a medium for the publication of new developments in computer science and information technology research, teaching, and education.

LNCS enjoys close cooperation with the computer science R & D community, the series counts many renowned academics among its volume editors and paper authors, and collaborates with prestigious societies. Its mission is to serve this international community by providing an invaluable service, mainly focused on the publication of conference and workshop proceedings and postproceedings. LNCS commenced publication in 1973.

Luís Correia · Aiala Rosá · Francisco Garijo
Editors

Advances in Artificial Intelligence – IBERAMIA 2024

18th Ibero-American Conference on AI
Montevideo, Uruguay, November 13–15, 2024
Proceedings

 Springer

Editors
Luís Correia (ID)
Universidade de Lisboa
Lisbon, Portugal

Aiala Rosá
Universidad de la República
Montevideo, Uruguay

Francisco Garijo
Universidad Complutense de Madrid
Madrid, Spain

ISSN 0302-9743   ISSN 1611-3349 (electronic)
Lecture Notes in Computer Science
ISBN 978-3-031-80365-9   ISBN 978-3-031-80366-6 (eBook)
https://doi.org/10.1007/978-3-031-80366-6

This Springer imprint is published by the registered company Springer Nature Switzerland AG
The registered company address is: Gewerbestrasse 11, 6330 Cham, Switzerland

If disposing of this product, please recycle the paper.

Preface

We are glad to introduce the Proceedings of the 18th Ibero-American Conference on Artificial Intelligence (IBERAMIA 2024). This corresponds to the latest edition of a recognised international conference taking place every other year (except for the Covid pandemic period) since 1988, organised by the Iberamia Association and researchers from Ibero-American countries, bringing together researchers in artificial intelligence (AI) from all over the world. This series of conferences has been continuously supported by the main Ibero-American AI societies.

IBERAMIA 2024 took place in Montevideo, Uruguay for the first time, on 13–15 November, organized by the Facultad de Ingeniería de la Universidad de la República. This resulted from a drive by the Iberamia Association and Ibero-American AI societies to increase the coverage of Ibero-American countries and foster research in AI in all of them.

As the result of a long-lasting collaboration between IBERAMIA conferences and Springer, these proceedings are being published in the Springer Lecture Notes in Computer Science series. We followed the highest standards in the paper selection process. Namely every submitted paper was reviewed by at least three reviewers in a double-blind review process. The Program Committee (PC) was formed by reputed researchers from all over the world in different topics of AI working both in academia and industry. In this edition we received a total of 96 papers including those from a specific call for late-breaking papers. 36 articles were accepted as full papers and 9 as short papers, plus 5 late-breaking papers, which are all collected and published in these proceedings.

Following the IBERAMIA 2024 call for papers we organised the proceedings following the corresponding scientific topic of each accepted paper. We notice a dominance of AI Engineering and Applications, and of Machine Learning (ML), reflecting well the importance of these topics as AI applications grow steadily in almost every area of human intervention with ML playing an important role in this. Other topics of the published papers include Bio-inspired & Soft Computing models, Computer Vision & Robotics, Knowledge Representation and Reasoning, Multi-agent Systems, Natural Language Processing, and Social AI, also highlighting the relevance of AI in human society.

We would like to express our gratitude to the ones that helped to make IBERAMIA 2024 happen. Namely we thank all the authors for sharing their research with us in this event and for the effort in coping with the requirements and deadlines of the production of the proceedings. We also are indebted to the PC members who, with their pro bono work, helped us to guarantee the highest quality of the review process and subsequently of the published papers. We thank Springer for support with the publication of the proceedings and for providing the conference management portal EquinOCS that we used throughout the process, from the paper submission to the editing of the proceedings. We acknowledge the initiative of the Iberamia Association in challenging us to take on this endeavour and the support of all the Ibero-American AI societies involved. Our final word of appreciation is a warm thank you to the colleagues of the organising committee from the Facultad de Ingeniería of Universidad de la República in Montevideo, whose dedication was what ultimately allowed us to convene in such an attractive venue.

We wish all the participants of IBERAMIA 2024 fruitful work and stimulating scientific discussions, and we hope that all the readers of these proceedings can find new advancements and insights on AI, the scientific area we all cherish so much.

November 2024   Luís Correia
                                                                 Aiala Rosá
                                                              Francisco Garijo

Organization

Program Committee Members

Diana Francisca Adamatti	Universidade Federal do Rio Grande, Brazil
Helena Aidos	Universidade de Lisboa, Portugal
Matias Alvarado-Mentado	Centro de Investigacion y de Estudios Avanzados del IPN, Mexico
Luis Antunes	Universidade de Lisboa, Portugal
Federico Barber	Universidad Politécnica de Valencia, Spain
Ana Bazzan	Universidade Federal do Estado do Rio Grande do Sul, Brazil
Reinaldo A. C. Bianchi	Centro Universitario FEI, Brazil
Ana Cristina Bicharra Garcia	Universidade Federal do Estado do Rio de Janeiro, Brazil
Juan Carlos Burguillo	University of Vigo, Spain
Amílcar Cardoso	Univ. Coimbra, Portugal
Luis Chiruzzo	UdelaR, Uruguay
Luís Correia	Universidade de Lisboa, Portugal
Paulo Cortez	Universidade do Minho, Portugal
Fabio Cozman	EP-USP, Brazil
Luciano Antonio Digiampietri	University of São Paulo, Brazil
Lorena Etcheverry	UdelaR, Uruguay
Ramon Fabregat	Universitat de Girona, Spain
María Inés Fariello	UdelaR, Uruguay
Nuno Garcia	Universidade de Lisboa, Portugal
Francisco Garijo	Universidad Complutense de Madrid, Spain
Daniela Godoy	ISISTAN Research Institute, Argentina
Agustín Gravano	UBA, Argentina
Paulo T. Guerra	Federal University of Ceará, Brazil
Stella Heras	Universitat Politècnica de València, Spain
Alípio Jorge	FC-UP, Portugal
Vicente Julian	Universitat Politècnica de València, Spain
Fernando Koch	IBM Global Services, USA
Henrique Lopes Cardoso	FE-UP, Portugal
José Machado	University of Minho, Portugal
Ricardo Marcondes Marcacini	ICMC/USP, Brazil
José Fco. Martínez-Trinidad	Instituto Nacional de Astrofísica, Óptica y Electrónica, Mexico

Nestor Dario Mendez Duque	Universidad Nacional de Colombia, Colombia
Ivan Vladimir Meza Ruiz	Universidad Nacional Autónoma de México, Mexico
José Manuel Molina Lopez	Universidad Carlos III de Madrid, Spain
Guillermo Moncecchi	UdelaR, Uruguay
Manuel Montes-Y-Gomez	National Institute of Astrophysics, Optics and Electronics, Mexico
Eduardo Morales	Instituto Nacional de Astrofísica, Optica y Electrónica, Mexico
Paulo Novais	Universidade do Minho, Portugal
Pedro Nuñez	Universidad de Extremadura, Spain
José Luis Oliveira	University of Aveiro, Portugal
Aline Paes	UFF, Brazil
Thiago Pardo	University of São Paulo, Brazil
João Paulo Papa	UNESP, Brazil
Juan Pavon	Universidad Complutense de Madrid, Spain
Martín Pedemonte	UdelaR, Uruguay
João Carlos Pereira Da Silva	UFRJ, Brazil
Ramon Pino Perez	Universidad de Los Andes, Venezuela
Rui Prada	INESC-ID and Instituto Superior Técnico, Universidade de Lisboa, Portugal
Maurício Reis	Universidade da Madeira, Portugal
Rita Ribeiro	Universidade do Porto, Portugal
Rui P. Rocha	University of Coimbra, Portugal
Ricardo O. Rodriguez	Universidad de Buenos Aires, Argentina
Aiala Rosá	Universidad de la República, Uruguay
Fernando Santos	Universidade do Estado de Santa Catarina, Brazil
Silvia Schiaffino	Instituto Superior de Ingeniería de Software Tandil, (CONICET – UNCPBA), Argentina
Ivan Serina	University of Brescia, Italy
Onn Shehory	Bar-Ilan University, Israel
Guillermo Simari	Universidad Nacional del Sur, Argentina
Elaine Sousa	University of São Paulo – ICMC/USP, Brazil
Luis Enrique Sucar	Instituto Nacional de Astrofísica, Óptica y Electrónica, Mexico, Mexico
Stefano Tedeschi	Università degli Studi di Torino, Italy
Flavio Tonidandel	Centro Universitario da FEI, Brazil
Flavio Varejao	Universidade Federal do Espírito Santo, Brazil
José Viterbo	Universidade Federal Fluminense, Brazil
Leandro Krug Wives	UFRGS, Brazil

Contents

AI Engineering and Applications

Adjusting Convolution Blocks of U-Net to Improve Sugarcane Crop Line
Segmentation 3
   João Batista Ribeiro and André Ricardo Backes

Mapping Motion: A Cognitive Approach to Dyspraxia Multimodal
Analysis 14
   David Muvdi, Mateo R. Hernández, Eiker J. Cifuentes,
   Andrea Di Somma, Francesco Chiappone, Vicenzo Di Maro,
   Juan Carlos Martínez-Santos, and Edwin Puertas

Improving Assembly Lines with Digital Twins: A Synthetic Case Study 26
   Maria Gabriela Juarez Juarez, Adriana Giret, and Vicente Botti

Peak Ground Acceleration Prediction for Earthquake Early Warning
with Multivariable Long Short-Term Memory Networks and Temporal
Transformers 38
   Yhon Fuentes, Yessenia Yari, Aurea Soriano-Vargas, and Anderson Rocha

Green Security Along Trails 50
   Nicolas Betancourt and Mauricio Velasco

Identification of Dating Violence with Machine Learning Algorithms:
Analysis and Results 62
   Mariana-Carolyn Cruz-Mendoza, Juana Canul-Reich,
   Roberto Ángel Meléndez-Armenta, and Carolina Malvaez-Hernández

Capturing the Essence of Plankton: A Gradient-Weighted Class Activation
Mapping Analysis 74
   Sofía Callejas, Hernan Lira, Andrew Berry, Luis Martí,
   and Nayat Sanchez-Pi

Agent-Based Model and Machine Learning for the Analysis of Barter
Regulated by Lotka-Volterra Equations in an Artificial Society 87
   Juan Villacrés

A Geometric Attention Mechanism to Classify Parkinsonism Smooth
Pursuit Patterns 99
   Luis Fernando Celis, Juan Olmos, and Fabio Martínez

A Mixed Audio-Video SPD Network for Online Classification
of Parkinsonian Speech Patterns 110
John Archila, Antoine Manzanera, and Fabio Martínez

Transformers for Genomic Prediction 122
*María Inés Fariello, Graciana Castro, Romina Hoffman,
Mateo Musitelli, Diego Belzarena, and Federico Lecumberry*

AI-Assisted Bronchoscopy in the Intensive Care Unit: Corpus
Development and an Application to Anatomic Position Identification 132
*Luciano Tarsia, Nicolas Mastropasqua, Indalecio Carboni Bisso,
Marcos Las Heras, Valeria Burgos, Marcelo Risk,
María Florencia Courtois, Ignacio Fernández Ceballos,
Carolina Lockhart, Daniel Acevedo, and Viviana Cotik*

Bio-inspired and Soft Computing

Evaluation of Mares Uterine Health Based on Endometrial Biopsies Using
Image Processing and Machine Learning Techniques 147
*Sofía Zimmer, Agustina Díaz, Nicolás Aguilera, María José Estradé,
Federico Lecumberry, and Pablo Musé*

Computer Vision and Robotics

X-COVNet: Externally Validated Model for Computer-Aided Diagnosis
of Pneumonia-Like Lung Diseases in Chest X-Rays 161
*Jorge Felix Martínez Pazos, Arturo Orellana García,
David Batard Lorenzo, and Jorge Gulín González*

FairTrees: A Deep Learning Approach for Identifying Deforestation
on Satellite Images 173
Hernan Lira, Taco de Wolff, Luis Martí, and Nayat Sanchez-Pi

Ensembling Convolutional Neural Networks for Human Skin Segmentation 185
Patryk Kuban and Michal Kawulok

Risk Assessment for UAV Autonomous Landing in Urban Environments
Using Semantic Segmentation 197
*Jesús Alejandro Loera-Ponce, Diego A. Mercado-Ravell,
Israel Becerra, and Luis Manuel Valentin-Coronado*

Visual SLAM in Underground Environments: Preliminary Results 209
Bliman Federico, Monzon Pablo, and Llofriu Martin

Sound-Based Parakeets Detection System 221
    Ernesto Rován, Pablo Monzón, and Facundo Benavides

Knowledge Representation and Reasoning

AI-Based Medical Education: Coping with Clinical Decisions
in GLARE-Edu 237
    Alessio Bottrighi, Antonio Maconi, Stefano Nera, Luca Piovesan,
    Erica Raina, and Paolo Terenziani

Generating Contrastive Explanations from Gradual Semantics Rankings 250
    Mariela Morveli-Espinoza and Juan Carlos Nieves

Automatic Classification of Secondary and High School Students Dropout
Risk via Knowledge Graphs and Machine Learning 262
    Daniel Zapata-Medina, Albeiro Espinosa-Bedoya,
    and Jovani Alberto Jiménez-Builes

Machine Learning

Semi-supervised Hierarchical Bayesian Multi-label Classification 275
    Jonathan Serrano-Pérez and L. Enrique Sucar

Advancing Photovoltaic Forecasting with Neural Networks: Integrating
N-Beats and Sequential Models with Fourier Analysis 287
    Gonzalo Surribas-Sayago, Jose David Fernández-Rodríguez,
    and Enrique Dominguez

Efficiency of the Transformer Model in Time Series Forecasting: A Case
Study in Wastewater Treatment Plants 298
    Gonçalo Medeiros, Francisco S. Marcondes, Pedro Oliveira,
    José Machado, and Paulo Novais

Bayesian Regularized Iterative Soft Thresholding Algorithm 310
    Nicolas Cutrona and Dominique Guillot

Performance Evaluation of Data Analysis Techniques in Dry Bean Seed
Classification Using kNN and MLP 323
    Victor Hugo Schneider Lopes, Alessandro Bof de Oliveira,
    Patricia Bof, and Dante Augusto Couto Barone

Integrating Convolutional Neural Networks and Omics to Promote
Precision Medicine in Atopic Dermatitis 335
    Ana Duarte and Orlando Belo

Multi Agent Systems

Reliability Analysis of Organization-Based Multiagent System Designs 347
Juan C. García-Ojeda

Emotions Identification in Exchanges of Messages Between Agents 360
Thiago Dantas, Giovani Farias, Cleo Billa, Eder Gonçalves,
and Diana Adamatti

Explaining Task Delegation Through Argumentation Debates with Votes 372
Jeferson José Baqueta and Cesar A. Tacla

Natural Language Processing

Te Ahorré Un Click: A Revised Definition of Clickbait and Detection
in Spanish News 387
Gabriel Mordecki, Guillermo Moncecchi, and Javier Couto

Entrainment-Metrics: An Open-Source Toolkit for Quantifying
Acoustic-Prosodic Entrainment in Spoken Dialogue 400
Erik Ernst, Ramiro H. Gálvez, and Agustín Gravano

Derivation Prompting: A Logic-Based Method for Improving
Retrieval-Augmented Generation 412
Ignacio Sastre, Guillermo Moncecchi, and Aiala Rosá

Information Extraction from Electronic Health Records Written in Spanish
for Epidemic Intelligence 424
Javier Petri, Pilar Barcena Barbeira, and Viviana Cotik

Social AI

From AI Act to Public-Private-People Partnerships: Building AI
as a Global Public Good 439
Migle Laukyte

Posters

Optimization of Generalized Assignment Problem for a Machinery-Aided
Composting Process 453
Yael Andrade-Ibarra, Uriel Trejo-Ramirez, Oliver Cuate, Adriana Lara,
and Lourdes Uribe

Configuring an LLM Chatbot as Practice Partner for Language Learning 458
Pablo Gervás, Carlos León, Mayuresh Kumar, Gonzalo Méndez,
and Susana Bautista

Personalizing Learning with Intelligent Tutoring Systems: Leveraging
GenAI and Predictive AI for Adaptive Education 463
Juan C. Zuluaga-Morillo, Eduardo J. Tous-De la Ossa,
Álvaro J. Giraldo-Cadavid, Carlos M. González-McMahon,
Gustavo A. Moreno-López, Néstor D. Duque-Méndez,
Jaime A. Restrepo-Carmona, and Jovani A. Jiménez-Builes

Improving Efficiency of QBF Planning with Mixed Linear Compact Tree
Encodings 468
Frédéric Maris

Adversarial Attacks in a Shallow Neural Network to Classify Android
Malware 472
Leidy Marcela Aldana and Jorge E. Camargo

Harnessing Deep Learning for Detection of Violence and Vandalism 476
Tiago Ribeiro, Juan Pavón, José Machado, Paulo Novais,
and Manuel Rodrigues

Magic Matching: A Virtual Assistant for Financial Trading Using Machine
Learning 481
Melisa Arena, Belén Olivera, Santiago Pérez, Paola Romay,
and Victor Sabbia

Design of a Machine Learning Algorithm for Precipitation Prediction
in Riobamba Using Neural Networks 486
Gladys Urquizo, Cristina Ramos, Juan Villacr, and Alex Pozo

Comparative Evaluation of Algorithms for Adaptive Learning Analytics:
A Systematic Review of Higher Education Applications 490
Pablo Andres Quijano-Cabezas, Néstor Duque-Méndez,
and Jovani Alberto Jiménez-Builes

Audio-Based Violence Detection Using Spectrograms and Deep Learning 496
Bruno Campos, Carlos Rodríguez-Domínguez, Miguel J. Hornos,
and Manuel Rodrigues

Capturing Collective Memories of the Disappeared with Artificial
Intelligence 501
Tomas Laurenzo

Using Deep Learning Models in Clinical Histories in Order to Find
Precursors of Diseases 505
    Juli Climent Querol, Gonzalo Hernández Ortega,
    and Ernest Valveny Llobet

Evaluating State-of-the-Art Extractive Summarization Methods
for Brazilian Portuguese 510
    Germano Antonio Zani Jorge, Davi Alves Bezerra, Clarissa Castell,
    and Thiago Alexandre Salgueiro Pardo

Assessing AI's Persuasive Power: Can an AI Agent Influence Belief
in News? 515
    Jean Gabriel Nguema Ngomo and Ana Cristina Bicharra Garcia

Author Index 521

AI Engineering and Applications

Adjusting Convolution Blocks of U-Net to Improve Sugarcane Crop Line Segmentation

João Batista Ribeiro[1] and André Ricardo Backes[2]([✉])

[1] School of Computer Science, Federal University of Uberlândia, Uberlândia, Brazil
[2] Department of Computing, Federal University of São Carlos, São Carlos, Brazil
arbackes@yahoo.com.br

Abstract. In recent years, using Unmanned Aerial Vehicles (UAV) has enabled the development of many precision agriculture applications. Among these applications is the location of crop lines at low or medium altitude imagery, which allow us to estimate many attributes of a crop, such as a crop yield, number of plants, and failures in the sowing process. Since crop lines always show an almost constant appearance (greenish plants against reddish soil), this work proposes to evaluate the U-Net under different configurations of the numbers of filters and convolutional blocks. We also tested how different training sets affect its training. Results show that U-Net is a feasible approach to segment crop lines using fewer blocks and filters than traditional U-Net, being the first more important than the latter.

Keywords: Image Registration · Image Segmentation · Deep Learning · Precision Agriculture · UAV

1 Introduction

Precision Agriculture (PA) seeks to meet future demand for food with the smart use of natural resources. This area includes a vast number of technics, such as spatial analysis of the planted area, and soil and plant information. The information used to feed these technics comes from several sources, mainly images taken by cameras attached to Unmanned Aerial vehicles (UAV) and satellites [3,13]. This allows producers to plan and monitor their plantations, and it helps the decision-making process in the many stages of cropping.

The UAV has facilitated the availability of detailed world observation data. They made possible images with Ground Sampling Distance (GSD) in centimeters and with good quality. It is also possible to acquire new images when it is needed, different from satellites and other ways. The many images of one single UAV flight can be combined, through a mosaicking process, resulting in a big image called mosaic [6,12].

After capturing these images, many approaches are useful to obtain the desired information. We can divide the image into various regions, a process

L. Correia et al. (Eds.): IBERAMIA 2024, LNCS 15277, pp. 3–13, 2025.
https://doi.org/10.1007/978-3-031-80366-6_1

called image segmentation, to separate crop from weed, for example [4]. A sub-type of segmentation, called semantic segmentation, is used in many PA applications. In this approach, all image pixels are labeled according to a set of objects categories (e.g., tree, weed, crop line). Numerous algorithms were developed (e.g., thresholding, region-growing, Watershed) for image and semantic segmentation. Recent advances in the Deep Learning methods have improved the performance in segmentation tasks, often achieving the highest accuracy in various scenarios [5, 20].

Many fields use deep learning (DL) techniques, e.g., analysis of medical and agricultural images. Among these fields, image segmentation is one of the most challenging problems, and many networks architectures were developed to accomplish this task. The U-Net is one of the most known and used segmentation network. Because of his success, many variants/modifications have been proposed [5, 18].

One of the main applications of PA is crop row detection. It is a crucial step for other applications (e.g., weed detection, crop production estimation, detection of fault in the culture) [10]. Also, autonomous vehicles use this information to guide themselves in the plantation, thus avoiding trampling the crop, and it helps the selective spraying and mechanical weed removal [7].

This paper addresses the problem of crop row segmentation in UAV images. We evaluated the U-Net under different configurations of the numbers of filters and convolutional blocks and how it impacts the segmentation results. The remaining paper is structured as follows. In Sect. 2, we present some recent works related to our proposal. In Sect. 3 we present an introduction to Deep Learning and the U-Net architecture with details. We describe our experiments, the dataset, and the metric used for evaluation in Sect. 4. In Sect. 5 we present and discuss the results of our experiments. Finally, Sect. 6 concludes this paper.

2   Related Work

In [1], the authors investigated the use of transfer learning to improve U-Net to ultrasound image segmentation. Among the strategies tested, two have provided the best results: fine-tuning the whole network and all network except for the last block of the contract path. Also, the authors claim that the common practice of fine-tuning the final layers in transfer learning from one domain to another does not work well for their application of the U-Net, with fine-tuning the firsts layers performing better. Therefore, the U-Net architecture requires distinct transfer learning approaches. [18] made a literature review of U-Net and its variants with a focus on medical image segmentation. Among many works analyzed, they list a vast use of the U-Net and its improvements with most recent strategies developed by the deep learning scientists (e.g., residual, attention, recurrent strategies).

The authors in [8] compared different U-Net models for building extraction from high-resolution aerial imagery. They analyzed the U-Net with the contract path replaced by a Convolutional Neural Network (CNN). They tested three CNNs: VGG16, InceptionResNetV2, and DenseNet121. The models have made above 85% of F1 score, although they acquired the best result by the majority voting method, which combined the characteristics of three U-Net models. Unfortunately, the authors didn't test the traditional U-Net (i.e., without any modification), making it impossible to analyze the results produced only by the U-Net and the proposed improvements.

[5] compared the performance of the U-Net tuning some parameters, such as the activation function, pooling strategies, filter size, dropout, and batch normalization. This work's main goal was brain tumor segmentation. They noted that the use of batch normalization and dropout (implemented individually or in combination) do not enhance the performance of the U-Net. The kernel size 3×3 made the best result. The pooling strategies had made no significant change in the results. On the other hand, the models with activation function Rectified Linear Unit (ReLU) performed better.

Some works use a modified version (i.e., more suitable with the problem to be solved) of the U-Net architecture to segment images to Precision Agriculture, such as [2,7,11,14,21], and most of them use UAV images. [2] proposed a method using the U-Net architecture in a reduced manner (i.e., with lesser number of convolution filters) to classify weed and crop. Their approach achieved better accuracy than the standard U-Net and has lower number of parameters, although the test took place in a dataset with few images. [11] applied the U-Net and searched for best combination of spectral band, and analyzed the use of vegetation's indices to crop classification. They obtained the best result with RGB bands and textural features, although the tests were carried on a very limited dataset.

[14] applied a modified U-Net to detect crop rows on a dataset of images of different cultures taken by the ground. The works focus on supplying the crop rows to autonomous vehicles used in the agricultural field. They claim that the network performed well with the dataset's images, but poorly in a real-world video, where a clustering pixel strategy was used to improve the results. [7] proposed a weed mapping method, that uses a modified U-Net to segment the crops, later used to obtain the weed part by thresholding with vegetation indices. [21] used the U-Net to segment the emergence point of plaints of cereal fields.

Despite many works using the U-Net model, and some analyze its parameters to enhance performance, we didn't find any work that analyzed the number of blocks or the number of filters/kernels in each block. Also, and mainly, we didn't find any work focusing on improving the performance of crop row (for any culture) by tuning these parameters in the U-Net models. This gap and the vast use of the U-Net, motivates our work.

3   Theoretical Background

3.1   Deep Learning

Deep Learning (DL) methods work by learning information and representing the data by the application of a series of semantic layers in sequence, thus generating a useful model. After training, this model can be used to receive input data and return a valid response to a specific problem. DL methods have gained a lot of interest in the last decade, mainly because of the increase in computer power and the more availability of data, thereby becoming the state-of-the-art in many computer vision and image processing problems [15].

DL covers a range of methods, where many types of layers can be used, the CNN is one of the most used. The CNNs have some building blocks (e.g., convolution and transposed convolution, pooling and max-pooling, activation functions) that can be used to make networks for specific problems. CNN employs the mathematical operation convolution with a filter as the main part, given name to the networks [16].

In the convolutional layers, the convolution operation uses different filters (also called kernels) to extract features from the input data useful for the network. After a convolutional layer, it is usual for the network to apply a pooling operation (normally max-pooling, where we select the maximum value from the local area). The number of filters and their size relays on the problem to be solved. The CNNs commonly use filter sizes with width and height of the same size and odd values (e.g., 1, 3, 5, or 7) [20].

3.2 U-Net

The U-Net architecture, initially proposed to segment medical images in 2015, is a "U" shaped network from where it comes its name. It consists of a contraction (left side, also called the encoder) and an expansive path (right side, also called the decoder). The original network has in the contraction path layers of two 3×3 convolutions (unpadded convolutions, defined as "valid") followed by a ReLU and a 2×2 max-pooling with stride 2. Every downsampling doubles the number of feature channels [17].

On the other side, the expansive path have 2×2 transposed convolution (called "up-convolutions") for upsampling, that halves the number of feature channels. Then, it applies a concatenation with the correspondent cropped feature maps from the contraction path and two 3×3 convolutions layers, each followed by ReLU. The final layer has 1×1 convolution to map each 64 feature vectors with the desired number of classes (dimensionality reduction step). Thereby, the network has 23 convolutions layers (also counting the transposed convolutions) as it can be seen in the Fig. 1. This network don't use any other layers, such as dense or flatten.

This network receives as input an image ($572 \times 572 \times 1$) and make an output (the segmentation image) a little small, $388 \times 388 \times 2$, because the convolution used was unpadded. The U-Net have skip connections (line in gray, as show in Fig. 1), used on the concatenation step, and cropped for the same reason cited above. This skip connections are crucial to insert important feature maps from the previous layers [9, 17].

There are many variants and modifications of U-Net architecture due to its success in image segmentation. Primarily for medical image analysis (e.g., brain tumor, breast cancer, blood vessels) [18], and after for many other fields, like

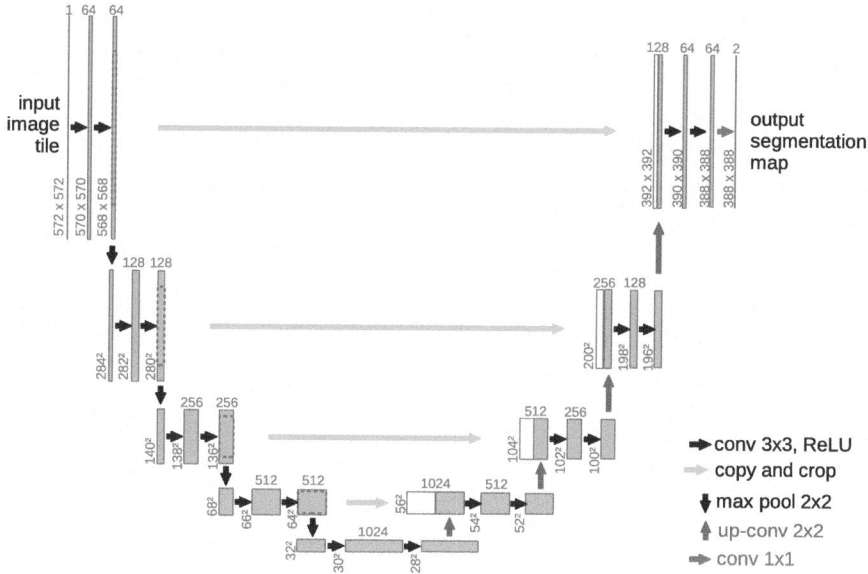

Fig. 1. U-Net architecture. Source: [17]. (Color figure online)

remote sensing [19] and precision agriculture [11]. The vast and numerous use of the U-Net elucidates its importance and motivates our experiments.

4   Proposed Experiments

In this work, we aimed to study the influence of the various convolutional blocks and filters in the U-Net segmentation results. We created 12 configuration (see Table 1) of blocks and filters. The configuration N° 01 (i.e., [16, 32, 64, 128, 256]) is used as standard configuration in the comparison. It has 5 blocks of filters starting with 16 filters and doubling every time, as well as the others configurations. This configuration differs from the standard U-Net, which also has 5 blocks, but starting with 64 filters. We opted to change the number of filters in each block due to the small size of our images. Additionally, a large number of filters may be prohibitive for some computers to handle.

We must emphasize that the U-Net uses the blocks of filters in sequence order on the contraction path and reverse order on the expansive path, although the last block (e.g., 256 in configuration number 1) is not used on the expansive path. Each block of filters is used two time, in the 3 × 3 convolution (blue arrow in the Fig. 1).

The main goal here is to investigate which configuration (i.e., arrangement of the number of filters in each block and the numbers of blocks to use) can lead to a better result on the U-Net and why. Accordingly, we have these research questions:

Table 1. U-Net configurations used in the experiments.

Nº	Nº of blocks and filters	Nº of parameters	Percentage
01	[16, 32, 64, 128, 256]	3, 331, 697	100.00%
02	[16, 32, 64, 128]	823, 921	24.72%
03	[16, 32, 64]	196, 721	5.90%
04	[16, 32]	39, 793	1.19%
05	[16]	2, 785	0.08%
06	[16, 16]	14, 369	0.43%
07	[16, 16, 16]	25, 953	0.77%
08	[16, 16, 16, 16]	37, 537	1.12%
09	[16, 16, 16, 16, 16]	49, 121	1.47%
10	[32]	10, 177	0.30%
11	[64]	38, 785	1.16%
12	[128]	151, 297	4.54%

- What is more important to the network to learn about the planting lines? The number of blocks or the numbers of filters?
- More parameters to train is always the better choice, or a configuration with low number of parameters can create scenarios with less time to train and better accuracy?
- The dataset used to train the network has influence in the network learning? Of how much?

For our experiments, we used four mosaics of aerial images of sugarcane crop areas with crop lines of different ages and width. We acquired these images' mosaics with an eBee SenseFly mapping drone with a SenseFly S.O.D.A. camera of 1 in Sensor, that has 5472×3648 pixels of resolution and RGB lens F/2.8-11, 10.6 mm. The GSD of each pixel is about 5 cm (0.053 m).

The four mosaics can be visualized in the Fig. 2, named Dataset A, B, C, and D, respectively. Each mosaic has been segmented his crop lines by an expert, as some examples can be seen in the Fig. 3. The expert have marked with a uniform line the sugarcane crop lines where the culture is present and where it should exist (also marked regions where the crop line should exist).

We crop each mosaic in slices of 256×256 pixels with no overlapping (i.e., stride of 256 pixels). After the crop, we discard samples with less of 80% of useful information (i.e., pixels with values different of zero). After discarding some samples, the datasets A, B, C, and D contained a total of, respectively, 678, 3291, 1552 and 2162 images.

To compare our segmentation results to the one provided by the expert, we used the Dice coefficient, a measurement of how similar two binary images are. Let A and B binary images, we compute Dice coefficient as follows:

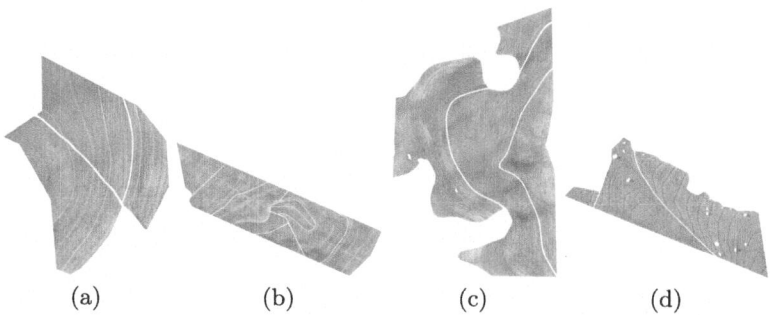

Fig. 2. Mosaics and their sizes: (a) 11180×8449; (b) 19833×30255; (c) 17497×10771; (d) 16677×24181.

Fig. 3. Examples of planting lines and the segmentation provided by an expert.

$$D = 2\frac{|A \cap B|}{|A| + |B|}. \tag{1}$$

This result is the image similarity D, $0 \leq D \leq 1$, where the higher the value D, the more similar are the images A and B.

5   Results and Discussion

For each configuration presented in Table 1 we trained a U-Net to segment crop lines from images. We performed the training for 50 epochs using the Adam optimizer with a 0.001 learning rate and 80% for training and 20% for test and a batch size of 8. We used a personal computer with Intel® Core™ i7-7700 CPU @ 3.60 GHz, 32 GB RAM, 64-bit Windows OS, and GPU NVIDIA GeForce GTX 1050 Ti, 4 GB GDDR5. We also used Python 3.6 and Keras 2.1.6-tf with TensorFlow 1.10.0 and CUDA Toolkit 9.0 to implement train and test the U-Net.

As we can see in Fig. 3, crop rows in each dataset can vary in width due to several factors, such as the age of the crop, and the level of success of the planting process. Thus, we trained each U-Net configuration using each of the four datasets. After training, we made predictions (segmented the sugarcane images) for the whole dataset used in the training step and the others. With the predictions, we calculated the Dice coefficient for each combination.

Test 01: [16, 32, 64, 128, 256]

	A	B	C	D
A	0.96	0.68	0.84	0.73
B	0.90	0.97	0.90	0.84
C	0.90	0.91	0.96	0.86
D	0.87	0.91	0.89	0.94

Test 02: [16, 32, 64, 128]

	A	B	C	D
A	0.94	0.69	0.82	0.70
B	0.88	0.95	0.89	0.80
C	0.87	0.91	0.94	0.79
D	0.84	0.87	0.85	0.92

Test 03: [16, 32, 64]

	A	B	C	D
A	0.92	0.69	0.81	0.69
B	0.64	0.93	0.85	0.65
C	0.61	0.88	0.90	0.49
D	0.83	0.86	0.81	0.88

Test 04: [16, 32]

	A	B	C	D
A	0.86	0.60	0.74	0.65
B	0.54	0.89	0.75	0.57
C	0.65	0.79	0.85	0.57
D	0.78	0.79	0.74	0.81

Test 05: [16]

	A	B	C	D
A	0.69	0.14	0.42	0.59
B	0.27	0.69	0.61	0.24
C	0.50	0.60	0.65	0.41
D	0.64	0.30	0.49	0.63

Test 06: [16, 16]

	A	B	C	D
A	0.84	0.53	0.74	0.63
B	0.55	0.88	0.73	0.56
C	0.69	0.82	0.85	0.61
D	0.77	0.76	0.72	0.80

Test 07: [16, 16, 16]

	A	B	C	D
A	0.90	0.67	0.79	0.69
B	0.69	0.92	0.86	0.67
C	0.70	0.87	0.88	0.52
D	0.77	0.81	0.80	0.85

Test 08: [16, 16, 16, 16]

	A	B	C	D
A	0.92	0.63	0.82	0.71
B	0.84	0.94	0.89	0.76
C	0.84	0.89	0.92	0.70
D	0.85	0.89	0.85	0.89

Test 09: [16, 16, 16, 16, 16]

	A	B	C	D
A	0.93	0.67	0.83	0.71
B	0.91	0.94	0.91	0.83
C	0.91	0.91	0.92	0.80
D	0.84	0.90	0.86	0.90

Test 10: [32]

	A	B	C	D
A	0.75	0.25	0.48	0.62
B	0.29	0.71	0.61	0.28
C	0.51	0.61	0.67	0.43
D	0.65	0.34	0.50	0.64

Test 11: [64]

	A	B	C	D
A	0.76	0.38	0.54	0.63
B	0.36	0.75	0.63	0.37
C	0.54	0.63	0.69	0.45
D	0.62	0.24	0.50	0.63

Test 12: [128]

	A	B	C	D
A	0.74	0.29	0.47	0.59
B	0.26	0.71	0.62	0.25
C	0.49	0.60	0.68	0.40
D	0.69	0.39	0.57	0.68

Fig. 4. Influence of the number of filters and convolutional layers in the crop line segmentation task. For each test, the row identify the dataset used for training while columns are the evaluated datasets. Red shades highlight the deterioration in the Dice coefficient in comparison to the result in the diagonal of Test 01 (in blue). (Color figure online)

Figure 4 shows the average Dice coefficient results for each U-Net configuration present in Table 1, using the datasets A, B, C and D. They are organized by each configuration/test and within a matrix of training and test/prediction, where the row dataset indicates the dataset used for training, while the columns are the tested ones. For example, $A \times B$ in Test 01 means that the U-Net with configuration number one ([16, 32, 64, 128, 256]) was trained in dataset A and tested on dataset B, resulting in an average Dice coefficient of 0.68. The main diagonal represents the training and test performed in the same dataset. Therefore, we expect the main diagonal to present better results in contrast with other combinations.

Firstly, we notice that reducing from 5 to 3 blocks affects little the performance of the U-Net, except for some cases where the training and test sets are different. In general, the average Dice coefficients remain high as we reduced the number of convolutional blocks and, consequently, the number of parameters to only 5.90% of the original parameters (Test 03). Any attempt to further reduce the number of blocks compromises the network performance, especially when training and test occur in a different dataset (Testes 04 and 05).

Any reduction in the number of blocks affects the total number of parameters of U-Net. For example, Test 05 uses a single convolutional block and only 0.08% of the parameters (2,785). The drop in performance could be a result of the reduction in the number of trainable parameters instead of the decrease in the number of convolutional blocks.

To test which hypothesis is the most likely we evaluated some U-Net configurations using a single block and different numbers of filters in it (Tests 10, 11 and 12), and different number of blocks with a constant number of filters in each block (Tests 06, 07, 08 and 09).

Results show that a single convolutional block is not capable to learn the attributes of the image to properly segment it. Contrarily, as we increase the number of blocks, and keep a constant and small number of filters, we obtain results similar to the original U-Net architecture. A single block with 128 filters contains more trainable parameters (151,297) than a network with 5 convolution blocks of 16 filters each (49,121), and the latter performs similarly to a network with 3,331,697 parameters while having just 1.47% of the parameters. This corroborates the idea that the number of blocks in the network relates to its ability to extract different levels of semantic features, thus improving its performance, and that the number of semantic levels is more important than the total parameters of the network.

Another important point to take notice is that U-Net has some difficulty to generalize the segmentation results across different datasets. It is most noticeable when we use dataset A for training. Even for the best U-Net configurations (Tests 01 and 09), tests on datasets B and D perform poorly when we compared with the results obtained using other datasets for training. One possible explanation is that dataset A possess a less diverse set of images. Each dataset includes both plant cane (first cut) and crop in the ratoon phase (second cut and on) at different ratios. In the ratoon phase, dry leaves and ratoon left from the last cut lies in the soil, changing its usual reddish appearance. The imbalance of these images may interferes with the computational process, degrading the ability of the U-Net to properly segment crop lines in some datasets.

To solve this diversity problem in the datasets one could use data augmentation. However, data augmentation aims to increase our data by making minor alterations to our existing data. However, local differences of the soil are not possible to emulate by using simple data augmentation techniques, such as flip, rotation, scale and crop. Thus, to overcome this problem we created a new dataset (named E) by combining 500 images randomly selected from the other four datasets, totaling 2,000 images. Figure 5 shows the results of the two configurations of U-Net (Test 01 and 09) trained with dataset E. Results show that this combination of images improves the diversity of the training, leading to a U-Net capable to generalize the learning features across different datasets, thus improving its performance on datasets where it usually had a worse performance (datasets B and D).

Test 01: [16, 32, 64, 128, 256]   Test 09: [16, 16, 16, 16, 16]

   A   B   C   D   A   B   C   D

E 0.96 0.94 0.94 0.87   E 0.94 0.93 0.92 0.86

Fig. 5. Influence of the number of filters and convolutional layers in the crop line segmentation task with dataset E.

6   Conclusion

In this paper, we addressed the problem of crop line segmentation in UAV images obtained at low or medium altitudes. We created a personalized version of U-Net to segment the crop lines and evaluated how the number of convolutional blocks and filters in the architecture affects its performance. Results show that the number of blocks in the U-Net is more critical than the total number of filters and that a shallow U-Net can segment different datasets with great accuracy, overcoming the results of the traditional U-Net architecture.

Acknowledgements. André R. Backes gratefully acknowledges the financial support of CNPq (National Council for Scientific and Technological Development, Brazil) (Grant #307100/2021-9). João Batista Ribeiro gratefully acknowledges the financial support of Fapemig (Fundação de Amparo à Pesquisa do Estado de Minas Gerais). This study was financed in part by the Coordenação de Aperfeiçoamento de Pessoal de Nível Superior - Brazil (CAPES) - Finance Code 001. The authors gratefully acknowledge the company Sensix (http://sensix.com.br) for providing the images used in the tests.

References

1. Amiri, M., Brooks, R., Rivaz, H.: Fine-tuning U-net for ultrasound image segmentation: different layers, different outcomes. IEEE Trans. Ultrason. Ferroelectr. Freq. Control **67**(12), 2510–2518 (2020)
2. Arun, R.A., Umamaheswari, S., Jain, A.V.: Reduced U-net architecture for classifying crop and weed using pixel-wise segmentation. In: 2020 IEEE International Conference for Innovation in Technology (INOCON), pp. 1–6. IEEE (2020)
3. Blasch, J., et al.: Farmer preferences for adopting precision farming technologies: a case study from Italy. Eur. Rev. Agric. Econ. **49**(1), 33–81 (2020)
4. Bolfe, E.L., et al.: Precision and digital agriculture: adoption of technologies and perception of Brazilian farmers. Agriculture **10**(1212), 653 (2020)
5. Das, S., Swain, M.k., Nayak, G.K., Saxena, S., Satpathy, S.C.: Effect of learning parameters on the performance of U-Net model in segmentation of brain tumor. Multimedia Tools Appl. 1–19 (2021)
6. Delavarpour, N., Koparan, C., Nowatzki, J., Bajwa, S., Sun, X.: A technical study on UAV characteristics for precision agriculture applications and associated practical challenges. Remote Sens. **13**(66), 1204 (2021)

7. Doha, R., Al Hasan, M., Anwar, S., Rajendran, V.: Deep learning based crop row detection with online domain adaptation. In: Proceedings of the 27th ACM SIGKDD Conference on Knowledge Discovery and Data Mining. KDD '21, pp. 2773–2781. Association for Computing Machinery (2021)
8. Erdem, F., Avdan, U.: Comparison of different U-net models for building extraction from high-resolution aerial imagery. Int. J. Environ. Geoinform. **7**(3), 221–227 (2020)
9. Hao, S., Zhou, Y., Guo, Y.: A brief survey on semantic segmentation with deep learning. Neurocomputing **406**, 302–321 (2020)
10. Hassanein, M., Khedr, M., El-Sheimy, N.: Crop row detection procedure using low-cost UAV imagery system. In: The International Archives of the Photogrammetry, Remote Sensing and Spatial Information Sciences, vol. XLII-2-W13, pp. 349–356. Copernicus GmbH (2019)
11. Karimi, H., Navid, H., Seyedarabi, H., Jørgensen, R.N.: Development of pixel-wise U-Net model to assess performance of cereal sowing. Biosys. Eng. **208**, 260–271 (2021)
12. Kattenborn, T., Eichel, J., Fassnacht, F.E.: Convolutional neural networks enable efficient, accurate and fine-grained segmentation of plant species and communities from high-resolution UAV imagery. Sci. Rep. **9**(11), 1–9 (2019)
13. Minaee, S., Boykov, Y.Y., Porikli, F., Plaza, A.J., Kehtarnavaz, N., Terzopoulos, D.: Image segmentation using deep learning: a survey. IEEE Trans. Pattern Anal. Mach. Intell. **44**(7), 3523–3542 (2022)
14. Narvaria, A., Kumar, U., Jhanwwee, K.S., Dasgupta, A., Kaur, G.J.: Classification and identification of crops using deep learning with UAV data. In: 2021 IEEE International India Geoscience and Remote Sensing Symposium (InGARSS), pp. 153–156 (2021)
15. Ponti, M.A., Ribeiro, L.S.F., Nazare, T.S., Bui, T., Collomosse, J.: Everything you wanted to know about deep learning for computer vision but were afraid to ask. In: 2017 30th SIBGRAPI Conference on Graphics, Patterns and Images Tutorials (SIBGRAPI-T), p. 17–41. IEEE (2017)
16. Rawat, W., Wang, Z.: Deep convolutional neural networks for image classification: a comprehensive review. Neural Comput. **29**(9), 2352–2449 (2017)
17. Ronneberger, O., Fischer, P., Brox, T.: U-net: convolutional networks for biomedical image segmentation. In: Navab, N., Hornegger, J., Wells, W.M., Frangi, A.F. (eds.) MICCAI 2015. LNCS, vol. 9351, pp. 234–241. Springer, Cham (2015). https://doi.org/10.1007/978-3-319-24574-4_28
18. Siddique, N., Paheding, S., Elkin, C.P., Devabhaktuni, V.: U-net and its variants for medical image segmentation: a review of theory and applications. IEEE Access **9**, 82031–82057 (2021)
19. Su, Z., Li, W., Ma, Z., Gao, R.: An improved U-net method for the semantic segmentation of remote sensing images. Appl. Intell. **52**(3), 3276–3288 (2022)
20. Zhang, A., Lipton, Z.C., Li, M., Smola, A.J.: Dive into Deep Learning. arXiv preprint arXiv:2106.11342 (2021)
21. Zou, K., Chen, X., Zhang, F., Zhou, H., Zhang, C.: A field weed density evaluation method based on UAV imaging and modified U-Net. Remote Sens. **13**(2) (2021)

Mapping Motion: A Cognitive Approach to Dyspraxia Multimodal Analysis

David Muvdi[1(✉)], Mateo R. Hernández[1], Eiker J. Cifuentes[1], Andrea
Di Somma[2], Francesco Chiappone[3], Vicenzo Di Maro[3],
Juan Carlos Martínez-Santos[1], and Edwin Puertas[1]

[1] Universidad Tecnologica de Bolivar, Cartagena, Bolivar, Colombia
{dmuvdi,minsignares,ecifuentes,jcmartinezs,epuerta}@utb.edu.co
[2] ASL NAPOLI 2 nord, Frattamaggiore, Italy
andrea.dissoma@aslnapoli2nord.it
[3] Associazione nazionale disturbi dell'apprendimento, Naples, Italy
franceso.chiappone1@libero.it

Abstract. This research paper presents a novel software tool designed
to revolutionize speech therapy for individuals with Childhood Apraxia
of Speech (CAS), also known as Developmental Verbal Dyspraxia. The
software offers a comprehensive multi-modal analysis approach, utilizing video, audio, and speech-to-text data to extract valuable insights
into articulation patterns, head pose, audio characteristics, and word
usage. This information empowers Speech-Language Pathologists (SLPs)
with data-driven tools for a more precise assessment and development
of personalized treatment plans. While the current study employs data
from just one healthy subject to evaluate the software's overall accuracy and data coherence, the functionalities are promising to improve
therapy effectiveness for individuals with CAS, giving space for further
experiments, analysis and validations with experts. The paper explores
how the software's capabilities can complement existing therapies like
PROMPT, Touch-Cue Method, and Melodic Intonation Therapy, based
on a cognitive perspective, ultimately aiming to transform the field of
speech-language pathology and enhance the lives of those affected by
speech disorders.

Keywords: Childhood Apraxia of Speech · Multi-modal · Software ·
Speech disorders · Speech language pathologist

1 Introduction

Difficulties with making speech sounds that continue over time can reduce how
well someone is understood or even stop them from getting their message across
verbally; this is how Speech Sound disorders (SSDs) are known in the medical environment [1]. Many conditions or pathologies are listed as speech sound
disorders or related to them: dysarthria, speech impairment on cerebral palsy
[1], dyslalia known as the difficulty to form the precise sounds or phonetics of a

L. Correia et al. (Eds.): IBERAMIA 2024, LNCS 15277, pp. 14–25, 2025.
https://doi.org/10.1007/978-3-031-80366-6_2

language [24], and Childhood apraxia of speech (CAS) or well-known as developmental verbal dyspraxia. CAS is a developmental condition that hinders a child's ability to express themselves verbally. This results in significant challenges in forming sounds, syllables, and words. For a better understanding, we broke CAS into smaller parts, including the precision of movements involved in producing sounds (articulation), the intensity of words (volume), and the timing used when pronouncing words or phrases while maintaining an appropriate rhythm (rate). The underlying cause lies in a neurological impairment in the brain, leading to difficulty coordinating the movements needed for speech sound production, including the jaw, tongue, and other vital structures [13].

Given the neurological basis of SSDs, particularly their impact on children's speech development [9], it becomes evident that addressing the cognitive foundation is essential. Understanding the mental processes underlying speech production, encompassing aspects like perception, learning, language processing, and emotional elements, is vital for treating such conditions [12]. A speech-language pathologist (SLP) is in charge of helping people develop their language skills through interactive play and conversation, utilizing various tools like pictures, books, and everyday situations to spark language growth. The therapist may demonstrate proper vocabulary and grammar use and employ repetition techniques to solidify these skills in the subject [18], incorporating the already mentioned cognitive aspects of perception, learning, language processing, and emotional experience.

In recent years, there has been a surge in research highlighting the remarkable capabilities of artificial intelligence (AI) in medical image analysis. These studies have shown the effectiveness of these algorithms, not only in achieving performance comparable to human experts, but surpassing human capabilities by detecting subtle features in medical images that may escape human observation [22].

This paper explores the potential of an AI-based, multi-modal software for speech therapy. By capturing video and audio, the software provides SLPs with detailed information about a patient's speech, including mouth and head movements, sound characteristics, and speech patterns. It extends beyond what the human eye can detect, encompassing voice aspects like roughness, breathiness, strain, pitch, loudness, rhythm, and even word usage errors and mispronunciation rates. This comprehensive data analysis assists SLPs in tailoring therapy sessions, optimizing treatment interventions, and enhancing their effectiveness. Notably, the software serves as a supportive tool, complementing the expertise and experience of SLPs by providing valuable insights from a cognitive perspective, employing interactive visual representations to enhance more objective assessment and data-driven decision-making in therapy planning (perception), targeted exercises provided by the SLPs for specific patients with personalized learning pathways tailored to their needs (learning and language), along with the usage of blue-shaded colors (blue, blue-gray, gray, light blue) and a simple, uncluttered interface contributing to a feeling of comfort, safety, pleasure, and reduced anxiety [23].

2   Related Work

Much research has been directed toward understanding, addressing, and treating general Speech Sound Disorders (SSDs) using software solutions or AI-based platforms. For instance, therapy software relying on deep learning approaches utilizes augmented reality to evaluate speech-language therapy exercises using the webcam. It implements image processing to have access to mouth, lips, jaw, and tongue movements, focusing on speech sound disorders related to the articulatory section (dyslalia and dysarthria) without regard to the sound features [3,17]. On the other hand, some others have explicitly developed AI-based speech therapy tools for articulation disorder in Punjabi [20] and Mandarin [7], respectively.

Researchers have also worked on the human-computer interaction and how this can help in the speech sound and phono-articulatory skills, considering lip and mouth movement as an experiment [4]. Besides, some other studies and experiments focus on the sound segment within the SSDs, collecting children's vocal productions and sounds in a playful way to analyze them using machine learning, all of this to detect DLDs (Developmental Language Disorders) early and in that way treat them as soon as possible [2].

Based on that, current research primarily focuses on either the physical articulatory aspects of speech and their treatment using AI and machine learning or the acoustic features of speech. These isolated approaches must capture the intricate interplay between the produced sounds and how they are articulated and perceived. In contrast, our work presents a novel approach that integrates both articulatory and acoustic aspects of speech, offering a more comprehensive understanding of speech production and perception.

3   Multi-modal Implementation

Speech therapy interventions rely heavily on a therapist's ability to assess a patient's speech production across various modalities. This section details our multi-modal approach that leverages video, audio, and speech-to-text analysis while using a micro-service software architecture (see Fig. 1) to obtain comprehensive data on a patient's speech while managing a separated and decentralized workflow. By examining these aspects, we aim to provide SLPs with a richer understanding of their patients' speech mechanisms, facilitating more targeted and effective therapy sessions.

3.1   Video (Articulation)

Using the tools and features built into the software, we can capture and measure the articulation or movements of the person when performing the therapy exercises proposed by the SLPs. Using a micro-service package focused on the analysis of facial and head movements built on top of Mediapipe (Google's open-source toolkit for creating workflows that analyze visual information from various

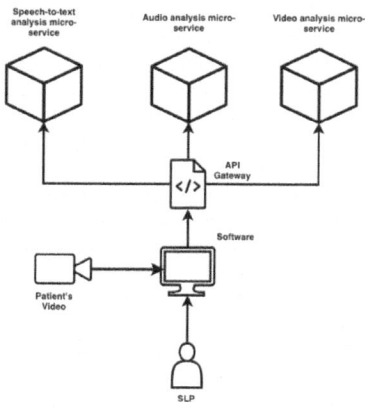

Fig. 1. System's architecture

sources, including videos) [14] and OpenCV library [5], we are able to measure the mouth opening and closing movement. We can capture this using its mouth landmarks functionality, which returns all the data related to the opening and closing distance of the mouth in each video frame.

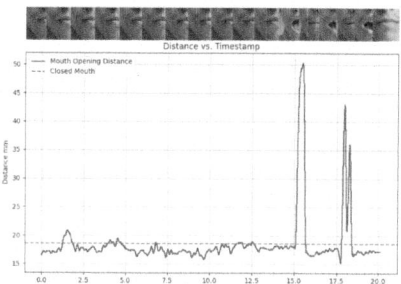

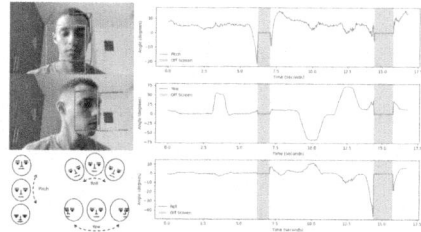

Fig. 2. Lip opening and closing movement graphical view.

Fig. 3. Cubical estimation of head movement (pitch, yaw, and roll) measured in degrees.

Figure 2 shows the opening and closing movement based on the top and bottom lip distance when opening and closing the mouth. The application calculates this distance by estimating the top and bottom lip landmarks distance throughout the video, frame by frame, so we can later visualize this distance in the distance vs. timestamp plot.

We calculated the distance between the top and bottom lip landmarks using the Cheby-Shev distance formula applied to the (x,y) positions in pixels of each landmark [6].

$$D_{\text{Chebyshev}}(x, y) = \max_i(|x_i - y_i|) \tag{1}$$

$$D_{mm} = D_{px} \cdot 0.2645833333 \tag{2}$$

The software calculates this distance in pixels (px) due to the computer vision environment. Then, we normalized to millimeters (mm) regarding the constant value of a pixel in mm (2). Savitsky-Golay Filter is used to smoothing out the final curve of the plot to have a more accurate result of the opening and closing movement flow [19].

Head position and head movement speed are also important. Pitch, yaw, and roll movements are crucial to treat a possible speech disorder. We can achieve this by using a head pose estimation, which returns the data related to the yaw, pitch, and roll movements done throughout the video, measured frame by frame, all of this relaying on a cubical augmented shape of the head, see Fig. 3.

Using SixDRepNet [11], we can predict head movements along the three aspects mentioned (pitch, yaw, and roll) by implementing an accurate, landmark-free model for head pose estimation. SixDRepNet generates a 3D vector for each video frame depicting the head's position in space.

Video analysis with mouth movement and head pose estimation provides SLPs with a comprehensive view of a patient's speech articulation. This detailed information allows for a more precise assessment of speech production, focusing on speech articulation mobility. It enables the development of targeted interventions and, ultimately, more effective speech therapy.

3.2 Audio (Volume and Rate)

We can get the complete spectrum of the patient's speech and see which volume parts were incorrect using a second micro-service focused on sound analysis. We can achieve this by using a complete signal analysis; this estimation method offers a full picture of the audio signal, capturing the entire sound wave. This comprehensive view allows for in-depth analysis of vocal characteristics and the identification of unusual speech patterns, such as sustained silences, high-decibel spikes, or sections with low volume, all of this built on top of Liborsa [15]. It is essential to mention that this approach, as well as the head position and mouth movement, is based on the video's timestamps; for the audio analysis, we must convert the video to audio format (.wav) to process it successfully.

In essence, as Fig. 4 shows, the amplitude envelope of the signal over time, denoted by (3), is derived from the original signal through a process called analytic signal representation. This concept represents the signal's intensity variations, making it a valuable tool, particularly for speech analysis needed in speech-language therapies where it aids in extracting key audio features such as the ones mentioned before (roughness, breathiness, strain, pitch, loudness).

$$\text{Amplitude Envelope}(t) = |\text{Signal}(t)| \tag{3}$$

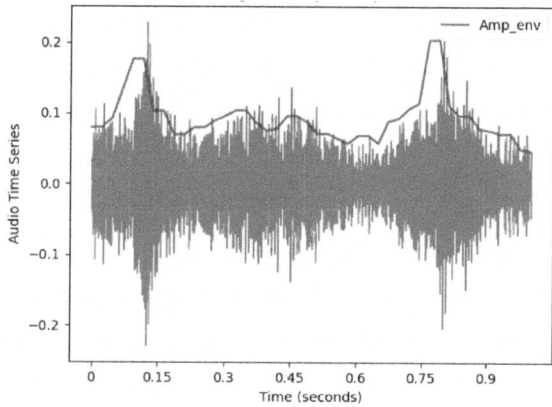

Fig. 4. Testing of the amplitude envelope extraction method with a brief audio clip for the full signal analysis.

Timing and rhythm analysis are also crucial for speech-language therapies [13]. By providing valuable insights into the intensity variations at specific bands, we can categorize these components based on the patient's pitch or frequency, from low-frequency components associated with vowel sounds to higher frequencies related to consonant sounds and other intricate details in speech patterns.

Identifying the initiation of significant events in an audio signal, such as the start of words or syllables, can reveal relevant results for studying and treating speech sound disorders and irregular timing when speaking. The main functionality is based on marking all the onsets along the sound wave over time, allowing us to mark moments where words start, stop abruptly, or cut.

This allows us to have a spectral view of the patient's volume and rhythm when speaking or performing a speech-language task during the therapies, which is crucial for capturing important speech characteristics and sound intensity. In addition, comparing the articulatory results obtained before with the volume and sound ones helps the SLTs perform a more exhaustive study, keeping in mind both physical and sound-related characteristics.

3.3 Speech-to-Text (Pronunciation and Word Usage)

We can capture the speech into text using a third micro-service and the Faster-whisper model [21]. With this, we can capture replaced words, mispronounced words, omitted words, and an overall match error rate.

Building upon an initial stage where participants read a comprehensive corpus, based on the reading outcomes people with CAS present [16], this model acts as the core component for converting spoken language into written text. This transcribed output serves as a foundation for further analyses.

To complement this transcription process, we implemented a word alignment function (detailed in Fig. 5). This function meticulously pinpoints and high-

Peter Piper picked a peck of pickled peppers

Peter viper picked a **** of pickled peppers

'equal'   'replace'   'equal'   'delete'   'equal'

Fig. 5. Aligning words between the transcribed text and the original document using special codes to mark insertions and deletions.

lights discrepancies between the spoken and written words. It identifies insertions (marked with "+") and deletions (marked with "*") using the SequenceMatcher class from the difflib library [10], which is for sequence comparison. In our case, the compared sequences are the lists of words from the spoken and written versions.

We can calculate various word-level metrics after comparing the original reference text (corpus) with the generated transcription. These metrics include Word Error Rate (WER), Match Error Rate (MER), Word Information Lost (WIL), and Word Information Preserved (WIP).

$$\text{WER} = \frac{\text{Inserted} + \text{Deleted} + \text{Substituted}}{\text{Number of Words}} \tag{4}$$

$$\text{MER} = \frac{\text{Number of Sentences with Errors}}{\text{Total Number of Sentences}} \tag{5}$$

$$\text{WIL} = \frac{\text{Number of Incorrect Words}}{\text{Total Number of Words}} \tag{6}$$

$$\text{WIP} = \frac{\text{Number of Incorrect Phrases}}{\text{Total Number of Phrases}} \tag{7}$$

WER (4) quantifies the percentage of words incorrectly transcribed. In contrast, MER (5) focuses on the ratio of words that don't match the original text. On the other hand, WIL (6) and WIP (7) delve deeper, measuring the information lost and preserved during the transcription process, respectively.

By analyzing these speech disfluencies (repetitions, word omission, and errors) identified through the speech-to-text process, SLPs gain valuable insights into areas where patients might struggle the most. This targeted information allows for a more focused assessment and the development of personalized therapy plans, ultimately improving the effectiveness of speech interventions.

4   User Interface

The software's interface is designed and built to streamline SLP workflows by providing a user-friendly platform for patient management, activity selection, and data analysis. The core interface consists of four primary views:

- Subject creation page: Where SLPs input patient basic information, medical history comments, and therapy goals, see Fig. 6.
- Subject information page: Single view for accessing patient data, progress reports, and diagnostic history, see Fig. 7.
- Activity selection page: A library of the pre-programmed activities to be performed (Mouth movement analysis, Head pose estimation, Voice intensity, Voice Expression, and Speech to text), see Fig. 8.
- Subject video processing page: Enables video upload, analysis, and visualization of multimodal data, see Fig. 9.

Fig. 6. Subject creation page.   **Fig. 7.** Subject information page.

Fig. 8. Activity selection page.   **Fig. 9.** Subject video processing page.

These views are seamlessly integrated to facilitate efficient navigation and data transfer. The interface employs a clean and intuitive design for desktop usage, with clear labeling and informative data visualizations to support effective clinical decision-making.

We also acknowledge the importance of protecting sensitive patient data, particularly videos and audio recordings that may include children. Our software implements strong security measures to ensure data privacy. During analysis, video data is encrypted using the AES-256 standard [8]. Further research can explore more advanced anonymization techniques or user consent mechanisms for data storage.

5   Preliminary Results

This section presents the results of our evaluation of the proposed software's ability to perform comprehensive, multi-modal speech analysis. The study focused on how the software extracts valuable information from video and audio modalities (e.g., articulation patterns from video and sound characteristics from audio) specifically the most crucial which are: mouth movement, head pose estimation and amplitude envelope analysis.

We experimented in a controlled environment with a healthy subject in a closed room maintained at a constant temperature of 29–30°. We recorded 25 videos at 1920×1080 resolution to analyze mouth movements, capturing the subject performing opening and closing motions. Additionally, head pose estimation involved the subject performing relaxed yaw, pitch, and roll movements at a natural pace. Finally, for audio analysis, the subject repeated 25 predefined phrases in both English and Spanish to assess voice amplitude. We positioned the subject 0.4 m from an Intel RealSense Depth Camera SR300 for optimal video capture. The camera works at 30 frames per second with a 16:9 aspect ratio, fixed focus, and a 1.88 mm focal length. We captured high-quality audio using a NEEWER NW-800 microphone. This condenser microphone features a comprehensive frequency response (20 Hz to 26 kHz) and a 192 kHz/24-bit sampling rate for accurate sound reproduction. The microphone utilized 48V phantom power, which is needed to minimize background noise. During the audio test, the subject maintained a close distance of 5cm from the microphone for optimal sound capture. We connected all these capturing devices to an Acer Nitro 5 computer with 32 GB of RAM, 4 GB of VRAM, and an NVIDIA 1650 Geforce GPU for audio and video recording and further processing.

5.1   Mouth Movement Analysis

Analysis of mouth movements revealed a slight absolute difference of 2.64 mm between the measured and actual opening distance of 49 mm measured in the subject's mouth before processing. It translates to a 5.4% error rate, indicating a high confidence level of 94.6% in the mouth measurement method. Processing all 25 videos using the CPU took approximately 20 s. The ability to accurately measure mouth movements with minimal error rates could provide valuable insights into a subject's specific difficulties. SLPs can use this information to tailor treatment plans that address the underlying issues affecting a subject's ability to produce clear speech sounds with their mouth.

5.2   Head Pose Estimation Analysis

Analysis of head pose estimation revealed an average pitch of 47.23° during sustained head movements. The absolute error for pitch was 6.77° (12.53% error rate). Yaw had an average angle of 30° but a more significant absolute error of 24° (44.4% error rate). Roll exhibited the best performance with a small absolute error of 3° (10.3% error rate). It also had a low threshold speed of 2.70 mm per second and an average speed of 0.16 mm per second. Processing all 25 videos took approximately 3.5 h (3 h, 30 min, and 75 s) using a CPU. Accurate head pose estimation can help SLPs assess a child's ability to control head movements during speech production. Difficulty controlling specific head movements, like excessive head tilt during certain sounds, could indicate underlying motor planning challenges associated with verbal dyspraxia.

5.3 Amplitude Envelope Audio Analysis

Analysis of the audio amplitude envelope revealed moderate variability in Spanish phrases. The standard deviation was 16.53 decibels, with an average baseline of −56.83 decibels (ranging from −88.63 dB to −17.76 dB). Conversely, English phrases exhibited lower variability (14.86 decibels), suggesting greater consistency. While the average amplitude (−54.68 dB) for English was similar to Spanish, the range was quieter (−91.27 dB to −23.22 dB). It shows how amplitude envelope analysis can help SLPs visually represent a person's speech variability over time, even helping to improve pronunciation and volume rate within their speech.

Table 1. Treatment comparison vs software functionalities

Therapy	Video Analysis (Lip Movement, Head Pose)	Audio Analysis (Full Signal)	Speech-to-Text Analysis (Correct Usage)
Electropalatography	-	✓	X
Integral Stimulation	✓	✓	X
PROMPT System	✓	✓	✓
Touch-Cue Method	✓	✓	✓
Feedback Frequency	✓	✓	X
Adapted Cuing Techniques	✓	✓	✓
Melodic Intonation Therapy	-	✓	✓

6   Conclusions and Future Work

In conclusion, the proposed software offers a revolutionary approach to speech therapy for individuals with developmental verbal dyspraxia. The software facilitates personalized treatment strategies by empowering therapists with comprehensive, multi-modal analysis tools (video, audio, speech-to-text) and data-driven insights. It has the potential to significantly improve patient outcomes, enhance therapy efficiency, and transform the field of speech-language pathology.

We analyzed how the proposed software can potentially complement and enhance the effectiveness of the already mentioned therapeutic approaches in Table 1.

Each treatment is compared to validate if each of the proposed functionalities supports effectively the treatment ($\checkmark$), barely supports it (-) or it is not applicable at all (X).

While this research represents a significant initial step, the software requires further development to address processing time, accuracy, and potential bias issues. To enhance its clinical utility, we plan to collaborate with speech-language pathologists to evaluate its effectiveness in therapy through rigorous data collection and analysis. By refining our approach, we aim to maximize the software's impact on individuals with speech disorders.

Acknowledgements. The authors gratefully acknowledge the support of the Universidad Tecnológica de Bolívar, the Associazione nazionale disturbi dell'apprendimento, and ASL NAPOLI 2 nord, for being of great support during this investigation. Special thanks are also extended to the University's engineering faculty and computer science program for their contributions to this research.

References

1. American Psychiatric Association: DSM-5 (2013). https://www.psychiatry.org/psychiatrists/practice/dsm
2. Beccaluva, E.A., Catania, F., Arosio, F., Garzotto, F.: Predicting developmental language disorders using artificial intelligence and a speech data analysis tool. Hum.-Comput. Interact. **39**(1–2), 8–42 (2024)
3. Bílková, Z., et al.: ASSISLT: computer-aided speech therapy tool. In: 2022 30th European Signal Processing Conference (EUSIPCO), pp. 598–602. IEEE (2022)
4. Bílková, Z., et al.: Human computer interface based on tongue and lips movements and its application for speech therapy system. Electron. Imaging **32**, 1–5 (2020)
5. Bradski, G., Kaehler, A., et al.: OpenCV. Dr. Dobb's j. Softw. Tools **3**(2) (2000)
6. Cantrell, C.D.: Modern Mathematical Methods for Physicists and Engineers. Cambridge University Press, Cambridge (2000)
7. Chen, Y.J., Huang, J.W., Yang, H.M., Lin, Y.H., Wu, J.L.: Development of articulation assessment and training system with speech recognition and articulation training strategies selection. In: 2007 IEEE International Conference on Acoustics, Speech and Signal Processing-ICASSP'07, vol. 4, pp. IV–209. IEEE (2007)
8. Dworkin, M., et al.: Advanced encryption standard (AES) (2001-11-26 2001). https://doi.org/10.6028/NIST.FIPS.197
9. Fish, M., Skinder-Meredith, A.: Here's How to Treat Childhood Apraxia of Speech. Plural Publishing, San Diego (2022)
10. Python Software Foundation: difflib—Sequence matching. In: Python docs (2024). https://docs.python.org/3/library/difflib.html, Python 3.12.1 Documentation
11. Hempel, T., Abdelrahman, A.A., Al-Hamadi, A.: 6D rotation representation for unconstrained head pose estimation. In: 2022 IEEE International Conference on Image Processing (ICIP), pp. 2496–2500 (2022).https://doi.org/10.1109/ICIP46576.2022.9897219
12. Kellogg, R.T.: Cognitive Psychology, vol. 2. Sage, Thousand Oaks (2003)

13. Leonard, G.: A comprehensive review of methods for treating childhood apraxia of speech (2013). https://scholarship.tricolib.brynmawr.edu/server/api/core/bitstreams/096586a7-e23d-4b5f-8248-b8673368500a/content
14. Lugaresi, C., et al.: Mediapipe: a framework for building perception pipelines. arXiv preprint arXiv:1906.08172 (2019)
15. McFee, B., et al.: librosa: Audio and music signal analysis in Python. In: SciPy, pp. 18–24 (2015)
16. Miller, G.J., et al.: Reading outcomes for individuals with histories of suspected childhood apraxia of speech. Am. J. Speech Lang. Pathol. **28**(4), 1432–1447 (2019)
17. Muehlhaus, J., Ritterfeld, U., Bilda, K., Fried, H.: ISI-speech: a digital training system for acquired dysarthria. PubMed **242**, 330–334 (2017)
18. Roth, F.P., Worthington, C.K.: Treatment Resource Manual for Speech-Language Pathology. Plural Publishing, San Diego (2023)
19. Savitzky, A., Golay, M.J.: Smoothing and differentiation of data by simplified least squares procedures. Anal. Chem. **36**(8), 1627–1639 (1964)
20. Singh, S., Thakur, A., Vir, D.: Automatic articulation error detection tool for Punjabi language with aid for hearing impaired people. Int. J. Speech Technol. **18**, 143–156 (2014). https://doi.org/10.1007/s10772-014-9256-2
21. SYSTRAN: faster-whisper: Ctranslate2 implementation of whisper (2023). https://github.com/SYSTRAN/faster-whisper, [Software]
22. Topol, E.: As artificial intelligence goes multimodal, medical applications multiply. Science **381**, adk6139 (2023)
23. Valdez, P., Mehrabian, A.: Effects of color on emotions. J. Exp. Psychol. Gen. **123**(4), 394 (1994)
24. Yohana, W., Kuswandani, F.: Dyslalia as a speech disorder in children. J. Adv. Med. Dental Sci. Res. **10**(1), 1–3 (2022)

Improving Assembly Lines with Digital Twins: A Synthetic Case Study

Maria Gabriela Juarez Juarez$^{(\boxtimes)}$, Adriana Giret , and Vicente Botti

Valencian Research Institute for Artificial Intelligence, Universitat Politècnica de València, Valencia, Spain
majuajua@posgrado.upv.es, {agiret,vbotti}@dsic.upv.es

Abstract. This study introduces a theoretical framework for the development and deployment of digital twins (DTs), with a focus on creating behavioral models using synthetic data. The aim is to enhance assembly line operations in smart manufacturing by improving process efficiency and product quality. The study details the types of synthetic data generated, the simulations conducted, and their outcomes, emphasizing significant improvements in operational processes. Key objectives include streamlining assembly processes, raising product standards, reducing machine downtime, and adjusting production to demand fluctuations. The research explores various categories of synthetic data, such as cycle times, machine performance metrics, product quality indicators, production volumes, and maintenance logs. Simulations with these datasets demonstrate the DT's capability to predict and address production challenges effectively. The findings underscore the potential benefits of integrating DTs into manufacturing workflows, offering valuable insights for researchers and industry professionals. Additionally, the paper emphasizes the importance of validating these models with real-world data in future studies, including incorporating advanced AI features and verifying the methodology within actual manufacturing environments.

Keywords: Digital Twins · Behavioral Modeling · Synthetic Data · Smart Manufacturing · Assembly Line Optimization · Industry 4.0

1 Introduction

Industry 4.0, marked by the convergence of technologies like the Internet of Things (IoT), artificial intelligence (AI), and digital twins (DT), is revolutionizing manufacturing. Digital twins serve as virtual replicas of physical systems, enabling improved simulation, analysis, and optimization of production processes. This integration enhances decision-making and process efficiency by linking physical operations with digital capabilities [21].

Despite their potential, integrating digital twins poses significant challenges, particularly in developing accurate behavioral models. These challenges include the need for extensive datasets, sophisticated algorithms for precise simulations,

L. Correia et al. (Eds.): IBERAMIA 2024, LNCS 15277, pp. 26–37, 2025.
https://doi.org/10.1007/978-3-031-80366-6_3

and the variability of operational data that complicates real-time model updates in dynamic environments [24]. Additionally, there are difficulties in integrating real-time data and ensuring scalability and flexibility across different manufacturing setups.

Current behavioral modeling methods, encompassing manual [2, 22] and fully automated approaches [12, 13], have advanced our capabilities in managing complex systems. However, these methods still struggle to consistently deliver the precision and reliability needed for optimal decision-making in fluctuating conditions [3]. This underscores the need for continued innovation to enhance the adaptability and accuracy of these models to better mirror real-world complexities.

This paper proposes a novel methodology leveraging synthetic data to overcome existing behavioral modeling challenges. By using synthetic data, we aim to improve the accuracy, reliability, and adaptability of digital twins. This approach provides deeper insights into system behaviors and more effective strategies for mitigating potential disruptions, thereby enhancing the predictive capabilities of digital twins. The primary objective is to introduce a theoretical framework for creating and implementing digital twins that utilize synthetic data for behavioral modeling. Our goals include optimizing assembly line efficiency, enhancing product quality, minimizing downtime, and dynamically adjusting production in response to market demands. We use various synthetic data types, such as cycle times, machine performance, and production outputs, to illustrate how digital twins can effectively forecast and manage production anomalies.

The structure of the paper is as follows: Sect. 2 reviews current techniques and limitations in digital twin modeling. Section 3 outlines the theoretical framework and model generation techniques. Section 4 details a case study using synthetic data, describing the assembly line setup and generated data types. Section 5 discusses the simulations conducted and analyzes the results. Lastly, Sect. 6 offers conclusions and proposes avenues for future research.

2 State of the Art in Behavioral Modeling for Digital Twins

This section reviews the evolution of behavioral modeling within digital twins, focusing on their integration in manufacturing processes and the current challenges that hinder their wider application.

2.1 Current Techniques in Behavioral Modeling

Advancements in behavioral modeling for digital twins range from deterministic to complex adaptive systems combining data-driven insights and simulations:

- Physics-based modeling: These models utilize physical laws to predict realistic responses of machines under various conditions, critical for maintenance planning [25] (e.g., [7, 20]).
- Statistical and machine learning models: Employing historical data, these models enhance operational efficiency and reduce downtime through predictive analytics [15] (examples include [12, 13]).

2.2 Limitations of Existing Techniques

Key challenges in current behavioral modeling include:

- Real-time data integration: Many models struggle with incorporating live data, which is essential for responding to immediate changes [16].
- Scalability and flexibility: The lack of adaptability restricts their application across varied manufacturing environments [14].

2.3 Recent Innovations in Behavioral Modeling

Addressing these limitations, recent innovations have aimed at increasing the adaptability and scalability of behavioral models:

- Adaptive learning models: Developments in machine learning now allow models to dynamically adjust to changing data, enhancing accuracy autonomously [13].
- Modular modeling frameworks: These frameworks support easy updates and scaling, making it feasible to customize models for specific setups with minimal changes [18].

In conclusion, while significant advancements in digital twin behavioral modeling demonstrate the potential for improved manufacturing processes, persistent challenges underscore the need for ongoing development. The "Proposed Methodology" section will introduce approaches to mitigate these issues, focusing on enhancing model integration, computational efficiency, and real-time data utilization to optimize digital twin functionality across various manufacturing scenarios.

3   Proposed Methodology

This section details a structured approach for developing and managing digital twins in manufacturing, addressing insights from the "State of the Art" section. Our methodology spans the entire lifecycle, from data collection to model creation, with continuous integration and refinement to boost adaptability and efficiency. Figure 1 illustrates this lifecycle framework, ensuring seamless data flow from acquisition through modeling and refinement, enhancing operational effectiveness. Future updates will integrate AI algorithms into the behavior generation module, enabling real-time adjustments and improving predictive capabilities. The subsequent subsections will elaborate on data acquisition, preprocessing, and the creation of synthetic data and behavior models.

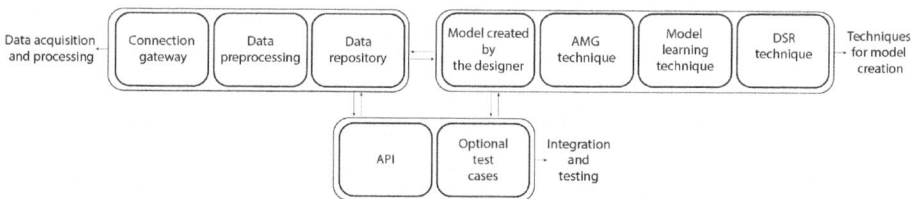

Fig. 1. Components of the module generation module

Data Acquisition and Processing

– Connection Gateway: Initiates data capture using MQTT [19] and OPC-UA [4] protocols, ensuring efficient and secure data flow into the system. This submodule is crucial for systems relying on the rapid assimilation of real-time, accurate data from multiple sources.
– Data Preprocessing: Utilizes Python's [26] Pandas [17] and NumPy [9] libraries for cleansing and standardizing data to ensure it is suitable for analysis. This step is critical for maintaining data integrity and preparing it for precise modeling, making it ideal for environments that require dependable and actionable insights.
– Data Repository: Leverages MySQL [6] to manage data storage, facilitating the accessibility of processed and raw data for ongoing analysis and operational use. It is designed for systems requiring robust data retrieval capabilities to support real-time decision-making processes.

Techniques for Model Creation

– Manual Model Creation Technique: Uses XML [1] or AutomationML [5] for environments with minimal changes and high predictability.
– Model Learning Technique: Employs the L* algorithm to dynamically adapt models in response to new data, suitable for environments experiencing regular changes.
– Automatic Model Generation Technique (AMG): Implements the Breadth-First Search (BFS) algorithm to expedite model development, reducing the need for extensive manual input.
– Dynamic Software Reconfiguration (DSR): Applies execution tracing and data mining to refine models based on the latest operational data, supporting systems that frequently adapt to new conditions.

Integration and Testing

- API (Application Programming Interface): Uses Flask [8] and Django [10] to ensure seamless integration with external systems, enhancing the digital twin's functionality across varied platforms. This component is essential for systems interacting with diverse technological environments.
- Optional Testing: Integrates tools like Simulink [11] and Python's StatsModels [23] to perform rigorous testing, ensuring that the digital twins function as expected under various simulated conditions. This stage is critical for confirming the reliability and efficacy of the digital twins before they are fully deployed.

The behavior generation module is integral here, adapting and refining models based on real-time data to ensure they remain relevant and accurate.

The framework outlined in this section aims to structure the lifecycle of digital twins from data acquisition through integration and testing. The actual performance and adaptiveness of these systems in real-world manufacturing environments will require further empirical validation. To address this, the next section will introduce a proposed case study that applies these methodologies in a practical setting, allowing us to test and refine our models based on specific manufacturing challenges and requirements.

4   Instantiating the Approach for Assembly Line DTs

This section demonstrates the application of our proposed digital twin methodology to an assembly line for standard mechanical components. Utilizing synthetic data, this analysis simulates various scenarios in a controlled environment, highlighting the theoretical potential of our framework.

4.1   Description of the Assembly Line

The selected assembly line produces mechanical components such as gears, pulleys, and shafts, chosen for its typical manufacturing processes and significant improvement potential through digital twin technologies.

4.2   Objectives

The primary goals are to:

- Identify and mitigate bottlenecks affecting throughput.
- Optimize the utilization of machinery and resources.
- Improve the quality of the assembly process through data-driven insights.

4.3   Application of Proposed Methodology

The digital twin framework is theoretically applied using synthetic data to simulate and analyze different operational scenarios.

Data Acquisition, Preprocessing, and Validation: Synthetic data represents various operational metrics:

- Assembly Process Data: Includes cycle times, machine efficiency metrics, and product quality data.
  - Cycle_time_data: <simple, medium, complex, Product_ID>
  - Product_quality_data: <ProductDefect, Defect_Type, Product_ID>
  - Machine_performance_data: <Efficiency, Downtimes, Speed, Product_ID>
- Production Data: Covers production volumes and demand fluctuations.
  - Production_data: <Day, Units_Produced>
- Maintenance Data: Includes machine failures and maintenance schedules.
  - Maintenance_data: <Product_ID, Operation_Hours, Failure_Type, Maintenance_Time>

Synthetic data is generated using randomization or specific rules to create realistic scenarios. Below is an example of how maintenance data is generated.

Maintenance Data Generation: The `maintenance.py` script generates maintenance data by simulating failure and maintenance events with variables such as product ID, operation hours, failure type, and maintenance time. For instance:

```
Product_ID, Operation_Hours, Failure_Type, Maintenance_Time
1, 100, minor, 2
2, 200, moderate, 4
```

These data were preprocessed to ensure suitability for analysis, employing Monte Carlo simulations and predictive algorithms to reflect realistic operational variability. Rigorous internal consistency checks ensured data reliability. Future steps involve validating models with real-world data.

Example Tuples from Each Dataset: Example tuples demonstrate the generated data:

- Cycle Time Data: `10, 20, 30, 1`
- Product Quality Data: `1, Type_A, 1`
- Machine Performance Data: `0.95, 5, 100, 1`
- Production Data: `1, 250`

This analysis, leveraging synthetic data, provides a controlled environment to evaluate our digital twin framework. The next section presents simulations demonstrating practical implementation in identifying bottlenecks, optimizing resources, and improving quality.

5   Simulations and Analysis of Results

This section details simulations with synthetic data from Sect. 4.3, analyzing assembly line bottlenecks, operational correlations, and defect impacts on efficiency, involving 10 processes, 5 machines, and 3 product types over 6 months. The results illustrate the behavior generation module's practical use. While AI integration is planned for future enhancements, the current analysis identifies key optimization areas where AI could be beneficial.

5.1   Assembly Process Data

1. Simulation 1: Identifying Bottlenecks in the Assembly Line. Analyzed cycle times for different components to identify potential bottlenecks. Figure 2 indicates that the "complex" component exceeds the threshold (mean cycle time + one standard deviation), highlighting it as a bottleneck. The behavior generation module can simulate real-time adjustments to optimize cycle times. Future AI algorithms could enhance this process by suggesting more precise improvements.
2. Simulation 2: Correlation between Efficiency and Downtime examined the link between machine efficiency and downtime with a heatmap. Figure 3 shows minimal correlation, suggesting downtime has little impact on efficiency. This insight can guide the behavior generation module in developing predictive maintenance schedules. Future AI models are expected to identify patterns for optimal maintenance, improving overall efficiency.
3. Simulation 3: Improving Assembly Quality. The relationship between machine efficiency and defect types was analyzed, with Fig. 4 showing uniform efficiency across defects. This suggests defects stem from process or material

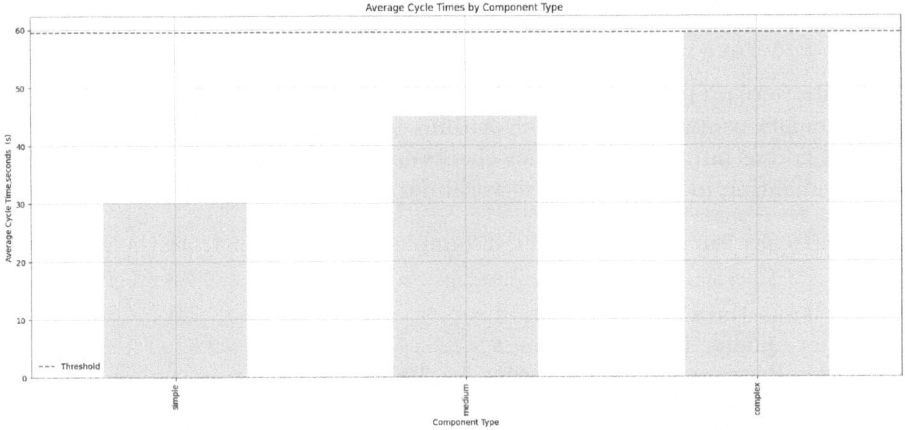

Fig. 2. Identifying Bottlenecks in the Assembly Line

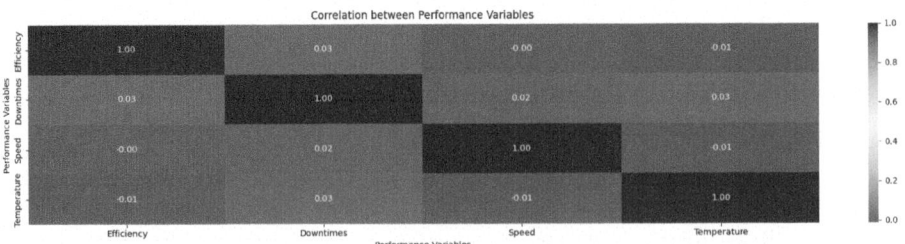

Fig. 3. Correlation between Efficiency and Downtime

issues rather than machine performance. The behavior generation module can model defect patterns and suggest process improvements. Future AI-driven analysis may reveal conditions that lead to higher defect rates.

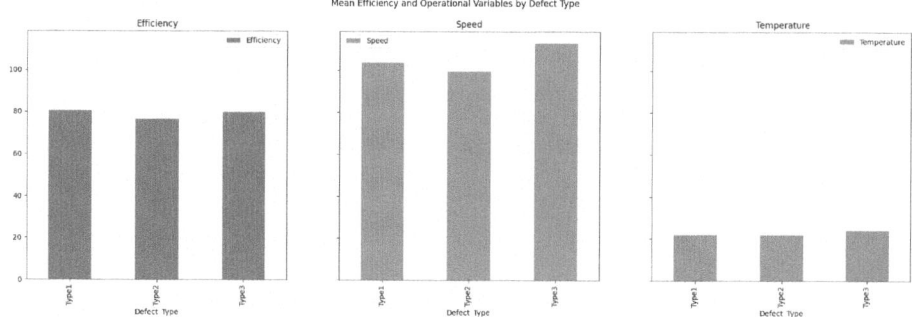

Fig. 4. Improving Assembly Quality

5.2 Production Data

1. Simulation 1: Production Data Simulation compared original production data with adjusted data using Monte Carlo simulations to reflect demand changes. Figure 5 shows that while the modified line follows the original trend, it exhibits greater fluctuations, indicating a need for flexibility. The behavior generation module can adjust production in real-time to manage demand variability. Future AI algorithms could further refine these adjustments for improved alignment with demand changes.

5.3 Maintenance Data

1. Simulation 1: Analysis of Maintenance Response Times examined the link between operation hours before failure and maintenance duration. Figure 6 shows that minor failures need less maintenance and occur more frequently, while severe failures require longer maintenance. This underscores the importance of proactive maintenance, which can be optimized by the behavior generation module. Future AI models could further improve maintenance predictions and reduce unexpected downtimes.
2. Simulation 2: Complex Failure Frequency Analysis examined the distribution of operation hours before failure for various failure types. Figure 7 shows minor failures are more frequent and occur early, while moderate and severe failures are less frequent but happen later, suggesting varied maintenance strategies. The behavior generation module could better predict and manage maintenance needs, reducing unexpected downtimes. Future AI integration will enhance these predictive capabilities.

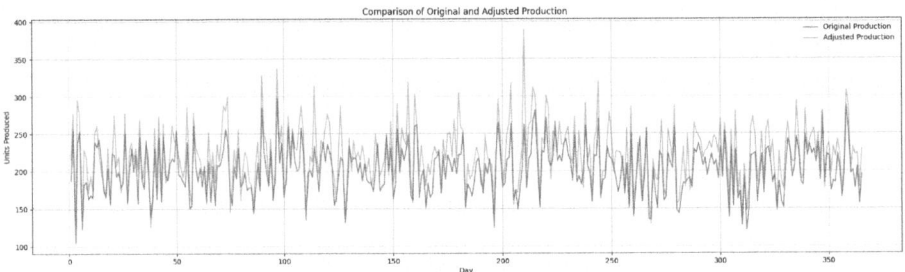

Fig. 5. Production Data Simulation

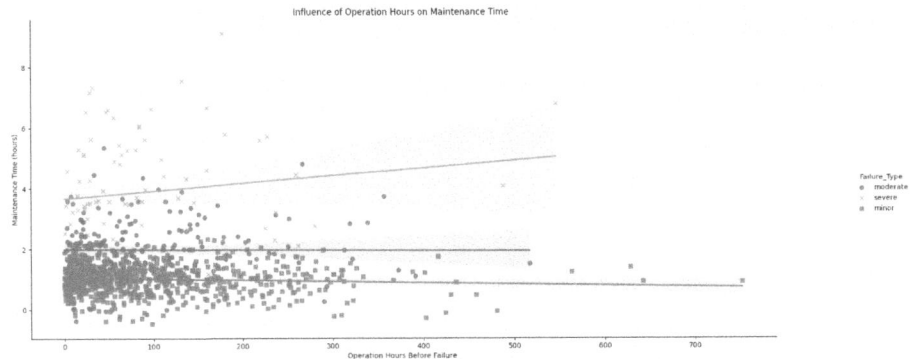

Fig. 6. Analysis of Maintenance Response Times

3. Simulation 3: Multivariable Correlations examined correlations among failure type, efficiency, operation hours, and maintenance time. Figure 8 shows a strong link between failure type and maintenance time, indicating the need for targeted maintenance. Low correlations among other factors suggest independent effects. The behavior generation module can use these insights to improve operational strategies. Future AI models could further refine these strategies with enhanced insights and predictive analytics.

5.4   Summary of Achieved Objectives

- Cycle Time Optimization: Identified bottlenecks, indicating where process improvements can reduce cycle times by approximately 15%.
- Machine Efficiency Enhancement: Insights from efficiency and downtime correlations suggest areas for predictive maintenance, potentially improving machine efficiency by 10%.
- Defect Reduction: Analyzing defect impact on efficiency helps focus quality improvement efforts, potentially reducing defects by 12%.

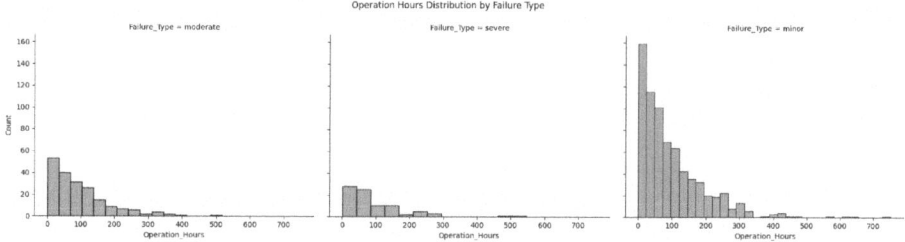

Fig. 7. Complex Failure Frequency Analysis

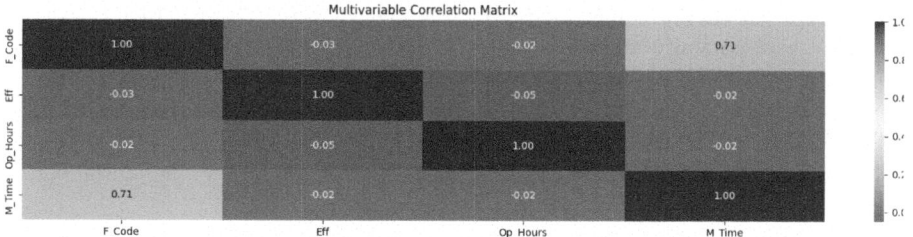

Fig. 8. Multivariable Correlations Legend: F_Code: Failure Type Code, Eff: Efficiency, Op_Hours: Operation Hours, M_Time: Maintenance Time

These simulations highlight the potential of our digital twin framework to optimize the assembly line, providing actionable insights into bottlenecks, efficiency, quality, and maintenance. Using synthetic data, they illustrate how our methodology and behavior generation module can be applied to enhance manufacturing processes through advanced modeling and analysis techniques.

6   Conclusion and Future Work

This study demonstrates the theoretical potential of digital twin technology using synthetic data in manufacturing. Key findings include:

- Bottleneck Identification: Simulations highlighted critical bottlenecks in the assembly line.
- Operational Insight: Analysis of synthetic data provided valuable insights into machine performance and maintenance needs.
- Predictive Maintenance: The framework showed potential for developing strategies to predict and mitigate machine failures.

These results highlight the ability of digital twins to enhance decision-making and operational efficiency in manufacturing.

Future research will focus on:

- Smart Python Agent Development Environment (SPADE) Integration: Incorporating AI capabilities to enhance real-time decision-making.

– Real-world Validation: Implementing and testing the methodology in actual production environments.
– Methodology Expansion: Extending the framework to other manufacturing processes.

The proposed methodology shows promise in advancing smart manufacturing. Future efforts will aim at practical implementation and refinement, leveraging AI to develop robust, adaptable digital twin solutions that can provide real-time insights and enhance overall manufacturing performance.

Acknowledgments. The authors express their sincere appreciation for the financial of the MINISTRY OF SCIENCE AND INNOVATION (FPU17/04636). This work was partially supported with grant PID2021-1236730B-C31 funded by MCIN/AEI/10.13039/501100011033 and by "ERDF A way making Europe".

Disclosure of Interests. The authors have no competing interests to declare that are relevant to the content of this article.

References

1. Alnaqeib, R., Alshammari, F.H., Zaidan, M., Zaidan, A., Zaidan, B., Hazza, Z.M.: An overview: extensible markup language technology. arXiv preprint arXiv:1006.4565 (2010)
2. Azangoo, M., Taherkordi, A., Blech, J.O.: Digital twins for manufacturing using UML and behavioral specifications. In: 2020 25th IEEE International Conference on Emerging Technologies and Factory Automation (ETFA), vol. 1, pp. 1035–1038. IEEE (2020)
3. Dave, D.M.K., Mittapally, B.K.: Data integration and interoperability in IoT: challenges, strategies and future direction (2024)
4. Drahoš, P., Kučera, E., Haffner, O., Klimo, I.: Trends in industrial communication and OPC UA. In: 2018 Cybernetics and Informatics (K&I), pp. 1–5. IEEE (2018)
5. Drath, R.: AutomationML: A Practical Guide. Walter de Gruyter GmbH & Co KG (2021)
6. Gilmore, W.J.: Introducing mySQL. Beginning PHP and MySQL: From Novice to Professional, pp. 621–633 (2008)
7. Glatt, M., Sinnwell, C., Yi, L., Donohoe, S., Ravani, B., Aurich, J.C.: Modeling and implementation of a digital twin of material flows based on physics simulation. J. Manuf. Syst. **58**, 231–245 (2021)
8. Grinberg, M.: Flask Web Development. O'Reilly Media, Inc., Sebastopol (2018)
9. Harris, C.R., et al.: Array programming with NumPy. Nature **585**(7825), 357–362 (2020)
10. Hillar, G.C.: Django RESTful Web Services: The Easiest Way to Build Python RESTful APIs and Web Services with Django. Packt Publishing Ltd, Birmingham (2018)
11. Ibrahim, M., Rjabtšikov, V., Gilbert, R.: Overview of digital twin platforms for EV applications. Sensors **23**(3), 1414 (2023)
12. Ko, T., Lee, J.H., Cho, H., Cho, S., Lee, W., Lee, M.: Machine learning based anomaly detection via integration of manufacturing, inspection and after sales service data. Indust. Manag. Data Syst. **117**(5), 927–945 (2017)

13. Kulkarni, V., Barat, S., Clark, T.: Towards adaptive enterprises using digital twins. In: 2019 Winter Simulation Conference (WSC), pp. 60–74. IEEE (2019)
14. Liu, Q., Liu, B., Wang, G., Zhang, C.: A comparative study on digital twin models. In: AIP Conference Proceedings, vol. 2073. AIP Publishing (2019)
15. Madni, A.M., Madni, C.C., Lucero, S.D.: Leveraging digital twin technology in model based systems engineering. Systems **7**(1), 7 (2019)
16. Martinez, G.S., Sierla, S., Karhela, T., Vyatkin, V.: Automatic generation of a simulation based digital twin of an industrial process plant. In: IECON 2018-44th Annual Conference of the IEEE Industrial Electronics Society, pp. 3084–3089. IEEE (2018)
17. McKinney, W., et al.: pandas: a foundational python library for data analysis and statistics. Python High Perform. Sci. Comput. **14**(9), 1–9 (2011)
18. Mykoniatis, K., Harris, G.A.: A digital twin emulator of a modular production system using a data driven hybrid modeling and simulation approach. J. Intell. Manuf. **32**(7), 1899–1911 (2021)
19. Rausch, T., Nastic, S., Dustdar, S.: Emma: distributed QoS aware MQTT middleware for edge computing applications. In: 2018 IEEE International Conference on Cloud Engineering (IC2E), pp. 191–197. IEEE (2018)
20. Reifsnider, K., Majumdar, P.: Multiphysics stimulated simulation digital twin methods for fleet management. In: 54th AIAA/ASME/ASCE/AHS/ASC Structures, Structural Dynamics, and Materials Conference, p. 1578 (2013)
21. Rosen, R., Von Wichert, G., Lo, G., Bettenhausen, K.D.: About the importance of autonomy and digital twins for the future of manufacturing. IFAC-Papersonline **48**(3), 567–572 (2015)
22. Schroeder, G.N., Steinmetz, C., Rodrigues, R.N., Henriques, R.V.B., Rettberg, A., Pereira, C.E.: A methodology for digital twin modeling and deployment for industry 4.0. Proc. IEEE **109**(4), 556–567 (2020)
23. Seabold, S., Perktold, J.: Statsmodels: econometric and statistical modeling with python. SciPy **7**, 1 (2010)
24. Segovia, M., Garcia-Alfaro, J.: Design, modeling and implementation of digital twins. Sensors **22**(14), 5396 (2022)
25. Stary, C., Elstermann, M., Fleischmann, A., Schmidt, W.: Behavior centered digital twin design for dynamic cyber physical system development. Complex Syst. Inform. Model. Q. (CSIMQ) **30**, 31–52 (2022)
26. Van Rossum, G., et al.: Python programming language. In: USENIX Annual Technical Conference, vol. 41, pp. 1–36. Santa Clara, CA (2007)

Peak Ground Acceleration Prediction for Earthquake Early Warning with Multivariable Long Short-Term Memory Networks and Temporal Transformers

Yhon Fuentes[1]([⊠]) [iD], Yessenia Yari[1] [iD], Aurea Soriano-Vargas[2] [iD],
and Anderson Rocha[2] [iD]

[1] National University of San Agustin of Arequipa (UNSA), Arequipa, Peru
{yfuentesh,yyarira}@unsa.edu.pe
[2] State University of Campinas (UNICAMP), Campinas, Brazil
{asoriano,arrocha}@unicamp.br

Abstract. Peru's seismic history, characterized by devastating earthquakes resulting in significant loss of life and property damage, underscores the urgency of effective early warning systems. Notably, events like the 1746 and 2007 Pisco earthquakes highlight the vulnerability of the region to seismic activity. In this context, this work presents a novel approach to earthquake early warning systems using deep learning architectures, specifically Long Short-Term Memory (LSTM) networks, and Temporal Fusion Transformer (TFT) networks. The study focuses on predicting Peak Ground Acceleration (PGA), a crucial parameter for issuing timely alerts to mitigate earthquake hazards. Using a comprehensive dataset comprising 5045 seismic records from various locations in Peru, the study employs LSTM and TFT networks to predict PGA values. Data preprocessing involves homogenizing acceleration records and dividing them into training, validation, and testing sets.

Results indicate that both LSTM and TFT networks demonstrate promising performance across different time windows (5, 30, and 60 s). For LSTM networks, the 60-s time window yields the most accurate predictions, with validation accuracy reaching 98.015% and testing accuracy at 88.89%. Meanwhile, TFT networks achieve competitive results, particularly with 30-s time windows, showing validation accuracy of 96.03% and testing accuracy of 91.32%. The findings underscore the potential of deep learning architectures in enhancing early warning systems, contributing to more effective disaster preparedness and response strategies in earthquake-prone regions.

Keywords: Earthquake Early Warning · Seismic Data · Deep Learning · Peak Ground Acceleration · PGA Prediction

L. Correia et al. (Eds.): IBERAMIA 2024, LNCS 15277, pp. 38–49, 2025.
https://doi.org/10.1007/978-3-031-80366-6_4

1   Introduction

Earthquakes are devastating natural disasters causing loss of life and property. With over one million earthquakes occurring annually worldwide, or an average of two every minute, major seismic events in urban areas rank among the most catastrophic natural disasters. Over the past four decades (1970–2017), earthquakes have claimed over a million lives across various regions, including Armenia, China, Ecuador, Guatemala, Haiti, Iran, India, Indonesia, Japan, Mexico, Pakistan, Peru, and Turkey [9].

Peru, with its lengthy history of seismic activity, stands as a region particularly vulnerable to the destructive forces of earthquakes. Throughout the centuries, Peru's seismic records have chronicled numerous impactful earthquakes, with fatalities ranging from 74 to 6000 between 1740 and 2019. Notable among these events is the 1746 earthquake, claiming approximately 1300 lives, while El Callao witnessed a staggering toll of around 3800 fatalities [2].

One of the most significant seismic events in Peru's history is the "Pisco" earthquake of August 15, 2007, which registered a magnitude of 7.0 on the Richter scale (ML) and 7.9 on the moment magnitude scale (Mw). Spanning an area of approximately 250 km, the Pisco earthquake left a profound impact on the region, resulting in extensive damage and loss of life. Reports from the National Institute of Civil Defense (INDECI) and the National Institute of Statistics and Informatics (INEI) highlighted the grim aftermath, with 32,000 individuals affected, 595 fatalities, and severe damage in 12 locations. Particularly hard-hit were Ica, Pisco, and Chincha, where over 230,000 houses sustained damage, with 52,150 destroyed due to structural deficiencies [13, 14].

In this context, early warning systems play a crucial role in mitigating the effects of earthquakes by providing valuable seconds of warning before strong ground shaking arrives [6]. The (PGA) is a key parameter used to assess earthquake intensity and trigger emergency protocols. Recognizing the urgency of effective earthquake early warning systems, this study introduces a novel approach leveraging the power of deep learning architectures. Deep learning offers promising avenues for addressing complex prediction tasks [3]. Its ability to autonomously learn intricate patterns from vast datasets makes it particularly suited for analyzing seismic data and forecasting critical parameters such (PGA).

Deep learning techniques have shown remarkable capability in handling the complexities and uncertainties inherent in temporal data analysis. Traditional methods often struggle to capture the nonlinear relationships and temporal dependencies present in seismic signals, limiting their predictive accuracy. In contrast, deep learning models, such as networks (LSTM) and (TFT), excel at learning from sequential data and extracting meaningful features, offering significant potential to improve prediction performance in earthquake early warning systems [7].

The motivation driving this research is multifaceted, rooted in the imperative to mitigate the impact of seismic events in Peru. Our main contributions encompass the prediction of (PGA) to facilitate timely early warning systems, prevention of civilian deaths resulting from high-magnitude earthquakes, early detection of earthquakes in vulnerable areas to safeguard lives, and utilization of seismic data and historical records to inform predictive models. Through the adoption of advanced deep learning architectures, we aim to improve prediction accuracy and response time in earthquake early warning systems, ultimately fostering greater resilience and disaster preparedness in earthquake-prone regions like Peru.

2   Related Work

Numerous studies have explored methodologies and technologies aimed at improving the effectiveness and reliability of early warning systems in mitigating the impact of seismic events.

Graizer and Kalkan [12], introduced a novel approach centered on the maximum ground acceleration (PGA) to devise a predictive model for 5% damped pseudospectral acceleration (S_A) ordinates [4]. Their model is parameterized within an approximation function leveraging momentum magnitude, fault distance (closest distance to the fault), and VS_{30} (average shear wave velocity in the upper 30 m) as independent variables. Saffari et al. (2012) also consider the VS_{30} in a model to estimate response spectra by considering real PGA, earthquake magnitude, distance, and VS_{30}. However, Zhu et al. (2022) introduced the HydriNet model, employing physics-based time series analysis focusing on amplitude (Pd), maximum velocity (P_V), and maximum velocity acceleration (P_a) [17].

Additionally, various technologies, such as IoT systems, XGB models, and Machine Learning (ML) algorithms, have been utilized to calculate maximum PGAs, achieving high accuracy rates like 98.59% [1] [11]. These models leveraged data from 386 stations within the Italian national seismic network, trained on 50,000 occurrences, comprising 150,000 two-second three-component (3C) seismic windows [1].

In Wang et al.'s [16] proposal, the model incorporates 8 sequential characteristics from stations, serving as indicators of energy and other physical parameters, as input to predict the Peak Ground Acceleration (PGA) recorded at the station. The study utilizes a dataset comprising 5,961 records from 119 earthquakes recorded by the Japan Strong Motion Earthquake Network (K-NET) for training purposes. Subsequently, 3,433 records from 73 earthquakes are reserved for validation to assess the model's generalizability. The model achieved an accuracy of 67.88% with Support Vector Machine (SVM) and 64.91% with Convolutional Neural Network (CNN), as measured by the Average Similarity Measure (ASM) [16].

In our proposed paper, we work with a substantial dataset comprising 31150987 seismic acceleration records derived from 5045 earthquakes in Peru. To optimize our analysis, we refined the dataset to 1119710 seismic records associated with these earthquakes, focusing on prediction with time windows of 5, 30, and 60 s. This proposed model is capable of issuing earthquake alerts 1 min in advance.

A comparison of the results obtained by the proposed neural networks (LSTM and TFT), the reading of the sensors and the mathematical model proposed by [8] was made, which shows the following equation.

$$log_{10}PGA = -3.93 + 0.78M_L - 1.5log_{10}R - \epsilon\sigma_{log}PGA \qquad (1)$$

Figure 1, shows the comparison of traditional models, AI, and data read by sensors, which are very similarly distributed.

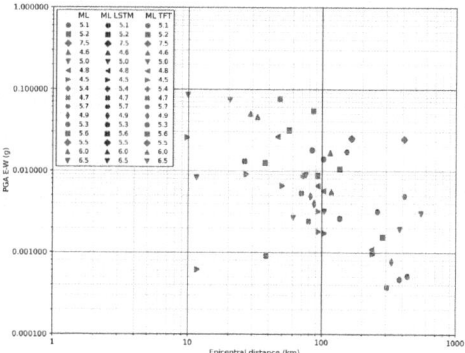

Fig. 1. Correlation between the empirical model proposed by Marin, PGAs taken by sensors and artificial neural networks

3 Methodology

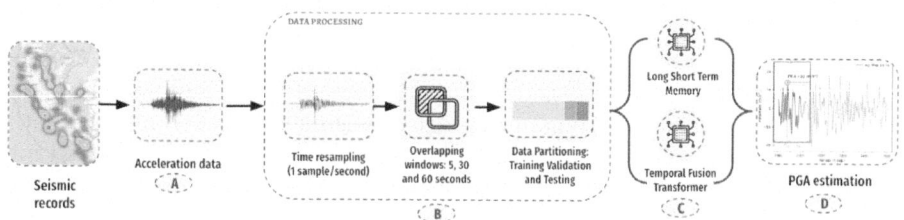

Fig. 2. Estimation methodology (PGA) involves three main stages: seismic record data collection (A), time series preprocessing encompassing resampling, windowing, and data partitioning (B), and model training and evaluation (C).

As illustrated in Fig. 2, the methodology for the PGA estimation encompasses data collection from seismic records (Fig. 2.A), time series preprocessing including resampling, windowing, and data partitioning (Fig. 2. B), model training and evaluation (Fig. 2. C).

3.1   Acceleration Data

Seismic data from diverse locations in Peru were gathered from reliable sources, notably the Geodetic Institute of Peru (IGP) and other seismic monitoring networks. This dataset encompasses accelerograms captured during seismic events throughout history, including earthquakes of varying magnitudes and intensities.

From the collected data, we organize the latitude, longitude, epicentral distance, magnitude, and ground acceleration. Ground acceleration data exhibit a range of formats and time resolutions, with sampling rates including 10, 50, 100, and 200 samples/second.

3.2   Data Preprocessing

The seismic data collected from the weather stations underwent a preprocessing consisting of: First, the preprocessing consisted of reading the plain text files downloaded from the stations that have different records of (time and seismic acceleration) and a single parameter of (magnitude, epicentral distance, latitude, longitude and depth), then these data were placed in a matrix repeating the data that are unique per seismic record then was saved in CSV files by earthquake. Second, the CSV records were homogeneous over time at a sampling rate of 1 sample per second. The reason is because the data is taken by different stations located at different points and each station takes the data at different times, for example: 10,20,50,100,200 samples per second. Then resampling was applied, which consisted of homogenizing the amount of data in a single time (to one sample per second), for example, from 10 samples per second I took windows of 10 samples and selected the maximum of that interval, if my records were at 50 samples per second, take windows of 50 samples and select the maximum of those 50 samples. With this procedure, the number of records was reduced and it can also be predicted with more future time. The next step involved the time window, a process by which ground acceleration data was segmented into discrete windows. This segmentation allowed the incorporation of temporal dependencies in the model by including past observations within each window as input characteristics. We consider time windows of 5, 30, and 60 s. This step was crucial in capturing the dynamic nature of seismic events and improving the model's predictive capabilities. Finally, the preprocessed data were divided into training, validation, and testing sets according to a 70-15-15 split ratio. This partitioning scheme ensured that the model was trained on a sufficient amount of data while also allowing for a robust assessment of its performance on unseen data.

3.3   Model Training and Evaluation

From the ground acceleration data windows and the tabular data (latitude, longitude, epicentral distance, and magnitude), two deep learning architectures, namely the multivariable (LSTM) network and (TFT) network, were proposed for PGA estimation at 3 different time windows prediction: 5, 30, 60 s.

Long Short-Term Memory. Networking (LSTM) is a specialized type of recurrent neural network (RNN) architecture designed to address the disappearing gradient problem. By integrating a memory cell and a series of gates, LSTMs effectively manage the flow of information, allowing long-term dependencies to be captured within sequential data. This attribute makes LSTMs very effective for tasks that involve sequential data processing, such as time series forecasting.

In our proposal, we introduce a novel architecture that leverages 128 initial layers sourced from four tabular input data: Latitude, Longitude, Magnitude, and Epicentral Distances, along with ground acceleration time windows. This model aims to predict seismic accelerations within time windows of 5, 30, and 60 s, akin to methodologies utilized in previous studies like [5,16]. Additionally, we employ the RMSprop optimizer and conduct training over 30 epochs to optimize model performance.

Temporal Fusion Transformer Network Model. The (TFT) Network Model is a deep learning architecture designed for time series forecasting tasks. It combines the strengths of both Transformer networks and traditional time series forecasting methods to effectively capture temporal dependencies and make accurate predictions.

In this model, the Transformer architecture is adapted to handle temporal sequences by incorporating positional encodings to represent the time order of input data. The model consists of multiple layers of self-attention mechanisms, which allow it to attend to relevant temporal patterns across different time steps in the input sequence. Additionally, the model utilizes feedforward neural networks and residual connections to capture non-linear relationships within the data.

The "fusion" aspect of the model refers to its ability to integrate information from multiple sources or features. This allows the model to leverage diverse types of data, such as time series measurements, categorical variables, and external factors, to improve forecasting accuracy.

We also propose a TFT from four tabular input data: Latitude, Longitude, Magnitude, and Epicentral Distances, along with ground acceleration time windows. This model aims to predict seismic accelerations within time windows of 5, 30, and 60 s.

For model evaluation, we employ Mean Absolute Percentage Error (MAPE) and Symmetric Mean Absolute Percentage Error (SMAPE) metrics to assess the accuracy of time series forecasting for ground accelerations.

MAPE quantifies the average percentage difference between predicted and actual values, calculated as the mean of the absolute percentage errors across all observations. SMAPE measures the relative accuracy of forecasts, considering both underestimation and overestimation errors.

In addition to these metrics, we introduce an evaluation criterion for earthquake early warning systems based on a classification problem. Specifically, we utilize an average PGA threshold of $20 \, \text{cm/s}^2$ to classify earthquakes as either warranting an early warning or not. This threshold serves as a benchmark for assessing the effectiveness of the two forecasting models in predicting seismic events. To address the imbalanced nature of earthquake warnings, we employ the F1-score and balanced accuracy metrics for evaluation. The F1-score represents the harmonic mean of precision and recall, offering a comprehensive measure of model performance. On the other hand, Balanced Accuracy provides a balanced assessment of model accuracy by considering both sensitivity (true positive rate) and specificity (true negative rate) of the classifier.

3.4   Peak Ground Acceleration (PGA)

The PGA represents the maximum seismic acceleration recorded by an accelerograph within a specific period during an earthquake event. This value, as described by Thenhaus et al. [15], can fluctuate between positive and negative extremes, but its significance lies in its magnitude, which aids in determining the severity of an earthquake.

High values of PGA within a seismic record are indicative of intense ground shaking and can provide valuable insights into the potential damage caused by the earthquake. Utilizing PGA, in conjunction with other pertinent parameters, allows for the estimation of earthquake magnitude and helps in assessing the potential impact on structures and infrastructure.

Figure 3 illustrates the computation of PGA across time windows of 5, 30, and 60 s. When PGA values surpass certain thresholds, it triggers alerts signaling high seismic magnitudes, highlighting the importance of monitoring and analyzing PGA data for earthquake early warning systems.

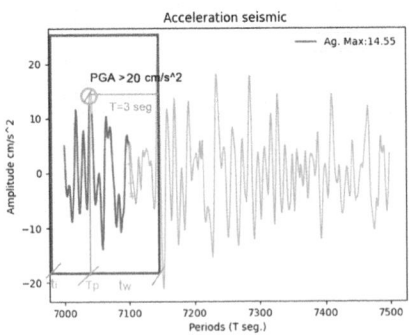

Fig. 3. Prediction of seismic accelerations, PGA greater than $20 \, \text{cm/s}^2$ with time windows of 5,30,60 s

4   Experimental Results and Discussion

4.1   Dataset

For validation purposes, we utilized seismic data collected from accelerographic stations situated across various locations in Peru, encompassing a total of 5045 seismic events recorded between 2021 and 2024. These stations are operated by three distinct institutions: the Instituto Geofísico del Perú (IGP) [10], Colegio de Ingenieros del Perú, and Sencico. Each station is equipped with accelerographs that sample seismic movements at different rates, ranging from 10 to 200 samples per second.

From the initial pool of 5045 seismic records, we extracted a total of 31150987 seismic acceleration data points. Subsequently, all samples were standardized to a uniform sampling rate of 1 sample per second, resulting in a consolidated dataset comprising 1119710 seismic acceleration records. These standardized data were then utilized for testing purposes, employing time windows of 5, 30, and 60 s for analysis. This rigorous validation process ensures the robustness and reliability of our predictive models in forecasting seismic events.

From the data, which consists of a total of 1119710 records, 70%, 15%, and 15% were divided for training, validation, and testing. They were organized according to the dates of occurrence.

4.2   Ground Acceleration Forecasting

Table 1 presents (MAPE) and (SMAPE) of the LSTM and TFT models in different time windows: 5, 30, and 60 s. These metrics provide insight into the accuracy and performance of models in forecasting ground accelerations.

Table 1. MAPE and SMAPE of the LSTM and TFT models across different time windows: 5, 30, and 60 s.

Window	LSTM		TFT	
	MAPE	SMAPE	MAPE	SMAPE
5 s	0.0024	0.1206	0.0981	0.2780
30 s	0.0027	0.1396	0.0952	0.2649
60 s	0.0017	0.0889	0.0953	0.3125

Based on Table 1, the best performance of LSTM was evident with a 60-s window, achieving a 0.0889 SMAPE and a 0.0017 MAE. Conversely, TFT demonstrated its strength with a 30-s window, yielding a 0.0952 MAE and 0.2649 SMAPE, showcasing its effectiveness under different temporal contexts.

In Figs. 4 and 5, we visualize the RMSE loss of our LSTM and TFT models during training and testing phases across time windows of 5, 30, and 60 s. The red line represents the model loss on the training data, indicating how effectively it learns from the provided data. Meanwhile, the blue line illustrates the loss of the test data, reflecting the model's performance on unseen examples of earthquakes, which were not part of the training set.

Our experimentation initially involved testing the model with various numbers of epochs, ranging from 10 to 30. Ultimately, we found that the optimal results were achieved with 30 epochs, as demonstrated by the plot. It's noteworthy that the network's performance could potentially continue to improve with additional epochs, as evidenced by the trends observed in the experiments.

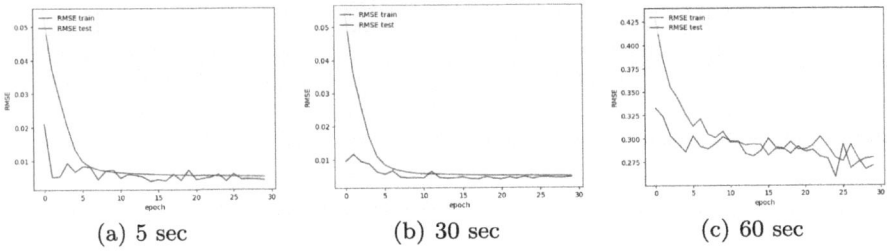

|(a) 5 sec|(b) 30 sec|(c) 60 sec|

Fig. 4. RMSE loss of our LSTM model during training and testing phases across time windows of 5, 30, and 60 s.

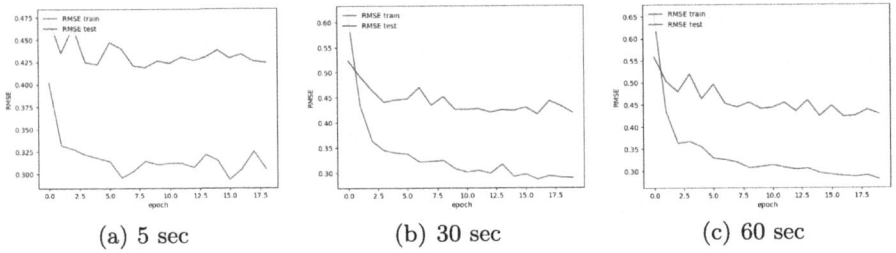

|(a) 5 sec|(b) 30 sec|(c) 60 sec|

Fig. 5. RMSE loss of our TFT model during training and testing phases across time windows of 5, 30, and 60 s.

4.3 PGA for Early Earthquake Warning Classification

Table 2 showcases the F1-score and balanced accuracy metrics for both the LSTM and TFT models, evaluated across various time windows: 5, 30, and 60 s, treating early warning as a classification problem. We utilize an average (PGA) threshold of 20 cm²/s as a criterion for classifying earthquakes, determining whether they necessitate an early warning. These metrics offer insights into the classification performance of the models in predicting early earthquake warning events.

Table 2. F1-score and Balanced accuracy of the LSTM and TFT models across different time windows: 5, 30, and 60 s.

Window	LSTM		TFT	
	F1-score	B. Accur	F1-score	B. Accur
5 s.	0.9259	0.9623	0.9434	0.9639
30 s.	0.9434	0.9630	0.9630	0.9815
60 s.	0.8727	0.9444	0.9999	0.9808

Based on the findings presented in Table 2, for the early warning classification of earthquakes with PGAs exceeding 20*gal*, LSTM showcased its prowess with a 30-s window, achieving a balanced accuracy of 0.9630 and an F1-score of 0.9434. Conversely, TFT obtained the best result with the same window, attaining a balanced accuracy of 0.9815 and an F1-score of 0.9630. Notably, across all experimental protocols, there was consistent evidence of strong performance in early warning earthquake classification.

Figures 6 and 7 provide insights into the classification performance of both the LSTM and TFT models, employing an average (PGA) threshold of $20\,cm^2/s$ for earthquake classification across time windows of 5, 30, and 60 s.

Throughout the experiments, a total of 27 earthquakes surpassing the $20\,cm^2/s$ PGA threshold were identified. In the 5-s and 30-s experiments, the LSTM model accurately classified 25 out of 27 predicted warnings, achieving an effectiveness of 92.59% according to the proposed model criteria. Similarly, in the 60-s experiment, out of the 27 alert data points, the LSTM model correctly predicted 24 alarms, demonstrating an effectiveness of 88.89%.

Furthermore, in both the 5-s and 60-s experiments, the TFT model exhibited comparable performance, accurately classifying 25 out of 27 predicted warnings. This resulted in an effectiveness of 92.59% according to the proposed model criteria.

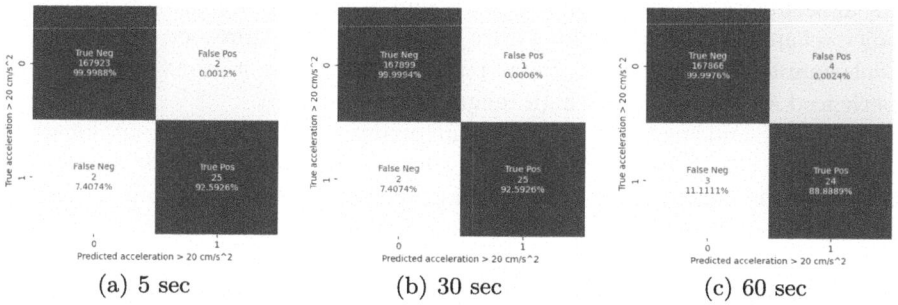

(a) 5 sec   (b) 30 sec   (c) 60 sec

Fig. 6. Confusion matrices for our LSTM model are depicted when utilizing the classification protocol across time windows of 5, 30, and 60 s.

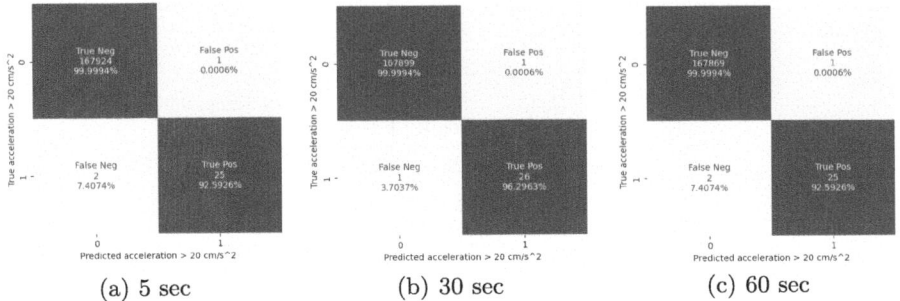

(a) 5 sec   (b) 30 sec   (c) 60 sec

Fig. 7. Confusion matrices for our TFT model are depicted when utilizing the classification protocol across time windows of 5, 30, and 60 s.

5   Conclusions and Future Work

In this study, we introduce a methodology for earthquake early warning systems utilizing deep learning architectures. We compare the performance of (LSTM) and (TFT) networks across three different time windows: 5 s, 30 s, and 60 s.

We assess the models' performance through two protocols: predicting ground acceleration data evaluated with MAPE and SMAPE metrics, and classifying early warnings evaluated with F1-score and balanced accuracy to address the imbalanced nature of the data.

The LSTM model achieves the best performance in predicting ground acceleration data with a 60-s window, yielding an SMAPE of 0.0889 and an MAPE of 0.0017. For early warning classification, the TFT model with a 60-s window achieves the highest result, with an F1-score of 0.9999 and a balanced accuracy of 0.9808.

The utilization of deep learning models necessitates significant computational resources. Nonetheless, this research lays the groundwork for future advancements in early warning prediction. By adjusting the seismic acceleration parameters in seismic acceleration blocks with uniform times, it becomes feasible to predict seismic accelerations for entire seismic events. Moreover, the proposed models enable the generation of new response spectra in stations that have not experienced seismic events of high magnitudes previously.

For future endeavors, experiments involving data from different locations play a pivotal role in demonstrating the generalization capability of the models. Additionally, exploring data augmentation techniques based on time series and varying time windows from historical data is imperative.

Acknowledgments. This work was financed by CONCYTEC – FONDECYT, under the "Program for Doctorates in Peruvian Universities" [Contract No. 173-2020-FONDECYT]. Special thanks to the Universidad Nacional de San Agustín de Arequipa and the University of Campinas for making it possible to carry out the research proposed in this article.

Disclosure of Interests. The authors have no competing interests to declare that are relevant to the content of this article.

References

1. Abdalzaher, M.S., Soliman, M.S., El-Hady, S.M.: Seismic intensity estimation for earthquake early warning using optimized machine learning model. IEEE Trans. Geosci, Remote Sens (2023)
2. Carcelén, C., Morán, D., Amador, L.: El terremoto de 1746 y su impacto en la salud en la ciudad de lima. Rev. Peru. Med. Exp. Salud Publica **37**, 164–168 (2020)
3. Geng, Y., Su, L., Jia, Y., Han, C., et al.: Seismic events prediction using deep temporal convolution networks. J. Electr. Comput. Eng. **2019** (2019)
4. Graizer, V., Kalkan, E.: Prediction of spectral acceleration response ordinates based on PGA attenuation. Earthq. Spectra **25**(1), 39–69 (2009)
5. Hsu, T.Y., Pratomo, A.: Early peak ground acceleration prediction for on-site earthquake early warning using LSTM neural network. Front. Earth Sci. **10**, 911947 (2022)
6. Irwansyah, E., Winarko, E., Rasjid, Z., Bekti, R.: Earthquake hazard zonation using peak ground acceleration (PGA) approach. J. Phys. Conf. Ser. **423**, 012067. IOP Publishing (2013)
7. Li, Q., et al.: PGA-net: polynomial global attention network with mean curvature loss for lane detection. IEEE Trans. Intell. Trans. Syst. (2023)
8. Marin, S., Avouac, J.P., Nicolas, M., Schlupp, A.: A probabilistic approach to seismic hazard in metropolitan France. Bull. Seismol. Soc. Am. **94**(6), 2137–2163 (2004)
9. Pan American Health Organization: Earthquakes. https://www.paho.org/en/topics/earthquakes, Accessed 12 Apr 2024
10. IG of Peru: IGP. https://www.igp.gob.pe/servicios/aceldat-peru/reportes-registros-acelerometricos. Accessed 28 June 2024
11. Saad, O.M., Helmy, I., Mohammed, M., Savvaidis, A., Chatterjee, A., Chen, Y.: Deep learning peak ground acceleration prediction using single-station waveforms. IEEE Trans. Geosci. Remote Sens. **62**, 1–13 (2024)
12. Saffari, H., Kuwata, Y., Takada, S., Mahdavian, A.: Updated PGA, PGV, and spectral acceleration attenuation relations for Iran. Earthq. Spectra **28**(1), 257–276 (2012)
13. Tavera, H.: El terremoto de la región del sur del perú del 23 de junio de 2001. Informes Técnicos (2001)
14. Tavera, H.: El terremoto de pisco (perú) del 15 de agosto de 2007 (7.9 mw). Informes Técnicos (2008)
15. Thenhaus, P.C., Campbell, K.W., Chen, W., Scawthorn, C.: Seismic hazard analysis. Earthq. Eng. Handb. **8**, 1–50 (2003)
16. Wang, A., Li, S., Lu, J., Zhang, H., Wang, B., Xie, Z.: Prediction of PGA in earthquake early warning using a long short-term memory neural network. Geophys. J. Int. **234**(1), 12–24 (2023)
17. Zhu, J., Li, S., Song, J.: Hybrid deep-learning network for rapid on-site peak ground velocity prediction. IEEE Trans. Geosci. Remote Sens. **60**, 1–12 (2022)

Green Security Along Trails

Nicolas Betancourt$^{(\boxtimes)}$ and Mauricio Velasco

Departamento de Informática, Universidad Católica del Uruguay (UCU), Av. 8 de Octubre, 2738 Montevideo, Uruguay
nicolasbetancourt80@gmail.com

Abstract. Park rangers worldwide are tasked with patrolling protected areas to prevent illegal activities, a challenge compounded by the vastness of the land compared to the available manpower. This is exacerbated by the dense vegetation, the rugged terrain, and the adaptability of lawbreakers, compounded by a lack of historical data. To address these constraints, collaborative efforts between park rangers and the computer science community have led to theoretical advances and novel practical implementations. However, the unique features of tropical parks, including vegetation, geography, and the behavior of illegal actors, require adaptation of currently available tools. In this work, we propose two key contributions: employing combinatorial multi-armed bandits for sequential patrol route suggestions, tailored to the trail networks common in Latin American parks, and integrating our algorithm with an acoustic monitoring system. We provide theoretical performance guarantees as well as computational simulations, using information provided by the Jama-Coaque Ecological Reserve (JCR) in Ecuador.

Keywords: Combinatorial multi-armed bandits · Green security · Combinatorial optimization over networks · Green and Sustainable Artificial Intelligence · AI for social good

1 Introduction to the Rangers' Problem

The Jama-Coaque Ecological Reserve (JCR) is one of the last significant remnants of tropical moist forest and premontane cloud forest in the world. It is constantly under threat by poachers and illegal loggers. The reserve is patrolled by a small group of rangers on a daily basis. To make walking through the rugged terrain and the thick vegetation easier, rangers have established a system of $n = 154$ trails. These trails are the edges of a *graph of trails* which we denote by G.

A set of B available rangers visits some of the trails looking for illegal activity. Every spot of the reserve other than the rangers' shelter is unsuitable for resting. This forces every patrol route to start at the rangers' shelter and to eventually return there, even if camping is considered. As a result, the set $\mathcal{C}_G$ of feasible patrolling routes for the JCR staff consists of paths on G that are cycle-shaped and go through the rangers' shelter.

© The Author(s), under exclusive license to Springer Nature Switzerland AG 2025
L. Correia et al. (Eds.): IBERAMIA 2024, LNCS 15277, pp. 50–61, 2025.
https://doi.org/10.1007/978-3-031-80366-6_5

The *rangers' problem* is to choose B different circuits on $\mathcal{C}_G$ every day so that the success of the resulting patrol policy (the number of encounters between rangers and illegal actors) is maximized. This problem is particularly difficult because it simultaneously involves a learning problem, namely trying to estimate the unknown distribution of illegal activity, and a combinatorial optimization problem, which results from having to select routes from the enormous space of available options: In the particular case of Jama-Coaque, among the more than 130.000 available circuits, 10.216 of which cross the rangers' shelter. Currently, there are $B = 4$ rangers in Jama-Coaque and therefore the problem is to appropriately choose quadruples of patrol routes among the $\binom{10.000}{4} \approx 10^{14}$ alternatives daily.

The main contribution of this article is to propose a novel route recommendation algorithm for solving the rangers' problem. The key idea is to observe that the framework of combinatorial multi-armed bandits is very well suited for it. We prove theoretical success guarantees for the algorithm and illustrate its performance on practical simulations built on data from the JCR. We believe these results serve as a representative sample of a wide variety of protected areas of tropical rainforest and thus give relevant evidence in favor of the practicality of the proposed approach.

As a second contribution, we suggest a method for incorporating real-time alerts into our system, enabling rangers to respond to threats by considering critical regions identified by the route suggestion algorithm. This approach not only utilizes the information provided by our system but also enhances the responsiveness of the route suggestion algorithm to sophisticated illegal actors, as further discussed in later Sections.

This algorithm has not been implemented on the field yet. This would require developing an app for rangers to receive route recommendations, monitor related statistics and for us to measure the resulting effectiveness. Field implementation and the related technological development are part of the next research steps of this project.

Related Work. The task of equipping rangers with technological tools giving them an advantage over illegal actors has led to various advances. Most of these contributions rely on supervised learning models to automatically detect illegal activity on images (see [2]) or to predict the next attack's location and timing (see [6,12,15]). In some cases, the availability of a predictive model is assumed and used as input for a decision-making algorithm, resulting in an optimal patrol policy (see [12,13]) assuming the validity of the model. Although some works acknowledge that predictive models are not always available and propose alternatives (see [7]), almost all approaches depend directly or indirectly on the availability of significant historical data.

To the best of our knowledge there is only one work which develops an assistant tool for park rangers which does not rely on historical data. The authors of [14] developed a unique assistant tool for park rangers that efficiently collects real-time data to inform future models. Their algorithm utilizes a combinatorial multi-armed bandit framework, selecting effort levels for each cell in a discretized grid model of the protected area. In this instance, the arms are tuples of effort levels and cells, each equipped with a data vector. These vectors contain relevant information such as the distance from the cell to water bodies and reserve boundaries, providing the model with a notion of similarity between arms. Super arms are collections of cells together with effort levels that do not exceed a maximum threshold. By optimizing convergence rates based on similar characteristics, their algorithm surpasses other bandit algorithms, particularly under the assumption of Lipschitz continuity of the reward function.

These theoretical contributions stem from questions arising in practical fieldwork, shaped by collaborative efforts with personnel from national parks in Uganda and Cambodia. While the understanding of the daily tasks of park rangers, the operating methods of poachers, and the physical constraints are applicable to the African Savannah and the dry forests of Indochina, we find these observations less relevant for many Latin American contexts. Regions like the Amazon and the Tumbes-Chocó-Magdalena hotspot, feature rugged terrain and dense vegetation, limiting movement to trail networks for both park rangers and illegal actors. Furthermore, illegal activities in these areas, such as hunting with firearms and logging, occur persistently, reducing the need for extensive search for traps and other hunting devices. The need for area discretization and effort level determination is thus often unnecessary in addressing similar issues in other regions.

As far as we know, our work is the first to consider the trail structure for a decision-making model that does not rely on historical data. By not requiring a discretization of the terrain, our model does not suffer from the dilemma of choosing between low precision and low dimensionality. The number of super arms depends exponentially on an invariant feature over time which already gives geographical accuracy: the number of trails n. Similarly, not having to consider a discrete set of effort levels avoids increasing the set of super arms. We also leverage the combinatorial structure of the optimization set to reduce the initialization phase. Our algorithm optimizes directly over the set of feasible routes, ensuring that possible patrols occur only through valid routes avoiding the task of deciding how to connect the critical areas to be visited.

While there are several initiatives focused on using automatic acoustic detection for preventing poaching and logging [4,5,8,10,11], none, to our knowledge, addresses the critical issue of how park rangers should respond to threats detected by the system with imprecise location pinpointing. This is particularly important in tropical areas where foot patrol is the primary means of reserve traversal. Our research fills this gap by considering real-time alerts to augment existing patrol strategies, offering a novel approach to countering the adaptability of intruders.

1.1   Multi-armed Bandits Preliminaries

With the aim of describing our results with greater precision, we begin with some preliminaries. An n-armed bandit is a process determined by a fixed amount of trials or times $t = 1, \ldots, T$ and a set of random variables $X_1, \cdots, X_n$ supported on $[0, 1]$, called arms, with finite means $\mu_i := \mathbb{E}[X_i]$ $(1 \leq i \leq n)$. The distributions and the means of the arms are unknown. At each time t our agent is allowed to sample one of the arms perceiving a reward denoted $x_t^{i_t}$.

If $\mu^* := \max\{\mu_1, \cdots, \mu_n\}$ and $i^* := \operatorname{argmax}\{\mu_1, \cdots, \mu_n\}$ the decision rule that maximizes the expected cumulative reward of the agent after T stages is to always choose the i^*−th arm. This results in a expected cumulative reward of $T\mu^*$. The difficulty lies in the fact that the μ_i's are unknown. The agent thus needs to allocate some of its trials for estimating these quantities. Regardless of the picked rule, the expected reward will always under-perform choosing i^* consistently. A way of measuring the success of a given rule after t stages is by comparing it to this ideal rule, by computing the difference between $t\mu^*$ and the expected return up to stage t of the given rule.

Definition 1. *The regret at stage t of a rule A that has chosen arm indices $i_1, i_2, \cdots, i_t$ and perceived outcomes $x_1^{i_1}, \cdots, x_t^{i_t}$ is defined by*

$$R(t) := t\mu^* - \mathbb{E}\left[\sum_{s=1}^{t} x_s^{i_s}\right] \tag{1}$$

The best course of action in estimating i^* is to approximate the expected value of all of the arms using the sample mean. The Law of Large Numbers states that for every arm X_i the sample mean $\overline{X}_{i,t} = \frac{1}{t}\sum_{s=1}^{t} x_s^i$ is a consistent estimator of μ_i i.e. for every $\epsilon > 0$

$$\lim_{t \to \infty} \mathbb{P}\left\{\left|\overline{X}_{i,t} - \mu_i\right| > \epsilon\right\} = 0 \tag{2}$$

However, several samples might be required for each arm in order to make satisfactory estimations. It is reasonable then to consider not only the sample means $\overline{X}_{i,t}$, but also the size of the sample up to stage t for every arm index i, $T_i(t)$. Moreover, a decision rule considering this should incorporate the information on whether or not the sample size is enough for a good estimation as summarized by the Chernoff inequality.

Theorem 1 (Chernoff inequality). *Let $\overline{X}_{i,t}$ be the mean of the first t trials of the i−th arm. Then for every $A > 0$*

$$\mathbb{P}\{\mu_i - \overline{X}_{i,t} \geq A\} \leq \exp(-2kA^2) \tag{3}$$

$$\mathbb{P}\{\overline{X}_{i,t} - \mu_i \geq A\} \leq \exp(-2kA^2) \tag{4}$$

In 2002 Auer, Cesa-Bianchi and Fischer propose a decision rule considering exactly these three variables [1], the *Upper Confidence Bound (UCB)*. At each stage, the UCB estimates the expected values of the arms using the available sample means and adds a distortion that measures how big the size of the sample is with respect to the number of elapsed stages

$$\overline{\mu}_{i,t} := \overline{X}_{i,T_i(t-1)} + \sqrt{\frac{3\ln(t-1)}{2T_i(t-1)}} \tag{5}$$

$$i_t^* = \mathrm{argmax}\{\overline{\mu}_{1,t}, \cdots, \overline{\mu}_{n,t}\} \tag{6}$$

The UCB selects the arm with maximal $\overline{\mu}_{i,t}$ at each stage and updates the parameters for the next stage. The formula for the distortion is designed by considering the rate of convergence of the Law of Large Numbers to attain a desired asymptotic behavior. The UCB's regret increases at most logarithmically with the number of stages.

Theorem 2 (Auer, Cesa-Bianchi). *If the upper confidence bound policy is implemented then*

$$R(t) \le 8\ln(t) \sum_{i \neq i^*} \frac{1}{\mu^* - \mu_i} + \left(1 + \frac{\pi^2}{3}\right) \sum_{i \neq i^*} (\mu^* - \mu_i) \tag{7}$$

Furthermore, Lai and Robbins proved that the regret $R(t)$ of any decision rule at stage t is at least $O\left(\log(t)\right)$, which shows that the UCB is asymptotically optimal [9, Theorem 1, pg 6]. The UCB also applies to more general MAB frameworks. In the combinatorial Multi-armed Bandits (CMAB) setup, a subset of the arms (known as super-arm) from a given collection can be selected at each stage. The resulting reward is a function of the sampled values of the arms belonging to the super-arm S which we denote by $Rw(S)$.

This increases the number of alternatives at each stage and thus the framework would seem to require more samples for estimation. Nonetheless, [3, Section 2, pg 4] proved that if the function $r_\mu(S) := \mathbb{E}\left[Rw(S)\right]$ satisfies mild analytical conditions then the UCB can be adapted to this general setting while keeping its optimal behavior. On its combinatorial modification, called Combinatorial Upper Confidence Bound (CUCB), the means are still estimated using the sample means and the distortion is defined by

$$\overline{\mu}_{i,t} := \overline{X}_{i,T_i(t-1)} + \sqrt{\frac{3\ln(t-1)}{2T_i(t-1)}}. \tag{8}$$

The super-arm maximizing $r_{\overline{\mu}}(\cdot)$ is selected at each stage just as in the regular case

$$S_t^* = \mathrm{argmax}\{r_\mu(S) : S \text{ in super-arms}\} \tag{9}$$

The main difference lies in the optimization step for in CUCB this maximization may be difficult to solve. In many examples, the optimization step of [3, Section 4.1] is in fact NP-hard. Considering this, the authors proved that even when the optimization step is solved approximately, the regret keeps growing logarithmically when CUCB is implemented. For more information on how the quality of the approximation affects the performance of CUCB the interested reader should refer to [3, Section 3, Theorem 1].

2   The Rangers' Problem as a CMAB

The rangers' problem is to choose elements from $\mathcal{C}_G$ sequentially incorporating all constraints and maximizing the success of the patrol policy. Our main result is to model the rangers' problem as a stochastic combinatorial optimization problem. Acknowledging that some of the input parameters are unknown, the problem will be modeled as a CMAB.

Initially, we make two assumptions about poachers. First, we assume a fixed but unknown probability distribution of illegal activity. Specifically, we model their actions a collection of independent Bernoulli variables indexed by trails, each representing the likelihood of encountering illegal activity on that trail at any given day. This assumption is grounded in reality for two reasons. First, regarding logging, trails with higher timber concentrations typically experience increased illegal activity. Second, while poaching is more dynamic, the assumption may still apply as intruders often access reserves through fixed points along the boundary, such as near villages, which remain static over time. Additionally, proximity to rivers or roads may make certain trails more appealing for resource extraction.

The second assumption is that the Bernoulli variables will remain unchanged through time. This assumption will not hold in practice for the poachers will respond to unwelcome encounters with rangers by changing their behavior. In later sections of the article, we explore how a reactive poacher impacts the performance of our algorithm.

2.1   Theoretical Guarantees

Let t_i be the Bernoulli variable modeling the presence of illegal activity on the i−th trail ($i \in \{1, \cdots, n\}$). If $\eta \in \{0,1\}^{\#\mathcal{C}_G}$ encodes rangers' routes i.e. $\eta_c = 1$ if and only if some ranger picked circuit C, then, the number of trails (with multiplicity) in the intersection of the rangers' and the poacher's route is given by

$$Rw(t,\eta) = \sum_{C \in \mathcal{C}_G} \eta_C \sum_{i \in C} t_i \qquad (10)$$

If μ_i is the probability of finding illegal activity in the i−th trail then the expected number of trails in the intersection, given the route choices η, is

$$r_\mu(\eta) := \mathbb{E}[Rw(t,\eta)] = \sum_{C \in \mathcal{C}_G} \eta_C \sum_{i \in C} \mu_i \qquad (11)$$

Lemma 1. *If the team of rangers knows the values of the vector of probabilities μ, then the choice η that maximizes $r_\mu(\eta)$ is an optimal solution to the following integer linear programming (ILP) problem,*

$$(RP)_\mu \begin{cases} \max\limits_{\eta \in \{0,1\}^{\#\mathcal{C}_G}, y \in \mathbb{Z}^n} & \sum_{i=1}^n \mu_i y_i \\ & s.t.: \\ & \sum_{C \in \mathcal{C}_G} \eta_C = B \\ & y_i = \sum_{C:i \in C} \eta_C \\ & \eta_C \in \{0,1\}, y_i \in \{0,1,\cdots,B\} \end{cases} \qquad (12)$$

Proof. Note that

$$r_\mu(\eta) = \sum_{C \in \mathcal{C}_G} \eta_C \sum_{i \in C} \mu_i$$
$$= \sum_{i=1}^n \sum_{C \in \mathcal{C}_G} \eta_C \mu_i 1_{i \in C}$$

where $1_{i \in C}$ indicates whether or not i is in C.

$$r_\mu(\eta) = \sum_{i=1}^n \sum_{C \in \mathcal{C}_G} \eta_C \mu_i 1_{i \in C}$$
$$= \sum_{i=1}^n \mu_i \sum_{C:i \in C} \eta_C$$
$$= \sum_{i=1}^n \mu_i y_i$$

therefore, r_μ is upper bounded by the optimal value of Problem (12) and that optimal value is equal to $r_\mu(\eta^*)$ for some η^* satisfying the constraints of the rangers' problem.

Recall that the CUCB framework allows for approximate optimization while maintaining asymptotic guarantees. It is therefore natural to ask for efficiently solvable approximations to the ILP (12). The following Lemma shows that problem (12) is equivalent to a relaxation with random rounding, implying that we can perform the optimization step in polynomial time without requiring an approximation.

Theorem 3. *Consider the relaxation of* $(RP)_\mu$ *given by*

$$(\widetilde{RP})_\mu \begin{cases} \max\limits_{\tilde\eta \in \mathbb{R}^{\#\mathcal{C}_G}, \tilde y \in \mathbb{R}^n} & \sum \mu_i \tilde y_i \\ & s.t.: \\ & \sum_{C \in \mathcal{C}_G}^M \tilde\eta_C = B \\ & \tilde y_i = \sum_{C:i \in C} \tilde\eta_C \\ & \tilde\eta_C \in [0,1], \tilde y_i \in [0,B] \end{cases} \qquad (13)$$

Let $\tilde{\eta}^*, \tilde{y}^*$ be optimal solutions to problem 13. Consider the experiment of sampling B circuits (independently and with replacement) using $\tilde{\eta}^*/B$ and let $\eta^* \in \{0,1\}^M$ be the binary vector indicating the resulting selected sets. Then η^* is an optimal solution to problem 12.

Proof. Let Z_i be the random variable that counts the number of selected circuits crossing the i−th trail. For $1 \leq l \leq B$ let X_l^i be the random variable which is 1 when the l−th sampled cycle contains the i−th trail. Then $Z_i = \sum_{l=1}^{B} X_l^i$. Since the sample is independent and with replacement, $\{X_l^i\}_{l=1}^{B}$ is a collection of i.i.d Bernoulli random variables with success probabilities given by

$$\sum_{C:i\in C} \tilde{\eta}_C^*/B = \tilde{y}_i^*/B \tag{14}$$

This proves that the expected value of Z_i is $\tilde{y}_i^*$. Note that $r_\mu(\eta^*) = \sum \mu_i Z_i$ and then the expected value of $r_\mu(\eta^*)$ is $\sum \mu_i \mathbb{E}[Z_i] = \mathrm{Opt}(\widetilde{RP})_\mu$ where $\mathrm{Opt}\,(\widetilde{RP})_\mu$ the optimal value of problem (13).

The feasible region of problem (12) is a subset of the feasible region of problem (13), therefore

$$\mathrm{Opt}\,(\widetilde{RP})_\mu \geq \mathrm{Opt}(RP)_\mu \tag{15}$$

Besides, since η^* is in the feasible region of problem 12 then $r_\mu(\eta^*) \leq \mathrm{Opt}(RP)_\mu$ proving that $\mathrm{Opt}(RP)_\mu = r_\mu(\eta^*) = \mathrm{Opt}(\widetilde{RP})_\mu$.

In practice however, rangers ignore the exact values of μ_i and have very few or no historical poaching data for its estimation. We overcome this problem by modeling the rangers's problem as a CMAB. In this instance the arms are the Bernoulli variables indexed by the trails and the super-arms are collections of trails contained in any selection of B distinct circuits i.e. collections of arms induced by the vectors $\eta \in \{0,1\}^{\#C_G}$. Implementing CUCB in our problem requires us to solve problem (12) at every stage, using the estimation for the success probabilities provided by the algorithm. Specializing the regret bounds in [3, Section 3, Theorem 1] we obtain the following theoretical guarantee for this procedure,

Theorem 4. *The regret of the team of rangers up to stage t after following the CUCB policy $R(t)$, is at most*

$$n\Delta_{max}\left(\frac{6n^2 B^2 \ln(t)}{\Delta_{min}^2} + \frac{\pi^2}{3} + 1\right) \tag{16}$$

where

$$\Delta_{max}^i = \mathrm{Opt}_\mu - \min\{r_\mu(\eta) : \eta \text{ crosses } i, \ r_\mu(\eta) < \mathrm{Opt}_\mu\} \tag{17}$$

$$\Delta_{min}^i = \mathrm{Opt}_\mu - \max\{r_\mu(\eta) : \eta \text{ crosses } i, \ r_\mu(\eta) < \mathrm{Opt}_\mu\} \tag{18}$$

are the difference between the performance of the optimal route Opt_μ with the worst and the second best route respectively and the variables

$$\Delta_{min} = \min_i \Delta_{min}^{i,j} \tag{19}$$

$$\Delta_{max} = \max_i \Delta_{max}^{i,j} \tag{20}$$

are the minimum and maximum of those quantities over the trails

Running the CUCB strategy above requires an initialization: for every arm one should pull a super-arm containing it. Since the patrolling routes are chosen on a daily basis, pulling $n = 154$ super-arms would take more than 5 months. But leveraging the combinatorial structure of our super-arms, the initialization requires considerably less resources. We found out that a minimal collection of circuits in $\mathcal{C}_G$ crossing all the trails at least once consists of 6 simple circuits. Using this minimal covering collection results in a reduction of the initialization stage from more than 5 months to less than one week.

2.2   Computational Illustrations

In the computational illustrations shown in this Section, we used the actual map of trails of the JCR. It has $n = 154$ trails and $\#\mathcal{C}_G = 10216$ simple circuits crossing the rangers's shelter. We simulated an illegal actor by choosing a route uniformly between those shown in Figure 1. The poacher's behavior induces the poaching probabilities shown in Fig. 2a. These are not known to the rangers.

Fig. 1. Poacher's routes

In Fig. 2c we see a heat map showing the execution of our algorithm for a single ranger ($B = 1$). The heat map shows the proportion of times that the ranger has visited every trail after $t = 180$ stages with brighter colors corresponding to higher proportion of visits. Even though the ranger does not have initial information about the poacher, the algorithm gradually discovers her routes and the heat concentrates on the trails lying on the ideal optimal route shown in Fig. 2b.

3   Future Research

In the previous simulations the ranger was capable of discovering the optimal route because the poacher is not modifying her behavior reactively, keeping a fixed probability distribution among routes. In what's left of this document, we

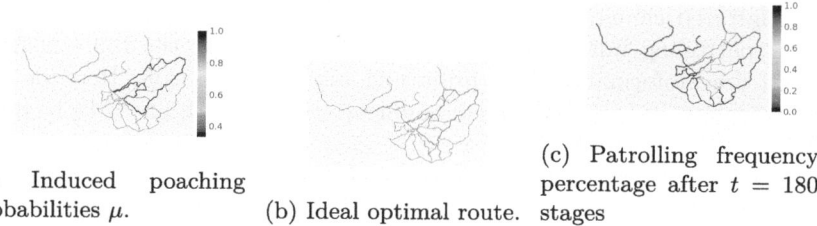

(a) Induced poaching probabilities μ.

(b) Ideal optimal route.

(c) Patrolling frequency percentage after $t = 180$ stages

Fig. 2. .

propose simple models of strategic poachers and provide an empirical demonstration of how the algorithm's performance diminishes when opposing them. We also point out alternatives to overcome this challenge.

A Π_1 poacher is one that moves according to a fixed probability distribution $\alpha \in \mathbb{R}^n$ with one exception: if in stage t the sampled route meets at least one of the trails of the ranger's chosen route, then the poacher will skip stage $t+1$ and return at stage $t+2$ sampling again from the fixed route distribution α. A Π_2 poacher is a poacher that has two unrelated patrolling distributions α and β. The Π_2 poacher will switch between the distributions after any encounter with the ranger.

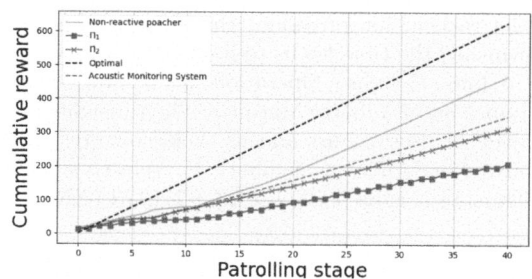

Fig. 3. The cumulative reward during 25 patrolling stages for each kind of ranger.

We simulated 40 stages for each kind of poacher using the CUCB a hundred times. Then, we computed the average cumulative reward at each stage (see Fig. 3). The straight black line is the optimal possible reward in the stationary scenario. As one would expect, faced against reactive poachers the performance of the CUCB worsens.

3.1 Acoustic Monitoring Systems

There are different ways in which rangers in protected areas use available information. Potentially they could use information provided by satellite imagery, thermal infrared cameras and acoustic monitoring systems (AMS) for enhancing their patrolling initiatives. Finding ways of coupling the route suggestion with real-time information is a promising research direction. Here we give a brief description of how our system could be coupled with an AMS to improve performance.

Acoustic identification is known to be effective for managing natural areas. Such systems provide rangers with real-time alerts by detecting sounds up to 1500 m away from the tree in which a microphone is installed. This covers vast areas but lacks precision in pinpointing the source. In this section we show a combination between AMS and the route-suggestion system giving rangers real time alerts and a route for heading towards the threat.

More concretely, we modified the CUCB strategy to incorporate synthetic real time alerts. The simulated AMS notifies the ranger whenever illegal activity is detected by informing her of a collection of possibly threatened trails. We assume this collection contains some trails in the poacher's route but also additional spurious trails. The algorithm is modified by solving problem (12) adding the constraint that the solution must cross at least one of the threatened trails.

To illustrate the enhancement we compare the improved CUCB algorithm against a Π_2 poacher again. As shown in Fig. 3, the green line outperforms plain CUCB against Π_2 poacher. This synthetic simulation might encourage future empirical research on the topic.

Acknowledgments. We are grateful to Third Millenium Alliance, specially Ryan Lynch and Shawn McCracken, for providing the shape file for the trails of Jama-Coaque. They also furnished the time for us to hold interviews with the reserve's staff. We are also grateful to Jama-Coaque's Operation Team comprised by Moises Tenorio, Dany Murillo, Sixto Lopez and Edilberto Marquez. Their insights and advice have been essential for this project. N. Betancourt and M. Velasco were partially supported by Proyecto INV-2018-50-1392 (UAndes, Colombia). M. Velasco is partially supported by ANII grant Fondo Clemente Estable FCE-1-2023-1-176172 (ANII, Uruguay).

References

1. Auer, P., Cesa-Bianchi, N., Fischer, P.: Finite-time analysis of the multiarmed bandit problem. Mach. Learn. **47**(2–3), 235–256 (2002)
2. Bondi, E., et al.: Spot poachers in action: augmenting conservation drones with automatic detection in near real time. In: Proceedings of the AAAI Conference on Artificial Intelligence, vol. 32 (2018). https://doi.org/10.1609/aaai.v32i1.11414, https://ojs.aaai.org/index.php/AAAI/article/view/11414
3. Chen, W., Wang, Y., Yuan, Y.: Combinatorial multi-armed bandit: general framework and applications. In: Dasgupta, S., McAllester, D. (eds.) Proceedings of the 30th International Conference on Machine Learning, pp. 151–159. no. 1 in Proceedings of Machine Learning Research, PMLR, Atlanta, Georgia, USA (17–19 June 2013). https://proceedings.mlr.press/v28/chen13a.html
4. Connection, R.: Guardian platform. https://rfcx.org/guardian. Accessed 03 Sept 2024

5. Czuni, L., Varga, P.Z.: Time domain audio features for chainsaw noise detection using WSNs. IEEE Sens. J. **17**(9), 2917–2924 (2017)
6. Gholami, S., et al.: Taking it for a test drive: a hybrid spatio-temporal model for wildlife poaching prediction evaluated through a controlled field test. In: The European Conference on Machine Learning and Principles and Practice of Knowledge Discovery in Databases (ECML PKDD 2017 Applied Data Science Track) (2017)
7. Guo, R., Xu, L., Cronin, D., Okeke, F., Plumptre, A., Tambe, M.: Enhancing poaching predictions for under-resourced wildlife conservation parks using remote sensing imagery (2020)
8. Katsis, L.K., et al.: Automated detection of gunshots in tropical forests using convolutional neural networks. Ecol. Indicat. **141**, 109128 (2022)
9. Lai, T.L., Robbins, H., et al.: Asymptotically efficient adaptive allocation rules. Adv. Appl. Math. **6**(1), 4–22 (1985)
10. Suman, P., Karan, S., Singh, V., Maringanti, R.: Algorithm for gunshot detection using Mel-Frequency Cepstrum Coefficients (MFCC). In: Maringanti, R., Tiwari, M., Arora, A. (eds.) Proceedings of Ninth International Conference on Wireless Communication and Sensor Networks. LNEE, vol. 299, pp. 155–166. Springer, New Delhi (2014). https://doi.org/10.1007/978-81-322-1823-4_15
11. Mporas, I., Perikos, I., Kelefouras, V., Paraskevas, M.: Illegal logging detection based on acoustic surveillance of forest. Appl. Sci. **10**(20), 7379 (2020)
12. Silvestro, D., Goria, S., Sterner, T., Antonelli, A.: Improving biodiversity protection through artificial intelligence. Nat. Sustain. **5**(5) (2022). https://doi.org/10.1038/s41893-022-00851-6
13. Xu, H., et al.: Optimal patrol planning for green security games with black-box attackers. In: Conference on Decision and Game Theory for Security (GameSec) 2017 (2017)
14. Xu, L., Bondi, E., Fang, F., Perrault, A., Wang, K., Tambe, M.: Dual-mandate patrols: multi-armed bandits for green security. CoRR abs/2009.06560 (2020). https://arxiv.org/abs/2009.06560
15. Xu, L., et al.: Stay ahead of poachers: illegal wildlife poaching prediction and patrol planning under uncertainty with field test evaluations (short version). In: 2020 IEEE 36th International Conference on Data Engineering (ICDE), pp. 1898–1901 (2020). https://doi.org/10.1109/ICDE48307.2020.00198

Identification of Dating Violence with Machine Learning Algorithms: Analysis and Results

Mariana-Carolyn Cruz-Mendoza[1,3] (ID), Juana Canul-Reich[2(✉)] (ID),
Roberto Ángel Meléndez-Armenta[1] (ID), and Carolina Malvaez-Hernández[3] (ID)

[1] División de Estudios de Posgrado e Investigación, Tecnológico Nacional de México/
Instituto Tecnológico Superior de Misantla, 93821 Misantla, Veracruz, Mexico
ramelendeza@itsm.edu.mx

[2] División Académica de Ciencias y Tecnologías de la Información de la Universidad
Juárez Autónoma de Tabasco, Av. Universidad s/n, Magisterial, 86040 Villahermosa,
Tabasco, Mexico
juana.canul@ujat.mx

[3] División de Ingeniería en Sistemas Computacionales, Tecnológico Nacional de
México/Tecnológico de Estudios Superiores de Valle de Bravo, Km. 31 de la carretera
Monumento-Valle de Bravo ejido de San Antonio de la Laguna,
Valle de Bravo 51200, Mexico
1202007009@vbravo.tecnm.mx

Abstract. This study evaluated three classification models: Naïve Bayes, Random Forest, and Neural Networks, to predict dating violence. Random Forest was superior, achieving an accuracy of 86% and effectively classifying various forms of violence, although with limitations in less represented categories. Naïve Bayes showed limited effectiveness for infrequent classes, and Neural Networks required a large amount of data to generalize effectively. The results indicate that implementing Random Forest in digital platforms can offer a 4% higher accuracy and a 1% higher F1-Score compared to Neural Networks, and a 60% higher accuracy, 28% higher sensitivity, and 1% higher F1-Score compared to Naïve Bayes. In conclusion, Random Forest is suitable for robust and varied datasets in the analysis of violence.

Keywords: Naïve Bayes · Random Forest · Neural Network · Mujer Segura

1 Introduction

Violence against women is a violation of human rights and a global issue of gender and health. Intimate partner violence is common and still often perceived as a private matter; however, observers can take active or passive actions to intervene promptly [17].

L. Correia et al. (Eds.): IBERAMIA 2024, LNCS 15277, pp. 62–73, 2025.
https://doi.org/10.1007/978-3-031-80366-6_6

In Latin America and the Caribbean, a woman is a victim of homicide every two hours. This alarming statistic underscores the severity of the problem and reveals the limitations of public records, which do not always capture the complexity of the phenomenon and its monitoring challenges [7].

To address this problem, it is crucial to understand and prevent the different forms of violence in intimate relationships. Liston R., *et al.* highlight the importance of prevention, categorizing the causes of violence into three dimensions: primary, which addresses the root causes; secondary, focused on at-risk groups; and tertiary, which aims to reduce recurrence through timely intervention [13].

Violence causes significant emotional impacts, inducing vulnerability and fear that can restrict the freedom of movement and everyday actions of victims [6]. These effects underscore the need for early and effective interventions to prevent the escalation of violence and its severe consequences.

Machine learning improves the detection and prevention of dating violence [14]. This study evaluates models to predict violence and proposes the integration of these models into a web platform to intervene with at-risk students. Three models will be evaluated for the early identification of violence, with a literature review in Sect. 2 and the methodology in Sect. 3.

In the selection phase, data is collected from a survey of university couples and variables from the Institutional Management System[1]. In preprocessing, missing and outlier values are handled, and necessary attributes are normalized. Then, the data is transformed for the algorithms. In the modeling phase, machine learning models such as Neural Networks, Random Forest, and Naïve Bayes are trained with the survey data to identify the type of violence in the couples.

In Sect. 4, the experiments are developed, and the accuracy percentages of the models are demonstrated. In the results and analysis Sect. 5, the integration of the models with greater confidence is proposed to be incorporated into a web platform that performs real-time data analysis. Finally, in Sect. 6, future work is presented.

This study compares three machine learning models: Naïve Bayes, Neural Network, and Random Forest, to identify the model with the best performance in detecting violent behaviors during dating. Surveys were conducted with university students in Valle de Bravo, Mexico, as the State of Mexico ranks among the highest in femicides in the country [4].

2   Literature Review

Intimate partner femicide is a global social challenge. Anticipating and understanding risk factors are essential for its prevention. Several studies have utilized AI and data analysis, including models such as Bayesian classifiers and decision trees, along with NLP, in legal documents, national databases, and social networks to detect patterns and signals that help prevent gender-based violence.

---

[1] System that manages student information at the Technological Institute of Higher Studies of Valle de Bravo.

García-Vergara *et al.* developed a predictive AI analysis on legal documents concerning intimate partner femicide. They used Bayesian classifiers based on functions, instances, trees, and rules. The variables included criminal background, sanctions, characteristics and frequency of violence, and the context of the crime. They correctly detected at least 75% of lethal and non-lethal cases [8].

AbiNader *et al.* used national databases from the United States to highlight the obstacles in the investigation of deaths due to intimate partner violence, pointing out the lack of operational definition of variables and the presence of numerous missing data [1].

González-Prieto *et al.* highlighted the importance of identifying whether a victim was assaulted again to ensure immediate protection, given their high risk of femicide. They examined reports from the VioGen system in Spain and used the Nearest Centroid (NC) model to identify and characterize archetypes of aggressors, classifying the cases into No, Low, and High risk [10].

Al-Garandi *et al.* used social media data to capture public reports of violence. They developed a natural language processing model to automatically identify tweets with content related to intimate partner violence. This model achieved human-like performance and did not exhibit specific gender or racial biases [2].

Ye *et al.* developed a project using motion sensors and AI to detect school violence, implementing Relief-F and Dempster-Shafer (DS) algorithms. They used sensors on the waist and leg to extract temporal and frequency features. They proposed a decision tree and an RBF network to classify activities. In simulations, the method improved recognition accuracy to 89.6% in school violence and 95.1% in daily activities. This approach is useful for monitoring activities of violence victims and preventing femicides [18]. In the study, Mhlanga *et al.* used Big Data to analyze interpersonal violence in the USA, mapping historical and current patterns to predict future trends. Data from violent crimes and physical altercations in public high schools were analyzed, focusing on variables like antisocial behavior, aggressiveness, hyperactivity, parental criminality, poor family management, poverty, and delinquent peers. [15].

3   Materials and Methods

In the process of evaluating the models (Naïve Bayes, Random Forest, and Neural Networks) to predict dating violence, these were chosen for their distinctive capabilities. Naïve Bayes was selected for its simplicity and computational efficiency, although it has limitations with imbalanced data. Random Forest was chosen for its robustness, accuracy, ability to handle variability, and identification of complex interactions. Neural Networks were justified for their ability to learn complex representations and their potential for continuous improvement, making them suitable for expanding data.

To obtain reliable and replicable results, a structured methodology such as CRISP-DM (Cross-Industry Standard Process for Data Mining) is recommended, which is widely used in machine learning. This methodology consists of six stages: business understanding, data understanding, data preparation, modeling, evaluation, and deployment [16].

3.1   Dataset Description

A sample of 250 students was selected from a subpopulation of 714 students with a romantic partner, representing 35% of this subpopulation. This sample size was chosen to ensure adequate representation and statistically significant results with a 95% confidence level and a 5% margin of error. The student population registered in the Institutional Management System at the National Technological Institute of Mexico, Valle de Bravo campus, includes 2000 university students, of which 50% have a romantic partner. The sample specifically included students who indicated having a violent partner, which was a crucial factor in the selection process. To ensure representativeness, stratified random sampling based on the enrollment of students with a romantic partner was used with the following formula [5]:

$$n_0 = \frac{1.96^2 \cdot 0.5 \cdot (1 - 0.5)}{0.05^2} = 384.16$$
$$n = \frac{384}{1 + \left(\dfrac{384 - 1}{714}\right)} = 250 \tag{1}$$

Table 1. Description of variables and data type

Field	Variable	Description	Data Type
1	gender	Subject's gender (M = Male, F = Female)	Binary
3	age	Subject's age between 18 and 25 years	Categorical
4	edo_civil	Subject's marital status	Categorical
5	career	Subject's field of study	Categorical
6	semester	Subject's current semester (from first to fifth semester)	Categorical
7	disability	Whether the subject has a disability (Yes or No)	Binary
8	language	Whether the subject belongs to an ethnic group (Yes or No)	Binary
9	apoyo_familiar	Whether the subject receives financial support (Yes or No)	Binary
10	percepcion_fami	Perception of the family environment (United, Very united, Dysfunctional)	Categorical
11	condicion_vive	Whether the subject lives alone or with parents	Binary
12	tipo_violencia	Classification of the type of violence (11 categories)	Categorical

This information is supported by the data protection law, which in its *Article 2 establishes the purpose of guaranteeing the right to the protection of personal data* [9] (see Table 1).

Establishing 11 classes instead of 4 allows for a nuanced understanding of various combinations and overlaps of violence types [15]: Class 0 (physical), Class 1 (physical and psychological), Class 2 (physical, psychological, and sexual), Class 3 (psychological), Class 4 (sexual), Class 5 (verbal), Class 6 (physical), Class 7 (verbal, physical, and psychological), Class 8 (verbal, physical, psychological, and sexual), Class 9 (verbal and psychological), and Class 10 (verbal and sexual). These classes enable a detailed analysis of different types of violence.

3.2 Description of the Machine Learning Models to Be Implemented

In this section, we provide a comprehensive overview of the chosen machine learning models for classifying dating violence. These models include Naïve Bayes, Random Forest, and Neural Networks, each with its own strengths and approaches to analyze and predict patterns of violence in romantic relationships.

3.3 Naïve Bayes algorithm

Bayes' theorem is of utmost importance in both inferential statistics and numerous advanced machine learning models, see Fig. 1. Bayesian reasoning constitutes a logical approach to updating the probability of hypotheses in light of new evidence, thereby playing a crucial role in science [3].

Naïve Bayes Theorem
Bayes' theorem establishes the relationship between the conditional probability of a class given a set of features and the probability of those features given the class. The general formula for Bayes' theorem is:

$$P(y|x) = \frac{P(x|y) \times P(y)}{P(x)} \tag{2}$$

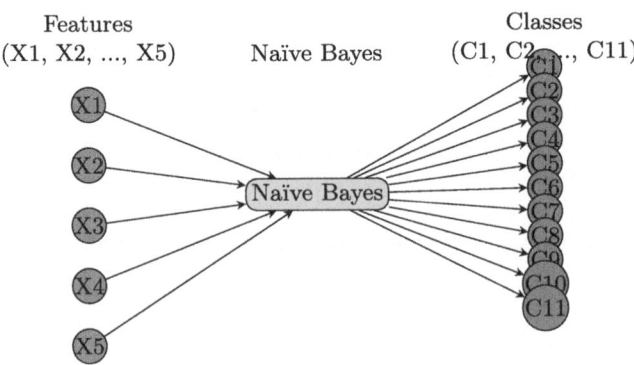

Fig. 1. Graphical representation of the Naïve Bayes model

3.4   Random Forest

The Random Forest model constructs multiple decision trees using random subsets of the training data, see Fig. 2. The final classification is decided by voting from these trees, each trained to minimize node impurity using the Gini index or entropy [14].

$$IG(p) = 1 - \sum_{i=1}^{K} p_i^2 \tag{3}$$

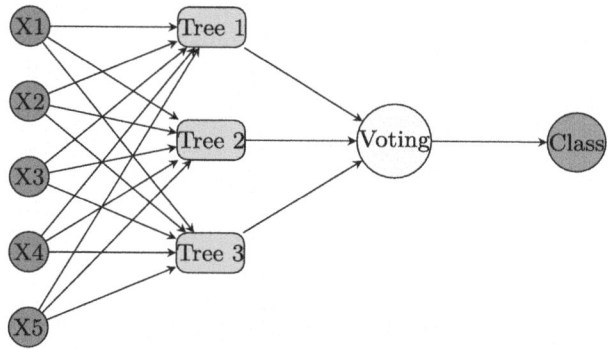

Fig. 2. Graphical representation of the Random Forest model

3.5   Multi-layer Perceptron (MLP) Neural Network

Neural networks are computational models inspired by the structure and function of the human brain, see Fig. 3. They use layers of interconnected artificial neurons to process information and make predictions [12].

The output of a neural network with L layers can be expressed as:

$$y = f_L(f_{L-1}(...f_2(f_1(X \cdot W_1 + b_1) \cdot W_2 + b_2)... \cdot W_{L-1} + b_{L-1}) \cdot W_L + b_L) \tag{4}$$

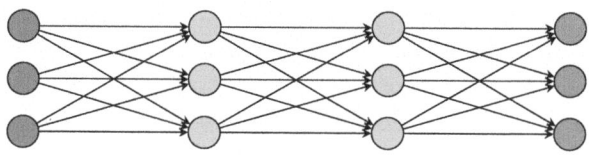

Fig. 3. Multi-Layer Perceptron (MLP) Neural Network used in the study

4   Experiments

In the preprocessing phase, a ColumnTransformer was used to transform numerical and categorical features. Irrelevant columns 'subject' and 'account' were removed, and categorical features such as 'gender' and 'marital status' and numerical features like 'age', 'edo_civil', 'semester', among others, were selected. Numerical features were standardized using StandardScaler, and categorical features were encoded using OneHotEncoder. A complete pipeline integrated these steps. Finally, the data was prepared by removing unwanted columns and separating the target variable 'type', which was encoded using OneHotEncoder for the classification model.

The experiments were conducted using Google Colab[2], Google Compute Engine, utilizing Python 3, system RAM of 1.7/12.7 GB, and disk space of 24.4/107.7 GB [11].

4.1   Neural Network

The data was split and transformed for model training and evaluation with an 80/20 data partitioning:

- **Training Set (X_train_prepared)**: 199 records with 16 features.
- **Test Set (X_test_prepared)**: 50 records with 16 features.
- **Training Set Classes (y_train)**: 199 records in 11 classes.
- **Test Set Classes (y_test)**: 50 records in 11 classes.

The dimensions ensure that each input has the same number of features and each output the same number of classes.

During the implementation of the neural network model to predict the type of dating violence, an overall accuracy of 82% was achieved, with an average precision of 78.04%. The average recall was 75.26% and the average F1-score was 76.46%, showing a good balance between precision and recall.

The analysis of the confusion matrix revealed issues with underrepresented classes being misclassified. It is recommended to increase the number of respondents and improve data completeness. While some classes showed excellent prediction, others, such as class 5, performed poorly, indicating areas for improvement.

The area under the ROC curve (AUC) varied between classes, with some reaching 1.0 and others, such as class 5, having an AUC of 0.8277. This suggests the need to improve the model's sensitivity and a more balanced approach in training to enhance fairness across classes.

Figures 4 and 5 graphically display the results of the Neural Network experiments.

---

[2] Google Colab is a development platform based on Jupyter Notebook that allows code execution in the cloud. Resources are available at https://github.com/MarianaCarolynCruz/Violence.

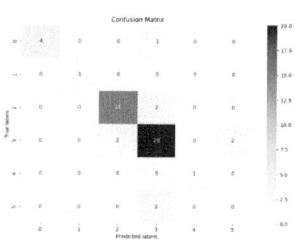

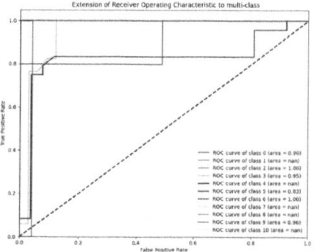

Fig. 4. Neural Network Confusion Matrix   **Fig. 5.** Neural Network ROC Curve

4.2   Naïve Bayes

The Naïve Bayes model was applied to predict the type of dating violence using demographic and personal variables. Hyperparameter optimization found that the optimal value for *var_smoothing* is 1.23×10^{-7}, improving accuracy to 84.34%. However, the mean cross-validation accuracy was 74.64%, suggesting a possible overfitting to the training data.

The model evaluation showed an overall accuracy of 28%, with an average precision of 50% and a sensitivity of 52%. The confusion matrix and ROC curves revealed significant variability in the model's ability to correctly identify the various categories of violence.

Figures 6 and 7 graphically display the results of the Naïve Bayes experiments.

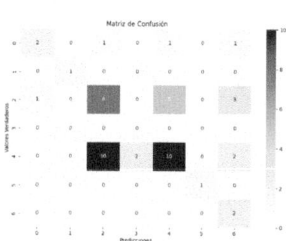

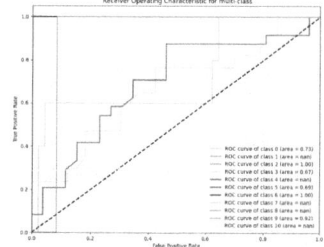

Fig. 6. Naïve Bayes Confusion Matrix   **Fig. 7.** Naïve Bayes ROC curve

4.3   Random Forest

The Random Forest model was optimized using GridSearchCV, identifying hyperparameters: max_depth of 10, max_features set to 'auto', min_samples_leaf of 1, min_samples_split of 2, and n_estimators of 100. Max_depth limits tree depth to avoid overfitting, max_features ('auto') selects all features to split a node, min_samples_leaf specifies the minimum samples in a leaf node, and min_samples_split defines the minimum samples to split a node. N_estimators indicate the number of trees, improving model robustness and accuracy.

These parameters achieved an average cross-validation accuracy of 74.36% using 3-fold cross-validation (cv = 3). The test set accuracy was 86%, with perfect precision and recall (100%) for 'Physical, Psychological, Sexual' and 'Verbal, Physical' categories, resulting in F1-Scores of 100%. However, 'Verbal, Psychological' had 0% precision and recall, indicating areas for improvement. The ROC curve for each class showed strong class discrimination, with AUCs near 1.00 for the best-represented classes. In Figs. 8 and 9, the results of the Random Forest experiments are graphically depicted.

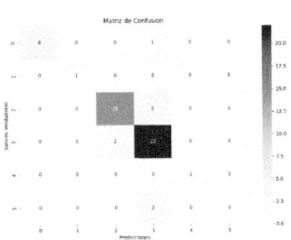

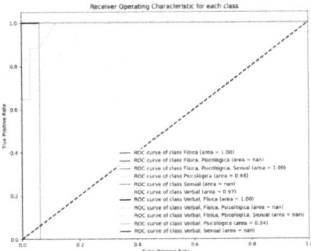

Fig. 8. Random Forest Confusion Matrix   **Fig. 9.** Random Forest ROC Curve

5   Results and Discussion

5.1   Results

Three classification models were evaluated: Naïve Bayes, Random Forest, and Neural Networks, each with its strengths and limitations. Naïve Bayes, known for its simplicity and effectiveness in high-dimensional data, had an average accuracy of 74.36%. However, it was less effective in handling classes with few data, showing low *recall* and *precision* in mixed categories.

Random Forest, with a robust structure to handle variability in data through multiple decision trees, achieved an accuracy of 86% on the test set. It excelled in managing various categories of violence, although it faced difficulties in classes with low representation. The ROC curve reflected an excellent discriminative ability for most classes.

Neural Networks, leveraging their ability to learn complex representations, could model non-linear interactions between features, essential in data with multiple interdependent attributes. However, hyperparameter tuning and the need for ample data to avoid overfitting are significant challenges to address to maximize their effectiveness (see Table 2).

Table 2. Comparison of performance of each model.

Model	Accuracy	Precision	Sensitivity	F1-Score
Random Forest	0.86	0.78	0.77	0.77
Neural Network	0.82	0.78	0.75	0.76
Naïve Bayes	0.28	0.50	0.52	0.43

5.2   Discussion

The comparison of these models shows that there is no one-size-fits-all solution for classifying dating violence. The choice of model depends on the dataset and the available resources. Naïve Bayes is useful for preliminary studies; Random Forest is preferable for practical applications; and Neural Networks are ideal for advanced research with extensive data.

These findings underscore the importance of carefully evaluating and fine-tuning machine learning models according to the problem's characteristics. The precision and generalization of Random Forest make it the most suitable model for this study. Neural Networks, with appropriate adjustments and resources, could offer significant improvements in future research.

6   Conclusions and Future Work

6.1   Conclusions

Violence against women is a critical and prioritized issue. This study highlights the importance of machine learning in identifying and predicting patterns of dating violence, facilitating early intervention. Implementing these models in institutional management systems can enhance the detection and prevention of violence, offering crucial support to victims.

Although Random Forest was the most accurate in this study, the choice of model depends on the size and nature of the dataset. In the Institutional Management System, with less than two thousand transactional data and added historical data, both Random Forest and Neural Networks are suitable for implementation in a web system and transforming it into an expert system.

6.2   Future Jobs

The integration of the Random Forest model into a web platform that performs real-time data analysis represents a novel approach to early intervention. This platform not only enhances the detection of violent behaviors but also provides timely support to at-risk students, potentially preventing the escalation of violence. The research highlights the importance of machine learning in addressing complex social issues and offers a practical solution for educational institutions to monitor and intervene in cases of dating violence.

Additionally it is necessary to explore the following processes to determine their impact on the proposed models.

- Increase the sample size to adequately represent all violence classes, enhancing the models' ability to generalize.
- Implement class balancing techniques, such as undersampling for majority classes to investigate the detection and classification of violence improvement.
- Optimize hyperparameters and feature engineering to improve model performance.
- Explore advanced feature engineering approaches to increase sensitivity and specificity in less represented categories.

References

1. AbiNader, M.A., Graham, L.M., Kafka, J.M.: Examining intimate partner violence-related fatalities: past Lessons and future directions using U.S. National data. J. Family Violence **38**(6), 1243–1254 (2023). https://doi.org/10.1007/s10896-022-00487-2
2. Al-Garadi, M.A., et al.: Natural language model for automatic identification of Intimate Partner Violence reports from Twitter. Array **15** (2022). https://doi.org/10.1016/j.array.2022.100217
3. Berrar, D.: Bayes' Theorem and Naive Bayes Classifier (2019)
4. Blanco-Ruiz, M., Sainz-De-baranda, C., Gutiérrez-Martín, L., Romero-Perales, E., López-Ongil, C.: Emotion elicitation under audiovisual stimuli reception: should artificial intelligence consider the gender perspective? Int. J. Environ. Res. Publ. Health **17**(22), 1–22 (2020). https://doi.org/10.3390/ijerph17228534
5. Caille, A., Billot, L., Kasza, J.: Practical and methodological challenges when conducting a cluster randomized trial: examples and recommendations. J. Epidemiol. Populat. Health **72**(1), 202199 (2024). https://doi.org/10.1016/j.jeph.2024.202199
6. Contreras, H., et al.: Linking physical violence to women's mobility in Chile. EPJ Data Sci. **12**(1) (2023). https://doi.org/10.1140/epjds/s13688-023-00430-5
7. D'Ignazio, C., et al.: Feminicide and counterdata production: activist efforts to monitor and challenge gender-related violence. Patterns **3**(7) (2022). https://doi.org/10.1016/j.patter.2022.100530
8. Garcia-Vergara, E., Almeda, N., Fernández-Navarro, F., Becerra-Alonso, D.: Artificial intelligence extracts key insights from legal documents to predict intimate partner femicide. Sci. Rep. **13**(1) (2023). https://doi.org/10.1038/s41598-023-45157-5
9. Gobierno de México: Protección de Datos Personales (2024)
10. González-Prieto, A., Brú, A., Nuño, J.C., González-Álvarez, J.L.: Hybrid machine learning methods for risk assessment in gender-based crime. Knowl.-Based Syst. **260**, 110130 (2023). https://doi.org/10.1016/j.knosys.2022.110130, https://linkinghub.elsevier.com/retrieve/pii/S0950705122012266
11. Google: Google Colaboratory (2024). https://colab.research.google.com/notebooks/intro.ipynb
12. Huang, C.W., Chen, Y.N.: Adapting pretrained transformer to lattices for spoken language understanding. In: 2019 IEEE Automatic Speech Recognition and Understanding Workshop (ASRU), pp. 845–852 (2019). https://doi.org/10.1109/ASRU46091.2019.9003825
13. Liston, R., Hamilton, G., McCook, S.: How do representatives from sporting organisations understand primary prevention of violence against women? Crime Prevent. Community Saf. **25**(3), 243–257 (2023). https://doi.org/10.1057/s41300-023-00179-z

14. Liu, C., Gu, Z., Wang, J.: A hybrid intrusion detection system based on scalable K-Means+ Random Forest and Deep Learning. IEEE Access **9**, 75729–75740 (2021). https://doi.org/10.1109/ACCESS.2021.3082147, https://ieeexplore.ieee.org/document/9437227/
15. Mhlanga, F.S., Perry, E.L., Kirchner, R.: Toward a predictive model ecosystem for interpersonal violence (WIP). Simul. Ser. **46**(10), 455–462 (2014)
16. Schröer, C., Kruse, F., Gómez, J.M.: A systematic literature review on applying CRISP-DM process model. Procedia Comput. Sci. **181**, 526–534 (2021). https://doi.org/10.1016/j.procs.2021.01.199, https://www.sciencedirect.com/science/article/pii/S1877050921002416
17. Vázquez-González, L.I., Bosch-Fiol, E., Sánchez-Prada, A., Ferreiro-Basurto, V., Delgado-Álvarez, C., Ferrer-Pérez, V.A.: Bystander behavior in violence against women in Spain: a scoping review (2023). https://doi.org/10.1016/j.avb.2023.101861
18. Ye, L., Wang, L., Ferdinando, H., Seppänen, T., Alasaarela, E.: A video-based DT-SVM school violence detecting algorithm. Sensors (Basel, Switzerland) **20**(7), 2018 (2020). https://doi.org/10.3390/s20072018, https://www.mdpi.com/1424-8220/20/7/2018, http://www.ncbi.nlm.nih.gov/pubmed/32260274, http://www.pubmedcentral.nih.gov/articlerender.fcgi?artid=PMC7181151

Capturing the Essence of Plankton: A Gradient-Weighted Class Activation Mapping Analysis

Sofía Callejas[(✉)] , Hernan Lira , Andrew Berry, Luis Martí ,
and Nayat Sanchez-Pi

Inria Chile Research Center, Avenue Apoquindo, 2827 Las Condes, Santiago, Chile
{sofia.callejas,hernan.lira,luis.marti,nayat.sanchez-pi}@inria.cl

Abstract. Plankton is a vital component of marine ecosystems, integral to the biogeochemical cycles and climate regulation processes. Their
abundance fluctuations serve as key indicators of ocean health. Accurate identification of plankton is therefore essential. Traditional monitoring methods, however, struggle with the complex spatial and temporal dynamics of ocean environments. In response, modern computer
vision techniques have increasingly been applied to enhance plankton
identification in microscopic images. While existing studies have predominantly used small, controlled laboratory datasets, our research addresses
this gap by employing large, unbalanced datasets, thereby aligning more
closely with real-world conditions. We utilize advanced models, including Swin Transformers and DeiT 3, incorporating data augmentation and
diverse loss functions to boost classification accuracy. Additionally, we
perform a systematic comparison of six GradCAM algorithms to gain
explainability insights into the recognition process. Our results indicate
that Vision Transformers (ViTs) outperform traditional methods, showcasing their potential to revolutionize plankton image recognition.

Keywords: Plankton Recognition · GradCAM · Vision Transformer

1 Introduction

Modeling oceans and climate change represents a critical application of artificial intelligence, with its significance expected to escalate in the forthcoming
years. This research focuses specifically on studying plankton, a diverse group of
microorganisms that form the foundation of aquatic ecosystems. Plankton may
play a crucial role in understanding and monitoring climate change due to their
sensitivity to variations in ocean temperature, which affects their productivity,
abundance, and distribution [12]. These change can lead to the proliferation
of harmful plankton concentrations, which deplete oxygen, release toxins, and
negatively impact other marine communities such as fish and invertebrates. As

A. Berry—Contribution while at Inria Chile Research Center.

a consequence, enhancing global observation of plankton is crucial for comprehending their responses to oceanic shifts and their role in climate regulation. Furthermore, this knowledge is essential for policymakers to effectively mitigate the effects of climate-related stressors [27].

The study of plankton is complicated by the intricate spatial and temporal dynamics of ocean ecosystems. Marine biologists prioritize plankton recognition to understand its variations across different oceanic regions. Despite recent advancements in robotics and artificial intelligence, capturing high-quality images in the ocean's variable light conditions—compounded by noise from suspended particles and detritus[1]—remains a significant challenge. These images frequently lack detail, contrast, and definition. Traditional methods, often utilizing professional instruments such as electron microscopes, are limited in scope and can only identify a few valuable categories in costly sampling tasks. This highlights the urgent need for enhanced methods to automatically recognize plankton images, facilitating rapid assessment of plankton's taxonomic composition across various scales. Our study also aims to analyze and identify the factors that influence the effectiveness of these recognition methods.

In the task of plankton image recognition, we can highlight the application of a Convolutional Neural Network (CNN) ensemble, which performed exceptionally well in the National Data Science Bowl on the Plankton Kaggle-121 dataset [2]. This dataset comprises approximately 30,000 samples across 121 plankton classes. Further advancements were achieved with the implementation of Deep Residual Networks (ResNets) [26] and variations of GoogLeNet on this dataset. Sun et al. [22] developed an attention mechanism based on Gradient-weighted Class Activation Maps (GradCAM) to direct a CNN to focus on the most informative regions in an image, highlighting discriminative features.

Recent studies have proposed models using large and challenging plankton datasets like the Woods Hole Oceanographic Institute (WHOI) plankton dataset [18] and ZooScanNet [8]. The WHOI plankton dataset contains over 3.5 million samples categorized into 103 classes, while the ZooScanNet dataset includes 1.4 million images across 93 classes. Both datasets suffer from severe class imbalance; for example, in the WHOI dataset, 90% of the data is concentrated in just five classes, and generally poor image quality is also an issue. Due to these complexities, researchers often work with smaller subsets of these datasets. Some encouraging recent results have been achieved using CNNs [4,7] and Visual Transformers (ViT) [1,11].

In this study we provide two main contributions to the field. First, we have developed five different models based on Visual Transformers (ViT) and CNNs that can understand the complete plankton datasets from WHOI and ZooScanNet. Second, we focus on evaluating the performance of our plankton recognition model by employing several GradCAM techniques. GradCAM provides visual explanations for decisions made by neural networks, highlighting the regions of the input image that are most influential for the model's predictions. By applying

---

[1] Detritus in the context of plankton refers to dead organic matter that has been broken down and colonized by communities of microorganisms.

these techniques, we aim to gain deeper insights into the decision-making process of the recognition model, identify potential areas for improvement, and enhance the overall interpretability and reliability of plankton image analysis. This approach will help us address the research gap by not only improving recognition accuracy but also providing a clearer understanding of the factors influencing model performance.

Our findings indicate that the Swin transformer is the best performing model on the ZooScanNet dataset, achieving an F1-score of 0.896. Additionally, the DeiT 3 model recorded an F1-score of 0.877. On the WHOI dataset, the DeiT 3 algorithm excelled, achieving an F1-score of 0.953 and demonstrating superior classification performance. We also provide a detailed comparison of six types of GradCAM algorithms applied to both datasets for each recognition model. These results underscore the potential of Vision Transformer (ViT)-based models to significantly enhance the accuracy of plankton image classification across diverse and large-scale image datasets.

The structure of the paper is as follows: Sect. 2 provides an overview of the background knowledge and context relevant to this study. Section 3 describes our methodology and the experiments conducted. Section 4 presents and discusses the findings. Finally, Sect. 5 outlines the conclusions and potential directions for future research.

2   Background

2.1   Studies on Plankton Image Recognition

The classification and recognition of plankton images have been areas of extensive research, fueled by the accessibility of high-quality data from marine biology studies. Traditionally, plankton images were analyzed using manually extracted features from relatively small datasets, often generated in controlled laboratory settings. However, images captured in natural environments tend to be of lower quality, presenting additional challenges. The field experienced a major shift in 2015 with the introduction of deep learning techniques. Researchers began leveraging Convolutional Neural Networks (CNNs) to process both raw and manually enhanced images, significantly improving the accuracy and efficiency of classification systems [6].

In recent years, the advent of Vision Transformer (ViT) architectures has further advanced the field. Data-efficient Image Transformers (DeiTs) have been particularly effective in addressing complex ecological and classification challenges within plankton datasets [11]. This shift towards ViT-based models reflects a broader trend in machine learning and computer vision, highlighting the potential of these techniques to revolutionize plankton image classification.

2.2   Plankton Image Datasets

Our research utilizes extensive plankton-focused datasets, particularly ZooScan-Net [8] and the Woods Hole Oceanographic Institution (WHOI) dataset [18].

These datasets are widely regarded as benchmarks in plankton classification studies. Historically, however, only subsets of these datasets have been employed in prior research [6,11], limiting the exploration of their full potential. Both the ZooScanNet and WHOI datasets exhibit significant class imbalance. The imbalance factors[2] for ZooScanNet and WHOI are 13,120 and 650,000, respectively. These figures are markedly higher than those found in other long-tail distribution benchmarks such as *ImageNet-LT* [15] (imbalance factor of 256) and *iNaturalist2017* [24] (imbalance factor of 500).

The presence of detritus samples in these datasets creates a realistic yet challenging scenario for plankton classification, mirroring the true complexity and diversity of species in oceanic environments. This inherent imbalance provides a unique opportunity to develop and test classification algorithms that can handle extreme variations in class distribution, which is critical for accurately representing and understanding marine biodiversity.

WHOI Plankton Dataset. The Woods Hole Oceanographic Institution (WHOI), a non-profit organization, offers a comprehensive plankton dataset comprising 3,563,596 images of 103 taxa. The initial version, known as *WHOI22*, consists of 22 distinct classes with 300 samples each [16]. Additionally, *WHOI40* [19] includes data from 2011 to 2014, divided into 40 balanced classes, totaling 4,071 images. These datasets provide a robust foundation for developing and evaluating plankton classification algorithms.

ZooScanNet Plankton Dataset. The ZooScanNet dataset includes 1,433,278 images representing 93 distinct taxa, specifically designed for capturing images of zooplankton. The distribution of classes within the ZooScanNet dataset reveals that approximately 70% of the images are concentrated in just six classes, highlighting the dataset's imbalanced nature.

2.3   GradCAM

Deep neural networks have achieved remarkable success in a wide range of computer vision tasks. However, the inherent complexity of these models often renders their decision-making processes opaque and difficult to interpret. Understanding why a model makes a particular prediction is crucial for ensuring transparency, trust, and accountability. Gradient-weighted Class Activation Mapping (GradCAM) has emerged as a popular technique for visualizing the regions of an image that are crucial for prediction, thereby providing valuable insights into the model's decision-making process.

The concept of visualizing deep neural networks through class activation mapping originated from the seminal work of Zhou et al. (2016) [28], who introduced Class Activation Mapping (CAM). CAM enabled the localization of dis-

---

[2] The imbalance factor of a class distribution corresponds to the ratio between the largest and smallest classes and is widely used for its simplicity.

criminative image regions responsible for predicting a particular class by leveraging global average pooling to produce class activation maps. However, CAM was limited to networks with global average pooling layers, thereby restricting its applicability to a subset of architectures. To address this limitation, Selvaraju et al. (2017) [21] proposed GradCAM, which extended the idea of CAM to any convolutional neural network (CNN) architecture, making it broadly applicable across various models.

3   Approach for Plankton Recognition and Evaluation

Building upon an extensive literature review on plankton recognition and our previous research efforts, this study aims to develop a robust deep learning model with the following objectives:

1. To process and classify images from the ZooScanNet and WHOI datasets directly.
2. To accurately assign these images to predefined categories corresponding to different plankton species.
3. To compare various GradCAM methods to assess their suitability in this context.

Figure 1 outlines the methodology employed to accomplish these objectives. The plankton datasets utilized in our study exhibit significant disparities in class sizes, a common characteristic in biological research due to the varying frequency of observations among different entities. To address these imbalances, we implemented two data augmentation strategies: Basic Image Manipulation (BIM), which includes Random Vertical Flip, Random Horizontal Flip, and Random Rotation, and RandAugment (RA) [5]. RA streamlines augmentation by eliminating the need for a separate search phase on a proxy task. It operates with two parameters: N, denoting the number of transforms per image, and M, indicating the magnitude of transformations, thereby reducing computational burden. Next, we utilized two different algorithms based on Vision Transformer (ViT): Swin [14] and DieT [23]. In addition, we employed two types of Loss Functions: Cross Entropy, a standard loss function used in classification tasks, and Focal Loss [13]. The latter computes the final loss by weighing losses for each class based on model confidence and class sizes. This prioritizes difficult and less frequent samples, potentially enhancing accuracy for underrepresented groups but may reduce overall model accuracy.

We employ six types of (GradCAM techniques in our analysis:

1. *GradCAM*: Utilizes gradients from any target concept flowing into the final convolutional layer to generate a coarse localization map.
2. *GradCAM++*: Provides a mathematical derivation for a method that uses a weighted combination of the positive partial derivatives of the last convolutional layer feature maps with respect to a specific class score to generate a visual explanation for the corresponding class label [3].

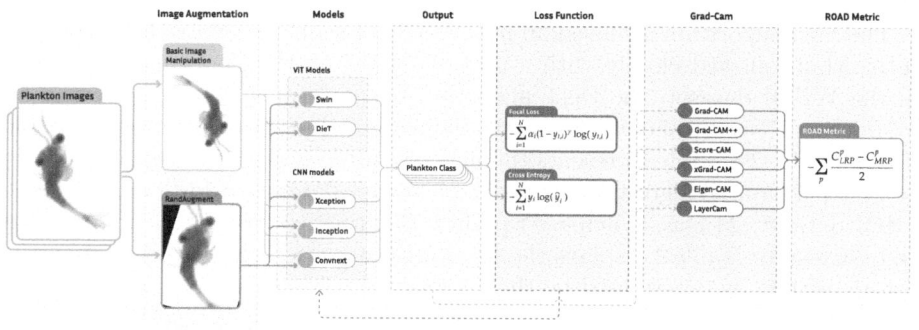

Fig. 1. Machine learning experimental methodology followed the experiments.

3. *ScoreCAM*: Eliminates the dependence on gradients by obtaining the weight of each activation map through its forward passing score on the target class. The final result is obtained by a linear combination of weights and activation maps [25].

4. *xGradCAM*: Introduces two axioms, Conservation and Sensitivity, into the visualization paradigm of CAM methods. Experiments show that xGradCAM enhances GradCAM in terms of conservation and sensitivity [9].

5. *EigenCAM*: Computes and visualizes the principal components of the learned features or representations from the convolutional layers [17].

6. *LayerCAM*: Extracts fine-grained object localization information from the class activation maps to locate the target objects more accurately [10].

Finally, to identify discrepancies among the methods and assess their effectiveness we employ the Remove and Debias (ROAD) metric [20] expressed as

$$ROAD = -\sum_{p} \frac{C^p_{\mathrm{LRP}} - C^p_{\mathrm{MRP}}}{2}, \tag{1}$$

as our evaluation standard. ROAD introduces noise through linear imputations and blurs image regions based on neighboring pixel values. The resulting changes in classification confidence scores when using the perturbed image, with the least relevant pixels (LRP) or most relevant pixels (MRP), are used to evaluate the accuracy of a CAM visualization. We combine both LRP and MRP to form the final metric. ROAD is evaluated at pixel perturbation thresholds of 20%, 40%, 60%, and 80%.

4 Experimental Results

Each experiment is conducted over 25 epochs training. The datasets were partitioned into training, validation, and test subsets using a randomized distribution ratio of 60%/20%/20%, respectively.

The Swin algorithm achieved an impressive F1-score of 0.896, while the DeiT 3 algorithm followed closely with a score of 0.877 on the ZooScanNet dataset. On the WHOI dataset, the DeiT algorithm outperformed others, achieving an outstanding F1-score of 0.953, while the Swin algorithm attained a commendable F1-score of 0.929. Table 1 presents the highest and lowest accuracies achieved by the best-performing algorithm on the ZooScanNet dataset. Additionally, for the WHOI dataset, 24 classes achieved perfect accuracy scores of 1.0, representing the best results. Table 2 displays the five lowest accuracy scores obtained by the best algorithm, as determined by the F1-score.

Table 1. Swin five top and bottom accuracy classes for the ZooScanNet dataset.

Top 5 classes	# Images	Accuracy
cyphonaute	9125	0.989
nauplii__cirripedia	6226	0.985
brachyura	5292	0.982
ostracoda	14392	0.981
obelia	1211	0.980
Bottom 5 classes	# Images	Accuracy
badfocus__copepoda	881	0.371
zoea__calatheidae	205	0.541
artefact	3849	0.544
nectophore__abylopsis_tetragona	208	0.681
cirrus	297	0.742

Table 2. DieT five worst accuracy classes for the WHOI dataset.

Bottom 5 classes	# Images	Accuracy
ditylum_parasite	258	0.563
hemiaulus	18	0.667
g_delicatula_external_parasite	433	0.675
diatom_flagellate	728	0.735
cochlodinium	14	0.750

For the GradCAM analysis, we conducted two experiments. In the first experiment, we computed the Remove and Debias (ROAD) metric for a single image across all classes and performed a permutation test to determine the superiority of one method over another. To evaluate algorithm performance, we used a RandomCAM method, which generates class activation maps (CAMs) with random uniform values in the range of $[-1, 1]$ for spatial activations. This method serves as a baseline comparison; an effective algorithm should outperform the RandomCAM. Figure 2 presents the p-value results for the ZooScanNet and WHOI datasets, respectively, using both the Swin and DeiT algorithms.

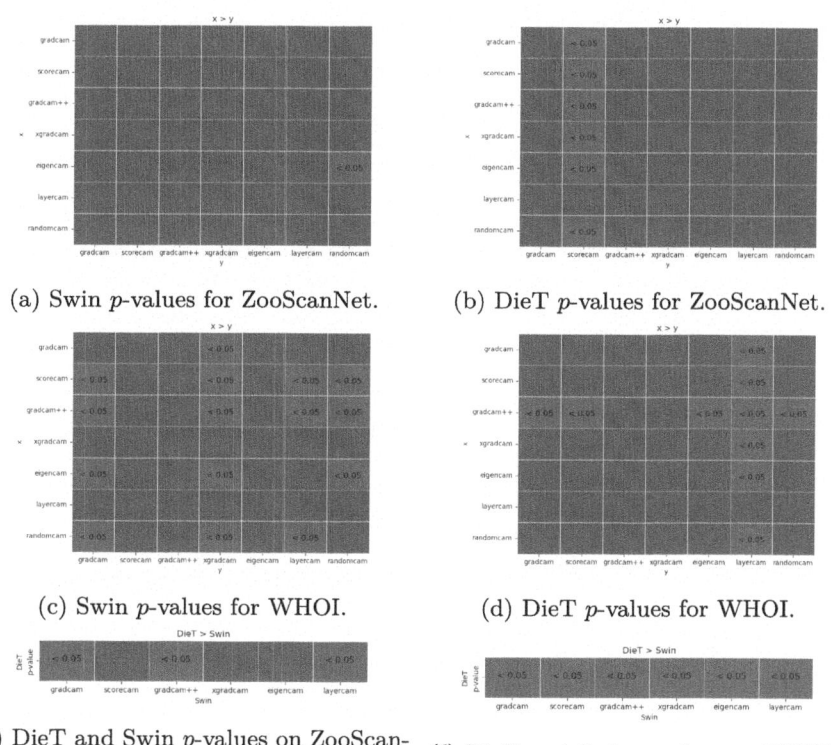

(a) Swin p-values for ZooScanNet.   (b) DieT p-values for ZooScanNet.

(c) Swin p-values for WHOI.   (d) DieT p-values for WHOI.

(e) DieT and Swin p-values on ZooScan-Net.   (f) DieT and Swin p-values on WHOI.

Fig. 2. p-values of the different experiments.

For the second experiment, we applied the ROAD metric to one image per class and conducted a permutation test. Our hypothesis posited that the DeiT algorithm would achieve a higher score than the Swin algorithm. We tested this hypothesis on both the ZooScanNet and WHOI datasets. Figure 2 presents the comparative results for the ZooScanNet and WHOI datasets.

4.1   Results Discussion

In the typical class-wise accuracy analysis of imbalanced datasets, it is generally expected that larger classes will yield better results, while smaller classes may perform less effectively. However, our investigation of the ZooScanNet and WHOI datasets revealed different outcomes. This suggests that, despite the dataset's imbalanced nature, the algorithms demonstrate a strong ability to generalize.

In a permutation test, a p-value less than the significance threshold of 0.05 indicates a significant difference. For the ZooScanNet dataset using the Swin algorithm, we observed only one significant difference, which was between Eigen-CAM and RandomCAM. This finding indicates that the score of EigenCAM is significantly higher than that of RandomCAM. Figure 3 illustrates that the EigenCAM method effectively highlights the plankton body in the image for both classes, while RandomCAM tends to focus more on the background.

For the DeiT algorithm, we observed that all methods, except LayerCAM, showed a significant difference compared to ScoreCAM. This implies that, according to the ROAD metric, ScoreCAM does not perform as well as the other methods. Figure 3 demonstrates that ScoreCAM does not effectively emphasize the plankton body in the image compared to the other methods, except for LayerCAM, which focuses attention on a specific area. As a result, the p-value exceeds 0.05 for these comparisons, indicating that we cannot draw definitive conclusions about the relative performance of LayerCAM and ScoreCAM.

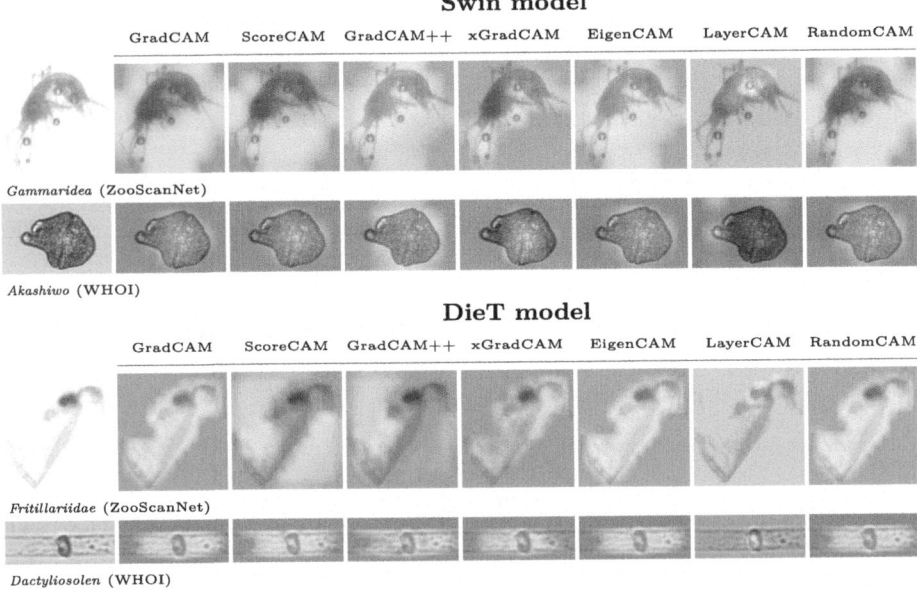

Fig. 3. Apply seven GradCAM methods to two classes of ZooScanNet and WHOI images using the Swin and DieT algorithms.

For the WHOI dataset using the Swin algorithm, the scores of RandomCam are significantly higher than those of the GradCAM, xGradCAM, and Layer-CAM methods based on the ROAD metric. This suggests that these methods may not be well-suited for this dataset when used with the Swin algorithm.

However, the ScoreCAM, GradCAM++, and EigenCAM methods exhibit higher ROAD metric scores compared to RandomCAM indicating their effectiveness. These methods outperform GradCAM, xGradCAM, and LayerCAM, except for EigenCAM and LayerCAM, where the p-value is not less than 0.05, preventing definitive conclusions. In Fig. 4, for the class *Akashiwo*, LayerCAM focuses solely on one specific area, suggesting that the other methods provide more accurate predictions for the plankton body.

Within the DeiT algorithm, GradCAM++ appears to be the most effective method, with the exception of xGradCAM, while LayerCAM consistently performs the poorest across all methods. In Fig. 3, LayerCAM shows a tendency to focus on a specific area of the plankton body in all images, whereas the GradCAM++ method displays clearer and more defined predictions.

In the final hypothesis test for the ZooScanNet dataset, the GradCAM, GradCAM++, and LayerCAM methods demonstrate superior performance when paired with the DeiT algorithm. This is evident in Fig. 4, where these methods consistently yield better predictions for the class *Haloptilus*. In the WHOI dataset, all methods exhibit significant differences according to the ROAD metric. However, this does not necessarily imply that the DeiT algorithm performs better in all images and classes, as seen in Fig. 4 for the class *Chaetoceros*.

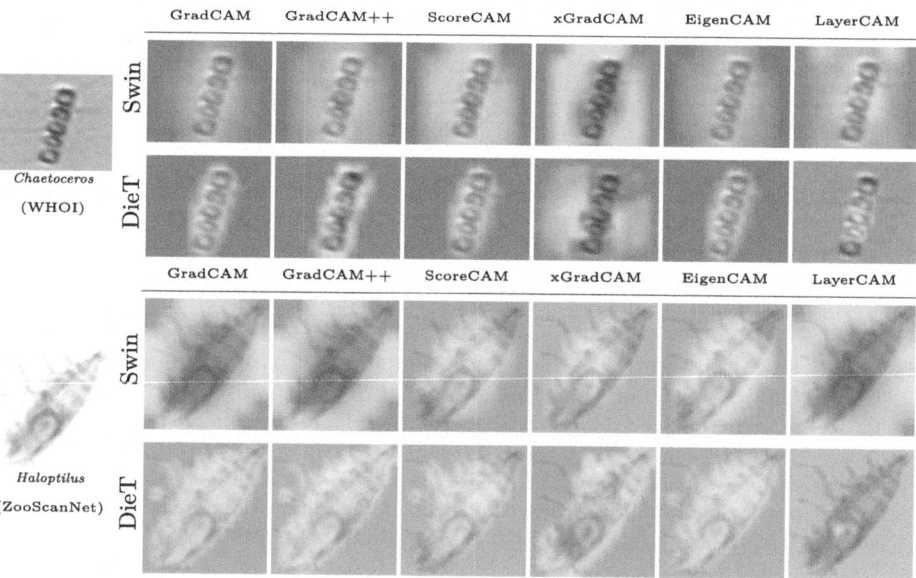

Fig. 4. Application of GradCAM methods to WHOI and ZooScanNet images using the Swin and DeiT algorithms.

5   Conclusions

Developing an algorithm for the ZooScanNet and WHOI plankton image datasets poses significant challenges due to their large size and imbalanced class distributions. Transformer-based algorithms, Swin for ZooScanNet and DeiT, achieved F1-scores of 0.894 and 0.953, respectively. Utilizing Grad-CAM methods requires hypothesis testing with RandomCAM, as some methods may yield lower scores than RandomCAM, indicating ineffectiveness for the dataset and algorithm. These methods might perform better under different conditions. Notably, strong algorithm performance does not always correlate with the highest ROAD metric scores in GradCAM analysis. For ZooScanNet, GradCAM, GradCAM++, and LayerCAM scored higher for DeiT compared to Swin.

In future research we plan to explore potential enhancements and alternatives during algorithm development. These included utilizing loss functions like LDAM, automatic augmentation techniques from the literature, and incorporating recent models. Additionally, hypothesis tests were conducted for different GradCAM methods, various loss functions, and data augmentation strategies to assess significant differences among them.

Acknowledgments. This work is funded by ANID Strengthening R&D capabilities Program CTI230007 Inria Chile and Inria Challenge OcéanIA. Authors wish to thank Mia Rose Elbo (Inria Chile) for the design of Fig. 1.

Disclosure of Interests. The authors have no competing interests to declare that are relevant to the content of this article.

References

1. Maracan, A., Pastore, V.P., Rosasco, L., Natale, L., Odone, F.: In-domain versus out-of-domain transfer learning in plankton image classification. Sci. Rep. **13**(1), 10443 (2023). https://doi.org/10.1038/s41598-023-37627-7
2. Aurelia, J., Luo, J., Allen, B., Sullivan, J., Mills, S., Cukierski, W.: National data science bowl (2014). https://kaggle.com/competitions/datasciencebowl
3. Chattopadhyay, A., Sarkar, A., Howlader, P., Balasubramanian, V.N.: Grad-CAM++: generalized gradient-based visual explanations for deep convolutional networks. CoRR abs/1710.11063 (2017). http://arxiv.org/abs/1710.11063
4. Cheng, K., Cheng, X., Wang, Y., Bi, H., Benfield, M.C.: Enhanced convolutional neural network for plankton identification and enumeration. PLoS ONE **14**(7), e0219570 (2019). https://doi.org/10.1371/journal.pone.0219570
5. Cubuk, E.D., Zoph, B., Shlens, J., Le, Q.V.: RandAugment: practical automated data augmentation with a reduced search space (2019)
6. Dai, J., Wang, R., Zheng, H., Ji, G., Qiao, X.: ZooplanktoNet: deep convolutional network for zooplankton classification. In: OCEANS 2016 - Shanghai, pp. 1–6 (2016). https://doi.org/10.1109/OCEANSAP.2016.7485680
7. Dai, J., Yu, Z., Zheng, H., Zheng, B., Wang, N.: A hybrid convolutional neural network for plankton classification. In: Chen, C.-S., Lu, J., Ma, K.-K. (eds.) ACCV 2016. LNCS, vol. 10118, pp. 102–114. Springer, Cham (2017). https://doi.org/10.1007/978-3-319-54526-4_8

8. Elineau, A., et al.: Zooscannet: plankton images captured with the zooscan. SEA-NOE (2017). https://doi.org/10.17882/55741
9. Fu, R., Hu, Q., Dong, X., Guo, Y., Gao, Y., Li, B.: Axiom-based Grad-CAM: towards accurate visualization and explanation of CNNs (2020)
10. Jiang, P.T., Zhang, C.B., Hou, Q., Cheng, M.M., Wei, Y.: LayerCAM: exploring hierarchical class activation maps for localization. IEEE Trans. Image Process. **30**, 5875–5888 (2021). https://doi.org/10.1109/TIP.2021.3089943
11. Kyathanahally, S., et al.: Ensembles of data-efficient vision transformers as a new paradigm for automated classification in ecology. Sci. Rep. **12**(1) (2022). https://doi.org/10.1038/s41598-022-21910-0
12. Li, Z., England, M.H., Groeskamp, S.: Recent acceleration in global ocean heat accumulation by mode and intermediate waters. Nat. Commun. **14**(1), 6888 (2023)
13. Lin, T.Y., Goyal, P., Girshick, R., He, K., Dollár, P.: Focal Loss for Dense Object Detection (2017). http://arxiv.org/abs/1708.02002
14. Liu, Z., et al.: Swin transformer: hierarchical vision transformer using shifted windows. CoRR abs/2103.14030 (2021). https://arxiv.org/abs/2103.14030
15. Liu, Z., Miao, Z., Zhan, X., Wang, J., Gong, B., Yu, S.X.: Large-scale long-tailed recognition in an open world. CoRR abs/1904.05160 (2019). http://arxiv.org/abs/1904.05160
16. Maracani, A., Pastore, V.P., Natale, L., Rosasco, L., Odone, F.: In-domain versus out-of-domain transfer learning in plankton image classification. Sci. Rep. **13**(10443), 1–10 (2023). https://doi.org/10.1038/s41598-023-37627-7
17. Muhammad, M.B., Yeasin, M.: Eigen-CAM: class activation map using principal components. In: 2020 International Joint Conference on Neural Networks (IJCNN). IEEE (2020). https://doi.org/10.1109/ijcnn48605.2020.9206626
18. Orenstein, E.C., Beijbom, O., Peacock, E.E., Sosik, H.M.: Whoi-plankton - a large scale fine grained visual recognition benchmark dataset for plankton classification. CoRR abs/1510.00745 (2015). http://arxiv.org/abs/1510.00745
19. Pastore, V.P., Zimmerman, T.G., Biswas, S.K., Bianco, S.: Annotation-free learning of plankton for classification and anomaly detection. Sci. Rep. **10**(1), 12142 (2020). https://doi.org/10.1038/s41598-020-68662-3
20. Rong, Y., Leemann, T., Borisov, V., Kasneci, G., Kasneci, E.: A consistent and efficient evaluation strategy for attribution methods (2022)
21. Selvaraju, R.R., Das, A., Vedantam, R., Cogswell, M., Parikh, D., Batra, D.: Grad-CAM: why did you say that? visual explanations from deep networks via gradient-based localization. CoRR abs/1610.02391 (2016). http://arxiv.org/abs/1610.02391
22. Sun, X., Xv, H., Dong, J., Zhou, H., Chen, C., Li, Q.: Few-shot learning for domain-specific fine-grained image classification. IEEE Trans. Industr. Electron. **68**(4), 3588–3598 (2020). https://doi.org/10.1109/TIE.2020.2977553
23. Touvron, H., Cord, M., Douze, M., Massa, F., Sablayrolles, A., Jégou, H.: Training data-efficient image transformers & distillation through attention. CoRR abs/2012.12877 (2020). https://proceedings.mlr.press/v139/touvron21a.html
24. Van Horn, G., et al.: The iNaturalist challenge 2017 dataset. CoRR abs/1707.06642 (2017). http://arxiv.org/abs/1707.06642
25. Wang, H., Du, M., Yang, F., Zhang, Z.: Score-CAM: improved visual explanations via score-weighted class activation mapping. CoRR abs/1910.01279 (2019). http://arxiv.org/abs/1910.01279
26. Xie, S., Girshick, R.B., Dollár, P., Tu, Z., He, K.: Aggregated residual transformations for deep neural networks. CoRR abs/1611.05431 (2016). http://arxiv.org/abs/1611.05431

27. Zemp, M., et al.: GCOS 2022 implementation plan. Glob. Climate Observ. Syst. GCOS (244), 85 (2022)
28. Zhou, B., Khosla, A., Lapedriza, A., Oliva, A., Torralba, A.: Learning deep features for discriminative localization (2015)

Agent-Based Model and Machine Learning for the Analysis of Barter Regulated by Lotka-Volterra Equations in an Artificial Society

Juan Villacrés$^{(\boxtimes)}$ (iD)

UTE University, Quito, Ecuador
jvillacresb@gmail.com

Abstract. In this paper, agent-based modeling and random forest are used to model the exchange of goods and services through barter. It is regulated by the Lotka-Volterra equation system–which is often used to analyze the dynamics of populations of different species. These equations, the ones that correspond to the case of mutualism between species, help to ensure that the amount of goods and services that a person can exchange does not accumulate during certain time among few individuals. This situation is repeated for the case in which empirical information is integrated into the modeling. Through such information, the coincidence of desires is modeled with the help of the random forest technique, which allows the modeling to be more closely related to a real situation.

Keywords: Agent-Based Model · Random Forest · Barter · Lotka-Volterra equations · Bio-inspired model

1 Introduction

Science can contribute to sustain society beyond the commodification of life. This observation is salient in the context of the multiple crises of the actual phase of capitalism [12]–as e.g. climate change. This is because, in order to face such crises we could need to look for alternative ways of exchange goods and services that differs from the ones proposed by capitalism. For instance, when the COVID-19 pandemic arrived–where many people with the capacity and willing to work, were left without it–there was a revival of barter powered by digital technology as e.g. using social media [4,7,8].

Certainly, a barter economy could function together or within a capitalist economy. However, for the former to collaborate with the latter for the benefit of humanity, the barter economy must overcome the problems attributed to capitalist economy. In this sense, it is worth to notice that, a criticized consequence of this phase of capitalism is inequality [9,10]. According to [1], by January 2023, the richest 1% of the entire population will hoard almost twice the wealth, generated in 2020, that was accessed by the 99% of humanity. Does a barter economy

© The Author(s), under exclusive license to Springer Nature Switzerland AG 2025
L. Correia et al. (Eds.): IBERAMIA 2024, LNCS 15277, pp. 87–98, 2025.
https://doi.org/10.1007/978-3-031-80366-6_8

also would give rise to monopoly? And in such case, Is there some regulation over the barter system that could decrease such a monopoly? In this work, I use agent-based modelling and a random forest model to analyze these questions. Furthermore, I propose a method to decrease monopoly in an artificial society that uses barter.

In the field of agent-based modeling, we have the paradigmatic case of the Sugarscape model [6] where societies are modeled in which the exchange of goods–in this case spices and sugar–takes place through barter regulated by a price. This is determined by the marginal rates of substitution[1] dictated by the people involved in the exchange.

More recent cases of agent-based barter modelling have occurred within The Village Ecodynamics Project, which models the settlement and livelihoods of Pueblo societies in Mesa Verde from 600 to 1300 [5]. For example, in [5] the amount of meat or corn to be exchanged is regulated based on a price. In this case, the price that person A assigns to a kilogram of protein is determined and quantified by the calories that he or she invested to obtain the protein. So if said person used $1000\,cal$ to obtain $10\,kg$ of protein, he or she will assign a value of $100\,cal$ to each kilogram. Thus, if a person B wants to exchange a good such as meat or corn with A, he or she can only exchange goods that add up to $100\,cal/kg$.

Although these models do not use machine learning techniques, such a techniques have been used lately to improve agent-based models. Thus, one of the most used techniques has been decision trees. This technique has been used to introduce empirical information to agent-based modeling, where, in addition, it allows the preprocessing of the data that will be used to determine the decision-making by the agents [2]. This has helped to improve the precision in the modeling/simulation of the behavior of the agents. For example, through this technique in [16] they use quantitative data about streets, taxis, weather, etc., to determine whether or not the agents will commit a crime.

In this paper, I will model a group of people who use bartering through online applications to exchange goods and services. Currently small businesses provide such applications [15]. They are in charge of finding matches between the offers and requirements of the users. Thus, we can imagine a scenario in which a person X offers O_x and expects E_x, while another Y offers Ex and expects Ey, and a third Z offers Ey and expects Ox. In this case, the application will identify the matches so that X can have what he wants through what Y offers, while Y does not get what he wants from X, but from Z, and it is X who satisfies Z's requirement, not Y. In other words, in this way the mentioned applications would help us to alleviate the problem of the coincidence of desires. This corresponds to a problem concerning goods exchange, namely: find a person Y that wants what X offers, which also gives to X what she or he wants [11].

---

[1] The marginal rate of substitution indicates how many units of one good a person is willing to give up in exchange for a unit of another good while keeping the level of utility constant.

Now, even if there were a community willing to barter, for which the problem of coincidence of desires is acceptably solved, this does not guarantee that the monopoly problem has been overcome. This problem is one of distribution of goods and services. To overcome it, I propose that this distribution be analogous to the way in which populations of species are distributed over time as dictated by the Lotka-Volterra (LV) system of equations.

This system can be written as:

$$\dot{x} = a_i x_i - b_i x_i^2 + \sum_{j=1, j \neq 1}^{N} c_{ij} x_i x_j$$

Here x_i corresponds to the population of the i-th species, a_i is the growth rate, b_i is the intraspecific competition, c_{ij} is the rate of interaction between species. In case the c_{ij} are positive we will be in the case of mutualism. One in which the interaction between species contributes to the growth of their population as in the case of bees and flowers [3]. While, for the case of species such as bears and fish, their population would obey a system of equations where the rate of interaction benefits the bears–here it will be positive–, and affects the population of fish–where said rate will be negative. This second case corresponds to the prey-predator interaction [3].

Certainly, Lotka-Volterra equations have already been used for modeling the dynamics of competition, and monopoly in market [13,14]. However, in the literature examined here, no works have been found that consider the mutualist case of the Lotka-Volterra equations to model economic dynamics, and even less to model barter as is done in this work.

To analyze whether monopoly exists in a barter-type system of exchange of goods and services through the aforementioned applications, and to examine the options for mitigating it if it does exist, I will proceed as follows. I will start by using NetLogo to model how a barter that is not regulated by the LV equations would turn out. For this I rely on the 'Simple Economy' [17] model from Netlogo's model library. I will then use Python to model barter based on LV equations in the following situations:

1. Barter regulated by LV equations for the mutualistic case.
2. Barter regulated by LV equations for a combined case of mutualism and predator-prey competition.
3. Integration of the random forest technique to mutualism-based barter

All documents created in the preparation of this work can be found at: https:// github.com/Juan-Villacres/TFM-ABM-Barter-Society

2   Methods

The Lotka-Volterra barter models that will be carried out are developed under the following assumptions:

1. There is a group of users using a barter digital application as described in the introduction, i.e. one that alleviates the problem of coincidence of desires.
2. Each person has a *barter capacity* for each instant of time. The *barter capacity* corresponds to the maximum amount of goods and hours of service provision that the barter application allows each person to offer for each *cycle*. A *cycle* is defined as a time interval that begins with the initialization of barter capacity based on individuals' wealth and concludes when the variance of the *barter capacity* reaches its minimum.
3. The initial *barter capacity* in each *cycle* is directly proportional to the wealth that each person has.
4. The *barter capacity* of a person A decreases by 1 if she gives a good or an hour of service to another person B. Whilst, the *barter capacity* of person A increases by 1 if she receives a good or an hour of service from another person B.
5. As the variance of the *barter capacity* decreases, its concentration among a few individuals diminishes, reducing the likelihood of a monopoly of the *barter capacity*.

2.1 Modeling Unregulated Barter

I considered the 'Simple Economy' model [17] from the NetLogo model library as a basis to model an unregulated barter system. This is a simple model in which at each instant a person gives a dollar to another person chosen at random. The modifications that I implement to this model are the following:

- I introduce a distribution of the initial barter capacity based on the distribution of wealth in 500 agents (NetLogo's turtles). Here, the wealth that each person has is assigned at random, and at most it can reach the value of 100. If the person has more wealth, he has a greater barter capacity.
- The transaction is completed when a randomly chosen person decreases his barter capacity by one, while another (chosen by chance) increases his barter capacity by one

2.2 Modeling Barter Regulated with Lotka-Volterra Equations

Here, I consider two cases for the distribution of goods and services. The first uses the LV equations for the mutualistic model, and the second uses a combined model of LV equations for mutualism and for predator-prey competition.

For the mutualistic case, I consider two groups of 12 people, where each can barter with the other 11 people in their own group. The initial distribution of *barter capacity* is randomly assigned with a maximum of 50. From each group of 12 people, 4 subgroups of three people are randomly chosen. In each subgroup, a person chosen randomly will be the one to provide a good or service. The mutualistic LV equations will be used to determine which of the other two individuals in the subgroup will receive the specified good or service. That is, when faced with the problem where two people want the same good or service, the aforementioned equations are used to determine who will actually receive it.

The *barter capacity* of each person would take the role of the population of a species in the LV equations. I use Python's odeint command to solve the LV system for the mutualist case, obtaining the values of the *barter capacity* of the two mentioned people. I use the barter capacity of the present time to replace it in, the right side of, LV equations. In this way, we are able to know the values of the derivative of the *barter capacity* for the pair of people interested in the same good or service for the present time. The person to which corresponds the highest value of the derivative will be the one to whom the good or service is assigned. Note that, here each unit of time is conceived as the realization of 8 barters; one for each subgroup taking into account the two groups of 12 people. This process is repeated over 200 units of time.

The described mutualistic barter completes its cycle when the variance of the *barter capacity* reaches its minimum. From here, I tried to combine this technique with a barter regulated by the LV equations for the prey-predator case. To do this, I obtained the total barter capacity of each group of 12 people by adding the barter capacities of each of them. The total capacity of one group took the role of the prey population, and that of the other group took the role of the predator population. Then, in a similar way to the mutualistic barter, the value of the derivative of these total capacities was used to determine which group gives, and which receives a good or service. Once it was determined which group receives the good or service, it is assigned to the person whose barter capacity is the lowest at the time when the exchange is made.

Finally, the random forest machine learning technique was integrated into the mutualistic model. This includes empirical information that helps determine who the three people are that would actually be involved in the mentioned barter dynamic (one offering a good or service and two interested in such an offering). In this case, the three people will not be chosen at random, but based on the coincidence of desires. For this, I use the Kaggle dataset 'Consumer Behaviour and Shopping Habits Dataset' that has collected real data on the preferences and behaviours of buyers when making purchases (on-line and off-line) of clothing and accessories in the United States.

I use the mentioned dataset and a random forest model during the barter, at each time step, to determine the particular item (e.g. boots) that people want to obtain in barter exchanges, based on their age, sex, the kind of item to be acquired (e.g. shoes), and the rating of the person who is going to give the item (this can be determined by 'likes' on online bartering applications). I will analyze a case in which I will hold certain characteristics constant (e.g. sex), while other characteristics (e.g. kind of item to be acquired) vary each unit of time.

The chosen random forest model uses the Gini criterion to determine the most relevant feature, the number of trees in the forest is 10, the minimum number of items in a node necessary for a classification (division) to be performed is 4, and the minimum number of items in a node for it to become a leaf is 2.

3   Results

3.1   Unregulated Barter

The initial distribution of *barter capacity* is associated with the initial random distribution of wealth as shown Fig. 1. Over time, the establishment of a monopoly can be noted, as e.g. in Fig. 2. This is because at time 732, the sum of the barter capacity of the richest 10%—which is 3926—exceeds that of the least rich 50%—which is 2541. That is, the aforementioned sum of the least rich 50% is the 64.7% of the richest 10%.

3.2   Mutualistic Barter and Combined Mutualistic Barter with the Prey-Predator Case

For the mutualistic barter case, 200 time units were used, where it can be seen that the variance of the *barter capacity* in each group of 12 people decreases until it reaches a minimum, and then begins to increase, as shown in Fig. 3. In addition, Fig. 4 shows, through a bar graph, how the *barter capacity* is distributed, for each group of 12 people, more homogeneously with respect to the initial distribution.

On the other hand, when implementing the combined case, no improvement was noted with respect to the mutualistic case. This is because, once the minimum variance was reached with the mutualistic exchange, the exchange regulated by the LV equations for the predator-prey case resulted in the variance only growing for a group of 12 people, while for the other group it reached zero, as shown in Fig. 5.

3.3   Mutualistic Barter with Random Forest

The models that we have examined do not include a criterion–based on empiric information–to stablish which should be the three people that get involved in the situation in which: two of them are intersted for the good or service that a third one offers. Here, a random forest model is incorporated into the mutualistic barter with the goal of determining the people who want the same goods and services. These people correspond to subgroups of three people that are formed based on shared desires. In the model, the predictor variables correspond to the following characteristics: sex, age, type of item to be acquired[2], qualification of the barterer, and size of the item. While, the objective variable corresponds to the desired *item* which can be: dress, pants, boots, sports shoes and belt.

It should be noted that the accuracy of the model varies depending on the type of 'desired items' that we choose from all the possible ones in the dataset. Thus, the ones I have mentioned were chosen because they allowed an accuracy of 0.62. This accuracy is sufficiently convenient for this modeling. On the other hand, the accuracy increases to 0.78 when the target variable *desired item* are

---

[2] Within the characteristic 'type of item to be acquired' are: clothing, shoes, accessories and street clothes, of which I only keep the first three.

the same as mentioned before except for boots. In this case, metrics as precision, recall, and f1-score that correspond to items 3 (sports shoes) and 4 (belt) are 1. This is because sports shoes are the only ones in the shoes category, and the belt is the only one in the accessories category. These categories correspond to the predictor variable *class of item to be purchased*. Since it is expected that the target variable is not determined solely or mainly by the variable *class of item to be purchased*, the *desired items* mentioned in the previous paragraph are preserved. The metrics and confusion matrix associated with this model can be seen in Fig. 6.

Once the model has been trained, it is used within the mutualistic barter. Such a model is applied to an array of characteristics, where some of them are kept constant: sex, age and qualification of the barterer. While the type of item to be acquired and the size of the item are varied each unit of time. In this case, it is again observed that the variance of the distribution of the bartering capacity does indeed decrease until it reaches a minimum, as shown in Fig. 7.

In order to analyze the process of reaching the minimum let us consider that the time until a person would be notified that he or she is in a position to receive, or to give a good or service[3]. I will refer as *time to receive* to the time that pass until a person be notified that she or he would receive a good or service. Also I will call *time to give* to the time that pass until a person be notified that she or he should give a good or service. Additionally, the ratio between the first and the second will be called *relative time*. After running the model for 25 cycles, the average times for *time to receive*, *time to give*, and *relative time* were 17.99, 6.45, and 2.78 respectively. That is, in each cycle the number of goods and services that each person gives does not necessarily coincide with the number that he or she receives. For example, for the case that corresponds to Fig. 7, in a cycle, nine persons receive more or equal number of goods and services than the others.

4   Discussion

As just mentioned, the mutualistic barter was successful in reducing the variance of the *barter capacity*, in all the cases analyzed, for the time it takes to reach its minimum. The decreasing of the variance is an emergent phenomenon of the mutualistic barter economy. This is because, such a decreasing cannot be reduced just to the sum of individual behaviours of people when they barter. That is, none of the possible behaviours provided to any trio of people involved in a barter includes explicitly (as e.g. by means of an equation) the decreasing of the *barter capacity*. This diminishing of monopoly arises from the collective dynamics in mutualistic barter economy; as expected of a complex system as the barter society presented here. One might initially think that even though the monopoly on the *barter capacity* has been reduced in each cycle, this has only occurred at the cost of several people giving more goods and hours of services than they receive in the mentioned cycle. However, let us note that in

---

[3] Here, the time mentioned runs from when the person registers his or her wishes and offers in the application.

the capitalist economy many employees with stable jobs, who do not work hourly, generally receive their salary 30 days after starting to provide their services. That is, capitalist economy shows a similar dynamics. Furthermore, the digital barter system could be improved by benefiting in each new cycle with a *relative time* between zero and one to those people who in the previous cycle had a *relative time* greater than one.

5   Conclusion

Throughout the work it was found that barter regulated by the LV equations for the mutualist case decreases monopoly. In economies based solely on this type of barter, the barter capacity would be equal to the wealth of each person. Hence, in such economies, although there would be moments in which wealth would accumulate in a few hands, at the beginning of each cycle, most of the time this accumulation will decrease. Hence, the proposed method for reducing monopoly in the examined artificial barter society has proven to be effective

While it is true that the use of the random forest technique was useful in bringing agent-based modeling closer to reality, there are many factors that can be fine-tuned and included. These comprise e.g. diversity of behaviors when buying or acquiring different goods and services beyond those related to clothing, cases when a barter involves exchanging not one good, but several goods simultaneously, or cases where people do not agree to give what they offer in the requested time. However, these improvements and other research projects suggested by this work should be pursued in future studies.

6   Figures

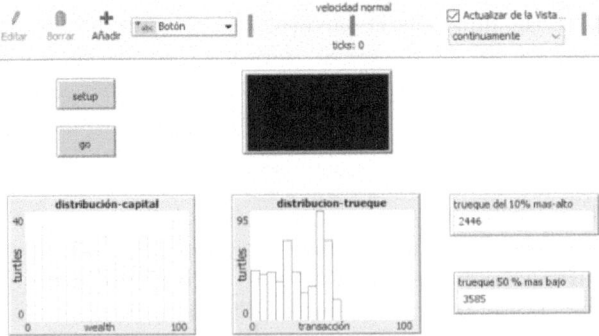

Fig. 1. Initial wealth and barter capacity distribution

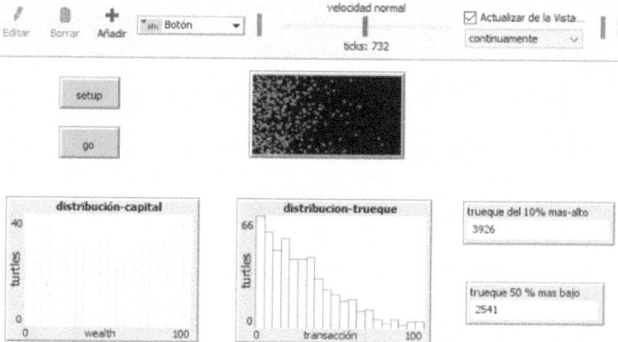

Fig. 2. Barter capacity distribution

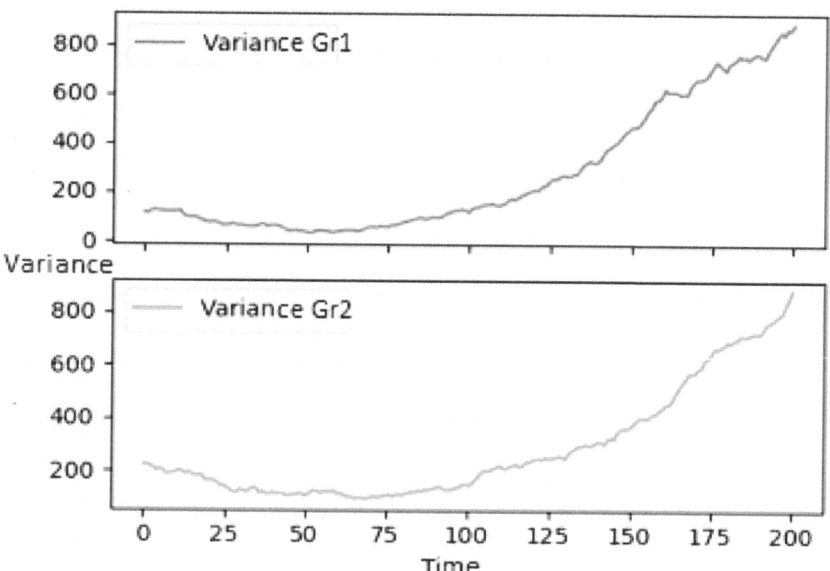

Fig. 3. Variance time evolution for the case of mutualism. The blue graph corresponds to first group of twelve people, whilst, the orange one to the other group. (Color figure online)

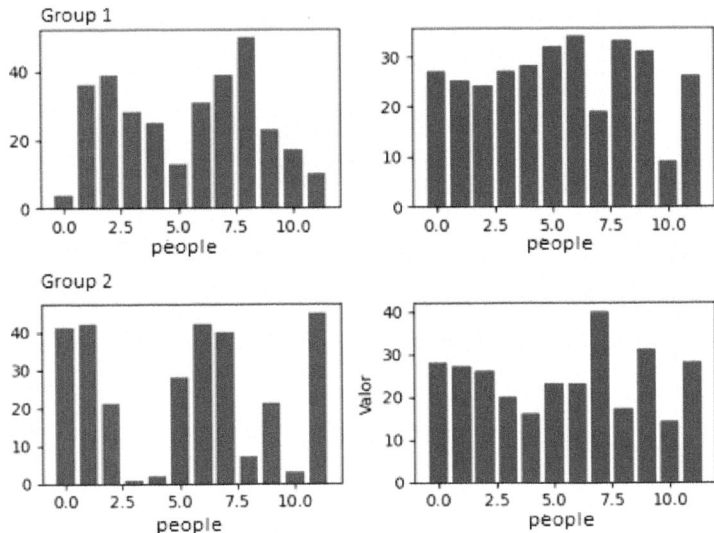

Fig. 4. Comparison of the barter capacity distribution. The graphs above correspond to the distribution of bartering capacity in the first group of twelve people. While the graphs below correspond to the second group. On the left side is the barter capacity at the beginning of the barter. While on the right side is the capacity at the end of a cycle.

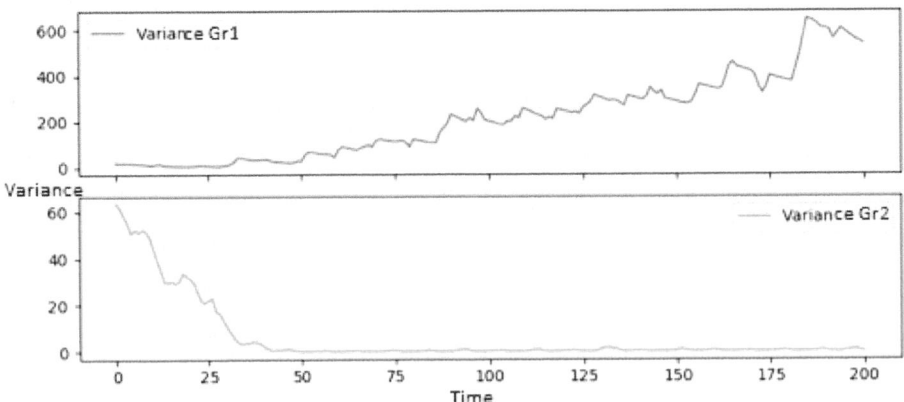

Fig. 5. Behaviour of the variance for the case that combine mutualism and prey-predator models. The graph above corresponds to the first group of twelve people. Whilst the graph below corresponds to the second one.

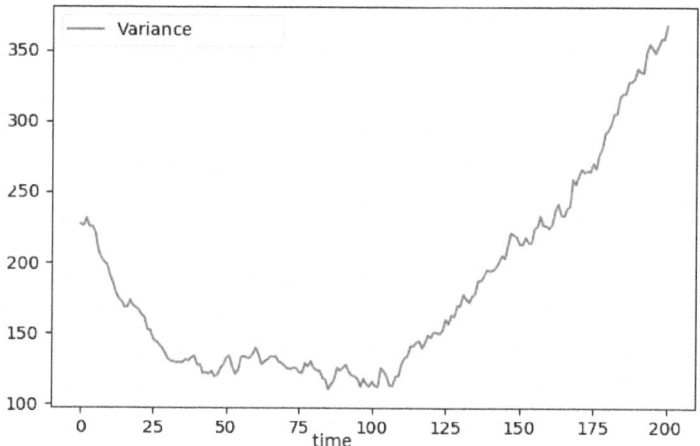

```
Accuracy: 0.620253164556962
Classification Report:
              precision    recall  f1-score   support

           1       0.49      0.45      0.47        38
           2       0.38      0.42      0.40        31
           3       1.00      1.00      1.00        37
           4       0.55      0.48      0.51        23
           5       0.62      0.69      0.66        29

    accuracy                           0.62       158
   macro avg       0.61      0.61      0.61       158
weighted avg       0.62      0.62      0.62       158
```

Confusion Matrix

```
array([[17, 18,  0,  0,  0],
       [21, 13,  0,  0,  0],
       [ 0,  0, 37,  0,  0],
       [ 0,  0,  0, 11,  9],
       [ 0,  0,  0, 12, 20]])
```

Fig. 6. Métrics and Confusion Matrix for random forest model

Fig. 7. Variance evolution in time for mutualist barter with random forest

References

1. Oxfam, cl 1 por ciento más rico acumula casi el doble de la riqueza que el resto de la población mundial en los últimos dos años (2023). https://www.oxfam.org/es/notas-prensa/el-1-mas-rico-acumula-casi-el-doble-de-riqueza-que-el-resto-de-la-poblacion-mundial-en
2. Ale Ebrahim Dehkordi, M., Lechner, J., Ghorbani, A., Nikolic, I., Chappin, E., Herder, P.: Using machine learning for agent specifications in agent-based models and simulations: a critical review and guidelines. J. Artif. Soc. Soc. Simul. **26**(1) (2023)
3. Brauer, F., Castillo-Chavez, C., Castillo-Chavez, C.: Mathematical Models in Population Biology and Epidemiology, vol. 2. Springer, New York (2012). https://doi.org/10.1007/978-1-4614-1686-9
4. Chin, F.M., Hastings, R.K., Tudy, R.A.: Reviving the ancient business model: the case of the Digos Barter trade during COVID-19 pandemic. Eubios J. Asian Int. Bioethics **31**(2) (2021)

5. Cockburn, D., Crabtree, S.A., Kobti, Z., Kohler, T.A., Bocinsky, R.K.: Simulating social and economic specialization in small-scale agricultural societies. J. Artif. Soc. Soc. Simul. **16**(4), 4 (2013)
6. Epstein, J.M., Axtell, R.: Growing Artificial Societies: Social Science from the Bottom Up. Brookings Institution Press, Washington (1996)
7. Finau, G., Kant, R.: Bartering as a form of resilience during COVID-19: case study of barter for better Fiji Facebook page. In: Ratuva, S., Ross, T., Crichton-Hill, Y., Basu, A., Vakaoti, P., Martin-Neuninger, R. (eds.) COVID-19 and Social Protection, pp. 193–204. Springer, , Singapore (2022). https://doi.org/10.1007/978-981-16-2948-8_11
8. Finau, G., Scobie, M.: Old ways and new means: indigenous accountings during and beyond the pandemic. Account. Audit. Account. J. **35**(1), 74–84 (2022)
9. Fisher, M.: Capitalist Realism: Is There No Alternative? John Hunt Publishing, London (2022)
10. Harvey, D.: Seventeen Contradictions and the End of Capitalism. Oxford University Press, Oxford (2014)
11. Jevons, W.S.: Money and the mechanism of exchange. In: General Equilibrium Models of Monetary Economies, pp. 55–65. Elsevier, Amsterdam (1989)
12. Lang, M., Bringel, B., Manahan, M.A.: Más allá del colonialismo verde: justicia global y geopolítica de las transiciones ecosociales (2024)
13. Liu, H., He, J., Chen, X.: Research on enterprise monopoly based on Lotka-Volterra model. In: Ahram, T., Karwowski, W., Pickl, S., Taiar, R. (eds.) IHSED 2019. AISC, vol. 1026, pp. 1018–1022. Springer, Cham (2020). https://doi.org/10.1007/978-3-030-27928-8_151
14. Marasco, A., Picucci, A., Romano, A.: Market share dynamics using Lotka-Volterra models. Technol. Forecast. Soc. Chang. **105**, 49–62 (2016)
15. Ozturan, C.: Network flow models for electronic barter exchanges. J. Organ. Comput. Electron. Commer. **14**(3), 175–194 (2004)
16. Rosés, R., Kadar, C., Malleson, N.: A data-driven agent-based simulation to predict crime patterns in an urban environment. Comput. Environ. Urban Syst. **89**, 101660 (2021)
17. Wilensky, U., Rand, W.: An Introduction to Agent-Based Modeling: Modeling Natural, Social, and Engineered Complex Systems with NetLogo. MIT Press, Cambridge (2015)

A Geometric Attention Mechanism to Classify Parkinsonism Smooth Pursuit Patterns

Luis Fernando Celis[1], Juan Olmos[1,2], and Fabio Martínez[1(✉)]

[1] Biomedical Imaging, Vision and Learning Laboratory (BIVL2ab), Universidad Industrial de Santander (UIS), 680002 Bucaramanga, Colombia
{luis2238327,jaolmosr}@correo.uis.edu.co, famarcar@saber.uis.edu.co
[2] Computer Science and Systems Engineering Laboratory (U2IS), ENSTA Paris, Institut Polytechnique de Paris, Palaiseau 91762, France

Abstract. Parkinson's disease (PD) is a neurodegenerative disorder that manifest progressive motor impairments. Smooth pursuit eye movement (SPEM) analysis has emerged as a potential PD biomarker, even at prodromal stages. Nonetheless, the standard protocols for SPEM analysis limit the study and discovery of abnormal eye movement patterns. Particularly, the protocols are invasive, and only recover global eye motion trajectories, losing kinematic information. This work introduces a video markerless representation that include a geometric attention mechanism with the capability to learn SPEM parkinsonism patterns under scenarios with limited training samples. This strategy first involves 3D convolutional layers to compute a volumetric bank of activations. Subsequently, the spatial and temporal relationships are synthesized into a symmetric positive definite (SPD) matrix. Then, a geometrical attention mechanism is introduced to identify the most significant feature relationships, even at specific moments in time, preserving the Riemannian geometry. The proposed approach was validated on a study with 15 patients with PD and 15 controls, achieving an average Recall and Precision scores of 0.80, and 0.89 respectively.

Keywords: Smooth pursuit eye movement · Parkinson's disease · Geometric attention · Riemannian deep learning

1 Introduction

Parkinson's disease (PD) is the second most common neurodegenerative disease, expressed as a progressive neurological disorder that includes locomotor disturbances, such as: the tremors, the rigidity, and the slowness [18]. These symptoms result from degeneration of dopamine-producing neurons in the substantia nigra [17]. Recently, oculomotor alterations have been remarked as potential early biomarkers, being sensible even before traditional motor symptoms, offering important insights into early disease detection or progression [1]. Predominantly, smooth pursuit eye movements (SPEM) have revealed abnormalities, including difficulties regulating voluntary eye movements (saccades), particularly reduced gain or speed of eye movement, involuntary saccades during tracking of moving stimuli, and impaired gaze convergence [6]. Nonetheless, the characterization, the detection, and the coding of such patterns require the use of specialized laboratory equipment, and sophisticated capture protocols. These technical requirements, together with the subsequent subjective analysis, hamper the efficiency of routine clinical attention for diagnosis and treatment [18]. To address these challenges, non-invasive methods have emerged using computer vision approaches based on video sequence analysis, leveraging advances in machine-learning-based technologies. Most of the state-of-the-art strategies are typically focused on coarse and common motor descriptors, analyzed from gait or tremor, which are typically assessed in clinical diagnosis in later stages of the disease [7]. Regarding eye movement, some studies have focused on the analysis of rapid eye movements during sleep [5], while others have analyzed ocular fixation tasks [15]. However, they often overlook the importance of spatio-temporal patterns in detecting specific abnormalities in SPEM, considering the substantial movement variability [20] (Fig. 1).

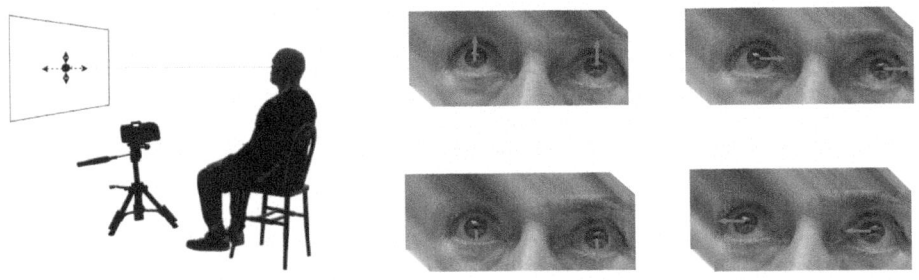

Fig. 1. A non-invasive capture configuration is depicted, showcasing a patient engaged in the task of smooth pursuit tracking a stimulus projected onto a screen.

The main contribution of this work is a novel end-to-end Riemannian self-attention mechanism to recover SPEM patterns, captured from a non-invasive setup, with limited training data information, and allowing to discriminate Parkinsonian from control patterns. The approach integrates 3D convolutional layers to extract deep spatio-temporal features, which thereafter are summarized in a compact symmetric positive definite (SPD) matrix, exploiting correlations among disease-related features. To stand out second-order patterns associated with the disease, the proposed geometric self-attention mechanism weights the most significant relationships within an SPD descriptor. In a retrospective study, the proposed approach was validated regarding the classification of Parkinson's videos, and also with the capability to recover online sequence predictions. The rest of the paper is organized as follows: the Sect. 2 presents the current strategies related with attention mechanisms and a brief background of Riemannian Learning. The Sect. 3 introduce the proposed approach and in the Sect. 4 is reported the achieved results. At the end, the Sect. 4.1 discuss and summarizes some conclusions of the work.

2   Geometric Learning and Attention Mechanisms

Geometric learning techniques have gained relevance in recent years, particularly for encoding high-dimensional representations into compact descriptors in many applications of computer vision. These techniques are effective for learning patterns while preserving pertinent information from local data affinity [2]. Particularly, similarity matrices, which are symmetric positive definite (SPD), form a Riemannian manifold, a topological space with local properties similar to Euclidean space and equipped with a metric structure. Hence, specialized modules have been developed to preserve the properties of the Riemannian space, progressively compressing information within the geometric space to facilitate nonlinear learning of covariate patterns [8].

Recent approaches leveraging SPD matrices have adopted state-of-the-art techniques to generalize and learn from non-Euclidean data, exploring the direct application of attention mechanisms over the Riemannian manifold, tailored to the intrinsic geometrical structure. Particularly, Konstantinidis et al. proposed an approach that computes linear multi-head attention, projecting the results into points in alternative geometric branches beyond Euclidean space, such as Grassmannian and SPD manifolds [9]. Such approach calculated similarities as distances within each respective space, subsequently concatenating and weighting the Euclidean value to capture intricate relationships and leverage geometrical and statistical properties for image classification. In the same way, Ma et al. proposed a cross-attention mechanism integrating structural and functional modalities from brain networks for Alzheimer's Disease analysis. This method calculates attention components over the same manifold, using geodesic distances to update information between modalities while ensuring coherent global geometrics [12]. Meanwhile, other approaches also utilize the SPD manifold for data analysis. Pan et al. developed a method for electroencephalographic decoding to track brain dynamics. This method divides the signal sequence, maps it from Euclidean space to the SPD manifold by computing the covariance for each segment, and then applies attention projections onto the manifold. The similarity function is based on the log-Euclidean distance between each pair of queries and keys, building a similarity matrix [16]. Similarly, Lu et al. integrated this similarity function with spatial and temporal convolution domains to create a symmetric matrix lying in the flat tangent plane [11]. In Euclidean space, similarity can be defined in several ways. However, attention components are SPD matrices instead of conventional vectors, requiring a definition in non-Euclidean contexts. The scalar measure obtained from distances is limited to a global approximation for weight, neglecting the latent information encoded in the matrix.

3   Proposed Approach

The proposed approach is in Fig. 2. Firstly, a bank of volumetric features is obtained to exploit spatio-temporal oculomotor information. Then, an SPD matrix is recovered as a geometrical SPEM descriptor. From this, novel SPD representation are calculated following a geometrical self-attention mechanism that allows to classify Parkinsonian patterns.

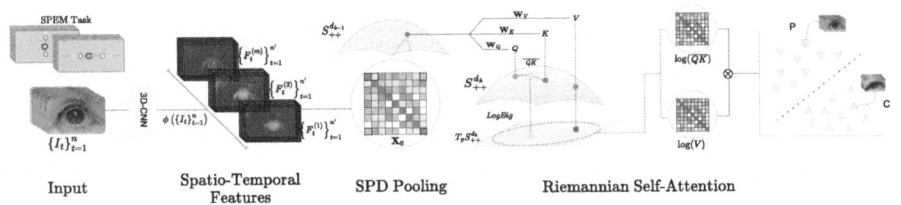

Fig. 2. Pipeline. First, a bank of deep representations is generated from an input smooth pursuit eye movement (SPEM) video, capturing the spatio-temporal patterns. Then, an SPD matrix is calculated to summarize the relationships between the volumetric deep representations. The subsequent step involves the implementation of riemannian module for spd learning, and a novel proposed attention, that allows to highlight relevant covariances. The scheme us trained end-to-end manner.

3.1   Extraction of Spatio-temporal Features

Oculomotor disturbances are often described as kinematic irregularities, displaying variations among PD patients. These disturbances exhibit variability in magnitude and temporal duration, making their characterization challenging. Even worse, in SPEM recordings, these disturbances are scarce, may appear spontaneously, and are often limited by recording frequency protocols [3,20].

In this work, a 3D convolutional representation (3D-Conv) extracts and process spatio-temporal features from the input video data. Particularly, the input video is represented as a sequence $\{I_t\}_{t=1}^{n}$ that is temporally composed of n frames with $W \times H$ spatial dimension. From this, the 3D-Conv can be seen as a function ϕ applied to the input video volume, recovering representations, such as $\phi\left(\{I_t\}_{t=1}^{n}\right) = \left\{F_t^{(i)}\right\}$, for $1 \leq i \leq m$, where m is the number of output volumetric representations, each one with a temporal length of n' frames and spatial dimension of $W' \times H'$, smaller than the input volumes ($W' \leq W$ and $H' \leq H$). This convolutional backbone serves as a 3D feature extractor, aiming to learn SPEM Parkinsonism-related patterns.

3.2   Riemannian Deep Representation

The bank of deep volumetric representations entails a set of several patterns that can be associated to the disease. In addition, these patterns entail high dimensional descriptors with complex and highly variable encoded information. For this, this work introduces a second-order descriptor from the last block of spatio-temporal convolutional activations. This descriptor encodes pairwise related to salient SPEM patterns. The output is a SPD matrix, which belongs to a Riemannian curved manifold S_{++}^n. Because their non-Euclidean character is required, a specialized module. To build the SPD descriptor, each volumetric feature $F_t^{(i)}$ is vectorized and rearranged into a matrix $\mathbf{M} \in \mathbb{R}^{m \times \ell}$, where $\ell = n' \times W' \times H'$. Following this, the SPD pooling operation calculates the matrix $\mathbf{X}_0 = \frac{1}{\ell^2} \mathbf{M}^\top \mathbf{M}$, where each entry $\mathbf{X}_0[i,j]$ indicates a similarity measure between the information captured by the feature $f_t^{((i)}$ and the information from $F_t^{(j)}$.

To learn new similarity features, we aim to encode novel descriptors from the SPD matrix $\mathbf{X}_0$. For this, and considering the Riemannian geometric structure of S_{++}^n, we implemented learning layers that preserve the geometric manifold structure [8]. In such case, a first layer defines a Bilinear Mapping (*BiMap*), which processes the input SPD descriptor via $\mathbf{X}_k = \mathbf{W}_k \mathbf{X}_{k-1} \mathbf{W}_k^\top$. Here, $\mathbf{X}_{k-1} \in S_{++}^{d_{k-1}}$ represents a previous SPD representation of dimension d_{k-1}. The weights $\mathbf{W}_k \in \mathbb{R}^{d_k \times d_{k-1}}$, are required to belong to the Stiefel manifold and being semi-orthogonal to still generating SPD matrices [8]. After each *BiMap* layer is applied, a rectification that prevents non-positivity errors (*BiRe* block). For this, the max function is calculated between each element of the diagonal of the eigenvalues matrix and a non-negative rectification threshold ε, such as $\mathbf{X}_k = \mathbf{U}_{k-1} \max(\varepsilon I, \mathbf{\Sigma}_{k-1}) \mathbf{U}_{k-1}^\top$, where $\mathbf{U}_{k-1}$ denotes the eigenvectors, and $\mathbf{\Sigma}_{k-1}$ is the diagonal matrix of eigenvalues of the matrix $\mathbf{X}_{k-1}$. This layer works as the rectified linear unit activation in common neural networks. Hence, $\mathbf{X}_k$ is a new SPD learned descriptor, which can be computed for independent parallel branches (with different learned $\mathbf{W}_k$) to enrich the PD related representations.

3.3 Riemannian Attention Mechanism

Nowadays, deep geometric learning has emerged as a promising alternative for data-sparse scenarios, being, among others, invariant to outliers artifacts and robust to noise data resulting in more robust alternative to capture patterns from high dimensional data [4]. Nonetheless, these representations report learning limitations. Also, there are no clear mechanisms to highlighting key relationships that may support the discrimination of spatiotemporal between Parkinsonian and control observations. This issue has been addressed by considering attention mechanisms, which have demonstrated prioritizing local information over a global context, effectively highlighting relevant image-level features [19]. Recently, these mechanisms have been extended in Riemannian architectures, by measuring the similarity of deep geometric representation as their distance in the geometric manifold where they lie [13]. However, these approximations are limited to a global scalar measure, overlooking the local information saved in the correlations of the SPD descriptor [12]. To address these limitations, the main contribution of this work is a novel Riemannian Self-Attention Mechanism (RSAM) that weights SPD descriptors according to global context information. First, we take advantage of *BiRe* projections to learn three parallel branches from the SPD descriptor $\mathbf{X}_k$, generating three independent Riemannian matrices within the same manifold using *BiRe* blocks. In correspondence with a common self-attention mechanism, these matrices are denominated the query $\mathbf{Q} = \mathbf{W}_Q \mathbf{X}_k \mathbf{W}_Q^\top$, the key $\mathbf{K} = \mathbf{W}_K \mathbf{X}_k \mathbf{W}_K^\top$, and the value $\mathbf{V} = \mathbf{W}_V \mathbf{X}_k \mathbf{W}_V^\top$, respectively. Hence, benefiting from the fact that this Riemannian manifold is geodesically complete with non-positive curvature and has a unique geodesic curve $\varphi(t)$ between any two points [21]. The mean can be calculated to integrate information from both $\mathbf{Q}$ and $\mathbf{K}$. The Riemannian mean is the midpoint on the geodesic curve defined as:

$$\overline{\mathbf{QK}} = \mathbf{Q}^{\frac{1}{2}}(\mathbf{Q}^{-\frac{1}{2}}\mathbf{KQ}^{-\frac{1}{2}})^{\frac{1}{2}}\mathbf{Q}^{\frac{1}{2}}.$$

The resulting $\overline{\mathbf{QK}}$ is an SPD matrix of the same dimensions as $\mathbf{Q}$ and $\mathbf{K}$, separate from both $\mathbf{Q}$ and $\mathbf{K}$, represents a novel descriptor that measures the similarity between $\mathbf{Q}$ and $\mathbf{K}$. In the classic attention mechanisms [14] after computing the product between $\mathbf{Q}$ and $\mathbf{K}$, a softmax operation is utilized for normalizes all the scores to a probability distribution. Here, given the nature of SPD data, we decided to weight the information captured on $\overline{\mathbf{QK}}$ by using the spectral information. For this, applying the softmax function to the eigenvalues weights the information in the matrix, to compute as follows $\overline{\mathbf{QK}}^* = \mathbf{U}_{k-1} softmax(\mathbf{\Sigma}_{k-1}) \mathbf{U}_{k-1}^\top$ This approach ensures a positive probability distribution, $softmax(\Sigma_{k-1}) > 0$ for each element in Σ_{k-1}, highlighting the relative importance of each eigenvalue in the matrix structure. This matrix is then projected into Euclidean space using the Riemannian logarithm (*LogEig*) layer [8]. The proposed RSAM employs the projected matrices $\log(\overline{\mathbf{QK}}^*)$ and $\log(\mathbf{V})$ weighting local relationships within the compact descriptor, as follows:

$$RSAM(\mathbf{X}_k) = \log(\overline{\mathbf{QK}}^*) \cdot \log(\mathbf{V}) = \mathbf{X}_{k+1}.$$

This weighted $\mathbf{X}_{k+1}$ incorporates the most relevant relationships during learning, quantifying those with the highest potential to capture oculomotor abnormalities, which is absent in classical approaches.

3.4   Data

In this study was recruited a total of 30 participants, 15 control subjects, with an average age of 74.6 ± 5.8 years, and 15 individuals diagnosed with Parkinson's disease with an average age of 74.4 ± 8.4 years. The PD patients were categorized with different disease level progression, including five in stage 2.5, five in stage 3, and two in stage 4, according to the Hoehn-Yahr scale. The data collection process involved a video recording of participants' eye regions while performing an SPEM task. First, participants were invited to stand in front of a screen positioned at a distance of 1 m and at the same height as the eyes. On this screen, a white dot (stimulus) was projected, performing a continuous smooth movement over a 15-s duration, first 8 s, horizontal movements were performed, followed by 7 s of vertical movements. During this SPEM task, participants were instructed to follow the stimulus with their gaze, while their faces were recorded by a camera with a resolution of 60 frames per second (FPS) positioned below the screen level. One video was recorded for each participant, and a template-matching technique was implemented to crop the area around the eyes, resulting in a rectangular region of dimensions 456×322 for each eye and each video sequence. In this way, for each participant, two video volumes were calculated after the pre-processing steps, each one corresponding to an independent eye. Written informed consent was obtained from all participants.

3.5   Experimental Setup

The convolutional backbone in this work was implemented with a total of three layers, with 32 volumetric features in the first layer, 64 in the second layer, and 128 in the third layer. Then, the SPD pooling is applied, resulting in a SPD descriptor of dimension of 128×128. Thereafter, the proposed RSAM was computed, specifically, the query (Q), the key (K), and the value (V) branches, resulting all of them in SPD matrices of size 32×32. For this work, we only consider one head of attention mechanism. In the ReEig layers, a rectification threshold of $\varepsilon = 10^{-4}$ was set. Regarding the training of the proposed approach, we run a total of 150 epochs with a learning rate of $\alpha_1 = 10^{-4}$ for SPD modules and $\alpha_2 = 10^{-5}$ for convolutional part, using a binary cross-entropy loss function, and following an Adam optimizer.

For validation, we employ a 5-fold cross-validation scheme. The dataset was divided into five folds, with each one having validation sets that included three control patients and three patients diagnosed with Parkinson's disease each. To evaluate model performance, average accuracy, recall, precision,F1 score and AUC-ROC scores were computed across folds. We compared the proposed approach with two approximation. The first, baseline consists of the 3D-Conv integrated with SPD modules (3D-ConvSPD) [15]. The second approach adapted a Cross-Modal Riemannian Network (CMRN) into a self-attention mechanism, computing the scalar attention as a Riemannian distance between the key and query $\lambda = d_R(Q, K)$ [12]. This attention mechanism is defined as $CMRN(\mathbf{X}_k) = \lambda \cdot \log(V)$.

4   Evaluation and Results

The proposed RSAM was evaluated on the classification of Parkinson sequences from complete SPEM video recordings, and compared with baseline strategies (3D-ConvSPD, CMRN). Table 1 summarizes the performance of the evaluated geometrical architectures.

Interestingly, despite challenges of SPEM classification patterns and the relatively limited set of data, all models achieved AUC-ROC scores exceeding 0.92. This result indicates the robustness of geometrical architectures to lead in clinical scenarios with observations with high variability and limited training conditions.

Table 1. Performance of the proposed method and comparison with baseline model and state-of-the-art method evaluation with Attention Mechanisms and the Baseline 3D-ConvSPD. Highest scores are highlighted in bold

Models	Scores (%)				
	Accuracy	*Recall*	*Precision*	*F1*	*AUC-ROC*
3D-ConvSPD	78.3 ± 12.5	73.3 ± 13.3	85.7 ± 18.1	77.4 ± 12.1	**93.3 ± 8.0**
CMRN [12]	80.0 ± 12.5	76.7 ± 8.2	85.7 ± 18.1	80.1 ± 10.8	92.2 ± 8.3
RSAM	**83.3 ± 9.1**	**80.0 ± 6.7**	**89.2 ± 14.8**	**83.3 ± 7.3**	92.2 ± 7.7

It should be also noted that the proposed RSAM method yielded the best score, with a significant improvement in Parkinson's patient discrimination, achieving a 80.0 ± 6.7 recall score and precision score of 89.2 ± 14.8. In addition, slightly improvements were observed when including the scalar attention to the baseline CMRN model. This underscores the limitation of the scalar attention approach, that loss local relationships saved in SPD matrices. In this regard, the proposed attention method entail a better integration of the most pertinent relationships by weighting SPD matrices considering the Riemannian geometry and the proposed way of performing the attention.

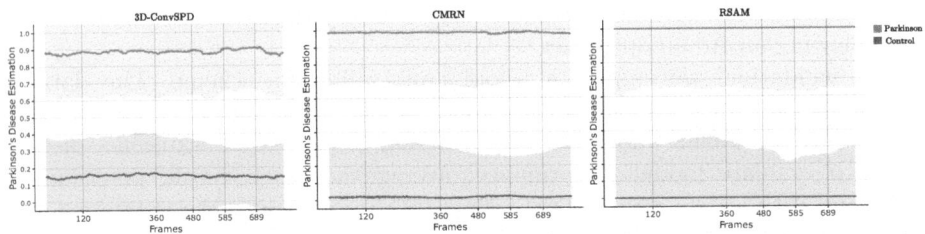

Fig. 3. Temporal Windowing Analysis. Temporal prediction comparison among evaluated models for Parkinson's class using a 120-frame window, differentiated by directional changes in SPEM in temporal (frames) axis. Each graph represents the baseline 3D-ConvSPD, the CMRN and the proposed RSAM respectively.

A main contribution of the proposed approach is the capability to carry out online prediction during a video recording, offering potential applications in clinical scenarios for identifying temporal abnormalities associated with PD. This is particularly relevant given existing literature that emphasizes challenges in specific eye movements within the PD population, such as directional changes [6]. Then, we study abnormal SPEM movements recovering an online and continuous prediction, computed from a sliding window approximation, *i.e.*, extracting multiple video segments for each time point (frame t) and throughout the interval $[t, t+120]$. Figure 3 summarizes the achieved results for all the test sequences, considering each instant by temporal sequence of medians, with the shadow being the standard deviation.

The evaluated models present outstanding performance for temporal predictions at all time instants. The baseline 3D-ConvSPD shows higher variability in the interval [120,360] for control patients, while Parkinson's cases exhibit similar behavior across all predictions. Adding an attention scheme considerably improves the median locations. Furthermore, the medians for the predictions of the models with attention consistently align with the obtained classification scores, demonstrating reduced variability between PD and control subjects, especially in the interval [480, 690], corresponding to changes in direction during vertical movements. These findings are consistent with patterns reported in the literature [6]. Additionally, the proposed model shows sensitivity in the same interval regarding the variability of control patients.

Table 2. Results of the ablation study of proposed method RSMA over different heads. Highest scores are highlighted in **bold**.

Number of Heads	Scores (%)				
	Accuracy	*Recall*	*Precision*	*F1*	*AUC-ROC*
1	83.3 ± 9.1	80.0 ± 6.7	$\mathbf{89.2 \pm 14.8}$	83.3 ± 7.3	92.2 ± 7.7
2	83.3 ± 12.9	83.3 ± 10.5	86.4 ± 17.6	84.0 ± 11.4	92.2 ± 11.6
3	81.7 ± 13.3	80.0 ± 12.5	84.6 ± 15.8	81.7 ± 12.4	91.7 ± 10.5
4	$\mathbf{85.0 \pm 14.3}$	$\mathbf{86.7 \pm 12.5}$	86.4 ± 17.6	$\mathbf{85.8 \pm 13.0}$	$\mathbf{92.8 \pm 11.7}$

The proposed method can be executed on multiple branches in parallel. These branches are joined at the end of the architecture through concatenation to initiate the linear layer. Up to four heads were evaluated, and the results are summarized in Table 2. Extending the architecture to a multi-head scheme with four heads shows benefits in most metrics, achieving outstanding results in RECALL with 86.7 ± 12.5 and in accuracy a 86.0 ± 14.3. However, in the evaluation of precision, its score decreases being outperformed by the single-head model, that has an advantage in the classification of control cases. Additionally, analyzing the F1 score, we see that the four-head model increased its performance, achieving an 85.8 ± 13.0.

4.1   Conclusions

This work proposes a novel Riemannian self-attention mechanism (RSAM) for PD classification using SPEM video recordings. Implemented within an end-to-end convolutional network, it involves the computation of SPD matrices. The efficacy of this methodology was evaluated with 15 PD patients and 15 control subjects. Current methods for SPEM analysis rely primarily on VOG and EOG, which require thorough analysis of time series data to identify patterns with high variability in frequencies and magnitudes among patients [10]. In contrast, attention mechanisms demonstrated notable capabilities to learn spatio-temporal relationships, effectively discriminating PD from control subjects. This method also allows for online predictions during specific time intervals, presenting a potential non-invasive alternative to support PD diagnosis and monitoring. Additionally, multi-head schemes demonstrated consistency, resulting in outstanding classification scores. However, the attention models evaluated, despite not surpassing the 3D-ConvSPD model's score of 93.3 ± 8.0 in AUC-ROC, do not show significant differences. This may be due to the use of geometric modules, which are robust to noise and data scarcity [4]. So, geometric attention mechanisms emerging as a promising technique to enhance covariate representations.

Future research should consider multimodal information, including different eye movement abnormalities and clinical data such as PD laterality. Furthermore, cross-attention mechanisms should be explored using larger datasets to confirm the generalization capability of the proposed approach

Acknowledgments. This work was supported by the Vicerrectoría de Investigación y Extensión (VIE) of the Universidad Industrial de Santander, project: *Cuantificación de lesiones de próstata comparando secuencias multi-paramétricas y bi-paramétricas de MRI usando herramientas de inteligencia artificial.* Code 3946.

Disclosure of Interests. The authors do not have any conflict of interest to disclose.

References

1. Antoniades, C.A., Spering, M.: Eye movements in Parkinson's disease: from neurophysiological mechanisms to diagnostic tools. Trends Neurosci. (2023)
2. Bronstein, M.M., Bruna, J., LeCun, Y., Szlam, A., Vandergheynst, P.: Geometric deep learning: going beyond Euclidean data. IEEE Sig. Process. Mag. **34**(4), 18–42 (2017)
3. Chang, Z., et al.: Accurate detection of cerebellar smooth pursuit eye movement abnormalities via mobile phone video and machine learning. Sci. Rep. **10**, 18641 (2020)
4. Congedo, M., Barachant, A., Bhatia, R.: Riemannian geometry for EEG-based brain-computer interfaces; a primer and a review. Brain-Comput. Interfaces **4**(3), 155–174 (2017)
5. Farashi, S.: Analysis of vertical eye movements in Parkinson's disease and its potential for diagnosis. Appl. Intell. **51**(11), 8260–8270 (2021)
6. Frei, K.: Abnormalities of smooth pursuit in Parkinson's disease: a systematic review. Clin. Parkinsonism Relat. Disord. **4**, 100085 (2021)
7. Guayacán, L.C., Rangel, E., Martínez, F.: Towards understanding spatio-temporal Parkinsonian patterns from salient regions of a 3D convolutional network. In: 2020 42nd Annual International Conference of the IEEE Engineering in Medicine and Biology Society (EMBC), pp. 3688–3691. IEEE (2020)
8. Huang, Z., Van Gool, L.: A Riemannian network for SPD matrix learning. In: Proceedings of the AAAI Conference on Artificial Intelligence, vol. 31 (2017)
9. Konstantinidis, D., Papastratis, I., Dimitropoulos, K., Daras, P.: Multi-manifold attention for vision transformers. IEEE Access (2023)

10. Larrazabal, A.J., Cena, C.G., Martínez, C.E.: Video-oculography eye tracking towards clinical applications: a review. Comput. Biol. Med. **108**, 57–66 (2019)
11. Lu, B., Huang, X., Chen, J., Fu, R., Wen, G.: Manifold attention-enhanced multi-domain convolutional network for decoding motor imagery intention. Knowl.-Based Syst. **296**, 111904 (2024)
12. Ma, J., Zhang, J., Wang, Z.: Multimodality Alzheimer's disease analysis in deep Riemannian manifold. Inf. Process. Manag. **59**(4), 102965 (2022)
13. Minh, H.Q., Murino, V.: Covariances in Computer Vision and Machine Learning. Springer, Cham (2022). https://doi.org/10.1007/978-3-031-01820-6
14. Niu, Z., Zhong, G., Yu, H.: A review on the attention mechanism of deep learning. Neurocomputing **452**, 48–62 (2021)
15. Olmos, J., Manzanera, A., Martínez, F.: Riemannian SPD learning to represent and characterize fixational oculomotor Parkinsonian abnormalities. Pattern Recogn. Lett. **177**, 157–163 (2024)
16. Pan, Y.T., Chou, J.L., Wei, C.S.: Matt: a manifold attention network for EEG decoding. In: Advances in Neural Information Processing Systems, vol. 35, pp. 31116–31129 (2022)
17. Shahed, J., Jankovic, J.: Motor symptoms in Parkinson's disease. Handb. Clin. Neurol. **83**, 329–342 (2007)
18. Tolosa, E., Garrido, A., Scholz, S.W., Poewe, W.: Challenges in the diagnosis of Parkinson's disease. Lancet Neurol. **20**(5), 385–397 (2021)
19. Vaswani, A., et al.: Attention is all you need. In: Advances in Neural Information Processing Systems, vol. 30 (2017)
20. Wu, C., et al.: Eye movement control during visual pursuit in Parkinson's disease. peerJ **6**, e5442 (2018)
21. Yair, O., Ben-Chen, M., Talmon, R.: Parallel transport on the cone manifold of SPD matrices for domain adaptation. IEEE Trans. Sig. Process. **67**(7), 1797–1811 (2019)

A Mixed Audio-Video SPD Network for Online Classification of Parkinsonian Speech Patterns

John Archila[1], Antoine Manzanera[2], and Fabio Martínez[1(✉)]

[1] Biomedical Imaging, Vision and Learning Laboratory (BivL2ab),
Universidad Industrial de Santander (UIS), Bucaramanga, Colombia
`famarcar@saber.uis.edu.co`
[2] U2IS/Robotics and Autonomous Systems, ENSTA Paris,
Institut Polytechnique de Paris, Palaiseau, France
`antoine.manzanera@ensta-paris.fr`

Abstract. Parkinson's disease (PD) is a neurodegenerative disease that produces progressive motor impairments. Dysarthria (speech disorders) and hypomimia (face rigidity) are two major Parkinsonism patterns observed even at the early stages of the disease. Nonetheless, the clinical diagnosis is mainly observational and dependent on the specialists' expertise. Besides, the categorization of each of these patterns is isolated, which may lead to delayed diagnosis and misplanning of treatments. This work introduces a non-invasive multimodal strategy that integrates video and audio modalities into the online characterization of speech exercises. Subjects were invited to pronounce sustained vowels while video and audio were recorded. Then, a temporal window is run along the sequence to build online covariance matrices of synchronized face landmarks position and characteristic voice frequencies. From these temporal covariance matrices are learned Riemannian descriptors that allow to discriminate between Parkinson's and control subjects. From a study with 14 subjects, the proposed approach achieved a mean accuracy of 70% in sustained vowel pronunciation. Considering online predictions, the proposed approach evidenced a consistent accuracy of 0.77 during pronunciation of close vowels.

Keywords: Mixed audio-video SPD networks · online Parkinson's Disease prediction

1 Introduction

Parkinson's disease (PD) is a chronic neurodegenerative disease with no cure, characterized by progressive degeneration of nerve cells, decreasing the production of dopamine, resulting in serious impairments regarding the control of movement and coordination [4]. Early motor impairments are usually manifested as dysarthria (speech affectation associated with rigidity of muscles) and

L. Correia et al. (Eds.): IBERAMIA 2024, LNCS 15277, pp. 110–121, 2025.
https://doi.org/10.1007/978-3-031-80366-6_10

hypomimia (facial expression affectation associated with movement slowness and rigidity) [18,21]. Patients with such symptoms may experience difficulties in articulating words or changing the tone of their voice, resulting in difficult and monotonous speech. These symptoms are manifested between 7 and 11 years before the definitive diagnosis of Parkinson [5,15]. Nowadays, these patterns are characterized only by observational tests, highly dependent on the specialist's expertise [2,6]. Additionally, they have low sensitivity in early stages, and researchers need to spend a significant amount of time developing the skills for an adequate evaluation [19].

The main contribution of this work is a geometrical online learning method to support Parkinson classification considering multimodal sources (audio and video). Thus, characterizing dysarthria and hypomimia, the proposed approach use a set of video landmarks that, together with fundamental frequencies, form a compact covariance descriptor. From this second-order representation, geometrical learning is herein implemented to learn covariative patterns associated to the disease at different temporal intervals. The paper is structured as follows: Sect. 2 provides an overview of the literature on Parkinson's disease focusing on methodologies to support hypomimia and dysarthria. Section 3 describes the proposed approach integrating audio and video modalities. Section 4 presents the classification results. Section 5 discusses the advantages and limitations of the proposed online geometrical representation.

2   Related Works

Communication is a fundamental daily life task, involving the coordination of multiple muscular, respiratory, and facial functions [11]. The facial expression during communication is based on the gesticulation of words, producing mouth movements and the coordination of the zygomatic muscles and the orbicular muscle. For patients affected by PD at early stages, there exist evidences of gesture limitations, which causing slowness and rigidity, known as mask face or hypomimia [18,21]. These persons may experience difficulties in articulating words or changing the voice tone, resulting also in speech difficulties known as dysarthria. Today, there are no significant advances on the characterization of such pattern and even worst, in the combination of dysarthria and hypomimia patterns, from multimodal approaches.

The quantification of hypomimia has been previously estimated using strategies to classify single images [7,17] or videos [14,23,26]. Approaches based on single-image classification consider the identification of facial landmarks whose spatial characteristics allow classification through classical machine learning methods [17] or statistical analysis [7]. Other proposed approaches have attempted to temporally characterize the most significant expressions during classification from activation maps of a 3D convolutional networks [23]. However, the retrieved activation maps only coarsely distinguish regions, so that differentiating patients and control subjects remains challenging. Recurrent networks have also been used to extract the temporal embedding to classify PD

[26]. Alternatively, landmarks have been located in face to associate emotions expressions with Parkinson patterns and carry out the classification [14].

Regarding dysarthria, the frequency analysis of voice has been used as descriptor to classify patients with PD, in particular harmonic analysis and signal-to-noise parameters [1,12]. Other works have incorporated deep learning stages using a CNN [9,25] or recurrent architectures [16], where these architectures learned new representations based on the frequency characteristics of emotional expression [9] and vowel pronunciation [16,25]. These computational approaches have evidenced remarked scores to classify Parkinson's disease, but their application is yet limited to operate in clinical scenarios, without complex setups of recording. Besides, to the best of our knowledge, there exists limited information about how to fuse hypomimia and dysarthria information to enhance Parkinson's representation.

3   Proposed Approach

This work introduces an online multimodal approach that fuses orofacial patterns, following an early fusion method based on covariance patterns. The covariance descriptor encodes both face landmarks trajectories and fundamental frequencies of the audio speech, aligned in intervals of time. Then, a geometrical representation is learned on the Riemannian manifold, to classify Parkinsonian patterns. The general pipeline of the proposed approach is illustrated in Fig. 1.

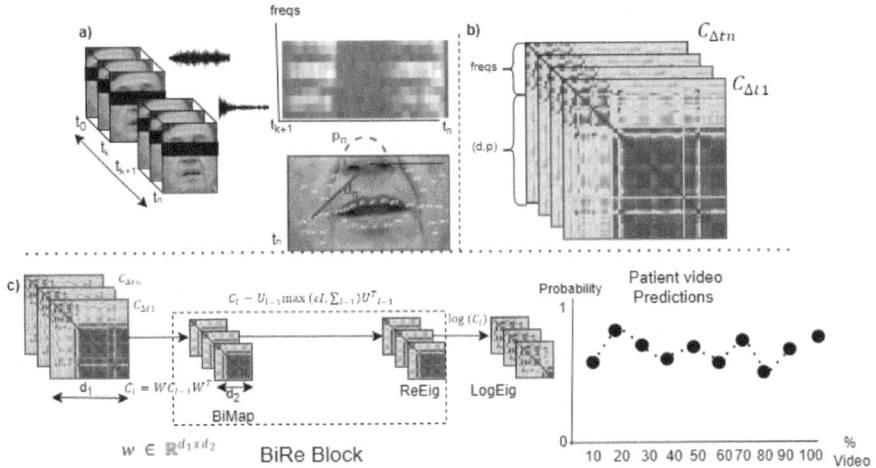

Fig. 1. Multimodal Architecture: a) The position of each key-point in polar coordinates $d(t,k)$ and $p(t,k)$ where d is the distance between the nose to the landmark and p is the angle, is combined with short time spectrogram $\sigma(t,f)$ through b) covariance matrices in time intervals $C_{\Delta t}$. c) Then, the model learns new representations more compact for quantification of PD, with the capacity to output a prediction for each video slice. *Riemannian geometry is form of BiRe blocks with a subsequent projection onto the tangent plane to carry out a classification.* Thus, this approach characterizes the patient's pronunciation temporally, by predicting the probability of PD during the vocalization (bottom right plot).

3.1   Facial and Audio Low-Level Features

In this work, we first computed low-level features, at each sequence time, to encode dysarthria and hypomimia disorders. For dysarthria, we computed short-time spectrograms $\sigma(t, f)$ as fundamental representations (Fig. 1(a), top right), capturing the essential frequency dynamics for frequencies f over sliding window at time t. Consequently, an audio sequence is represented by a spectrogram map with dimensions $N_f \times N_t$ where N_f is the number of frequencies and N_t is the number of time samples.

Regarding hypomimia, we computed the displacement of face key points in regions around the mouth because of the association with facial muscles involved in lip expression. The MediaPipe architecture was used to compute facial landmarks using only video information [10]. We selected 44 landmarks near the mouth and muscles involved in jaw movement during pronunciation. These landmarks allow summarizing the dynamics of the subject's face during various expressions and movements. Specifically, at each time synchronised with the audio spectrogram samples, we encode the position of each keypoint in polar coordinates, using as centre the tip of the nose (Fig. 1(a), bottom right), resulting in a sequence $\{d(t, k), p(t, k)\}$ of dimensions $2N_k \times N_t$, where d is the distance between the nose to the landmark and p is the angle. N_k is the number of keypoints and N_t the number of time samples. Using the nose as centre of coordinates allow to eliminate head movements and to focus on the motion of the mouth.

3.2   Temporal Covariance Computation

Now, for each time interval Δt, made of consecutive N_t time samples, we calculate the covariance matrix of the synchronised features $\Phi(t, i)$ composed of concatenated spectrogram frequencies $\sigma(t, f)$ and face keypoints $\{d(t, k), p(t, k)\}$:

$$C_{\Delta t}(i, j) = \mathbb{E}_{\Delta t} \left(\Phi(t, i)\Phi(t, j) \right) - \mathbb{E}_{\Delta t}\Phi(t, i)\mathbb{E}_{\Delta t}\Phi(t, j)$$

where $\mathbb{E}_{\Delta t}$ refers to the expectancy calculated over the N_t samples $t \in \Delta t$. This temporal covariance matrix, with dimension $(N_f + 2N_k)^2$ (Fig. 1(b)), encodes the dynamic relationships among integrated facial and speech features, providing a comprehensive description of their temporal dependencies. This representation helps with classification performance but also results self-explainable to support recognition of coordination patterns, which is crucial for unraveling the intricate temporal interplay between facial and voice features.

3.3   Covariance-Based Learning for Temporal Video Predictions

Covariance matrices are Symmetric Positive Definite (SPD) matrices that lie in Riemannian manifolds with particular geometry, and need to be processed in a dedicated framework. For each temporal covariance $C_{\Delta t}$, we then learn a geometrical representation, capturing the inherent temporal dependencies between the different modalities. To do so, we first code a BiMap Layer following a bilinear mapping in each layer l, as: $C_l = W_l C_{l-1} W_l^T$, with $C_{l-1} \in \mathbb{R}_*^{d_{l-1} \times d_{l-1}}$ being the SPD matrix output of the layer $l-1$ and $W_l \in \mathbb{R}_*^{d_l \times d_{l-1}}$ the weight matrix transformation [3]. Hence, to ensure SPD property, an eigenvalue rectification layer is carried out, as: $C_l = U_{l-1} \max(\varepsilon I, \Sigma_{l-1}) U_{l-1}^T$ where U_{l-1} and Σ_{l-1} are defined by the diagonal decomposition $C_{l-1} = U_{l-1} \Sigma_{l-1} U_{l-1}^T$. Here, $\varepsilon > 0$ is a rectification threshold value, I is the identity matrix and Σ_{l-1} the diagonal matrix of the eigenvalues of C_{l-1}. This operation adjusts the eigenvalues, avoiding negative values and improving discriminative performance. This specialized block facilitates the extraction of relevant information from the input data, contributing to the computation of effective covariation patterns.

Finally, to carry out the classification task, the learned matrix is projected onto a tangent plane (i.e. back to a Euclidean space), following a logarithm map $\log(C) = U \log(\Sigma) U^T$. Then, classical dense layers are implemented to achieve the classification of the multimodal pronunciation exercise input.

3.4 Dataset Description

This study involved 14 participants, consisting of 7 patients diagnosed with Parkinson's disease (PD) and 7 control patients. The PD group had an average age of 65 ± 4, while the control group had an average age of 61 ± 3. All PD patients were on (Levodopa) medication during data acquisition. Informed consent was obtained from each participant, and the study was approved by the ethics committee of the Universidad Industrial de Santander. The dataset captured synchronized audio and video modalities, with participants performing sustained vowel pronunciation used in the clinical routine. All recordings were conducted in the same environment using a Nikon D3500 digital camera with an integrated monaural microphone. Video was recorded at 1080p resolution and 60 fps, focusing on the face region, while audio was captured at a sampling rate of 48 kHz. Phonation patterns included the pronunciation of five vowels, each vowel being repeated three times, providing a comprehensive dataset for phonation and articulatory analysis. In the study, participants are asked to sustain the pronunciation of vowels for about 5 s. This exercise is incorporated into clinical routines to detect voice abnormalities and to observe the facial expressions of the individuals.

4   Evaluation and Results

The proposed approach was validated with the oral task of sustained vowels, which allows the identification of voice impairments such as dysarthria during PD diagnosis, but also to peculiar conditions such as strengthening vocal muscles and motor coordination during rehabilitation therapies. For validation was followed leave-one-patient-out cross-validation, where at each iteration, one patient is left out for testing and the remaining ones (13 subjects in our experiment) are used for training. To evaluate the performance of the multimodal prediction, the implemented model configurations were assessed for the sensitivity, specificity, accuracy, precision, and F1-score per video. A video was considered correctly predicted by majority vote of its temporal predictions. Specifically, table metrics were quantified by considering either 5, or 10 or 15 predictions during each video.

A first validation was carried out to establish the best video representation to classify PD according to hypomimia-encoded patterns. In this experiment, the temporal covariance matrices were built from landmarks information using only phase (dimension of 44 × 44), only distance (dimension of 44 × 44), and integrating both variables (dimension of 88 × 88). These experiments were also evaluated in different temporal intervals, by evenly dividing the video in five, ten, and fifteen slices respectively. Table 1 summarizes the achieved results, reporting the best performance with the covariance descriptor using only phase information. These results highlight a high sensitivity of 78%, with an accuracy of 65%, evidencing a capability to capture motor coordination changes, especially with 10 slices per video.

Table 1. Hypomimia video classification (facial expression alone) with different number of video slices and polar coordinates of landmarks.

Facial Features	Predictions per video	Ac	Pr	Sen	Spec	F1-s
Phase	5	0.5	0.5	0.69	0.3	0.58
	10	**0.65**	**0.62**	**0.78**	**0.52**	**0.69**
	15	0.4	0.4	0.41	0.39	0.41
Distance	5	0.59	0.58	0.64	0.54	0.61
	10	0.56	0.55	0.66	0.46	0.6
	15	0.55	0.55	0.56	0.53	0.56
Phase and Distance	5	0.58	0.58	0.56	0.6	0.57
	10	0.57	0.57	0.58	0.56	0.57
	15	0.41	0.4	0.39	0.43	0.4

In a second evaluation the audio branch was assessed concerning its capability to classify dysarthria patterns from temporal covariance matrices of spectrograms only, with 20 and 50 frequency bands. Each configuration was also evaluated with five, ten, and fifteen slices per video. Table 2 summarizes the achieved results, reporting a better score with the configuration of 20 frequencies and ten slices (sensitivity of 64%). The improvement in results with 20 frequencies in sustained vowel pronunciation could be attributed to a higher generalization capacity or efficiency in representing relevant features for detecting individuals with Parkinson. It is possible that the learning covariance model can extract more discriminative information with fewer dimensions, facilitating the identification of distinctive patterns in the case of 20 frequencies.

Then, in a third experiment, the proposed approach was evaluated by fusing vocal spectrogram frequencies with facial landmark phases and distances. In such

Table 2. Dysarthria Audio classification with different frequencies and different number of video slices

Freqs	Predictions per video	Ac	Pr	Sen	Spec	f1-s
20	5	0.52	0.52	0.6	0.45	0.55
	10	**0.62**	**0.61**	**0.64**	**0.6**	**0.62**
	15	0.57	0.57	0.58	0.56	0.57
50	5	0.54	0.54	0.5	0.57	0.52
	10	0.55	0.55	0.51	0.58	0.53
	15	0.53	0.54	0.5	0.56	0.52

Table 3. Multimodal (audio-video) classification with 20 speech frequencies, phase and distance facial features

Fusion Features	Predictions per video	Ac	Pr	Sen	Spec	f1-s
20 freqs, phase	5	0.44	0.44	0.44	0.45	0.44
	10	0.66	0.65	0.65	0.65	0.66
	15	0.65	0.64	0.64	0.62	0.67
20 freqs, Distance	5	0.6	0.59	0.69	0.56	0.61
	10	0.58	0.61	0.61	0.58	0.59
	15	0.64	0.63	0.63	0.6	0.65
20 freqs, Distance, Phase	5	0.58	0.58	0.58	0.54	0.6
	10	**0.70**	**0.69**	**0.73**	**0.68**	**0.71**
	15	0.62	0.62	0.62	0.64	0.61

cases, it was considered 20 frequency bands for audio, and whole facial configurations. Table 3 summarizes the achieved results with multimodal configurations, being the best performance achieved in the third experiment, where vocal frequencies were fused with both facial landmark phase and distance, improving accuracy to 70% (10 intervals). These results highlight the complementarity and synergy of features extracted from both modalities. Also, the temporal interval of ten frames shows an appropriate trade-off to capture pronunciation dynamics and avoiding excessive fragmentation of the task.

Besides, the probability for the multimodal approach was calculated for patients and control subjects, for each video percentage during the sustained vowel pronunciation (see Fig. 2). The performance remains stable for both Parkinson's and control groups, suggesting that all vocalization phases can yield similar predictions. Figure 4 (resp. Fig. 3) shows the probability predictions and accuracy for the pronunciation of open vowels, in Spanish: A, E and O (resp. closed vowels: I and U), for Control and Parkinson groups at each interval per video. Interestingly, this categorization is related to movement: Closed vowels are produced with minimal mouth cavity amplitude, while open vowels involve greater mouth cavity expansion with the tongue positioned low. The Fig. 3 of the closed vowels shows greater consistency in the control groups, maintaining the average probability and its stable variability. As for the Parkinson group, higher and more variable results were observed during the initial pronunciation of closed vowels. Similarly, in Fig. 4 of the open vowels, the Parkinson group presents greater variability in the initial intervals. But in contrast to the group of closed vowels for the Control group, the best-predicted values (closer to zero) are found in intermediate pronunciation stages. These results show different dynamics for each vowel group in patients and control subjects. The pronunciation is divided into three phases: initial, stabilization, and decay [22]. Figure 5 indicate in the initial phase (predictions at 10%, 20%, and 30%), there is significant effort, with pronounced facial muscle movements. The most discriminative predictions in this phase are 20% considering all vowels (blue line) with a mean accuracy of 72%. The stabilization phase (predictions from 40% to 70%) represents the maximum vocal production stability, with constant acoustic characteristics and minimal facial movement. The most discriminative intervals here are at 50% of videos with a mean accuracy of 70% (blue line). Finally, the decay phase (predictions at 80%, 90%, and 100%) shows a decline in vocal production and increased facial movement until the mouth closes. The most discriminative intervals in this phase are at 90% of video with a mean accuracy of 68% (blue line). The red and green line indicate that accuracy trends for both open and closed vowels remain relatively stable across video percentages, suggesting that prediction variability does not significantly change, indicating robustness in results. For control subjects, the most discriminative percentages are 20% for closed vowels (red line) with a mean accuracy of 76% (initial stage) and 60% for open vowels (green line) with a mean accuracy of 72% (stabilization stage). Future works will include the analysis of enriched representations with other input modalities, as well as an investigation toward an end-to-end processing of the complete information, since vowels are versatile and can combine with a variety of consonants

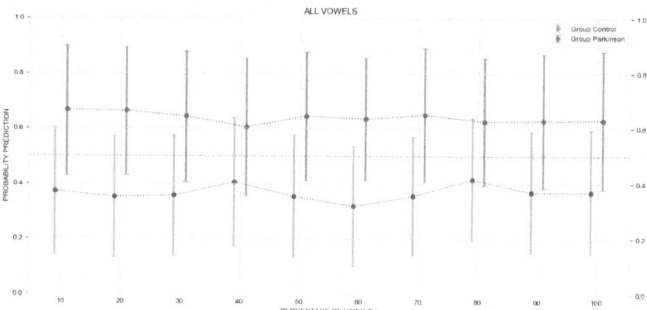

Fig. 2. Probability prediction per interval of video (red line and green line), for all vowels (Color figure online)

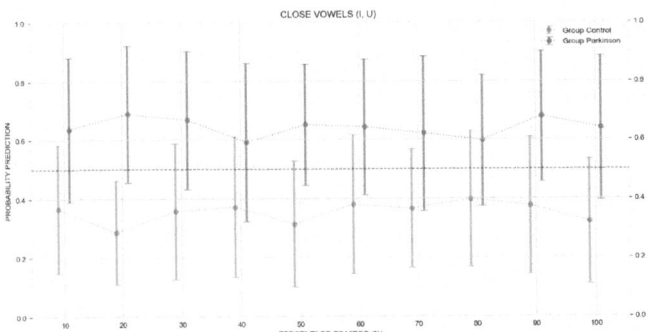

Fig. 3. Probability per interval of video (red line and green line), for close vowels (Color figure online)

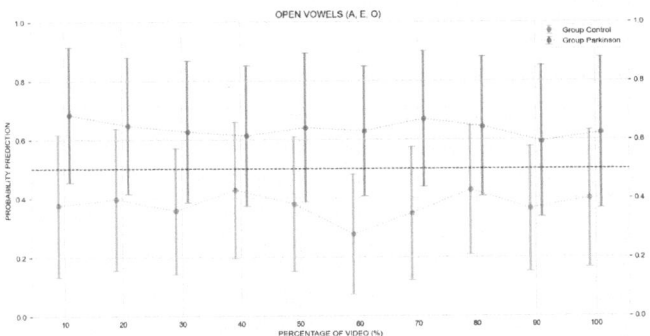

Fig. 4. Probability per interval of video (red line and green line), for open vowels (Color figure online)

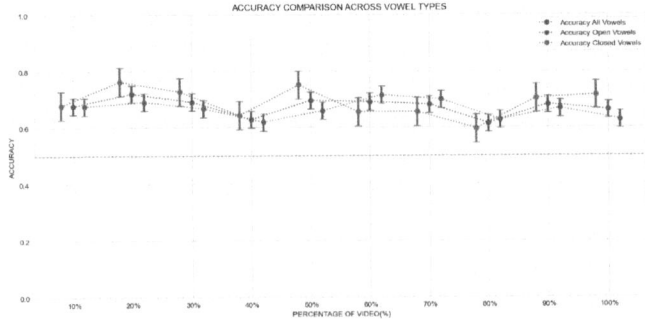

Fig. 5. Accuracy per interval of video for close vowels (red line), open vowels (green line) and all vowels (blue line). (Color figure online)

to create a wide range of sounds and words. Also, this study will be extended to other voice instructions to explore the capabilities of the proposed approach.

5 Discussion and Conclusive Remarks

This work introduced an online multimodal approximation to classify Parkinson disease from facial expression (hypomimia) and voice patterns (dysarthria). In the literature there exist evidences that dysarthria, through the pronunciation of sustained vowels, can identify speech difficulties, associated with early Parkinson's disease [20]. Additionally, treatments have been proposed that use vowel pronunciation in the attempt to improve these impairments [24]. Different studies have integrated voice with other modalities to achieve a broader range of motor impairments in patients and improve diagnostic prediction [8,13]. For example, they have integrated voice, gait, and tremor to extract kinematic features and classify between Parkinson's and Control using machine learning techniques [13]. Alternatively, the voice modality has been integrated with videos of smile expression and finger tapping. In this approach, landmarks are used to identify key points of movement, and models are trained in independent branches. An intermediate fusion is performed using a convolutional architecture, followed by classification [8]. However, this method integrates unsynchronized modalities, making it difficult to analyze and identify how the modalities and their associated symptoms are correlated. *Considering that, this work reported a multimodal approach that integrates visual and audio information to recover hypomimia and dysarthria-associated patterns. For doing so, the proposed approach captured face landmarks in video, and coded spectrograms from audio, which are integrated into temporal covariance descriptors, allowing to obtain a representation of bimodal vocalization. Then, this temporal covariance embedding is projected to a geometrical deep architecture to obtain a refined second-order representation with the ability to distinguish Parkinson patterns from control signals. Thanks to the sliding nature of time covariance descriptors, the geometrical net can bring a prediction at each time interval, allowing to detect abnormal patterns associated to PD, during the exercise, in clinical routine. The proposed geometrical representation was validated with respect to isolated video and audio patterns, and also with the integration of both modalities. Using only videos, the proposed approach encodes temporal covariance matrices using only the correlation among face landmarks. In such case, the proposed approach achieved 65% of accuracy, a f1-score of 69%, and a total of 4 Parkinson and 5 Control subjects were correctly classified. The mistakes in classification may be partially associated to instability of landmarks and recording conditions, but also to the limitation of visual information alone to determine Parkinsonian patterns. Regarding, an audio geometrical net, trained using only spectrogram voice information, was obtained an accuracy of 62% and a f1-score of 62%. These scores were achieved from a configuration of 20 frequencies an 10 intervals per video. Then, we conducted multimodal experiments using a geometrical net, learning from covariance matrices encoding the two modalities. In such case, the multimodal approximation has a gain of 5%*

and 8% in accuracy, and a gain of 2% and 9% in f1-score. The proposed app-roach, however, needs to be examined in a larger cohort of patients to determine statistical significance within the affected population. Additionally, it is crucial to design mechanisms that output disease stages based on observational scales, enabling their use in tracking disease progression.

Acknowledgment. To the Ministry of Science, Technology and Innovation of Colombia, through the project: *Caracterización de movimientos anormales del Parkinson desde patrones oculomotores, de marcha y enfoques multimodales basados en visión computacional* with code 92694.

References

1. Ahmed, I., Aljahdali, S., Khan, M.S., Kaddoura, S.: Classification of Parkinson disease based on patient's voice signal using machine learning. Intell. Autom. Soft Comput. **32**(2), 705 (2022)
2. Alegre-Ayala, J., et al.: The impact of Parkinson's disease severity on performance of activities of daily living: an observational study. Revista de Neurología **76**(8), 249 (2023)
3. Bronstein, M.M., et al.: Geometric deep learning: going beyond Euclidean data. IEEE Sig. Process. Mag. **34**(4), 18–42 (2017)
4. Feigin, V.L., et al.: Global, regional, and national burden of neurological disorders, 1990–2016: a systematic analysis for the global burden of disease study 2016. Lancet Neurol. **18**(5), 459–480 (2019)
5. Fereshtehnejad, S.M., et al.: Evolution of prodromal Parkinson's disease and dementia with Lewy bodies: a prospective study. Brain **142**(7), 2051–2067 (2019)
6. Friedman, J.H.: Misperceptions and Parkinson's disease. J. Neurol. Sci. **374**, 42–46 (2017)
7. Grammatikopoulou, A., Grammalidis, N., Bostantjopoulou, S., Katsarou, Z.: Detecting hypomimia symptoms by selfie photo analysis: for early Parkinson disease detection. In: Proceedings of the 12th ACM International Conference on PErvasive Technologies Related to Assistive Environments, pp. 517–522 (2019)
8. Islam, M.S., et al.: Accessible, at-home detection of Parkinson's disease via multi-task video analysis. arXiv preprint arXiv:2406.14856 (2024)
9. Khan, H., Ullah, M., Al-Machot, F., Cheikh, F.A., Sajjad, M.: Deep learning based speech emotion recognition for Parkinson patient. Electron. Imaging **35**, 298:1-298:6 (2023)
10. Lugaresi, C., et al.: Mediapipe: a framework for building perception pipelines. arXiv preprint arXiv:1906.08172 (2019)
11. Miller, N., Noble, E., Jones, D., Burn, D.: Life with communication changes in Parkinson's disease. Age Ageing **35**(3), 235–239 (2006)
12. Nayak, S.S., Darji, A.D., Shah, P.K.: Identification of Parkinson's disease from speech signal using machine learning approach. Int. J. Speech Technol. **26**(4), 981–990 (2023)
13. Orozco-Arroyave, J.R., et al.: Apkinson: the smartphone application for telemonitoring Parkinson's patients through speech, gait and hands movement. Neurodegener. Disease Manag. **10**(3), 137–157 (2020)
14. Pegolo, E., et al.: Quantitative evaluation of hypomimia in Parkinson's disease: a face tracking approach. Sensors **22**(4), 1358 (2022)

15. Postuma, R.B., et al.: How does Parkinsonism start? Prodromal Parkinsonism motor changes in idiopathic rem sleep behaviour disorder. Brain **135**(6), 1860–1870 (2012)
16. Quan, C., Ren, K., Luo, Z.: A deep learning based method for Parkinson's disease detection using dynamic features of speech. IEEE Access **9**, 10239–10252 (2021)
17. Rajnoha, M., Mekyska, J., Burget, R., Eliasova, I., Kostalova, M., Rektorova, I.: Towards identification of hypomimia in Parkinson's disease based on face recognition methods. In: 2018 10th International Congress on Ultra Modern Telecommunications and Control Systems and Workshops (ICUMT), pp. 1–4. IEEE (2018)
18. Ricciardi, L., De Angelis, A., et al.: Hypomimia in Parkinson's disease: an axial sign responsive to levodopa. Eur. J. Neurol. **27**(12), 2422–2429 (2020)
19. Rissardo, J.P., et al.: Parkinson's disease rating scales: a literature review. Ann. Movement Disord. **3**(1), 3–22 (2020)
20. Roland, V., Huet, K., Harmegnies, B., Piccaluga, M., Verhaegen, C., Delvaux, V.: Vowel production: a potential speech biomarker for early detection of dysarthria in Parkinson's disease. Front. Psychol. **14**, 1129830 (2023)
21. Rusz, J., et al.: Distinct patterns of speech disorder in early-onset and late-onset de-novo Parkinson's disease. NPJ Parkinson's Disease **7**(1), 98 (2021)
22. Tripathi, K., Rao, K.S.: Robust vowel region detection method for multimode speech. Multimedia Tools Appl. **80**(9), 13615–13637 (2021)
23. Valenzuela, B., et al.: A spatio-temporal hypomimic deep descriptor to discriminate Parkinsonian patients. In: 2022 44th Annual International Conference of the IEEE Engineering in Medicine and Biology Society (EMBC), pp. 4192–4195. IEEE (2022)
24. Wight, S., Miller, N.: Lee Silverman voice treatment for people with Parkinson's: audit of outcomes in a routine clinic. Int. J. Lang. Commun. Disord. **50**(2), 215–225 (2015)
25. Wodzinski, M., et al.: Deep learning approach to Parkinson's disease detection using voice recordings and convolutional neural network dedicated to image classification. In: 2019 41st Annual International Conference of the IEEE Engineering in Medicine and Biology Society (EMBC), pp. 717–720. IEEE (2019)
26. Xu, Z., Lv, D., Li, H., Li, H., Gao, H.: Application of resLSTM in hypomimia video detection for Parkinson's disease. In: 2023 International Conference on New Trends in Computational Intelligence (NTCI), vol. 1, pp. 243–247. IEEE (2023)

Transformers for Genomic Prediction

María Inés Fariello[1], Graciana Castro[2], Romina Hoffman[2],
Mateo Musitelli[1]([✉]), Diego Belzarena[2], and Federico Lecumberry[2]

[1] Instituto de Matemática y Estadística, Universidad de la República,
J. Herrera y Reissig 565, Montevideo 11300, Uruguay
{fariello,mmusitelli}@fing.edu.uy
[2] Instituto de Ingeniería Eléctrica, Universidad de la República,
J. Herrera y Reissig 565, Montevideo 11300, Uruguay
{gcastro,romina.hoffman,dbelzarena,fefo}@fing.edu.uy
https://iie.fing.edu.uy/

Abstract. AI is becoming state-of-the-art across scientific fields, giving novel solutions to age-old problems. In genomic prediction, Machine Learning methods could not outperform linear regressions in a general way yet, but are becoming closer. An important feature when working with genomic data, which is non other than a long sequence of information, is to account for the linkage disequilibrium, i.e. dependencies between genome variations that do not need to be close in the genome, and variate with respect to the reference genome. To explode this feature, we evaluate a *Transformer* trained in a small yeast dataset. Although it did not outperform the state-of-the-art results yet, the model got close achieving an R^2 score of 0.389 and 0.400 in Lactate and Lactose ambients, respectively, comparing to the R^2 score of 0.568 and 0.582 for Lactate and Lactose ambients, for the linear model of Lasso, proposed by [7]. This proves that there is still room for improvement.

Keywords: Genomic Prediction · SNPs · genotype · phenotype · Neural Networks · Transformers

1 Introduction

Genomic prediction involves using the information contained in the genome of an individual or a population to make inferences about phenotypes, such as specific traits or diseases. It is based on the premise that certain variations in DNA, such as nucleotide variations known as Single Nucleotide Polymorphisms (SNPs) are associated, due to linkage disequilibrium, with mutations responsible for the variation that certain traits present, or the presence or absence of diseases. In this context, improving the interpretation and prediction of data is a constant challenge due to significant differences in data sets, population structure, and sample size.

To continuously improve the results of linear models and seek alternatives to these models, we aim to apply *Transformers* [13] in the field of genomic

L. Correia et al. (Eds.): IBERAMIA 2024, LNCS 15277, pp. 122–131, 2025.
https://doi.org/10.1007/978-3-031-80366-6_11

Prediction, as they have demonstrated great capacity for capturing long-term relationships in sequences. Successfully adapting and training a model that can extract and learn biological dependencies from dependencies between positions in data sequences could lead to a major breakthrough [3]. The ability of *Transformers* to capture contextual information and model long-range dependencies makes them strong candidates for this task.

We propose to train a model based on the one proposed by Jubair et al. [8] to predict yeast growth in two different environments, Lactate and Lactose.

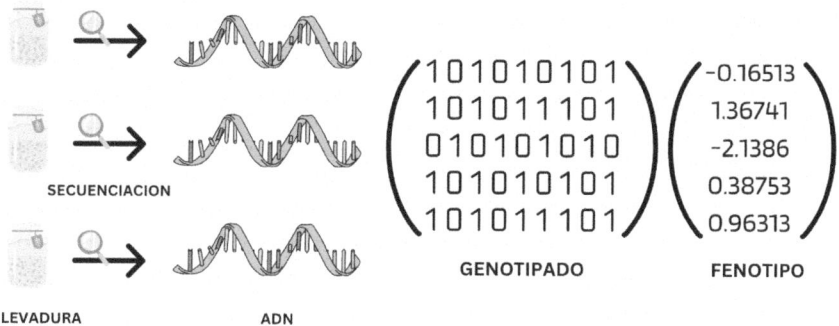

Fig. 1. Representation of the input data to our problem.

2 Problem Description

We have a database with information on yeast growth in forty-eight different environments. The yeast database contains growth information for 1,008 yeast strains in forty-eight different environments. Each strain includes information on 11,623 SNPs, encoded with values zero or one depending on whether the individual presents a variation at that position in their genotype. The phenotype value that quantifies its growth in that environment is associated with each individual.

The problem addressed is predicting yeast growth in each of the aforementioned environments. Specifically, we worked with the Lactate and Lactose environments. Yeast growth is a phenotype that is quantified numerically, having for each yeast genotype its corresponding growth phenotype, as illustrated in Fig. 1.

Although genomic prediction is a very promising approach in the field of genetics, increasing the *accuracy* of genomic predictions across various models remains a challenge. Multi-phenotypic models, that is, those that predict multiple phenotypes simultaneously, have shown promising results when evaluated according to the article *"Multi-trait multi-environment genomic prediction of agronomic traits in advanced breeding lines of winter wheat"* [5]. In light of the aforementioned, we therefore implement a multi-trait Transformer model and seek to compare its results to those of a single-trait Transformer model.

Additionally, a commonly used approach in multivariate genetics is index selection, which assigns different weights to each trait based on its economic importance. However, classical index selection only optimizes genetic gain in the next generation and requires experimentation to find the weights that lead to the desired outcomes, according to the article *"Multi-trait genomic selection methods for crop improvement"* [10].

3 Model

Transformers are particularly important because they revolutionized NLP by providing a more efficient way to process sequences compared to previous recurrent-based models. They excel at handling long-range dependencies, effectively understanding and modeling relationships between elements across entire sequences. This is achieved through the attention mechanism, which dynamically adjusts the importance of different elements based on their relevance to each other. Additionally, Transformers support parallelization during training, significantly enhancing both the performance and speed of training large models. These capabilities make Transformers a powerful and versatile tool for a wide range of applications beyond NLP, including genomics, where understanding complex dependencies within sequences is crucial.

In the field of genomics, the parallels between language and genetic sequences make the implementation of Transformers particularly appealing. The attention mechanism of Transformers can effectively model dependencies between different genomic regions, capturing the interactions that define linkage disequilibrium. An example of this occurrence is shown in Fig. 2, where the linkage disequilibrium is shown for a soybean protein genome.

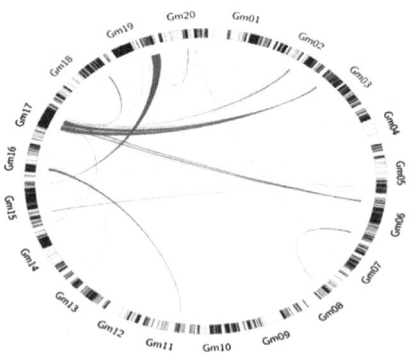

Fig. 2. Stable SNP interactions related to soybean protein content under multiple environments. The soybean genome is represented by a circle. The blue lines indicate the interactions between two markers or regions, presented by Chen et al. [2]. (Color figure online)

To predict a phenotype from the genotype, the model must learn the dependencies and semantics of the input data. In the *Transformer* algorithm, this task is performed by the *Encoder*, so the model used for this problem will not be a bidirectional *Encoder-Decoder* but will consist solely of the former.

The implemented model initially presents a linear layer functioning as an *Embedding* layer. It has as input dimension the number of SNPs (p) per individual and as output the hyperparameter of the dimension of the embedding space (`embed_dim`). Each of the positions that make up the individual's genotype is represented by a vector of dimension `embed_dim`, so when entering the *Encoder*, each individual is represented by a matrix of dimension `embed_dim` × p.

An explicit *Positional Encoding* module is not used since each position has a distinct representation at the output of the linear layer, thus preserving the positional information. The number of *Encoders* in the model is defined by the hyperparameter `NLayers`. The structure, in this case, is the same as the *Encoder* structure presented for the *Transformer*: a *Multi-Head Self Attention* block formed by h heads, a *Feed-Forward Neural Network* (FFN) of dimension `ff_dim`, and two *Add & Norm* layers at the output of each of the previously mentioned modules. Both h and `ff_dim` are hyperparameters of the model.

Finally, the model has a linear layer responsible for predicting the phenotype for each individual. The output dimension (`output_dim`) will be defined according to the number of phenotypes to be predicted with the same model (`output_dim` = 1 for predicting one phenotype and `output_dim` = 2 for predicting two). Figure 3 shows the diagram of the implemented model.

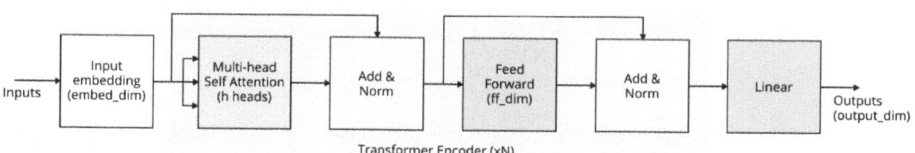

Fig. 3. Model trained with the Yeast database for predicting growth in different environments.

To implement the model, we use modules from the `Pytorch` library. In particular, the `Encoder` class *TransformerEncoder* and `nn.Linear` module for the *Embeddings* layer and output FFN.

4   Hyperparameter Search and Training

The model training was divided into two stages: first, a search for optimal hyperparameters was conducted, followed by the training of the model. All experiments were carried out on the ClusterUY [11] using a 40 GB GPU.

For the hyperparameter search, possible values for the *learning rate*, h, ff_dim, embed_dim, and dropout (used for regularization) were defined, where each training session used a different combination of these values. Finally, the combination that yielded the best value for the *Pearson Correlation Coefficient* (PCC), $r(\mathbf{x}, \mathbf{y})$, was selected. The result is a coefficient that measures the linear dependence between variables $\mathbf{x}$ and $\mathbf{y}$, with values ranging from $[-1, +1]$. The closer r is to the extremes of the interval, the greater the linear dependence between the variables, while the closer it is to the middle 0, the lesser the dependence. Its mathematical expression is as follows:

$$r(\mathbf{x}, \mathbf{y}) = \frac{\mathrm{cov}(\mathbf{x}, \mathbf{y})}{\sigma_{\mathbf{x}} \sigma_{\mathbf{y}}} = \frac{\sum_{i=0}^{n-1}(x_i - m_{\mathbf{x}})(y_i - m_{\mathbf{y}})}{\sqrt{\sum_{i=0}^{n-1}(x_i - m_{\mathbf{x}})^2 \sum_{i=0}^{n-1}(y_i - m_{\mathbf{y}})^2}}, \tag{1}$$

where $\sigma_{\mathbf{x}}$ and $\sigma_{\mathbf{y}}$ are the variances of $\mathbf{x}$ and $\mathbf{y}$, respectively, and $m_{\mathbf{x}}$ and $m_{\mathbf{y}}$ are the means of $\mathbf{x}$ and $\mathbf{y}$, respectively. Initially, tests were conducted to consider the number of *Encoders* as a hyperparameter, but due to computational limitations, it was decided to train the model with two. The implementation was done using the Optuna library [1].

The available dataset was divided into five folds to enable 5-fold cross-validation training with subsequent validation. Batches of eight individuals were taken, each with a sequence of 11,623 SNPs in length.

The *Mean Squared Error* (MSE) was used as the loss function, a measure of how accurate the machine learning model is in terms of predicting the values $\tilde{y}_i = g(\mathbf{x}_i)$. It is used in cases where high sensitivity to outlier values is desired due to being squared. Its mathematical expression is:

$$\mathrm{MSE}(\mathbf{x}) = \frac{1}{n} \sum_{i=0}^{n-1}(y_i - \tilde{y})^2. \tag{2}$$

The PCC (Eq. (1)) was used as the gain function. This meant that for each epoch, although the parameters were updated considering the MSE, once validation was done, the best model was chosen according to the epoch with the highest PCC. It is also noteworthy that the training MSE is computed during the forward propagation of the network and the validation MSE after backpropagation, resulting in better validation results than in training. The optimizer used was *Adam* [9]. The model parameters were initialized using the *Xavier Uniform* method [6].

Training was conducted for a maximum of one thousand epochs. *Early stopping* was used as a regularization method, stopping the training if no improvements in the validation PCC were observed after thirty-five consecutive epochs. Additionally, *Dropout* [12] stages were used both in the neural network within the *Encoder* and at its output, both with the same *dropout ratio*.

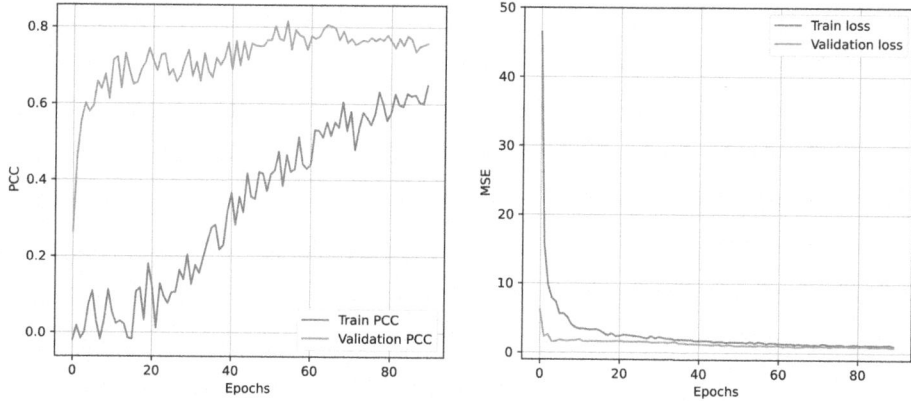

Fig. 4. Learning curves for data of yeast growth in Lactate environment. On the left is the PCC curve (used as gain function), and on the right is the MSE curve (used as the loss function). Training curves are in blue and validation curves are in orange. (Color figure online)

5   Results

Figure 4 shows the evolution of PCC and MSE for training and validation as a function of epochs for Lactate. It can be observed that as the epochs progress, both results improve, with PCC increasing and MSE decreasing, as expected. As in the case of the simulated data, the validation results are better than those for training. However, the training was stopped due to the *early stopping*. Additionally, the training PCC could have achieved a higher value if the validation PCC had not remained unchanged for approximately 30 epochs, which was the stopping condition.

In the case of Lactose, the behavior of MSE and PCC is similar to that of Lactate. It can be seen that there is some bias in the MSE and the presence of overfitting since around epoch 150. The MSE for validation surpasses that of training, as shown in Fig. 5. Again, it is observed that, despite no improvement being presented for more than approximately 50 epochs in validation PCC (causing the training to stop), the training PCC could have reached a higher value, as it shows an increasing trend up to this epoch. The same could have been manifested in the training MSE with respect to the decrease.

In Fig. 6, the PCC and MSE curves for each phenotype with the *multi-trait* model are presented. The results are similar to those obtained when training the model with a single phenotype: as the PCC increases, the MSE decreases. However, the MSE presents a smaller bias, and unlike in the case of Lactate, the model does not overfit.

In Fig. 7, the test results are presented. In both cases, the results obtained with *multi-trait* were better than training the model with a single phenotype. Although MSE does not show significant changes, the PCC increased moderately.

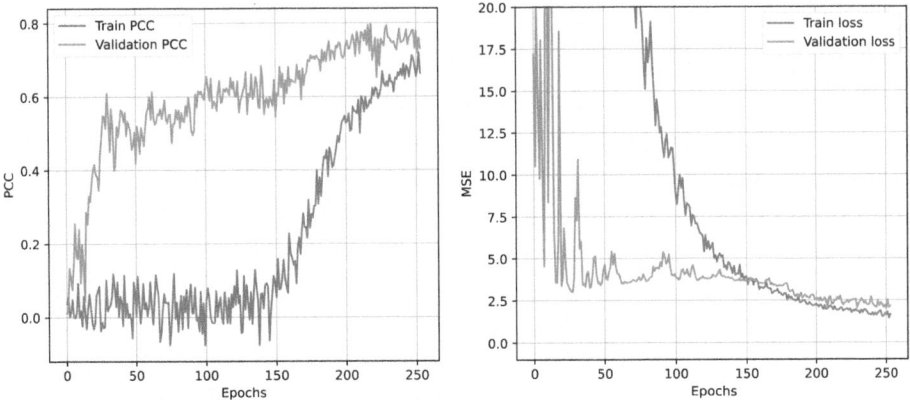

Fig. 5. Learning curves with Lactose growth data. On the left is the PCC curve (used as gain function), and on the right is the MSE curve (used as the loss function). Training curves are in blue and validation curves are in orange (Color figure online)

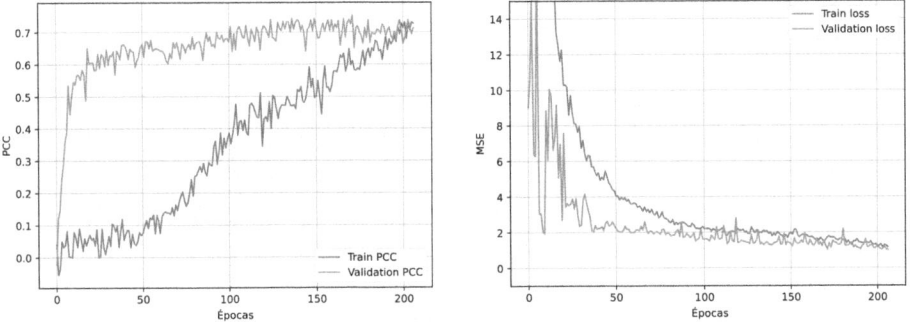

Fig. 6. Learning curves for yeast growth in Lactate and Lactose, predicted using the Multitrait model. On the left is the PCC curve which shows the average of the PCC obtained for the two phenotypes in training (blue) and validation (orange). On the right, the MSE curves, with training shown in blue and validation in orange. (Color figure online)

The relationship is not strongly linear in all cases, as the PCC is not as high as in the simulation cases.

To compare the results obtained with other models that have been trained for this dataset, the R^2 metric (*coefficient of determination*) is calculated. This metric is used to evaluate how well a model has performed on a dataset, with its best result being one and decreasing towards zero as the model's performance declines. A R^2 value of zero indicates that the model's predictions are as good as those of a random model, while if the result is outside the interval $[0, 1]$, the model has performed worse than random predictions. The R^2 metric results, obtained for the phenotype predictions of Lactate and Lactose *One trait* and *Multitrait*,

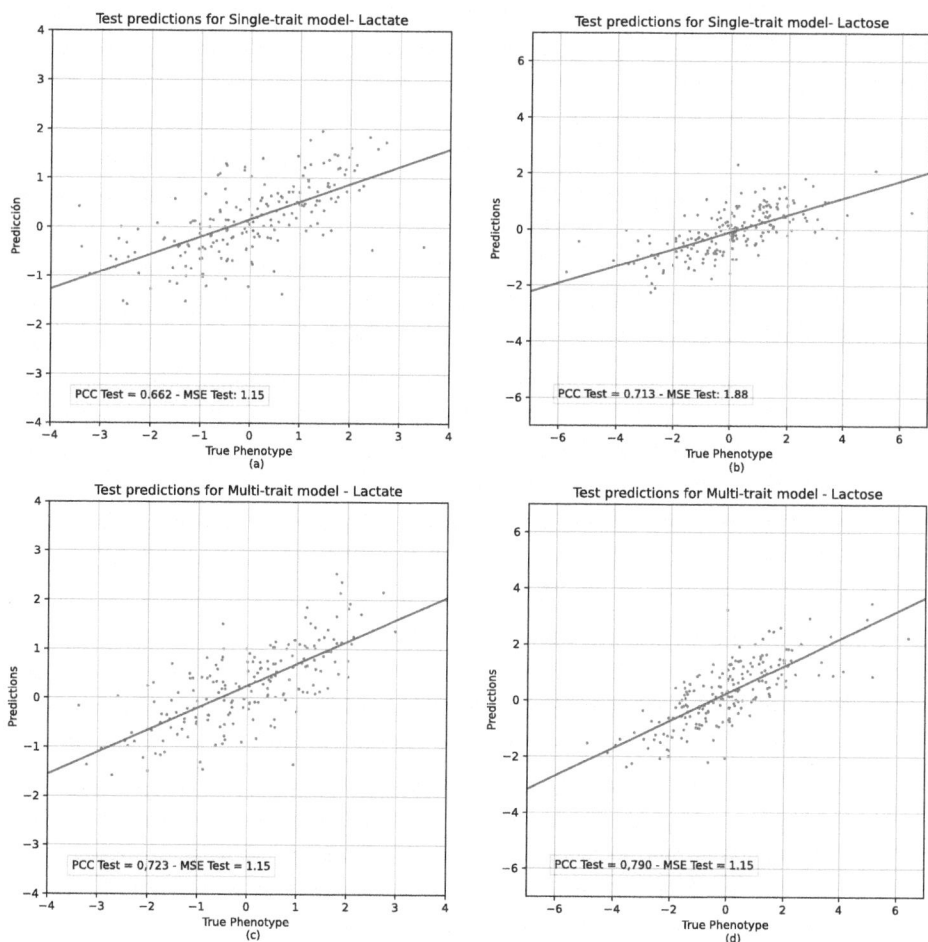

Fig. 7. Scatter plot of true phenotype values of the test set versus their predictions. In figure (a) the results of One-trait model used to predict phenotype in Lactate, in figure (b) One-trait model results for Lactose. Figures (c) and (d) show the results for the Multitrait model in Lactate and Lactose environments respectively. In all figures, the linear fit is shown in red (Color figure online)

are presented in Table 1, along with other results reported by Elenter et al. [4], compared for the same phenotypes predicted by other models. It can be observed from the table that the results for the *Multi-trait* models significantly outperform the *Single-trait* models, as previously indicated. On the other hand, while these results do not reach those of Gradient Boosting Machine (GBM) [4] presented in the table, they do achieve the order of magnitude of those of Grinberg et al. [7], confirming the robustness of the implemented algorithm.

Table 1. Comparison of R^2 metric results for yeast growth in Lactate and Lactose obtained with both our models, Transformer One-Trait and Transformer Multitrait, with the ones reported by Grinberg et al. [7], Elenter et al. [4] .

Environment	Grinberg	GBM	One trait	Multitrait
Lactate	0.568	0.830	0.389	0.478
Lactose	0.582	0.860	0.400	0.536

6   Conclusions

In this paper, we have experimented with Transformers applied to genomic prediction. We have described the different considerations taken to do the hyperparameter tunning and train the model.

The results obtained are promising and allow us to affirm that it is feasible to achieve satisfactory results by using models based on *Self-Attention* on genomic data sequences. However, there are modifications, validations, and new simulations that need to be explored, including:

1. More exhaustive hyperparameter searches.
2. Balancing choosing the best models according to PCC and MSE.
3. Investigating better parameter initialization methods.
4. Repeating the process on new datasets.

The first point is directly related to the computing power available. The maximum GPU memory accessible for us was 40 GB, and it was with this that hyperparameter searches were performed using *Optuna*. For all searches conducted, the best parameters obtained were always the maximum of the intervals studied, indicating that larger hyperparameter intervals must be studied. This could not be done as memory saturation was reached in all cases.

The second point on the list is due to the obtained results for Lactate, where overfitting is observed in the training curve with the MSE metric. This is related to the fact that the stopping condition and the choice of the best epoch were made based on the PCC coefficient, but a way should have been found to balance, and include, the results for each epoch of the MSE metric. Although the best epoch was tried according to the MSE metric, and the results were not better, other alternatives could have been considered, such as combining both PCC and MSE results.

In conclusion, although this work has shown positive results and promises great potential, the implementation of additional modifications and validations, as well as the exploration of new simulations and methods, are necessary to continue improving the accuracy and robustness of the model. Exploring these aspects will establish a solid foundation for future research and applications in the field of genomics using models based on *Transformers*.

References

1. Akiba, T., Sano, S., Yanase, T., Ohta, T., Koyama, M.: Optuna: A Next-generation Hyperparameter Optimization Framework. In: Proceedings of the 25th ACM SIGKDD International Conference on Knowledge Discovery and Data Mining (2019)
2. Chen, Q., et al.: Snp-snp interaction analysis of soybean protein content under multiple environments. Can. J. Plant Sci. **97**(6), 1090–1099 (2017)
3. Clauwaert, J., Menschaert, G., Waegeman, W.: Explainability in transformer models for functional genomics. Brief. Bioinform. **22**(5), 1–11 (2021)
4. Elenter, J., Etchebarne, G., Hounie, I.: DNAI: Machine learning for genome enabled prediction of complex traits in agriculture. Master's thesis (2021)
5. Gill, H.S., Halder, J., Zhang, J., Brar, N.K., Rai, T.S., Hall, C., Bernardo, A., Amand, P.S., Bai, G., Olson, E., et al.: Multi-trait multi-environment genomic prediction of agronomic traits in advanced breeding lines of winter wheat. Front. Plant Sci. **12**, 709545 (2021)
6. Glorot, X., Bengio, Y.: Understanding the difficulty of training deep feedforward neural networks. In: Teh, Y.W., Titterington, M. (eds.) Proceedings of the Thirteenth International Conference on Artificial Intelligence and Statistics. Proceedings of Machine Learning Research, vol. 9, pp. 249–256. PMLR, Chia Laguna Resort, Sardinia, Italy (13–15 May 2010). https://proceedings.mlr.press/v9/glorot10a.html
7. Grinberg, N.F., Orhobor, O.I., King, R.D.: An evaluation of machine-learning for predicting phenotype: studies in yeast, rice, and wheat. Mach. Learn. **109**, 251–277 (2020)
8. Jubair, S., et al.: Gptransformer: A transformer-based deep learning method for predicting fusarium related traits in barley. Front. Plant Sci. **12**, 2984 (2021)
9. Kingma, D.P., Ba, J.: Adam: A method for stochastic optimization. arXiv preprint arXiv:1412.6980 (2014)
10. Moeinizade, S., Kusmec, A., Hu, G., Wang, L., Schnable, P.S.: Multi-trait genomic selection methods for crop improvement. Genetics **215**(4), 931–945 (2020)
11. Nesmachnow, S., Iturriaga, S.: Cluster-uy: Collaborative scientific high performance computing in Uruguay. In: Torres, M., Klapp, J. (eds.) Supercomputing, pp. 188–202. Springer International Publishing, Cham (2019)
12. Srivastava, N., Hinton, G., Krizhevsky, A., Sutskever, I., Salakhutdinov, R.: Dropout: a simple way to prevent neural networks from overfitting. J. Mach. Learn. Res. **15**(1), 1929–1958 (2014)
13. Vaswani, A., et al.: Attention is all you need. In: Advances in Neural Information Processing Systems **30** (2017)

AI-Assisted Bronchoscopy in the Intensive Care Unit: Corpus Development and an Application to Anatomic Position Identification

Luciano Tarsia[1], Nicolas Mastropasqua[1,2(✉)], Indalecio Carboni Bisso[4],
Marcos Las Heras[4], Valeria Burgos[3], Marcelo Risk[3],
María Florencia Courtois[4], Ignacio Fernández Ceballos[4], Carolina Lockhart[4],
Daniel Acevedo[1,2], and Viviana Cotik[1,2]

[1] Departamento de Computación, FCEyN, UBA, Buenos Aires, Argentina
mastropasquanicolas@gmail.com
[2] Instituto de Investigación en Cs. de la Computación (ICC), CONICET-UBA,
Buenos Aires, Argentina
{dacevedo,vcotik}@dc.uba.ar
[3] Instituto de Medicina Traslacional e Ingeniería Biomédica (IMTIB), Buenos Aires,
Argentina
[4] Unidad de Terapia Intensiva, Hospital Italiano de Buenos Aires, Buenos Aires,
Argentina

Abstract. This study aims to create an artificial intelligence model capable of accurately identifying bronchial segments during broncho-endoscopic navigation. To achieve this, we analyzed 126 videos from bronchoscopic procedures conducted on critically ill patients at a university hospital in Buenos Aires, Argentina.

A dataset of consistently annotated videos, captured by bronchoscopists with varied expertise, was established. Inter-annotator agreement for image classification was evaluated using Cohen's kappa coefficient. Images of multiple bronchial segments were used as input to train a convolutional neural network in order to obtain a classification model. This paper presents the annotation schema, labeling guidelines, the developed corpus, and some preliminary results.

Keywords: neural networks · data annotation · bronchoscopy videos

1 Introduction

The exponential increase and availability of digital data in healthcare has raised the need to develop and adapt robust technologies for the management and analysis of medical aspects. In this sense, the resurgence of artificial intelligence (AI) took on a leading role in several data-generating fields of healthcare, such as bronchoscopy imaging.

© The Author(s), under exclusive license to Springer Nature Switzerland AG 2025
L. Correia et al. (Eds.): IBERAMIA 2024, LNCS 15277, pp. 132–144, 2025.
https://doi.org/10.1007/978-3-031-80366-6_12

A bronchoscopy is a minimally invasive procedure that looks into the airways to help in the diagnosis of lung diseases, collect mucus or tissue samples as well as removing blockages. During a procedure, an endoscope is inserted through the nose or mouth and navigates down the throat into the lungs. Identification of a specific airway segment may be a challenge to the non-expert eye, due to the particularly regular appearance and branching structure of the bronchial tree.

There is a broad range of applications for AI technologies in bronchoscopy, such as navigation assistance, prediction of anatomical areas, and targeting of lesions, among others [6,11,16]. From an educational perspective, the use of AI in assisted navigation during bronchoscopy procedures is a valuable addition to learning, acquiring, and integrating procedural skills in residency/fellowship training and post-graduate education.

A description of the steps to be followed in a bronchoscopy in a standarized way can be seen in the Step by Step Description[1] provided by the Bronchoscopy Education project.[2]

In this study, we developed a corpus of annotated images captured from bronchoscopy videos, tagged by expert bronchoscopists. These images were assessed for the detection of anatomical positions. To provide a baseline for this task, two Convolutional Neural Network architectures, namely ResNet18 [5], and ShuffleNetV2 [9], were trained to classify images into classes. We believe that the resulting model could serve as an accurate decision tool in clinical practice and as a valuable resource in training programs for physicians.

The scarce availability of publicly available corpora is a major limitation in the implementation of AI algorithms since annotated corpora is needed to train and test supervised machine learning models. Moreover, obtaining reliable annotated data is not a simple task: it requires the establishment of annotation schemas and criteria, input from specialists in the area, and, equally importantly, the anonymization of sensitive data. To the best of our knowledge, there are no publicly available annotated datasets of bronchoscopy images, nor schema and annotation criteria standards. For this reason, one important part of this work focuses on creating an annotated corpus of images corresponding to bronchoscopies navigation videos. Therefore, we developed an annotation criteria and schema, which was refined iteratively, following the MAMA cycle [10]. We describe the annotation process in detail to ensure that our corpus is accompanied by its Data Statements [1].

We generated a dataset of 126 annotated navigation bronchoscopy videos from 121 adult patients admitted to the intensive care unit (ICU). Videos were captured using bronchoscope equipment and the extracted images are in standard resolution. We present demographic information of the patients, including sex, age, height, weight, and body mass index (BMI).[3]

---

[1] https://www.bronchoscopy.org/wp-content/uploads/Step-by-Step-Description.pdf.

[2] The Step by Step guidance is provided by otolaryngologists, who perform it with awake patients, where 'left' refers to the right and vice versa. Conversely, bronchoscopies are performed with sedated patients, where 'right' is right and 'left' is left.

[3] We plan to release the annotated dataset in the future upon request.

The rest of the paper is organized as follows. Section 2 briefly describes previous approaches for corpus annotation and bronchoscopic image applications. Section 3 covers video collection criteria, labeling guidelines, and the annotation process. Section 4 presents a quantitative analysis of the resulting annotated dataset and the inter-annotator agreement. Section 5 describes the methodology used for classifying images into anatomical positions. Finally, Sect. 6 discusses the dataset analysis results and Sect. 7 presents the conclusions.

2   Previous Works

Corpus annotation has presented challenges for many years, making systems prone to inconsistencies where manual annotations cannot be avoided [15]. For instance, when dealing with text annotations in colonoscopy studies, Syed et al. [13] proposed a workflow for building a high-quality annotated corpus using specific domain taxonomies following standardized annotation guidelines. Similar shortcomings may be found in image segmentation works where several approaches are used to overcome imprecise or incomplete annotations [2].

Some recent studies on the detection of bronchial lumen segments have showed interesting results that could provide a complementary resource to practitioners during diagnostic bronchoscopy examinations. Keuth et al. [7] developed an airway guidance system based on the classification of video images through the combination of CNN (Convolutional Neural Network) and HMM (Hidden Markov Model). In this case, a synthetic dataset generated on simplified silicone phantoms of the bronchial tree was used.

Yoo et al. [16] introduced deep learning models for specific segment interpretation of bronchoscopy images that could provide an accurate navigation tool for anesthesiologists. They used 9793 video bronchoscopy images from 3216 patients and trained several CNNs designed to classify three bronchial segments (carina, left main bronchus and right main bronchus). They showed that the performance of the models was comparable with that of the most-experienced bronchoscopist.

Li et al. [8] presented the development and validation of an AI-based diagnostic model for bronchial lumen identification to enhance the quality of bronchoscopies and reduce variations in the skills of different bronchoscopists. The methodology involved collecting 342 bronchoscopy videos, segmenting them into 31 anatomical locations, training several CNN models on a large dataset of images, and evaluating the AI system's performance in assisting bronchoscopists in lumen recognition. The system first differentiates between *in vivo* and *in vitro* bronchoscopes, and then identifies 31 locations of the bronchial lumen with a dataset of 28441 images. Similarly, Chen et al. [3] collected a dataset of clear bronchoscopic images extracted from videos of 292 patients. Next, they trained CNN models to classify nine anatomical positions within the airway. During model training, they included at most one representative frame per class from each video, achieving AUC scores exceeding 0.98 for each class. However, these studies overlooked the temporal dependencies between these classes that might arise due to the standardized navigation procedure in bronchoscopic videos.

3   Annotation Process

We developed an annotation guideline through a three-iteration process involving expert annotations, inter-annotator agreement analysis, and annotation criteria refinement, by following the MAMA cycle [10]. In a fourth iteration, five bronchoscopists annotated the videos based on these guidelines. This section outlines the criteria for selecting and anonymizing the videos for annotation, the annotation schema and guidelines, and the manual annotation process.

3.1   Data

We collected videos of bronchoscopies performed on adult patients admitted to the intensive care unit of a university hospital in Buenos Aires, Argentina. The videos were recorded during bronchoscopic procedures between January 2022 and June 2024, using an aScope™ 4 connected to an aView™ 2 Advance (Ambu®, Denmark) equipment, coded with H264 codec in MP4 format with a maximum resolution of 560×480 pixels.

Selection of the Dataset. The videos selected for the study are from patients with preserved bronchial structures and no history of prior endoluminal lesions, obstructions, tumors, or any other interfering lesions that could hinder the identification of a specific position in the bronchial tree. These videos do not contain recordings of the segmental bronchus. Additionally, we selected videos without visual interferences or airway obstructions caused by typical bronchial secretions, and with controlled cough reflexes.

Since the demographic characteristics of individuals may influence the size of bronchial structures [14], we recorded the gender, age, height, and weight of the patients. Demographic information of the patients selected for the study is depicted in Table 1.

The videos were de-identified before annotations were made so it was not possible to determine any patient details from the videos. Only bronchoscopists involved in the study had access to a mapping that connects the video identification to the unique patient identifier assigned by the hospital. This unique identifier was used to present the demographics of the annotated data. Therefore, patient confidentiality and compliance with privacy standards are guaranteed.

3.2   Annotation Schema

We annotated thirteen classes, whose names and locations in the bronchial tree are depicted in Fig. 1. For the selection of the classes, we used as a reference the work by Li et al. [8], but decided to keep only those that the physicians considered relevant. We did not include the glottis as a category because the patients were intubated, and we also did not consider the distal segments.

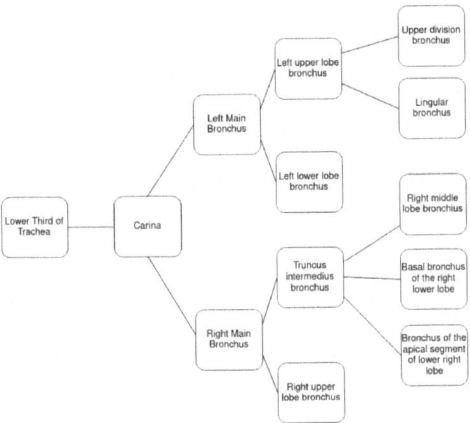

Fig. 1. Bronchus tree. Lingular bronchus is also called lower division bronchus.

3.3   Annotation Guidelines

The main annotation guidelines, established after the annotation process and discussion sessions with the annotators, are listed below:

1. Accurate annotation is crucial and must encompass the entire duration of each category, from the first second to the last.
2. During the annotation process, annotators should refrain from discussing their opinions, review criteria, or making corrections to the annotation with each other.
3. If the category cannot be determined with certainty, annotators should refrain from labeling it. Only categories that annotators are completely certain about must be registered.
4. In the same video, if a previously labeled category reappears during playback, it should not be labeled again, except for the carina.
5. Video segments should be labeled as "Lower 1/3 of Trachea" when both main bronchi are fully visible in the correspondent segment. (See Fig. 2a).
6. Video segments should be labeled as "carina" when both main bronchi are visible but one of them is not completely visible (See Fig. 2b).
7. The intermediate bronchus will be labeled once both distal bronchi are visible (for example, basal and apical; see Fig. 2c).
8. Only the distal progression is considered. Proximal retreat must not be annotated.
9. It is not necessary to tag the entire video. Irrelevant sections can be omitted (for example, very distal segments that do not fit any established categories).
10. The segment of the video where the bronchoscope is inside the orotracheal tube or tracheostomy tube should not be annotated.

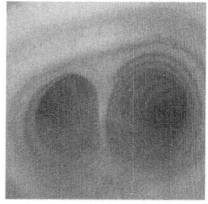
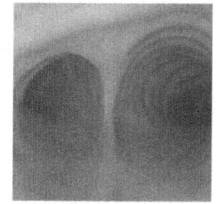
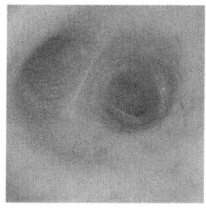

(a) Lower third of the trachea

(b) Carina

(c) Truncus intermedius bronchus

Fig. 2. Images of what is considered as Lower third of the trachea, Carina and Truncus intermedius bronchus.

3.4 Annotation

The manual annotation process was conducted by five physicians with varying levels of training in bronchoscopy procedures for critical care patients: one of the team members has over 10 years of experience, 3 have between 2 and 6 years, and the remaining member has 1 year of experience. Each practitioner performs approximately 350 bronchoscopies per year.

We conducted multiple meetings with the annotators to clarify any questions and resolve uncertainties. After the initial dataset was annotated (referred to as Annotation Iteration 1 in Table 3), we reviewed and addressed any doubts and differences in annotation criteria, which resulted in the development of more detailed annotation guidelines. After three annotation-revision iterations like the ones proposed in MAMA cycle [10], we obtained the final guidelines (described in Sect. 3.3). This annotated data was discarded because it was only intended to refine the annotation criteria.

Subsequently, the final round of annotations (referred to as Iteration 4) was carried out according to the revised guidelines. These annotations were then used to extract images and train our models.

To evaluate inter-annotator agreement, each annotation round involved multiple annotators analyzing a subset of videos. In the first three iterations, 7, 6, and 6 videos were annotated, respectively, with 2 annotators in the first round and 5 annotators in the second and third rounds, each tagging every video. The last iteration included 126 videos that were annotated by 5 annotators. Forty-five of the videos were annotated by two of the annotators to calculate the IAA. One of the annotators, who has 5 years of expertise, annotated a percentage of the videos that were also annotated by each of the other four annotators (see Fig. 3 to understand how data was divided among annotators). This setup ensured a portion of the videos intersected with the expert's set, while the remaining videos were uniquely assigned to individual annotators. This approach allowed us to monitor and control annotator agreement while scaling the number of annotated videos.

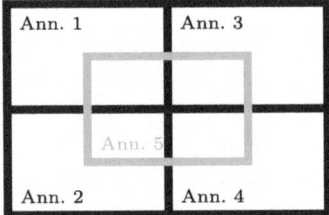

Fig. 3. Annotator number 5 (green rectangle), with 5 years of experience, labeled a percentage of videos annotated by the other annotators (black rectangles). (Color figure online)

To facilitate the annotation process, we utilized the open source data labelling platform LabelStudio[4]. An installation guide and a video demonstrating the annotation process were provided to ensure all annotators could efficiently use the tool. Each annotator had to log in to the annotation tool and could only see the videos assigned to them. Fig. 4 shows an example of a segment of an annotated video.

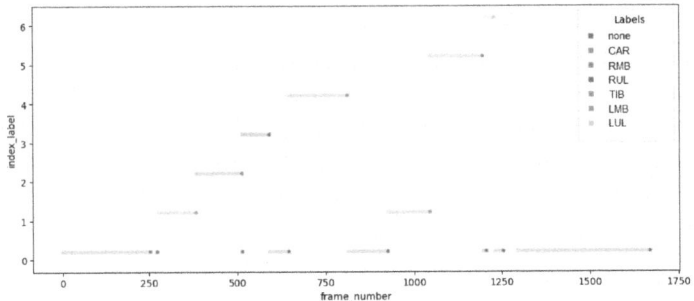

Fig. 4. Segment of an annotated video. Each class is displayed with a corresponding color and at equal height. Class 'none' corresponds to a segment of the video where the anatomical position could not be established according to guidelines. (Color figure online)

4   Dataset Analysis

A total of 126 bronchoscopy procedure videos, spanning from January 18, 2022, to June 6, 2024, were collected. Upon completion of the annotation process, the videos were processed alongside the annotations. Frames were extracted at a rate of 30 frames per second for each video. This approach yielded 193.398

---

[4] https://labelstud.io/.

images. Then, we examined the tags assigned by each annotator for every frame in order. This frame-by-frame comparison allowed us to evaluate the consistency and agreement of the annotations across different annotators (see Sect. 4.1). Tables 1 and 2 show the demographic of our annotated dataset, and the number of frames and videos obtained for each of the classes of interest.

Table 1. Demographics of 121 patients, whose data was annotated. For the sex attribute, the percentage of males was calculated. For the other attributes the median and interquartile range (IQR) are shown.

Variable	Unit	Median	IQR
Sex (Male)	quantity	79 (59,5%)	–
Age	years	61	48–70
Height	meter	1,68	1,60–1,75
Weight	kg	73	65–85
BMI	kg/m2	26,1	22,6–29,1

Table 2. Class distribution of the annotated dataset. For each class, the number of frames and the number of videos containing the class are shown.

Id	Abbrev.	Class	# Frames	# Videos
1	LTT	Lower third of the trachea	4287	52
2	CAR	Carina	31907	125
3	LMB	Left main bronchus	30820	125
4	RMB	Right main bronchus	21607	125
5	LUL	Left upper lobe bronchus	13099	118
6	LLL	Left lower lobe bronchus	14102	113
7	RUL	Right upper lobe bronchus	15538	124
8	TIB	Truncus intermedius bronchus	23599	126
9	UDB	Upper division bronchus	337	5
10	LDB	Lingular bronchus (Lower division bronchus)	611	6
11	RML	Right middle lobe bronchus	1111	12
12	RLL	Basal bronchus of the right lower lobe	1584	10
13	ASRLL	Bronchus of the apical segment of the lower right lobe	281	3

4.1 Inter-annotator Agreement

To assess the consistency among the annotations, the inter-annotator agreement (IAA) was calculated using the Cohen's Kappa coefficient (κ) on a frame level.

Given the inherent difficulty of having annotators start their annotations at the exact same millisecond, we considered incorporating a tolerance in the calculation of the IAA at class boundaries. However, it was not implemented.

Moreover, transitional segments between classes often lack a defined boundary, making it difficult to accurately assign a class. Consequently, there are segments of the videos where no class is assigned by a given annotator. To address this, we created a 'none' class, that was assigned to each frame in these videos where an annotator did not assign a class (relative to the annotator who did not make a class assignment). Therefore, we calculated κ for both: all frames and for only annotated frames. It was crucial that disagreements, where one annotator tagged a frame as one class and another annotator tagged it as a different class, occurred infrequently, as reflected in the metric value for only annotated frames. Conversely, to observe disagreements across the entire video, we examined the metric considering all frames.

IAA was calculated for every pair of annotations of a same video. Table 3 shows the average inter-annotator agreement (IAA) for all frames (considering 'none' class) and the average IAA only for annotated frames (not considering 'none' class) across each of the annotation iterations. It can be observed that in Iteration 3, the agreement value decreases. This is not due to a difference in annotator criteria, but rather to a technical error by one of them.

Table 3. Average IAA for all frames and only for annotated frames across each of the iterations, along with the number of videos and the percentage of videos annotated by more than one annotator, and the number of annotators.

Iteration	# of Videos (% of Videos)	# of Annotators	Average IAA (All Frames)	Average IAA (Annotated Frames)
1	7 (100%)	2	0.676	0.809
2	6 (100%)	5	0.818	0.939
3	6 (100%)	5	0.694	0.909
4	126 (38%)	5	0.800	0.946

5   Classification of Images Into Anatomical Positions

Given bronchoscopy procedure video streams, our main objective is to train a model capable of identifying all frames that depict any anatomical position of the bronchial tree defined in Fig. 1. We understand that previous works [8,16] only deal with images that show a clear view of bronchial lumen segments. Contrary to them, our proposed models should learn from frames as they come from the video feed. Thus, they do not necessarily provide a clear depiction of any of the anatomical positions of interest. This could be due to issues such as motion blur, over-exposed regions, partial occlusion due to camera position or even transitional sections between the well defined anatomical positions. To tackle this problem, we leveraged the PyTorch implementation of ResNet18 architecture, pretrained on the ImageNet dataset, solely to introduce a baseline on the collected dataset. Additionally, it is desirable for these models to achieve near real-time inference speed in a potential clinical application. Therefore, we have also selected ShuffleNetV2 as an interesting alternative.

For this preliminary analysis, we decided to subsample each video by extracting only even frame numbers. If any frame had been either annotated by two bronchoscopist and the annotations did not coincide, or if it was labeled as the none class, it was left out. Afterwards, we split the frames into training, validation, and test sets using a ratio of 6:2:2, ensuring that no two frames from the same video were kept in the same set. This strategy should help prevent the model to learn spurious correlations. At the same time, the split is made to maintain approximately the same class distribution across all sets. The aforementioned preprocessing step resulted in a refined subset of 49256 images which is tightly bounded to this particular analysis.

During the model selection process, the CNNs were trained for 50 epochs using weighted Cross Entropy Loss and Adam Optimizer, with the largest batch size that fit in memory (128 or 64 depending on the architecture). We searched for the best learning rate among 10^{-2}, 10^{-3}, 10^{-4}, and 10^{-5}, and assessed whether a weighted random sampler could improve model training. One-Cycle scheduler was used to adjust the learning rate during training. All hyperparameters were determined by evaluating each model's best performance on the validation set across all epochs. Images were later resized to a resolution of 224×224.

Additionally, to help reduce overfitting we performed online data augmentation using random rotation and random horizontal flip. Furthermore, to account for variability due to different random processes during training, four training iterations of the best model were carried out, picking a different seed each time. All models were trained using a Nvidia Geforce 1080 Ti GPU.

As shown in Table 2, the dataset is highly unbalanced. For this reason, F1-micro could be misleading, as it is equivalent to accuracy in single-label multi-class problems. Therefore, we monitored mean Average Precision (mAP) during training, along with F1-macro.

6   Results and Discussion

Our results, summarized in Table 4, highlight that the stated problem poses a significant challenge for the proposed models. We believe that there exists an inherent complexity in classifying certain types of frames that arise naturally during video bronchoscopy procedures: motion blur, over-exposed regions, partial occlusion due to camera position, or even the possible similarity between transitional sections and clearly defined anatomical positions. All of these factors might end up adding meaningful noise to the training process and degrade the learned feature representation, a well-known issue in the literature of CNNs [4].

Besides, we observe that using isolated frames as input, i.e., in the absence of temporal information, demonstrates that 2D CNNs alone are insufficient for this

Table 4. mAP, macro F1, and Micro F1 (equivalent to accuracy for this multiclass and single-label problem) on validation and test for the best models found of each CNN. For the validation column, median values out of 4 runs with different seeds are shown. Each best model was later tested on the test set.

Model	Validation			Test		
	mAP	macro-F1	micro-F1	mAP	macro-F1	micro-F1
ResNet18	0.43	0.40	0.70	0.50	0.46	0.76
ShuffleNetV2	0.40	0.39	0.66	0.44	0.43	0.71

problem. Our hypothesis is that modeling temporal dependencies, by adding an extra recurrent network trained on the CNNs' learned representations, should bring the performance closer to that of the simplified problem.

Direct comparison of our work with other bronchoscopy-related studies is challenging due to differences in datasets, number of classes, and image selection methods. For example, previous works [8, 16] demonstrated remarkable performance in identifying only clear frames of the anatomical positions. Nevertheless, it is noteworthy that a similar study [12], which classified 14 classes of anatomical landmarks and pathological lesions in endoscopies from raw frames, achieved a macro-averaged F1 score of 0.26.

7   Conclusions

In this work, we developed a corpus of bronchoscopy videos, with frames annotated by expert bronchoscopists into one of 13 anatomical positions. We then trained two CNNs to establish a baseline for the multiclass classification problem

We presented a careful design of the annotation schema and the resulting annotation guidelines. Iterations with experts, performing annotation, its evaluation, and discussion, were crucial in the process of defining the guidelines. As shown in Table 3, the IAA increased in each annotation round, except for one case explained in Sect. 4.1. Data selection was performed to reduce the demographic biases as much as possible.

We consider the macro F1 score of 0.46 achieved by our model to be reasonable given the difficulty of the problem.

Our work has some limitations: all generated data corresponds to individuals with conserved bronchial structure and with no visual interferences or obstruction of airways, and none of the patients presented anatomical variations. The main limitation is that videos have no register of the segmental bronchus. There is no record of patients' ethnicity, which is of interest for implementing fair-algorithms.

Besides, we are not incorporating any tolerance in the calculations of the IAA, which may make the results appear worse than they actually are. Furthermore, a small difference in the annotation of boundaries, which is expected, is weighted the same as a mismatch in the annotated class.

We plan as a future work, to compare how this AI-assisted tool performs against medical experts as well as to measure the system's ability to serve as an aid in the training of junior bronchoscopists. We also plan to expand this dataset to make it more diverse and challenging. On a final note, should an implementation like the one we have demonstrated be deployed, we believe it is fundamental for a multidisciplinary team to evaluate the implications of AI decision-making in clinical settings.

Acknowledgments. This study was funded by the PICT Project PICT-2021-GRF-TI-0067, granted by Agencia I+D+i through FONCyT.

Disclosure of Interests. The authors have no competing interests to declare.

References

1. Bender, E.M., Friedman, B.: Data statements for natural language processing: toward mitigating system bias and enabling better science. Trans. Assoc. Comput. Linguist. **6**, 587–604 (2018)
2. Cai, H., Li, S., Qi, L., Yu, Q., Shi, Y., Gao, Y.: Orthogonal annotation benefits barely-supervised medical image segmentation. In: IEEE Conference on Computer Vision and Pattern Recognition (CVPR), pp. 3302–3311 (2023)
3. Chen, C., et al.: Distinguishing bronchoscopically observed anatomical positions of airway under by convolutional neural network. Therapeut. Adv. Chronic Disease **14** (2023)
4. Dodge, S.F., Karam, L.: Understanding how image quality affects deep neural networks. In: 2016 Eighth International Conference on Quality of Multimedia Experience (QoMEX), pp. 1–6 (2016). https://api.semanticscholar.org/CorpusID: 12047850
5. He, K., Zhang, X., Ren, S., Sun, J.: Deep residual learning for image recognition. IEEE Conference on Computer Vision and Pattern Recognition (CVPR), pp. 770–778 (2016)
6. Keuth, R., Heinrich, M., Eichenlaub, M., Himstedt, M.: Weakly supervised airway orifice segmentation in video bronchoscopy. In: Medical Imaging: Image Processing, vol. 12464, pp. 58–65. SPIE (2023)
7. Keuth, R., Heinrich, M., Eichenlaub, M., Himstedt, M.: Airway label prediction in video bronchoscopy: capturing temporal dependencies utilizing anatomical knowledge. Int. J. Comput. Assist. Radiol. Surg. **19**, 713–721 (2024)
8. Li, Y., et al.: Development and validation of the artificial intelligence (AI)-based diagnostic model for bronchial lumen identification. Transl. Lung Cancer Res. **11**, 2261–2274 (2022)
9. Ma, N., Zhang, X., Zheng, H.-T., Sun, J.: ShuffleNet V2: practical guidelines for efficient CNN architecture design. In: Ferrari, V., Hebert, M., Sminchisescu, C., Weiss, Y. (eds.) Computer Vision – ECCV 2018. LNCS, vol. 11218, pp. 122–138. Springer, Cham (2018). https://doi.org/10.1007/978-3-030-01264-9_8
10. Pustejovsky, J., Stubbs, A.: Natural Language Annotation for Machine Learning: A Guide to Corpus-Building for Applications. O'Reilly Media (2012)
11. Shen, M., Gu, Y., Liu, N.: Context-aware depth and pose estimation for bronchoscopic navigation. IEEE Robot. Autom. Lett. **4**(2), 732–739 (2019)

12. Smedsrud, P.H., et al.: Kvasir-capsule, a video capsule endoscopy dataset. Sci. Data **8**(142) (2021)
13. Syed., S., et al.: Tax-corpus: taxonomy based annotations for colonoscopy evaluation. In: Proceedings of the 15th International Joint Conference on Biomedical Engineering System and Technology, pp. 162–169 (2022)
14. Ulusoy, M., et al.: Age and gender related changes in bronchial tree: a morphometric study with multidedector CT. Eur. Rev. Med. Pharmacl. Sci. **20**(16) (2016)
15. Xia, F., Yetisgen-Yildiz, M.: Clinical corpus annotation: challenges and strategies. In: Proceedings of Workshop on Building and Evaluating Resources for Biomedical Text Mining (2012)
16. Yoo, J.Y., et al.: Deep learning for anatomical interpretation of video bronchoscopy images. Sci. Rep. **11**(23765) (2021)

Bio-inspired and Soft Computing

Evaluation of Mares Uterine Health Based on Endometrial Biopsies Using Image Processing and Machine Learning Techniques

Sofía Zimmer[1]([✉]), Agustina Díaz[1], Nicolás Aguilera[1], María José Estradé[2], Federico Lecumberry[1][iD], and Pablo Musé[1][iD]

[1] Instituto de Ingeniería Eléctrica, Universidad de la República, J. Herrera y Reissig 565, Montevideo 11300, Uruguay
{sofia.zimmer,agustina.diaz,nicolas.aguilera, lecumberry,pmuse}@fing.edu.uy
[2] Facultad de Veterinaria, Universidad de la República, Ruta 8, km 18, Montevideo 13000, Uruguay
https://iie.fing.edu.uy/

Abstract. Animal fertility is a widely studied topic to promote the breeding of animals that inherit characteristics from their progenitors. Especially in equines, there are highly valued mares used for different equestrian sports whose genetic material is very valuable. Equines are not selected for their reproductive characteristics like other production animals but for their phenotype or sport aptitudes. Uterine health is fundamental for a healthy gestation, so it is of utmost importance to know the state of the mare's uterus. For this purpose, endometrial biopsies are performed, among other techniques. In these biopsies, pathologists can study the presence and disposition of the different structures in the endometrium and estimate the potential degree of fertility of the animal. This work integrates image processing and machine learning techniques to analyze endometrial biopsies from mares, with the aim of reducing sample evaluation times and providing quantitative data for diagnosis. Two models were used to segment the images: one for glands, adapted from a model pre-trained on human tissue, and one for fibrosis, trained on a database collected and labeled during the research. A learning-based color normalization technique was also applied to ensure the latter model's robustness to sample variations.

1 Introduction

Uterine health is fundamental for the establishment of gestation, so it is of utmost importance to know the state of the mare's uterus. For this purpose, among other techniques, endometrial biopsies are performed. In these biopsies, pathologists can study the presence and disposition of the different structures in the endometrium and estimate the potential degree of fertility of the animal.

L. Correia et al. (Eds.): IBERAMIA 2024, LNCS 15277, pp. 147–158, 2025.
https://doi.org/10.1007/978-3-031-80366-6_13

The sample is stained with hematoxylin and eosin (H&E) in order to visualize different structures of interest.

The most frequent and relevant pathologies observed in uterine biopsies are endometrosis, endometritis and lymphatic lacunae. In 1986, the Kenney-Doig [1] scale was introduced, which classifies specimens into four categories according to the degree of involvement by these pathologies. These categories have been correlated with the mare's expected ability to carry a gestation to term, known as foaling rate.

The Kenney-Doig scale serves as a reference for diagnosis, but it is subject to the interpretation and subjectivity of pathologists, depending on their experience and judgment. The absence of quantitative data on the sample leaves the criteria open to discussion. Analyzing these specimens is complex, as they contain a vast amount of information and require considerable time and meticulous attention to detail.

This work introduces image processing and machine learning techniques in mare endometrial biopsy evaluation, aiming to reduce the valuable time from experts and focus on significant regions of the samples analyzed.

The paper is structured as follows: Sect. 2 presents the literature background, including the framework used for data labeling and a histological image preprocessing technique. Section 3 focuses on the machine learning models used for structure segmentation; Sect. 4 describes the tests performed and presents the results obtained. Conclusions and future work directions are presented in Sect. 5. A more detailed presentation of this work can be found in [2].

2   Background

High-resolution histology images, known as *Whole Slide Images* (WSIs), are obtained from a complete scan of a histological preparation and are widely used in modern histology. These images can reach resolutions of up to $180,000 \times 180,000$ pixels, consuming up to 4 GB of memory. Special applications are needed to handle these images.

New software and tools are continually being developed and integrated into the physician's workflow to assist with and streamline specific tasks such as analysis and diagnosis. Graphical interfaces are available for viewing, annotating, and processing high-resolution tissue images. One such example is *FastPathology* [3], which can be used to run machine learning model inference on WSIs. *FastPathology* is built upon *FAST* library [4]. This tool is open source and was developed in *C++*. It was designed for high-performance processing, machine learning model inference, and visualization of medical images.

One of the most recurring tasks where deep learning practices have been applied in the medical field has been cancer diagnosis. Vekariya proposed [5] a five-step workflow to identify the presence or absence of tumors from breast sentinel node samples. This thesis was taken as a reference, as it performs an analysis of WSIs with H&E staining using deep learning techniques. The methodology applied by the author is commonly used when working with WSIs: first, the region of interest (ROI) where the tissue is located is selected; then, the WSI is divided into patches, with each patch processed separately. Finally, the global features of the sample are calculated as the sum of the features extracted from each patch.

Regarding the specific topic of mares endometrium, no prior research has applied machine learning techniques to analyze H&E-stained samples. Additionally, there are no available datasets that contain this tissue or any similar tissue that could exhibit comparable pathologies. Recent studies have investigated mare's endometrosis using genomics information, as reported in [6] and [7]. Furthermore, research by Yao [8] and Hu [9] has explored innovative techniques for the assessment of the female uterus. However, none of these studies are suitable as benchmarks for the current work as they do not address the same task.

2.1 Framework

Quick Annotator [10] is a project that aims to bring machine learning to people in biomedical areas. It allows the visualization of WSIs, streamlines the labeling process, facilitates the training of segmentation models, and performs inferences over new WSIs. It consists of a web interface where a database can be assembled from patches of WSIs, annotated, and used for training. The base architecture of the model to be trained is a U-Net [11].

In order to initialize the model weights, it is trained in an unsupervised way, as an autoencoder. Then begins a process of active learning, which consists of labeling a small amount of data, training the model, performing inference on new images, correcting the obtained inferences, and repeating these steps. Newly labeled patches (those for which inference was corrected) are added to the initial data during each iteration. The process concludes when the user is satisfied with the inference results or when some metrics indicate no further improvement.

At the end of the training stage, a database is generated containing annotations of the biological structures and additionally, the trained segmentation model is available.

2.2 Color Normalization

Color normalization procedures are a frequent practice in biological image preprocessing to address color variations that may arise between images of samples from the same tissue. These differences are inherent to the sample preparation process. Generally, color normalization is applied before training machine learning models to mitigate the effects of color variability within a set of images or to enhance the model's ability to learn from diverse color representations, depending on the approach. Its purpose is to adjust the color space of the images, aligning the chromatic distribution with a known or desired distribution.

HistAuGAN [12] is a Generative Adversarial Network (GAN) based augmentation technique for histopathological images. It recognizes two relevant aspects of the images to treat them separately. First, it works with a content encoder, which is invariant to different color domains and encodes information about tissue structure, such as sizes and shapes. Second, it uses an attribute encoder, which is different for each domain and learns color-specific features. The disaggregated representation of image information allows transfers between images, combining the content of one image with the attribute space of another.

This model was used to preprocess the images prior to fibrosis segmentation model training. The implementation was obtained from [13], and the results are presented in Sect. 4.2.

3 Segmentation Models

Of all the biological structures included in a mare's endometrial sample, the work concentrated on two structures of interest: glands and fibrous connective tissue. These classes were chosen because they allow the identification of the presence of some of the pathologies described in the Kenney-Doig scale in a sample.

3.1 Glands Segmentation

For gland segmentation, an epithelium segmenter is used. It was trained with colon mucosa samples developed by Pettersen et al. [14] and obtained from the FastPathology model repository. Its architecture is a U-Net, and the magnification of the input images is 10×.

Since the original purpose of the model is to detect glandular epithelium, it segments both glandular and luminal epithelium when used with mare endometrium images. A post-processing step was implemented in order to improve inference results. This process involves filling in the lumen of the most dilated glands and removing inferences over the luminal tissue. Figures 1 and 2 present the masks resulting from applying these filtering techniques.

3.2 Fibrosis Segmentation

Regarding fibrous connective tissue segmentation, there were no existing models capable of performing this task on endometrial images stained with H&E. Therefore, we chose to train our own model using our own data. Utilizing *Quick Annotator* (QA), we created a labeled dataset and employed a U-Net architecture implemented in PyTorch[1] as our model. The latter implementation was used, rather than the model provided by *Quick Annotator*, as it had a more accessible architecture, allowing greater flexibility for modifications and adjustments.

The U-Net model was chosen for the aforementioned task due to its widespread adoption in biomedical image segmentation since its development. Its encoder-decoder architecture, enhanced with skip connections, makes it an adequate design for segmenting images by considering both the relationship between observed structures and the target object, as well as the overall structures within the image. Additionally, more complex models risk overfitting due to the limited availability of data.

---

[1] https://github.com/milesial/Pytorch-UNet/tree/master.

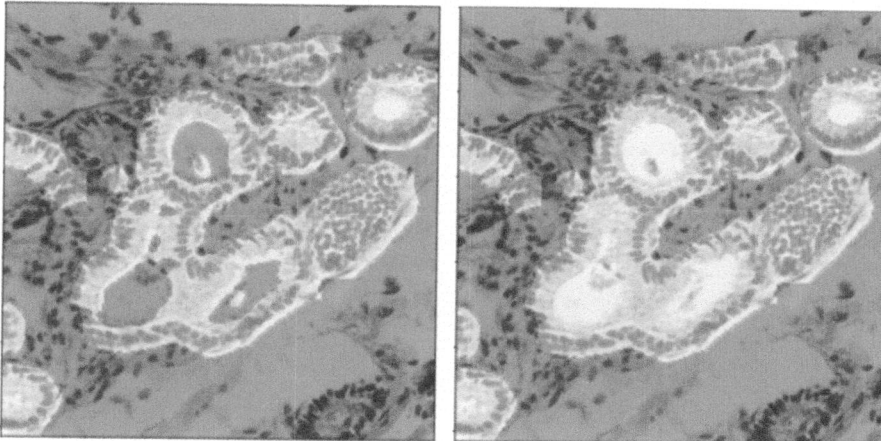

Fig. 1. Hole fill filter: The image on the left shows the raw inference of the model. The image on the right shows the mask resulting from applying the hole-filling algorithm.

The first step for using QA is to split the WSIs into patches of appropriate size and magnification. Following the guideline [15] a patch size of 1,000 ×1,000 was chosen, at a magnification of 10×. This configuration enables nests of dilated glands to be seen in a single patch. The selection of magnification size was influenced by the fact that it was impracticable to label the collagen fibers separately from the glands, considering their size and the image resolution. Thus, efforts were made to label the entire glandular nidus or dilated glands, including the collagen fibers in the segmentation.

After labeling all patches, the PyTorch U-Net model was trained using a PC equipped with an NVIDIA P100 GPU (12 GB VRAM), 4 CPUs, and 32 GB RAM.

Regarding the cost function, PyTorch U-Net employs the combination of cross-entropy with the Dice cost function during training. (Dice cost function was introduced by Milletari et al. in [16].) Whereas for weight initialization, a transfer learning technique was applied; the Pytorch U-Net model was initially trained to perform the H&E stained nuclei segmentation task.

The most extensive publicly available segmentation databases for H&E stained samples primarily focus on nuclei segmentation tasks. Hou et al. [17] produced semi-automatic cancer nuclei segmentation for over 5,000 WSIs stained with H&E, corresponding to diverse patients, laboratories, and tissue types. The segmented data belongs to *The Cancer Genome Atlas Program* (TCGA) sample repository[2], and each pixel is classified as positive (cancerous) or negative (non-cancerous). Even though the nuclei dataset used belongs to human organs, the knowledge learned by the model can be further applied to mares and other animals, since such as humans, animals are other large multicellular organisms

---

[2] The Cancer Genome Atlas Program (TCGA).

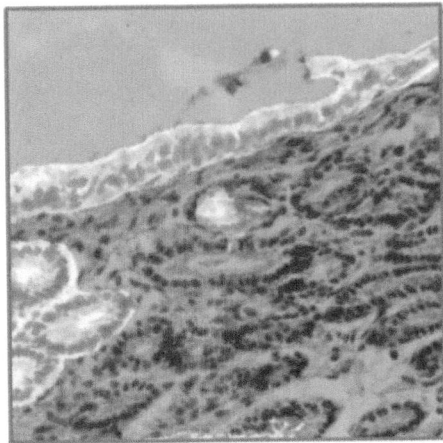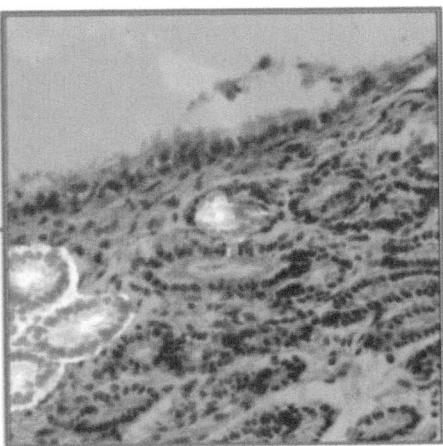

Fig. 2. Luminal epithelium removal filter: On the left is the inference of the original model, while on the right is the mask after applying the luminal epithelium removal filter.

that have the same four basic tissue types: epithelial tissue, connective tissue, muscle tissue, and nervous tissue [18].

From this database, 10 WSIs of urothelial cancer were chosen, generating 2,161 patches sized 1,000 ×1,000 pixels at a magnification of 10×. These images were divided into training and validation sets of 1,728 and 433 patches, respectively. This database will be referred to as BLCA-DB (*Bladder Cancer Database*), following the authors' convention. The patch size and magnification match those of the fibrosis database (hereafter referred to as F-DB).

In order to apply transfer learning, the PyTorch U-Net model was retrained using the fibrosis database. Amiri et al. [19] report that the results of retraining a U-Net by fixing the latent space weights are practically equal to retraining the entire network. Consequently, the latent space weights were frozen during the retraining process. Given that this section concentrates approximately half of the network parameters, such an approach represents a significant optimization when fine-tuning the model.

The model was retrained for 100 epochs using the F-DB training set. Section 4 describes how this dataset was obtained. Figure 3 shows the evolution of the cost function during training and the Dice coefficient on the validation data.

4   Experiments

The experiments require the execution of a workflow consisting of a series of sequential steps. It was implemented using *FAST* library. Initially, inference is performed over the entire WSI with each segmentation model, resulting in the generation of corresponding masks. Each mask is then trimmed into patches of equal size and magnification. The masks are post-processed at the patch level.

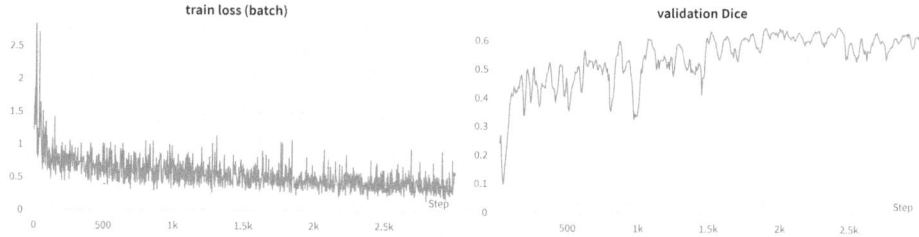

Fig. 3. PyTorch U-Net's metrics in the fibrosis dataset. Left: loss function evolution on the training set. Right: Dice coefficient evolution on the validation set.

Finally, features are computed across the entire WSI, and regions requiring further attention are pointed out on the WSI.

4.1   Dataset

The case study had an initial number of 34 WSIs of mare endometrial biopsies stained with H&E. These data have a resolution ranging around 60,000 ×12,0000 pixels, with individual sizes varying from 213 MB to 1.2 GB. All WSIs exhibit a maximum magnification of 40×. Of all the available samples, only 10 WSIs were carefully selected for labeling fibrosis. The quality of the sample was taken into account for this selection. Blur, torn tissue, folds, and invagination were some of the reasons why the other samples were disregarded.

As previously mentioned, a patch size of 1,000×1,000 pixels was chosen at 10× magnification, and a total 500 patches were labeled. However, after the initial training trail, 243 images were excluded from the dataset since they lacked positive labels and hindered the training process. Consequently, 257 patches of size 1,000 × 1,000 pixels remained available, all of them presenting fibrosis. Of this, 222 patches, corresponding to 8 of the 10 WSIs, were used for training, with 10% allocated for validation purposes. The remaining 35 patches from the other 2 WSIs were used for testing.

Three domains are recognized within the samples, characterized by the predominant colors of the staining they exhibit. The violet, red, and pink domains, which present 13, 12, and 9 WSIs, respectively, were obtained from the original set of 34 WSIs. The violet and red domains present the pathologies introduced in Sect. 1, with a mild to moderate degree of advancement. In contrast, the pink domain presents pathologies of greater severity, including dilated and nest-like glands (a sign of periglandular fibrosis), as well as edema and other alterations.

4.2   HistAuGAN Training Results

The HistAuGAN model was trained using the three aforementioned domains to apply a color normalization technique to the working data. Images sized 1,000 × 1,000 pixels with a magnification level of 10× were used. The training database for the HistAuGAN model did not require labeling; only recognition of different

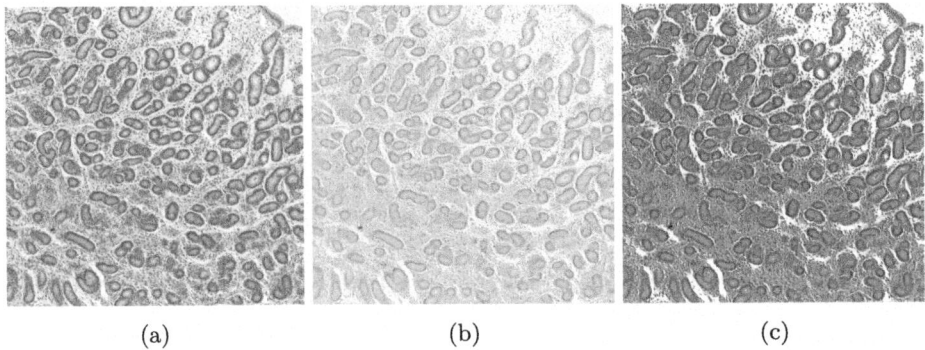

(a)   (b)   (c)

Fig. 4. (a) Test image obtained from the violet domain. (b-c) Generated images with the HistAuGAN model of the pink (b) and red (c) domains. (Color figure online)

domains was necessary. Consequently, this database is different from the one used for the fibrosis segmentation model. It comprised 241 images from the violet domain, 213 from the pink domain, and 226 from the red domain.

Four images from each domain were selected to create a test set. The inference results of the trained model can be observed in Fig. 4. The domain transfer outcomes achieved with the HistAuGAN model are satisfactory. The generated images are realistic and effectively capture the color space distribution of each domain.

4.3   Fibrosis Model Training Results

F-DB Augmented with HistAuGAN. In order to obtain a model that generalizes across a broader range of color domains, an augmented version of F-DB was generated. In this augmented dataset, each image was transformed from its original domain to the two remaining domains using the HistAuGAN network. Prior to this data augmentation, F-DB comprised 222 images, consisting of 128 belonging to the violet domain, 25 to the pink domain, and 69 to the red domain. After applying the color normalization algorithm, the total set of images tripled to 666, where each domain presents an equal distribution.

Transfer Learning Between BLCA-DB and F-DB. In the initial stage of this test, the segmenter was trained as described in Sect. 3.2 using the entire BLCA-DB for 30 epochs, with the data scaled to 25% of the original size (i.e., 500×500 pixel images). The input size reduction during training allowed for a larger batch size. During this training, each batch consisted of 8 images, constrained by computational limitations. This model achieved an average Dice coefficient of 0.69 on the BLCA-DB test set.

The model was then re-trained using the full F-DB, with images also scaled to 25% of the original size. The latent space weights were frozen, and the learning rate was fixed to 1×10^{-5}. Two variants of this experiment were conducted: one

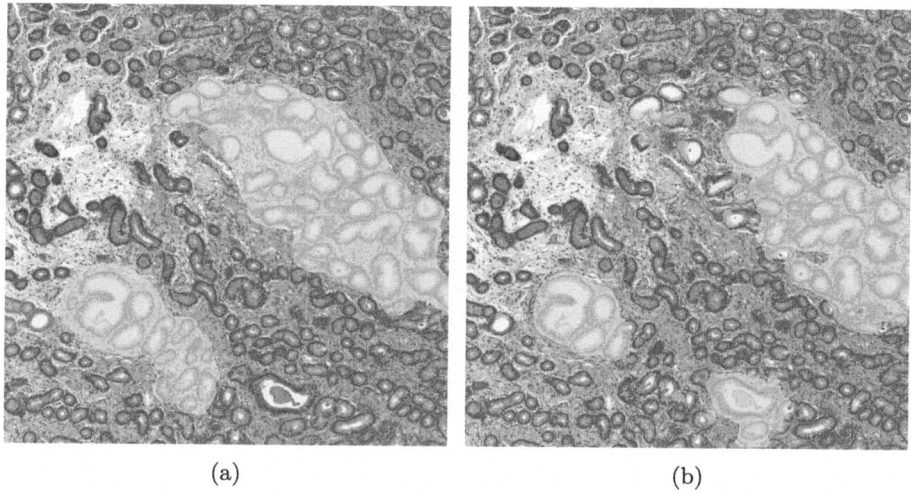

(a)   (b)

Fig. 5. (a) Labeled image from the test set. (b) Inference made with the fibrosis segmentation model that obtained the highest Dice coefficient over the test set.

Table 1. Details of the tests performed with the PyTorch U-Net fibrosis segmentation model. Dice coefficient reported over the test set.

Epochs	Data Augmentation	Batch size	Input size	Learning Rate	Dice	Training time (hours)
10	–	3	$1,000 \times 1,000$	1×10^{-5}	0.17	5
100	–	8	500×500	1×10^{-5}	0.52	5
130	**HistAuGAN**	**8**	**500×500**	**1×10^{-5}**	**0.54**	**7**

with F-DB in its original state, and another one with F-DB augmented using HistAuGAN.

In a third experiment, training with F-DB at the original resolution (i.e., 1,000×1,000) was tested. For this case, it was necessary to reduce the batch size to 3 images.

Experimental Results. As mentioned before, the test set consisting of 35 images, each sized 1,000×1,000 pixels and obtained at 10× magnification, was used. These images were labeled during the F-DB generation process and were not included in the training. All images exhibit fibrosis. The performance of the models was assessed by calculating the Dice coefficient of each model's predictions on the test set.

Table 1 presents the training results for the Pytorch U-Net model, while Fig. 5 illustrates the inference results of the best model on a test set image.

From the experiments that were performed, it is worth mentioning that initializing the network weights with a database containing a large and similar set of data was a good practice. This improved the training results.

Another technique that also contributed to the improved results was applying color normalization to the training images. By transforming each image to the other two working domains, the database was effectively tripled. HistAuGAN presented very surprising results, generating realistic images that accurately represent the tonalities of different domains. However, it can be concluded that the improvement introduced by this technique is not significant enough compared to the work involved, leading to only a 0.02 difference in the Dice test coefficient. Additionally, a notable increase in training times was observed. Table 1 shows that introducing data augmentation with HistAuGAN increased the training time by 2 hours due to the larger amount of data.

5   Conclusion and Future Work

This work presents an application of machine learning models and image processing techniques for analyzing biological samples of mare endometrium. Given the similarity to human tissue used to train the model for epithelium segmentation, these models could be applied without modifications to detect the same structure in mare endometrium images. Post-processing was performed to enhance the segmentation mask results.

A color normalization model was trained to balance the domains of the available images. Additionally, a fibrosis detection model was trained, requiring the generation of a labeled dataset. Various techniques were tested to improve the model's performance, with the initialization of network weights using labeled data for nuclei segmentation proving to be the most effective.

As a result of this work, specialists gain access to both quantitative and visual information for evaluating samples, facilitating objective and comparable results across specialists and between samples. However, while this investigation provides valuable insights, additional images are needed to better assess the model's generalization.

For further investigation, it would be interesting to extract quantitative information from the whole sample. For example, a gland density analysis could be performed. This information, together with the segmentation obtained by the fibrosis detection model, could be combined to improve the quality of the results obtained.

Code is available at the Mare FEst project on GitHub.

References

1. Kenney, R.M., Doig, P.A.: Equine endometrial biopsy. Current Therapy Theriogenol. **2**(3), 723–729 (1986)
2. Aguilera, N., Díaz, A., Zimmer, S.: Evaluación de la salud uterina en yeguas a partir de biopsias endometriales utilizando técnicas de procesamiento de imágenes y aprendizaje automático. Tesis de grado. Universidad de la República (Uruguay). Facultad de Ingeniería (2023)
3. Pedersen, A., Valla, M., Bofin, A.M., De Frutos, J.P., Reinertsen, I., Smistad, E.: FastPathology: an open-source platform for deep learning-based research and decision support in digital pathology. IEEE Access **9**, 58216–58229 (2021)
4. Smistad, E., Bozorgi, M., Lindseth, F.: FAST: framework for heterogeneous medical image computing and visualization. Int. J. Comput. Assist. Radiol. Surg. **10**(11), 1811–1822 (2015). https://doi.org/10.1007/s11548-015-1158-5
5. Vekariya, A.P.: A Deep Learning Based Pipeline For Metastatic Breast Cancer Classification From Whole Slide Images (WSI) (Doctoral dissertation) (2023)
6. Szóstek-Mioduchowska, A., et al.: Transcriptomic profiling of mare endometrium at different stages of endometrosis. Sci. Rep. **13**(1), 16263 (2023)
7. Wójtowicz, A., et al.: The potential role of miRNAs and regulation of their expression in the development of mare endometrial fibrosis. Sci. Rep. **13**(1), 15938 (2023)
8. Yao, Y., et al.: Polarization imaging feature characterization of different endometrium phases by machine learning. OSA Continuum **4**(6), 1776–1791 (2021)
9. Hu, Q., et al.: Noninvasive assessment of endometrial fibrosis in patients with intravoxel incoherent motion MR imaging. Sci. Rep. **11**(1), 12887 (2021)
10. Miao, R., Toth, R., Zhou, Y., Madabhushi, A., Janowczyk, A.: Quick Annotator: an opensource digital pathology based rapid image annotation tool. J. Pathol. Clin. Res. **7**(6), 542–547 (2021)
11. Ronneberger, O., Fischer, P., Brox, T.: U-Net: Convolutional Networks for Biomedical Image Segmentation. In: Navab, N., Hornegger, J., Wells, W.M., Frangi, A.F. (eds.) MICCAI 2015. LNCS, vol. 9351, pp. 234–241. Springer, Cham (2015). https://doi.org/10.1007/978-3-319-24574-4_28
12. Wagner, S.J., Khalili, N., Sharma, R., Boxberg, M., Marr, C., de Back, W., Peng, T.: Structure-preserving multi-domain stain color augmentation using style-transfer with disentangled representations. In: de Bruijne, M., et al. (eds.) MICCAI 2021. LNCS, vol. 12908, pp. 257–266. Springer, Cham (2021). https://doi.org/10.1007/978-3-030-87237-3_25
13. Wagner, S.: HistAuGAN. https://github.com/sophiajw/HistAuGAN Accessed 04 Nov 2023
14. Pettersen, H.S., et al.: Code-free development and deployment of deep segmentation models for digital pathology. Front. Med. **8**, 816281 (2022)
15. Rea, K.: How to select the correct magnification and patch size for digital pathology projects. https://andrewjanowczyk.com/how-to-select-the-correct-magnification-and-patch-size-for-digital-pathology-projects/ Accessed 04 Nov 2023
16. Milletari, F., Navab, N., Ahmadi, S.: V-net: fully convolutional neural networks for volumetric medical image segmentation. In 2016 Fourth International Conference on 3D Vision (3DV), pp. 565-571. IEEE (2016)
17. Hou, L., et al.: Dataset of segmented nuclei in hematoxylin and eosin stained histopathology images of ten cancer types. Sci. Data **7**(1), 185 (2020)

18. Khan Academy. Tissues, organs, & organ systems. https://www.khanacademy.org/science/biology/principles-of-physiology/body-structure-and-homeostasis/a/tissues-organs-organ-systems Accessed 24 July 2024
19. Amiri, M., Brooks, R., Rivaz, H.: Fine-tuning U-Net for ultrasound image segmentation: different layers, different outcomes. IEEE Trans. Ultrason. Ferroelectr. Freq. Control **67**(12), 2510–2518 (2020)

Computer Vision and Robotics

X-COVNet: Externally Validated Model for Computer-Aided Diagnosis of Pneumonia-Like Lung Diseases in Chest X-Rays

Jorge Felix Martínez Pazos[1,2]($\boxtimes$) (iD), Arturo Orellana García[1] (iD),
David Batard Lorenzo[2] (iD), and Jorge Gulín González[2] (iD)

[1] Medical Informatic Center, University of Informatics Science, Havana, Cuba
jorgefmp.mle@gmail.com
[2] Study Center On Computational Mathematics, University of Informatics Science Havana,
Havana, Cuba

Abstract. Since the appearance of COVID-19, the accurate diagnosis of pneumonia-type lung diseases by chest radiographs has been a challenging task for experts, mainly due to the similarity of patterns between COVID-19 and viral or bacterial pneumonia. To address this challenge, a model for the computer-aided diagnosis of chest X-Rays has been developed in this research, which might contribute to substantially increasing the accuracy and efficiency of the diagnosis. This approach is based on supervised learning using neural networks, where the quality of the result depends on the quality of the dataset used during training. Image data augmentation techniques, hyperparameter adjustments and dropout layer contributed to achieve high-performance values in a multi-class classification. The experiments conducted to evaluate the model yielded that it detects and classifies domain classes with an accuracy of 99.45% on training data, 99.27% on validation data and 99.06% on selected test data. The main contribution of this paper is a new Deep Convolutional Neural Network model using Deep Transfer Learning through the Xception architecture for the assisted diagnosis of COVID-19, pneumonia or healthy patients, trained on COVID-19 Chest X-Ray Database and evaluated through two external databases, which give the model novelty within the lack of external validation in all the reviewed literature.

Keywords: chest X-Rays · external validation · Deep Convolutional Neural Network · Deep Transfer Learning · Xception

1 Introduction

During the COVID-19 pandemic crisis, plenty of health institutions suffered periods of collapse and their workers had to deal with the burden of attending to each of the patients. Although other techniques were later used to diagnose the disease, resorting to a chest X-ray is the most common when a patient has signs of respiratory disease

L. Correia et al. (Eds.): IBERAMIA 2024, LNCS 15277, pp. 161–172, 2025.
https://doi.org/10.1007/978-3-031-80366-6_14

such as COVID-19 or pneumonia. In 2017, more than 808 000 children under five years of age deceased due to pneumonia, accounting for 15% of all deaths in children under five years of age. Individuals who are susceptible to developing risk pneumonia encompass those who are above the age of 65 as well as those with underlying medical conditions [1]. Accurately diagnosing pneumonia is a difficult task, requiring the review of a chest X-ray which must be performed by highly trained specialists. This disease usually manifests as an area of increased opacity on the chest X-ray [2]. Diagnosis in early 2019 is complicated by the similarity of patterns between pneumonia and COVID-19, as differences between pneumonia and COVID-19 in X-rays are difficult for the human eye to perceive. Intelligent system-assisted diagnosis could lead to greater accuracy, as they can examine the internal patterns in the pixel array of the image. Specialists are often faced with reading large volumes of images during each shift. Therefore, having a tool for computer-aided diagnosis could have a positive impact on the efficiency and effectiveness of the outcome. Intelligent systems based on deep learning have been used in several solutions for medicine and healthcare. Yu et al. [3], show a non-exhaustive list of the potential of AI applied in medicine grouped into, basic biomedical research, translational research and clinical practice.

The field of medical imaging for computer-aided diagnosis has been covered by several authors with excellent results, studies discussed below perform the classification of chest X-rays. Hashmi et al. [4], proposed to obtain the optimal weights for five deep convolutional neural network architectures - ResNet18, DenseNet121, Inception, Xception and MobileNetV2 - and feed the weighted predictions into a weighted classifier module to obtain the final weighted prediction to identify healthy patients or patients with pneumonia. Chouhan et al. [5] used an ensemble model consisting of multiple pre-trained deep neural network models. Toğaçar et al. [6], combined features from different deep learning models. Ayan et al. [7], applied transfer learning and fine tuning to VGG16 and Xception architectures. Civit-Masot et al. [8] also applied transfer learning to the VGG16 architecture to distinguish between COVID-19, pneumonia and normal classes. Owida et al. [9] proposed a method for extracting effective features from chest X-ray images using wavelet analysis and the Mel Frequency Cepstral Coefficients (MFCC) method. These features were then used in a classification process using a Support Vector Machine (SVM) classifier. In a study by Lee & Lim [10], fine-tuning techniques were applied to the DenseNet201 architecture to improve its performance in detecting COVID-19 cases from chest X-rays. Jyoti et al. [11], present a new approach to decompose chest radiographs from two different datasets using a two-dimensional (2D) tunable Q-wavelet transform (TQWT) based on a memristive crossbar array (MCA). The decomposed images were then classified as either COVID-19 or non-COVID-19 using convolutional neural network (CNN) models, specifically ResNet50 and AlexNet. Their results were achieved with less complexity, energy consumption and performance compared to conventional techniques. Dalvi et al. [12] proposed to apply transfer learning and tuning over DenseNet-169 architecture, where the data preprocessing is performed using the Nearest-Neighbors interpolation technique.

Considering the significant benefits of Intelligent systems based on deep learning, this research aims to create a predictive model based on deep learning techniques. This model enables computer-aided diagnosis of respiratory diseases in chest radiographs using a multiclass classification between normal/healthy, pneumonia and COVID-19. Image data augmentation techniques, hyperparameter adjustments, and dropout layer contributed to achieving high-performance values on test data and external datasets in multi-class classification.

2 Materials and Methods

The COVID-19 Chest X-Ray Database, which was officially published on Kaggle and developed by a team of researchers from Qatar University in Doha, Qatar and Dhaka University in Bangladesh, was deemed the most affordable option for the proposed approach [13]. The database contains chest X-ray images for healthy, COVID-19 positive, and Normal and Viral Pneumonia cases. The suitability of this database is given by the fact that the data were collected from various publicly available datasets, online sources, and published articles, and include images from several regions, including the Americas and Europe, since images were sourced from several reputable institutions, such as the Radiological Society of North America, the Institute for Diagnostic and Interventional Radiology at Hannover Medical School in Hannover, Germany, the Italian Society of Medical and Interventional Radiology (SIRM), and the Medical Imaging Databank of the Valencia Region (BIMCV). The inclusion of images from these different sources ensures that the training dataset is representative of a wide range of radiological data.

The database contains 2500 images of the normal class, 1345 of pneumonia, and 3616 of COVID-19. Originally, the database had more than 10,000 images belonging to the normal class, to balance the dataset, a random value between the maximum and minimum of the other classes was selected, resulting in 2500 images for the normal class. Figure 1 shows the first contact with the database.

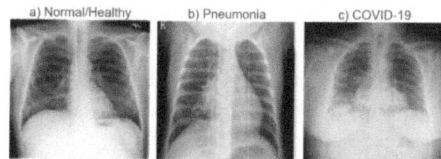

Fig.1. Chest radiographs of a) a healthy individual, b) an individual with pneumonia, and c) an individual with COVID-19.

Transfer learning is a technique that involves enhancing the learning process for a new task by transferring knowledge from a related task that has already been learned [15]. The use of pre-trained models as an initial foundation for tasks related to computer vision and natural language processing is a prevalent methodology within the field of deep learning [16]. Xception, the architecture used for feature extraction in our study is a deep convolutional neural network that employs depth-wise separable convolutions [17]. After data acquisition, exploratory data analysis is used to obtain insights from the dataset. The architecture of the deep learning model is defined by 3 blocks data preprocessing - data augmentation, model building - fine tuning and callbacks. The preprocessing is worked exhaustively because it has a direct impact on the model's performance, model building - fine tuning block and callbacks block are inter-connected due to the presence of a feedback mechanism between them as depicted in Fig. 2. To achieve improved optimization and more efficient handling of uncertainty, convolutional neural networks and transfer learning techniques were employed, which have demonstrated exceptional performance in computer vision tasks [18]. Accuracy, precision, recall, f1-score, AUC score, ROC curve and confusion matrix are the performance metrics used in this study [4].

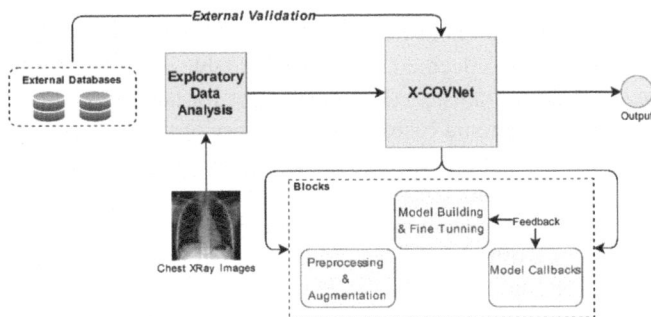

Fig.2. Workflow diagram of the proposed model.

The images were formatted to RGB with a size of 224 × 224 for optimal processing of the predictive model. Working with high resolution images would lead to a high level of processing and time cost. There is no significant dispersion of values concerning the total, 91.6% of the images have values distributed in (75 ≤ x ≤ 175) for mean and (30 ≤ y ≤ 80) for standard deviation. A detailed manual analysis of stochastic image sets was conducted to ascertain the number of radiographs with COVID-19 or pneumonia classification among patients admitted with visible medical wiring, which could lead to erroneous learning and misclassifications by the model, based on the assumption that all patients exhibiting these features belong to a given classification. Images with these features are sparsely represented in the dataset, obviating the need for their removal or treatment.

To increase the diversity of the training dataset and improve model performance, various image data augmentation techniques such as rotation with 25 degrade limit, horizontal flipping, random brightness adjustment, and RGB shifting are applied using Albumentations library [19, 20]. The proposed model is initialized and configured using the Xception architecture and adding new configurations, input layer 224 × 224 × 3, pretrained architecture weights were frozen and used the 'imagenet' weights which are pretrained on ImageNet database organized according to WordNet hierarchy [21]. In the output of the Xception architecture, was added a 2D Global Average Pooling layer. The parameter compression ratio is exponentially high in this type of layer resulting in a 2D dimensionality of the form (batch_dimension, n_channels). This is different to the Flatten layer commonly used to feed fully connected layers, which only restructures the matrix to a single dimension [22]. Followed by a Dropout layer with 75% probability to avoid possible overfitting mainly due to the extension of the pretrained model, and a Batch Normalization layer to normalize the inputs. This layer applies transformations that keep the mean output close to 0 and the standard deviation output close to 1 [23, 24]. It is concluded with a Dense output layer with 3 neurons referring to the three classes in the domain, using the softmax activation function.

The model is compiled using the Adaptive Moment Estimation (Adam) optimizer with a learning rate of 3e-4, the loss function categorical crossentropy, the metric accuracy and a batch size of 32. The values corresponding to learning rate and batch size were obtained from the fine tune process using keras tuner, the hyperparameters search was run on [0.03, 0.003 0.0003] and [16, 32, 64] for learning rate and batch size respectively [25]. The recommended values for the batch size parameter in low-performance computing were selected. The learning rate values were chosen because, although the default value for the Adam optimizer is 0.001, a widely used value is 0.0003 [26]. Therefore, an interval from 0.03 to 0.0003 was used, decreasing by a factor of 10. Four callbacks were used during the training process, early stop, learning rate reduction, tensorboard and checkpoint [22].

3 Results

After training the model on the training set for 15 epochs an accuracy of 0.9945 on training data and 0.9927 on validation data was obtained. The model exhibits robust generalization capabilities, as evidenced by its consistent performance across both the training and validation datasets. The benefits of transfer learning: a higher start, a higher slope and a higher asymptote are noticeable in the model training history [27]. After concluding the training, it is visible the considerable contribution of the callback learning rate reduction to the performance of the model on validation data. Also shows how the reduction of the learning rate stabilizes the model keeping it close to the best metrics, the decay of the learning rate is also translated in a smaller amplitude between the training and validation curve. This is because the lower the value of the learning rate used in the gradient descent implemented by the Adam optimizer, the smaller its displacement in search of the global minimum, therefore it can locate this or very close values with greater effectiveness, having less possibility of deviating to less favorable situations [20]. The best values for each metric are reached at epoch 15, so this is the saved configuration of the model via the checkpoint callback.

The model was evaluated on the test set, its performance was validated using previously described metrics, achieving 100% recall for the COVID-19 class, indicating that all instances of this class were correctly identified; such high sensitivity in detecting patients with COVID-19 suggests that there were no false negatives for this class. Results indicate that the model performed slightly better on the COVID-19 class than on the normal and pneumonia classes, suggesting that the model is more effective in identifying patients with this disease. A set of images were randomly selected from the test set for classification using the proposed model. Figure 3 provides a visual representation of each image, accompanied by its respective predicted classification from the X-COVNet model, indicated by the label on the x-axis (Predicted Label: PL), and its actual classification, indicated by the label on the y-axis (True Label: TL).

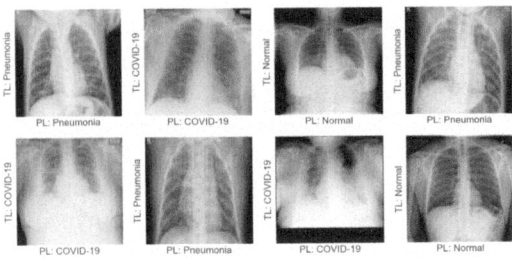

Fig.3. Classification results of chest radiographs using the proposed model.

It is important to evaluate predictive models on unseen data, especially in the context of medical imaging. If a model is not thoroughly evaluated with data external to the training process, there is a risk that it will not perform as well in real-life scenarios as it did during training. This could lead to incorrect diagnoses and potentially harmful therapeutic decisions. By evaluating the model with new data, practitioners can gain a more accurate understanding of its real-world performance and increase confidence in its use for assisted diagnosis. This can help ensure that the model is reliable and safe for use in clinical settings. This process involves the use of two external chest radiograph databases containing a large number of images of 6000 approximately. These databases allow for a comprehensive evaluation of the model's performance on both raw and preprocessed images. The evaluation process begins with an assessment of the model on the training database which serves as a reference point for further evaluation on external databases [13].

The first external database which will be called Raw Database, Chest X-ray (Covid-19 & Pneumonia), is derived from three previously published datasets and contains a total of 6432 images. One of these datasets is ieee8023 covid-chestxray-dataset, an official database approved by the University of Montreal's Ethics Committee and developed by Joseph Paul Cohen and his team at Mila, University of Montreal [28]. This database contains raw images in various formats and sizes, including images with figures and other artifacts. The second external database used is the Chest X-Ray Images Database, called Preprocessed Database, which has a total of 5228 images and contains preprocessed images resized to 232×232 in png format, including an even larger number of images with artifacts such as pointers and numbers. It uses information from medical websites such as eurorad.org, radiopedia.org and coronacases.org [29].

To gain a deeper understanding of the classification model on the three classes represented in the assessment databases, the ROC curve for One vs One and One vs Rest, as well as the AUC score was studied [30, 31]. The proposed model shows superior AUC values for the COVID19 class, with overall the highest results in the databases for both the OvO and OvR approaches. This shows that the model classifies radiographs of patients with COVID-19 more reliably, which is the best possible result in terms of domain classes because a higher performance in diagnosing COVID 19 not only implies more safety and care for patients with this disease but also limits the possibility of contagion by overlooking a patient with SarsCov2. Overall, the model has excellent values for all AUC scores with a minimum value of 0.9810 and 0.9875 for the OvO and OvR approaches respectively. In the context of the ROC Curve and AUC Score, it was found that there were no significant differences in the results obtained during the evaluation of the databases where the maximum difference in AUC score was 0.0083, observed between the training database with 0.9995 and the external preprocessed database with 0.9912.

The evaluation of the model on external databases yielded peculiar results. Contrary to expectations, the model performed best on the raw database, despite its previously described characteristics. Values for the accuracy metric obtained were 99.18%, 92.35%, and 96.78% for the training, preprocessed, and raw databases, respectively. These results are summarized in Table 1 indicating a good generalization of the model, as it was able to achieve high accuracy on external databases with a large number of chest X-rays images. When evaluating external databases, the lowest accuracy result was observed on the preprocessed database, with a value of 92.35%. However, this result is particularly noteworthy as it exceeds the accuracy values reported on the test sets of several models previously studied in the literature [7, 8].

Table 1. Metrics of the model evaluation on external databases.

Database	No Images	Loss	Accuracy (%)	AUC Score (%)
Training	7641	0.026	99.13	99.95
Preprocesed	6432	0.231	92.35	99.12
Raw	5228	0.103	96.78	99.43

These findings demonstrate the robust generalization capabilities of the proposed approach on real-world data, which not only achieves high accuracy on previously unseen data, but also outperforms other models in the field. A deeper evaluation of the model's performance on external datasets was only performed on the preprocessed dataset, as it had the lowest performance during the model evaluation. The proposed X-COVNet model showed an accuracy of 92.35% on this dataset, misclassifying a total of 400 chest X-rays out of a total of 5228. The recall for COVID-19 showed the best value among the three classes, which is very close to 1. This means that the model missed very few patients with COVID-19 in the preprocessed dataset. On the other hand, a high precision of 98.97% for the Normal class indicates that the model is very accurate in predicting cases as normal. This means that there are few false positives for this class, which is crucial in a medical context as it reduces the likelihood of misdiagnosing a patient with a condition such as pneumonia or COVID-19 as Normal. These results are very reliable as the dataset is not biased towards any class and has a very balanced total for each class, such as 1802, 1800 and 1626 for the Normal, Pneumonia and COVID-19 classes respectively. Table 2 below depicts the total number of misclassified images, categorized by each type of misclassification for further analysis.

As mentioned above, the model rarely classifies patients with any disease, including pneumonia or COVID-19, as normal, with a total of 16 misclassifications. It is more likely that the model will classify a healthy patient as having a disease such as pneumonia or COVID-19, with a total of 263 incorrect predictions, or misclassify pneumonia as COVID-19, with a total of 120 misclassifications. However, it is highly unlikely that the

Table 2. Distribution of a total 400 misclassifications by category.

True Label	Predicted Label	Amount
NORMAL	Pneumonia	107
NORMAL	COVID-19	156
Pneumonia	NORMAL	7
Pneumonia	COVID-19	120
COVID-19	NORMAL	9
COVID-19	Pneumonia	1

model would miss a patient with COVID-19 with a total of 10 misclassifications in this evaluation. As mentioned in the class breakdown of the evaluation dataset, this is not due to any bias in the evaluation dataset. Thus, it supports the statement in the article that the model classifies patients with COVID-19 with high performance.

4   Discussion

During the comparison of the developed model with those reported in the literature, it was observed that, in contrast to the proposed model, 8 of the total 9 studied models for diagnosing chest radiographs rely on binary classifications. Results used during the comparison process were reported by their respective authors in their studies. The comparison of the models is carried out based on the performance metrics obtained in the test set, as these are the ones most frequently collected in the research studied. While most of the literature reviewed utilized datasets with a magnitude of approximately 5000 images, the approach proposed in this study uses a larger dataset consisting of 7641 images.

Hashmi et al. [4] proposed a weighted classifier fed by 5 deep neural network architectures to classify normal or pneumonia in chest X-rays, achieving a recall value of 99.0%, higher than the 98.7% obtained by the proposed approach; however, their approach was slightly lower in terms of precision and accuracy metrics, with values of 98.3% and 98.4%, respectively. Ayan and Ünver [7] applied transfer learning and fine-tuning to the VGG16 and Xception architectures and obtained lower performance metrics than the proposed X-COVNet model, with accuracy, precision, and recall values of 87.0%, 87.0%, and 87.5%, respectively. The ensemble model proposed by Chouhan et al. [5] achieved recall and accuracy of 99.62% and 96.39%, respectively. Toğaçar et al. [6] achieved approximately 96.8% for recall, precision, and accuracy metrics. The approach proposed by Owida et al. [9] using the Mel Frequency Cepstral Coefficients method and a support vector machine classifier achieved 98.8% accuracy, unfortunately, others performance metrics were not reported in the research. The study conducted by Lee & Lim [10] achieved an accuracy of 99.90%, a precision of 98.99%, and a recall of 98.00%. Lee and Lim's approach outperforms the proposed model only in terms of accuracy, with a difference of 0.84%. Jyoti et al. [11] achieved accuracies of 98.82% and 95.67% on small and large datasets respectively, which are less accurate than the proposed X-COVNet model. The approach proposed by Dalvi et al. [12], where data preprocessing is performed using the Nearest-Neighbors interpolation technique, achieved 96.37%, 94.08%, and 98.89% for accuracy, precision, and recall, respectively, which tends to be less performing than the proposed approach.

All of the models discussed above perform a binary classification of the chest radiograph, either for the COVID19/non-COVID19 or pneumonia/normal classes. In contrast to the aforementioned models, Civit-Masot et al. [8] used a training set of 316 images for a multiclass classification approach similar to the proposed approach, distinguishing between normal/healthy, pneumonia, and COVID-19 using the VGG16 architecture; they obtained 86.0% for each macro average of the metrics used in the comparison. The model proposed in this paper generally showed better performance compared to the literature reviewed. The closest performance was achieved by Hashmi et al. [4], Lee & Lim [10], and Jyoti et al. [11], who used binary classification; however, the proposed approach outperformed these studies by using multi-class classification. Table 3 shows that the proposed X-COVNet model outperforms other approaches in general terms and demonstrates its superiority by achieving better metrics in a more complex classification task compared to most models in the reviewed literature.

A new computer-aided diagnostic model for chest X-ray has been achieved, capable of classifying patients with pneumonia-type lung disease. The model, which employs convolutional neural networks and transfer learning techniques, has demonstrated remarkable performance on validation and test data, achieving an accuracy of 0.9906 on test set. With its high accuracy and ability to differentiate between healthy patients, those with pneumonia and those with COVID-19, the model has the potential to improve the diagnosis of respiratory illnesses and patient care. The proposed model X-COVNet could serve as an adjunct to clinical decision making, it can assist radiologists in the decision-making process, but the final decision must be made by an expert. The proposed model is intended to support, not replace, the expertise of a trained radiologist in making a diagnosis [4]. The lack of external validation is a common problem in many research studies [4–12], which is critical since it allows researchers to determine how well the model generalizes to real-world cases and can help identify potential biases or limitations in the model.

In the present research, an external validation was conducted and it was found that our proposed model showed high performance on external data. This is an important finding because it provides evidence that our model can accurately diagnose pneumonia-like lung diseases using chest X-rays, even when applied to data outside of our training dataset, which helps ensure that the model can accurately diagnose patients in a clinical setting. It is also important to note that AI-based tools should be used with caution by trained professionals and not at the discretion of an individual. Misuse of these tools can pose risks to the health of patients and communities. Proper training and oversight are necessary to ensure that AI-based diagnostic tools are used safely and effectively in healthcare settings. It should be acknowledged that the model has certain constraints, the Grad-CAM technique has not yet been implemented to facilitate the interpretation of the model's classification decisions by medical specialists. X-COVNet constitutes a substantial contribution in the field of computer-assisted diagnosis and can profoundly influence both medical imaging and patient care. Through its ability to increase the accuracy and speed of diagnoses, reduce costs, and facilitate the use

Table 3. Comparison of the proposed model with published models.

Published Models	No. Of Images	Precision (%)	Recall (%)	Accuracy (%)	Classification Type	External Validation
Civit-Masot et al., [8]	316	86.00	86.00	86.00	**Multi-class**	No
Ayan & Ünver [7]	5856	87.00	87.50	87.00	Binary	No
Chouhan et al., [5]	5247	93.28	**99.62**	96.39	Binary	No
Toğaçar et al., [6]	5849	96.88	96.83	98.00	Binary	No
Hashmi et al., [4]	5856	98.26	99.00	98.43	Binary	No
Owida et al., [9]	5856	-	-	98.80	Binary	No
Lee & Lim [10]	5000	98.99	98.00	**99.90**	Binary	No
Jyoti et al., [11] – Small	2193	**99.16**	98.50	98.82	Binary	No
Jyoti et al., [11] – Large	5275	95.37	95.94	95.67	Binary	No
Dalvi et al., [12]	3010	94.08	98.89	96.37	Binary	No
Proposed Model (X-COVNet)	**7641**	98.99	98.73	99.06	**Multi-class**	**Yes**

of remote medical consultations and monitoring, this model can greatly improve the accessibility, convenience, cost-effectiveness, and quality of healthcare for patients.

5   Conclusions and Future Work

Recent advances in artificial intelligence have led to significant efforts in computer-aided diagnosis of radiographs and medical images. However, most of the existing models for computer-aided diagnosis of pneumonia-like lung diseases from chest X-rays do not perform external validation with databases completely separate from the one selected for training. This study presents X-COVNet, a model that achieves 99% accuracy on the test and validation sets, and demonstrates robust performance on two external databases with different characteristics, achieving approximately 96% and 93% accuracy, respectively. These results provide real insight into the potential application of the model in real-world scenarios and underscore the importance of external validation in assessing the generalizability of AI models in medical imaging. Future research aims to integrate an algorithm for Gradient-weighted Class Activation Mapping (Grad-CAM) into the model to allow specialists to understand how the model has been driven to make the classification decision. Further, it is suggested to integrate the model with the help of CESIM into the Xavia-Pacs medical imaging tool deployed in Cuban medical centers.

Disclosure of Interests. The authors have no competing interests to declare that are relevant to the content of this article

References

1. WHO. Pneumonia. https://www.who.int/health-topics/pneumonia. Accessed 06 Apr 2023
2. Cleverley, J., Piper, J., Jones, M.M.: The role of chest radiography in confirming covid-19 pneumonia. BMJ m2426 (2020). https://doi.org/10.1136/bmj.m2426
3. Yu, K.H., Beam, A.L., Kohane, I.S.: Artificial intelligence in healthcare. Nat. Biomed. Eng. **2**(10), 719–731 (2018). https://doi.org https://doi.org/10.1038/s41551-018-0305-z
4. Hashmi, M.F., Katiyar, S., Keskar, A.G., Bokde, N.D., Geem, Z.W.: Efficient pneumonia detection in chest xray images using deep transfer learning. Diagnostics **10**(6), 417 (2020). https://doi.org/10.3390/diagnostics10060417
5. Chouhan, V., et al.: A novel transfer learning based approach for pneumonia detection in chest x-ray images. Appl. Sci. **10**, 559 (2020). https://doi.org/10.3390/app10020559
6. Toğaçar, M., Ergen, B., Cömert, Z.: A deep feature learning model for pneumonia detection applying a combination of mRMR feature selection and machine learning models. IRBM (2019). https://doi.org/10.1016/j.irbm.2019.10.006
7. Ayan, E., Ünver, H.M.: Diagnosis of pneumonia from chest X-ray images using deep learning. In: 2019 Scientific Meeting on Electrical-Electronics & Biomedical Engineering and Computer Science (EBBT), pp. 1–5. IEEE (2019). https://doi.org/10.1109/EBBT.2019.874 1582
8. Civit-Masot, J., Luna-Perejón, F., Domínguez Morales, M., Civit, A.: Deep learning system for COVID-19 diagnosis aid using X-ray pulmonary images. Appl. Sci. **10**, 4640 (2020). https://doi.org/10.3390/app10134640
9. Owida, H.A., Al-Ghraibah, A., Altayeb, M.A.: Classification of chest X-ray images using wavelet and MFCC features and support vector machine classifier. Eng. Technol. Appl. Sci. Res. (2021). https://doi.org/10.48084/etasr.4123
10. Lee, C.P., Lim, K.M.: COVID-19 diagnosis on chest radiographs with enhanced deep neural networks. Diagnostics **12**, 1828 (2022). https://doi.org/10.3390/diagnostics12081828
11. Jyoti, K., Sushma, S., Yadav, S., Kumar, P., Pachori, R.B., Mukherjee, S.: Automatic diagnosis of COVID-19 with MCA-inspired TQWT-based classification of chest X-ray images. Comput. Biol. Med. **152**, 106331 (2023). https://doi.org/10.1016/j.compbiomed.2022.106331
12. Dalvi, P.P., Edla, D.R., Purushothama, B.R.: Diagnosis of coronavirus disease from chest X-Ray images using DenseNet-169 architecture. SN Comput. Sci. **4**,(2023). https://doi.org/10.1007/s42979-022-01627-7
13. Muhammad E.H., et al.: COVID-19 Radiography Database (2022). https://www.kaggle.com/datasets/tawsifurrahman/covid19-radiography-database
14. Ting, K.M.: Confusion matrix. In: Sammut, C., Webb, G.I. (eds.) Encyclopedia of Machine Learning. Springer, Boston, MA (2011). https://doi.org/10.1007/978-0-387-30164-8_157
15. Soria, E., Martinez, J.D., Martinez, M., Magdalena, J.R., Serrano, A.J.: Handbook of research on machine learning applications. ISBN-10: 1605667668, ISBN-13: 978–1605667669 (2009)
16. Brownlee, J.: A gentle introduction to transfer learning for deep learning (2019). https://machinelearningmastery.com/transfer-learning-for-deep-learning

17. Chollet, F.: Xception: Deep learning with depthwise separable convolutions (2016). https://doi.org/10.48550/arXiv.1610.02357

18. Usha Kingsly Devi, K., Gomathi, V.: Deep convolutional neural networks with transfer learning for visual sentiment analysis. Neural Proc. Lett. 1–34 (2022).https://doi.org/10.1007/s11063-022-11082-3

19. Buslaev, A., Iglovikov, V.I., Khvedchenya, E., Parinov, A., Druzhinin, M., Kalinin, A.A.: Albumentations: fast and flexible image augmentations. Information **11**(2), 125 (2020). https://doi.org/10.3390/info11020125

20. Geron, A.: Hands-on machine learning with Scikit-Learn, Keras and Tensorflow: concepts, tools, and techniques to build intelligent systems (2nd ed.). Chapter 11: Training Deep Neural Nets, pp. 275–312. O'Reilly (2019)

21. Fei-Fei, L., Deng, J., Russakovsky, O., Berg, A., Li, K.: ImageNet (2009). https://www.image-net.org/index.php

22. Keras. Keras API Reference. https://keras.io/api. Accessed 28 Mar 2023

23. Hinton, G.E., Srivastava, N., Krizhevsky, A., Sutskever, I., Salakhutdinov, R.R.: Improving neural networks by preventing co-adaptation of feature detectors. arXiv preprint arXiv:1207.0580 (2012). https://doi.org/10.48550/arXiv.1207.0580

24. Ioffe, S., Szegedy, C.: Batch normalization: accelerating deep network training by reducing internal covariate shift. In: International conference on machine learning, pp. 448–456. Pmlr (2015). https://doi.org/10.48550/arXiv.1502.03167

25. O'Malley, T., Bursztein, E., Long, J., Chollet, F., Jin, H., Invernizzi, L.: Keras tuner (2019). https://github.com/keras-team/keras-tuner

26. Sordello, M., He, H., Su, W.: Robust learning rate selection for stochastic optimization via splitting diagnostic (2019). https://doi.org/10.48550/arXiv.1910.08597

27. Torrey, L.A., Shavlik, J.W.: Handbook of research on machine learning applications and trends: algorithms, methods, and techniques. Chapter 11 Transfer Learning, pp. 242–264. IGI global (2009)

28. Patel, P.: Chest X-ray (Covid-19 & Pneumonia) (2020). https://www.kaggle.com/datasets/prashant268/chest-xray-covid19-pneumonia

29. Kumar, S., Shastri, S.: COVID19+PNEUMONIA+NORMAL Chest X-Ray Image Dataset (2021). https://www.kaggle.com/datasets/sachinkumar413/covid-pneumonia-normal-chest-xray-images

30. Trevisan, V.: Multiclass classification evaluation with ROC curves and ROC AUC. towards data science (2022). https://towardsdatascience.com/multiclass-classification-evaluation-with-roc-curves-and-roc-auc-294fd4617e3a

31. Scikit-Learn. Multiclass Receiver Operating Characteristic (ROC). https://scikit-learn.org/stable/auto_examples/model_selection/plot_roc.html. Accessed 15 Jul 2023

FairTrees: A Deep Learning Approach for Identifying Deforestation on Satellite Images

Hernan Lira[✉][iD], Taco de Wolff[iD], Luis Martí[iD], and Nayat Sanchez-Pi[iD]

Inria Chile Research Center, Av. Apoquindo, 2827 Las Condes, Santiago, Chile
{hernan.lira,luis.marti,nayat.sanchez-pi}@inria.cl
https://inria.cl

Abstract. Deforestation poses a significant threat to global ecological stability, particularly in the Latin American and Caribbean (LAC) region. This study aims to develop an advanced instance segmentation model using deep learning to monitor deforestation with satellite imagery. The model integrates spatial and temporal analysis to accurately identify deforested areas, addressing challenges such as data quality, class imbalance, and varying image exposures with advanced preprocessing, a robust training pipeline, and U-Net and Feature Pyramid Network architecture. A visualization dashboard tracks deforestation over time, enabling model performance evaluation across multiple LAC regions. Validated against known deforestation events, the model effectively detects and monitors forest loss, providing a valuable tool for policymakers and environmental managers.

Keywords: Deforestation Monitoring · Image Segmentation · Satellite Imagery

1 Introduction

The Latin American and Caribbean (LAC) region, comprising 23% of the world's natural forests, is crucial for global ecological stability, with the Amazon rainforest alone contributing 20% of the Earth's oxygen. However, these forests are increasingly threatened by agriculture, forestry, and livestock expansion, leading to habitat destruction and significant CO_2 emissions. Between 2010 and 2020, South America had the second-highest annual deforestation rate globally, losing 2.6 million hectares of forest annually, largely due to coffee and cocoa plantation expansion. In 2021, the LAC region's harvested areas for coffee and cocoa reached 5.2 million and 1.8 million hectares, respectively, representing 52% and 18% of global production [7].

To mitigate deforestation, it is essential to gather accurate, timely information. Developing a remote-sensing platform that provides comprehensive geographical and imagery data enables temporal comparisons and the establishment

---

T. de Wolff—Contributed while at Inria Chile Research Center.

L. Correia et al. (Eds.): IBERAMIA 2024, LNCS 15277, pp. 173–184, 2025.
https://doi.org/10.1007/978-3-031-80366-6_15

of deforestation metrics. This study aims to create a robust model using advanced instance segmentation techniques to detect and quantify deforested areas accurately, offering a valuable tool for monitoring and managing deforestation in the LAC region.

The primary goals of this study are to develop a remote-sensing platform that provides detailed geographical and imagery information for the LAC region, establish a reliable deforestation metric through temporal satellite image comparisons, and create a robust model using instance segmentation techniques to accurately detect and quantify deforested areas. This involves extensive data gathering from open sources and thorough reliability analyses to ensure accuracy. The study also aims to produce detailed deforestation maps using Geographic Information System (GIS) tools to enhance understanding of deforestation patterns and drivers. By equipping policymakers and environmental managers with a valuable monitoring tool, this study seeks to support the development of targeted conservation strategies.

To develop our deforestation model, we focus on processing satellite imagery as input for a deep learning model. Deep learning techniques enable the extraction of deep semantic features, facilitating automatic object identification in satellite images. Previous research has shown promising results using both classical machine learning ensembles and deep learning approaches. The current state-of-the-art involves deep learning models with architectures like U-Net and feature pyramids, which leverage Convolutional Neural Networks (CNNs) for automatic feature extraction and object identification. These models provide a highly effective end-to-end framework for remote sensing tasks, often incorporating pixel-wise analysis, vegetation indices, and masking to enhance performance.

The structure of the paper is as follows: Sect. 2 reviews relevant previous research. Section 3 provides background knowledge needed for the study. Section 4 details our approach. Section 5 presents and discusses the model results and web application. Finally, Sect. 6 outlines the conclusions and suggests potential directions for future research.

2   Preliminary Studies

Deep learning computer vision techniques, in conjunction with satellite imagery, enable the extraction of deep semantic features, facilitating the automatic identification of objects when trained on such images. Moreover, these techniques have enabled automatic remote sensing for numerous critical applications, including deforestation monitoring.

For example, in 2021, researchers developed a multi-feature random forests (RF) model to identify cocoa plantations in Côte d'Ivoire and Ghana using satellite imagery [2]. A similar approach was used to monitor mangrove loss in Indonesia, applying an RF model to satellite images from Google Earth Engine (GEE) [14]. These studies utilized various vegetation indices, including the Normalized Difference Vegetation Index (NDVI).

Most contemporary research on this issue has employed deep learning models using architectures such as U-Net or feature pyramids [3,12,15,18,21]. These methods leverage Convolutional Neural Networks (CNNs), wherein convolutional layers automatically extract and learn discriminative features. The primary advantage of these models lies in their end-to-end framework for automatic identification, which significantly simplifies the feature engineering process compared to traditional machine learning approaches.

Recent advancements in deep learning models include attention mechanisms that enhance segmentation accuracy by helping the network focus on the most relevant parts of an image [16]. These mechanisms dynamically prioritize regions with significant deforestation signals while ignoring irrelevant background noise. Integrating CNN-based models, such as U-Net and FPNs, with attention mechanisms provides a powerful toolset for deforestation monitoring, automating feature extraction and improving detection accuracy and reliability. These advancements make such models indispensable for large-scale environmental monitoring efforts.

3 Background

3.1 Definitions

To accurately monitor deforestation, it is crucial to define the terms forest, deforestation, and forest degradation. According to the Food and Agriculture Organisation (FAO), a *forest* is land with more than 10% tree crown cover and an area of over 0.5 hectares, where trees can reach at least 5 m in height. Forests include closed formations with dense tree cover and open formations with continuous vegetation exceeding 10% crown cover. This definition also covers young natural stands, forestry plantations, temporarily unstocked areas expected to revert to forest, forest nurseries, seed orchards, forest roads, cleared tracts, firebreaks, small open areas, and protected forest areas. Land primarily used for agriculture is excluded.

Deforestation refers to the change in land cover where tree crown cover is reduced to less than 10%, signifying a shift from forest to non-forest land use, primarily due to human activities. In contrast, *forest degradation* involves changes within the forest that negatively affect its health and productivity without necessarily decreasing woody vegetation area. This degradation can be due to over-grazing, over-exploitation for firewood or timber, repeated fires, natural events like cyclones, and unsustainable logging practices, which reduce biomass, alter species composition, and degrade soil.

3.2 Deep Learning Instance Segmentation Models

As highlighted in Sect. 2, deep learning instance segmentation models have proven highly effective in satellite monitoring. Instance segmentation, a key area in computer vision, involves detecting multiple object instances within complex visual environments by delineating their boundaries. These models have wide-ranging applications, including crop monitoring, medical imaging, and autonomous driving.

Deep learning offers an end-to-end architecture that automatically extracts and learns relevant features, eliminating the need for handcrafted features. Instance segmentation models create a mask and label for each object in an image, providing detailed understanding. Unlike pixel-wise classification, these models detect coherent yet irregular shapes using conglomerate shapes. Most recent studies in this field utilize an encoder-decoder deep learning architecture.

One of the most prominent models in this category is U-Net [23]. It aims to reconstruct an input image from a feature map vector obtained through convolutions. The encoder contracts the image with two convolutional layers followed by max-pooling, increasing feature maps while reducing spatial information to learn complex structures. The decoder then expands the feature vectors using up-convolutions and concatenates these with spatial information to create a segmented image. This design allows the model to train at various scales, learning both small and large features, significantly enhancing its segmentation and classification capabilities. The model's architecture resembles a "U," hence its name.

U-Net is trained on images formatted as tuples (channel, height, width), typically using three channels (red, green, blue) and 256×256 pixels. However, its flexibility allows adjustments to the number of channels and pixel dimensions based on the specific task. This adaptability makes U-Net suitable for large-pixel satellite images with varying channels, including near-infrared.

More recent models for segmentation tasks include Feature Pyramid Networks (FPN) [19], which offer greater flexibility compared to U-Net by allowing the encoder to be interchanged with any existing backbone model, such as ResNet [9], DenseNet [11], or EfficientNet [24]. The decoder can also be customized with different layers depending on the segmentation task. This flexibility is advantageous, as using EfficientNet can reduce training time due to fewer trainable parameters. Numerous extensions to these models, such as incorporating attention mechanisms, have been proposed to enhance their segmentation capabilities [5,10,17].

4   An Instance Segmentation Approach to Deforestation Identification

In this section, we outline our approach for developing FairTrees, our deforestation model in the LAC region that uses deep learning instance segmentation models in satellite images.

Effective monitoring necessitates a model capable of distinguishing forest areas above and below 0.5 hectares with over 10% crown cover. The model must learn features characteristic of both forested and deforested regions from satellite images. This task is complicated by the region's vast and diverse geography, persistent cloud cover, and varying seasons and climates. Additionally, the model must evaluate both temporal and spatial information within a polygon area of interest (AOI) to determine if deforestation has occurred over time.

Leveraging the robust capabilities of Convolutional Neural Networks (CNNs), we employ state-of-the-art architectures such as U-Net and Feature Pyramid

Table 1. The four datasets used in our experiments with a total of 4901 satellite images with their forest loss regions as separate masks. The resolution of the images is in meters per pixel. The common channels are red, green, blue, and infrared.

Dataset	Location	Resolution	Samples	Bands	Classes
Bragagnolo et al. 2021 [4]	Brazil	N/A	1084	4	2
ForestNet [12]	Indonesia	15 m	2757	6	4
Isaienkov et al. 2021 [13]	Ukraine	12 m	212	13	2
Kaggle deforestation [20]	N/A	N/A	848	13	2

Networks (FPNs) to automatically extract and learn discriminative features from satellite imagery. Our approach incorporates both temporal and spatial analysis, enabling precise identification and tracking of deforestation over time.

4.1 Data Sources and Processing

The instance segmentation model is trained using publicly available datasets that include labeled satellite images indicating forested and deforested areas. It processes multi-temporal 10-meter satellite data obtained from Sentinel-1 and Sentinel-2.

Sentinel-1 (S1) is equipped with a Synthetic Aperture Radar (SAR), which generates two-dimensional images or three-dimensional reconstructions of landscapes at a 10-meter spatial resolution, regardless of cloud cover. Sentinel-2 (S2) features an optoelectronic sensor capable of detecting vegetation state variations using visible, near-infrared (VNIR), and short-wave infrared (SWIR) spectral zones, with resolutions ranging from 10 to 60 m. The satellite coverage revisits the same area every 10 d, creating an extensive time-series dataset essential for this project.

Four publicly available datasets have been utilized Bragagnolo et al. [4], ForestNet [12], Isaienkov et al. [13], and the Kaggle deforestation competition [20] (see dataset details on Table 1). We employed the red, green, blue, and infrared channels. Images were cropped to 332 × 332 pixels to enhance training efficiency, despite the model not requiring uniform image dimensions. Data augmentation techniques, such as random cropping, rotations, and the introduction of salt-and-pepper noise, were employed to improve the model's generalization and robustness. These strategies effectively expanded the original dataset fourfold.

We propose to fuse the inputs of Sentinel-1 (S1) and Sentinel-2 (S2) data to create cloud-free images, facilitating continuous monitoring throughout the selected time period. Integrating optical and radar data from Sentinel satellites enables the development of spatially explicit machine learning models that can capture the fine details necessary to distinguish between forested and deforested areas. Additionally, we extract various satellite images over a span of time to incorporate temporal information.

Temporal data allows for the estimation and visualization of deforestation over a given time frame. By analyzing the same region over an extended period

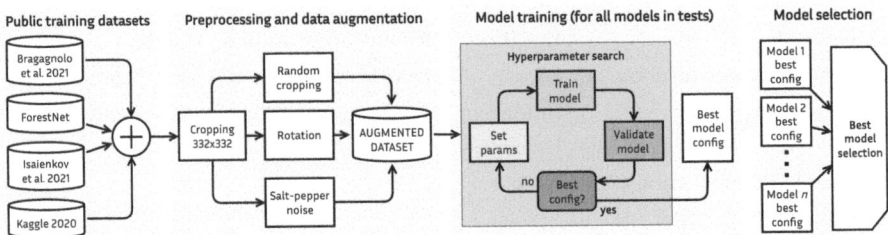

(a) *Training pipeline.* Publicly available datasets are merged, preprocessed, and subjected to data augmentation. A parameter search is then conducted to identify the optimal parameters for each model used in the experiments. The best configurations of each model are subsequently evaluated on a test dataset subset to determine the overall best-performing model, which is then implemented in the inference pipeline.

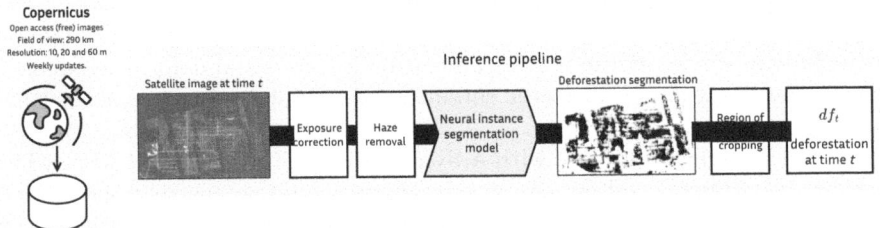

(b) *Inference pipeline.* Copernicus satellite images from the specified time interval, t_{min} to t_{max}, are downloaded and processed to correct exposure and eliminate noise and haze. The processed images are then input into the instance segmentation model. The focus polygon is selected, and the deforestation index is calculated.

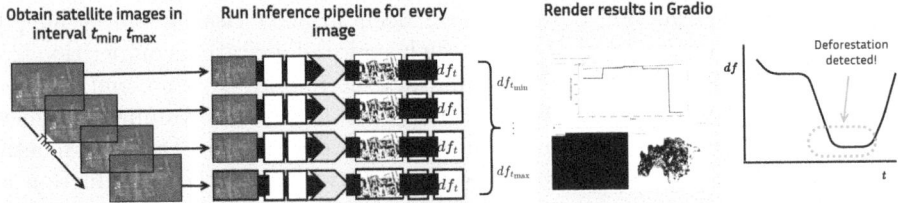

(c) *Deforestation detection pipeline.* Satellite images of the specified area and time interval are processed through the inference pipeline to obtain deforestation indexes. These indexes are rendered as a Gradio app [1], to detect deforestation events.

Fig. 1. FairTrees processing pipelines: deforestation segmentation training (1a), the deforestation segmentation inference (1b), and the deforestation deviation detection (1c) pipelines.

Training

Prediction

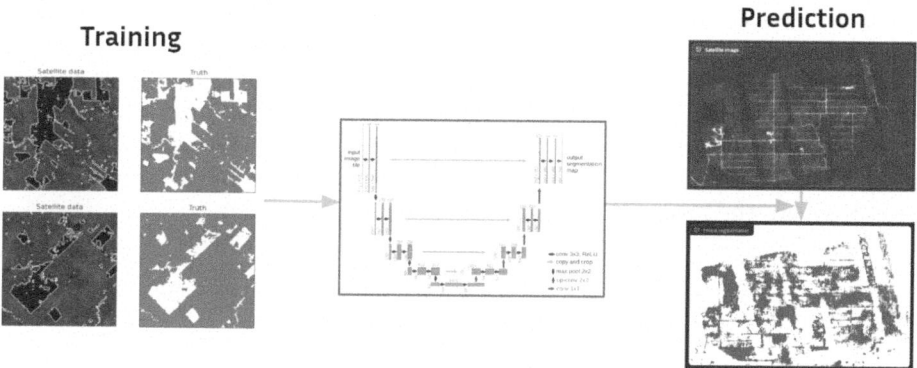

Fig. 2. Holistic view of the training and prediction processes within a deep learning segmentation model.

and measuring the reduction of forest cover in each image, we can plot the amount of deforestation that has occurred. An overview of this approach is presented in Fig. 1, where the detection pipeline processes satellite images over time, conducts segmentation analysis to identify deforestation, and plots this information onto a graph to provide a visual representation of deforestation trends.

4.2 Training and Evaluation Procedure

After processing the data, we train the model to learn the distinguishing features of both forested and deforested areas. Once trained, the model can predict whether a given area is forested or deforested. Figure 2 provides an overview of the training and prediction procedure.

We employed the standard U-Net model [23] as a baseline and further experimented with U-Net, Feature Pyramid Networks (FPN) [19], and Pyramid Attention Networks (PAN) [17] architectures, in combination with ResNet-34, ResNet-50 [9], and EfficientNet-B2 [24] backbones. We did not use pretrained networks, such as those trained on ImageNet [6], as those datasets significantly differ from our specific satellite imagery datasets.

Training was conducted on a node of the Inria HPC cluster, leveraging parallel GPUs. We utilized a random 80%/20% split for training and validation, employing the stochastic gradient descent optimizer with a cross-entropy loss function.

Once the models were trained, they were tested on several key areas in Latin America known for the presence of cacao and coffee plantations, which are particularly challenging to distinguish from forests. The definitive administrative boundaries for each region were retrieved from OpenStreetMap [22].

Each region was divided into a grid of 6144 m × 6144 m sections, resulting in images with a resolution of 12 m/px (i.e., 512 × 512 px). These images were downloaded from Copernicus Sentinel-2 (Copernicus Sentinel data 2022, processed by ESA) using the Google Earth Engine Python interface [8]. This process was repeated for T time periods between 2019 and 2022, selecting the image with the least cloud cover for each period.

For each image, a regional mask was generated from the GeoJSON file of the administrative boundary, along with a cloud mask using the S2 cloud probability dataset, where the cloud probability exceeded 0.6.

The image bands were normalized using the following equation:

$$\text{norm. band} = \frac{255}{0.3} \text{clip} \left(\frac{\text{band} - \min(\text{band})}{10000}, 0.0, 0.3 \right) \qquad (1)$$

This normalization ensured similar histograms across the bands and mitigated the impact of over-illuminated spots using the clip function.

Subsequently, the satellite images were processed by the model to classify each pixel as either deforested or not. A pixel was classified as deforested if the deforestation class probability was greater than or equal to 0.5. The optimal learning rate of 0.01 was determined through a hyperparameter grid search.

5   Results

Table 2 provides a comparison of the models we trained and tested, indicating that U-Net exhibited the best performance, albeit at the cost of increased execution time.

In order to assess the progress of deforestation in a given region we defined a *deforestation index*, $\text{DF}(r, t)$ of a territory r at time t as

$$\text{DF}(r, t) = \frac{\text{count}(\text{pixels\_deforested}(\text{image}(r, t)))}{\text{count\_pixels}(\text{image}(r, t))} \times \text{area}(r), \qquad (2)$$

where $\text{image}(r, t)$ represents a satellite image of the region r at time t.

These indexes can then be visualized over time to detect changes and, in the future, predict the evolution of this phenomenon.

5.1   Assessing Deforestation with Our Instance Segmentation Model

As demonstrated in the previous section, the developed model can effectively segment areas of deforestation when provided with high-resolution satellite images. A critical aspect of this process has been sourcing relevant data for training the models, which has proven challenging due to the limited availability of free data. Furthermore, a significant obstacle in the analysis has been the lack of validation and calibration data.

To address these challenges, we developed a visualization dashboard that facilitates the comparison of satellite images, instance segmentation results, and the progression of the deforestation index as defined in Eq. 2. This dashboard serves to illustrate the predictive performance of the models.

Table 2. Comparison between the image segmentation models feature pyramid networks (FPN) and pyramid attention networks (PAN). Each model was run for a total of 100 epochs.

Model	Validation Loss	Validation Accuracy	Execution time
U-Net	**0.06538**	**0.9773**	8h34m
ResNet-34 + U-Net	0.07431	0.9659	2h36m
ResNet-50 + U-Net	0.1086	0.9521	3h45m
EfficientNet-B2 + U-Net	0.09296	0.9627	3h53m
ResNet-34 + FPN	0.09503	0.9175	2h4m
ResNet-50 + FPN	0.1118	0.9214	2h56m
EfficientNet-B2 + FPN	0.08254	0.9362	3h28m
ResNet-34 + PAN	0.09208	0.9546	8h10m
ResNet-50 + PAN	0.3643	0.9466	8h1m
EfficientNet-B2 + PAN	0.1054	0.9583	5h48m

The dashboard was implemented as a web application using Gradio, allowing users to select from five municipalities: two located in Brazil and one each in Guatemala, Peru, and the Dominican Republic. These municipalities were chosen due to their high number of coffee and cocoa producers and documented evidence of deforestation. Figure 3 illustrates the model's deforestation predictions for a municipality in Brazil, demonstrating its efficacy in detecting deforestation in these regions. The visualization dashboard is available online as a HuggingFace space[1] in a restricted access mode.

By comparing Fig. 3a, which shows the municipality from satellite images, with the model's prediction output (Fig. 3b), it is evident that large portions of the image have been accurately segmented as deforested areas, including cities, towns, and agricultural land. Additionally, the prediction has correctly identified forested areas on the right-hand side of the municipality, corresponding to the satellite image. Beyond segmenting satellite images, the web application also features a time series graph that allows users to visualize the progression of deforestation over time, from July 2019 to July 2022, as illustrated in Fig. 4.

Our analysis revealed significant negative impacts resulting from poor satellite image quality. As shown in Fig. 4, deforestation initially increases in the region but then plateaus towards the end of 2020. This plateau can be attributed to the degraded quality of satellite images from 2022, which have compromised the model's performance. Furthermore, satellite images with extensive cloud cover presented significant challenges, despite the various mitigation strategies previously detailed. When clouds obscure the ground, the model can only access pixel information related to the clouds, which hinders its ability to accurately predict forested and deforested areas.

---

[1] https://huggingface.co/spaces/inriachile/fairtrees.

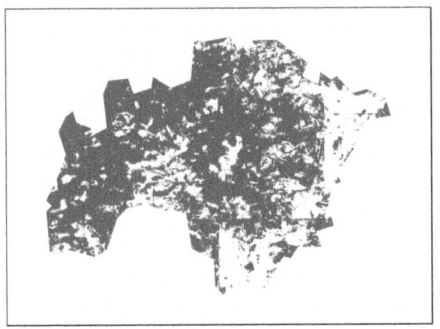

(a) A mosaic of images obtained from Sentinel2 around July 2019. This mosaic was selected for the low cloud coverage.

(b) Output of our trained segmentation model showing in red areas that are classified as deforestation.

Fig. 3. An example prediction of deforestation in Estêvão de Araújo, Araponga, Minas Gerais, Brazil.

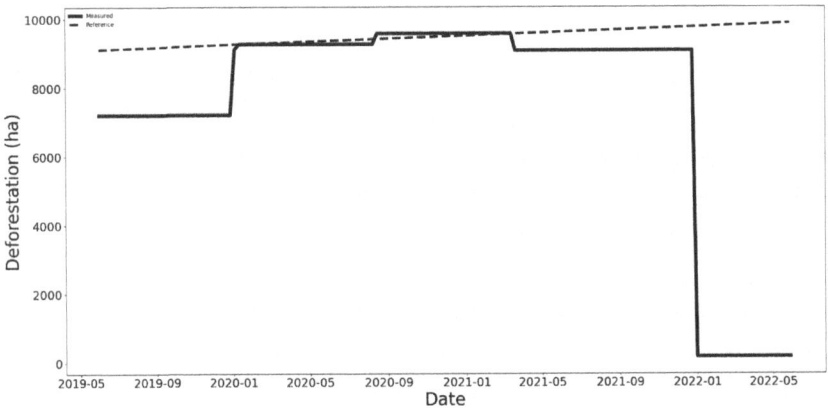

Fig. 4. Predicted deforested area (in red) over time for Minas Gerais, Brazil. As a reference, the reported deforested area is also shown (dotted line). Predictions from May 2019 to January 2022 are consistent with the reported values. (Color figure online)

Another issue identified is the impact of seasonality. In regions with pronounced dry and rainy seasons, deforestation predictions are sensitive to periods of low irrigation. A final issue is the phenomenon of tessellation in satellite images. When these images are merged into a single image and processed through the inference pipeline, they exhibit significant degradation and clear markings of the tile borders. This degradation is likely due to inconsistencies in image characteristics such as exposure and illumination.

6   Conclusion

In this study we developed FairTrees, a state-of-the-art instance segmentation model and a web application to detect deforestation. Our approach leverages high-resolution satellite imagery from Sentinel-1 and Sentinel-2, incorporating both spatial and temporal analyses to accurately detect and quantify deforested areas. The deep learning model implemented proved highly effective in distinguishing between forested and deforested regions.

The quality of satellite images is crucial for accurate deforestation detection, necessitating thorough preprocessing to address issues such as luminosity variations, cloud cover, and seasonal changes. Our visualization dashboard offers policymakers and environmental managers an intuitive tool for monitoring deforestation, aiding informed decision-making. While the model demonstrated strong performance, future improvements should focus on integrating additional data sources, enhancing image quality management, and addressing class imbalance to boost accuracy. FairTrees marks a significant advancement in deforestation monitoring, providing a reliable, scalable, and practical tool for conservation efforts in the LAC region. Ongoing development will be essential to tackle the persistent challenges of deforestation and protect vital forest ecosystems.

Acknowledgments. This work is funded by ANID Strengthening R&D capabilities Program CTI230007 Inria Chile, Inria Challenge OcéanIA, and the Latin American and Caribbean Network of Fair Trade Small Producers and Workers (CLAC).

Disclosure of Interests. The authors have no competing interests to declare that are relevant to the content of this article.

References

1. Abid, A., Abdalla, A., Abid, A., Khan, D., Alfozan, A., Zou, J.: Gradio: hassle-free sharing and testing of ml models in the wild (2019). https://doi.org/10.48550/arXiv.1906.02569
2. Abu, I.O., Szantoi, Z., Brink, A., Robuchon, M., Thiel, M.: Detecting cocoa plantations in côte d'ivoire and ghana and their implications on protected areas. Ecol. Ind. **129**, 107863 (2021)
3. Alzu'bi, A., Alsmadi, L.: Monitoring deforestation in jordan using deep semantic segmentation with satellite imagery. Eco. Inform. **70**, 101745 (2022)
4. Bragagnolo, L., da Silva, R.V., Grzybowski, J.M.V.: Amazon and atlantic forest image datasets for semantic segmentation. Zenodo (2021). https://doi.org/10.5281/zenodo.4498086
5. Cao, J., Chen, Q., Guo, J., Shi, R.: Attention-guided context feature pyramid network for object detection (2020). https://doi.org/10.48550/arXiv.2005.11475
6. Deng, J., Dong, W., Socher, R., Li, L.J., Li, K., Fei-Fei, L.: ImageNet: a large-scale hierarchical image database. In: 2009 IEEE Conference on Computer Vision and Pattern Recognition, pp. 248–255 (2009). https://doi.org/10.1109/CVPR.2009.5206848

7. Economic Commission for Latin America and the Caribbean of the United Nations: Forest loss in Latin America and the Caribbean from 1990 to 2020: The statistical evidence. ECLAC Statistical Briefings (2021). https://hdl.handle.net/11362/47152

8. Gorelick, N., Hancher, M., Dixon, M., Ilyushchenko, S., Thau, D., Moore, R.: Google earth engine: planetary-scale geospatial analysis for everyone. Remote Sens. Environ. (2017). https://doi.org/10.1016/j.rse.2017.06.031

9. He, K., Zhang, X., Ren, S., Sun, J.: Deep residual learning for image recognition (2015). https://doi.org/10.48550/arXiv.1512.03385

10. Hu, M., Li, Y., Fang, L., Wang, S.: A2-FPN: attention aggregation based feature pyramid network for instance segmentation. In: Proceedings of the IEEE/CVF Conference on Computer Vision and Pattern Recognition (CVPR), pp. 15343–15352 (June 2021)

11. Huang, G., Liu, Z., van der Maaten, L., Weinberger, K.Q.: Densely connected convolutional networks (2016). https://doi.org/10.48550/arXiv.1608.06993

12. Irvin, J., et al.: ForestNet: classifying drivers of deforestation in Indonesia using deep learning on satellite imagery (2020). https://doi.org/10.48550/arXiv.2011.05479

13. Isaienkov, K., Yushchuk, M., Khramtsov, V., Seliverstov, O.: Deep learning for regular change detection in Ukrainian forest ecosystem with sentinel-2. IEEE J. Selected Topics Appli. Earth Observat. Remote Sensing **14**, 364–376 (2021). https://doi.org/10.1109/JSTARS.2020.3034186

14. Jamaluddin, I., Chen, Y.N., Ridha, S.M., Mahyatar, P., Ayudyanti, A.G.: Two decades mangroves loss monitoring using random forest and landsat data in east luwu, indonesia (2000–2020). Geomatics **2**(3), 282–296 (2022)

15. John, D., Zhang, C.: An attention-based u-net for detecting deforestation within satellite sensor imagery. Int. J. Appl. Earth Obs. Geoinf. **107**, 102685 (2022)

16. Kaselimi, M., Voulodimos, A., Daskalopoulos, I., Doulamis, N., Doulamis, A.: A vision transformer model for convolution-free multilabel classification of satellite imagery in deforestation monitoring. IEEE Trans. Neural Netw. Learn. Syst. (2022)

17. Li, H., Xiong, P., An, J., Wang, L.: Pyramid attention network for semantic segmentation (2018). https://doi.org/10.48550/arXiv.1805.10180

18. Li, R., Wang, L., Zhang, C., Duan, C., Zheng, S.: A2-FPN for semantic segmentation of fine-resolution remotely sensed images. Int. J. Remote Sens. **43**(3), 1131–1155 (2022)

19. Lin, T.Y., Dollár, P., Girshick, R., He, K., Hariharan, B., Belongie, S.: Feature pyramid networks for object detection (2016). https://doi.org/10.48550/arXiv.1612.03144

20. Marin, M.: Deforestation: can you find deforestation on satellite imagery? Kaggle (2020). https://kaggle.com/competitions/deforestation

21. Masolele, R.N., et al.: Spatial and temporal deep learning methods for deriving land-use following deforestation: a pan-tropical case study using landsat time series. Remote Sens. Environ. **264**, 112600 (2021)

22. OpenStreetMap contributors: Planet dump retrieved from https://planet.osm.org (2017)

23. Ronneberger, O., Fischer, P., Brox, T.: U-net: Convolutional networks for biomedical image segmentation (2015). https://doi.org/10.48550/arXiv.1505.04597

24. Tan, M., Le, Q.V.: EfficientNet: rethinking model scaling for convolutional neural networks (2019). https://doi.org/10.48550/arXiv.1905.11946

Ensembling Convolutional Neural Networks for Human Skin Segmentation

Patryk Kuban and Michal Kawulok(✉)

Department of Algorithmics and Software, Silesian University of Technology,
Gliwice, Poland
michal.kawulok@polsl.pl

Abstract. Detecting and segmenting human skin regions in digital images is an intensively explored topic of computer vision with a variety of approaches proposed over the years that have been found useful in numerous practical applications. The first methods were based on pixel-wise skin color modeling and they were later enhanced with context-based analysis to include the textural and geometrical features, recently extracted using deep convolutional neural networks. It has been also demonstrated that skin regions can be segmented from grayscale images without using color information at all. However, the possibility to combine these two sources of information has not been explored so far and we address this research gap with the contribution reported in this paper. We propose to train a convolutional network using the datasets focused on different features to create an ensemble whose individual outcomes are effectively combined using yet another convolutional network trained to produce the final segmentation map. The experimental results clearly indicate that the proposed approach outperforms the basic classifiers, as well as an ensemble based on the voting scheme. We expect that this study will help in developing new ensemble-based techniques that will improve the performance of semantic segmentation systems, reaching beyond the problem of detecting human skin.

Keywords: Skin segmentation · Convolutional neural networks · Ensemble learning · Color features · Grayscale features

1 Introduction

Human skin detection consists in taking a binary decision on whether the skin is present in digital imagery data at different granularity levels (a pixel, an image region, a whole image, or a video sequence) [13]. Commonly, it is performed at a pixel level and therefore consists in segmenting skin regions [28]. Skin segmentation plays a pivotal role in gesture recognition [21], objectionable content filtering [10], privacy protection [26], selective image compression [22], skin rendering [20], and more [25]. The recent techniques underpinned with deep learning allow for extracting skin regions from color [14,28] and grayscale [18,29] images,

© The Author(s), under exclusive license to Springer Nature Switzerland AG 2025
L. Correia et al. (Eds.): IBERAMIA 2024, LNCS 15277, pp. 185–196, 2025.
https://doi.org/10.1007/978-3-031-80366-6_16

however it remains a challenging problem of computer vision that requires a thorough visual scene understanding, as many other problems related with semantic segmentation [15].

1.1 Related Work

The first attempts to detect skin in digital images relied on modeling the skin color in a variety of color spaces [16]. By applying a set of handcrafted rules [3] or a learned model [7], every pixel can be classified as skin or non-skin based on its position in the color space. The large amount of pixels in the training set coupled with low dimensionality of color spaces (commonly limited to three dimensions) made the Bayesian classifier (BC) highly effective in this case [7,19]. However, as skin appearance varies across different individuals and image acquisition conditions, its color alone is not a sufficiently discriminating feature for segmenting the skin regions effectively. Therefore, features extracted from a wider context, capturing the texture or geometrical properties, allow for enhancing the segmentation performance [8].

As demonstrated in an extensive experimental study by Lumini et al. [13], the handcrafted features have been surpassed by those learned with convolutional neural networks (CNNs). A network-in-network (NiN) architecture introduced by Kim et al. [9] outperformed the BC and the techniques based on handcrafted textural features, as well as the VGG network [27] trained for segmenting skin areas. Zuo et al. proposed to combine the CNN with a recurrent network [31], and Arsanal et al. demonstrated the benefits of applying residual connections for this purpose [1]. These methods were subsequently outperformed with the Skinny network introduced by Tarasiewicz et al. [28], which is a lightweight U-Net architecture with inception modules and dense blocks. The segmentation outcome can also benefit from appropriate postprocessing [2], including the use of morphological filters [12]. In [4], color attention mechanism was proposed to decrease the computational burden and improve the real-time performance.

There were also some efforts aimed at improving the training process, including coupling body and skin detection in a semi-supervised fashion [6] and exploring data augmentation techniques to improve the robustness against varying lighting conditions [30]. Appropriate data augmentation performed in the spectral dimension may also help in learning color invariant features [29]. Another possibility explored in the literature is to rely exclusively on grayscale images— this was motivated by the fact that human observers are capable to identify the skin regions using textural and context information [24]. Paracchini et al. proposed to train an encoder-decoder architecture with grayscale images, thus making the segmentation fully independent from color information. The grayscale image does not contain any additional information compared with the color one, so the networks trained from color images may also exploit the features present in grayscale images. However, the color-based features are definitely easier to extract and without appropriate guidance, the geometrical and texture features present in the grayscale image may not be fully exploited during training. While several image segmentation networks can be treated as an ensemble whose responses are averaged to improve the final segmentation [17], to our best knowledge it has not been attempted to combine networks trained from grayscale and color information.

1.2 Contribution

In the research reported here, we address the identified research gap related with extracting and coupling different types of skin-presence features. In particular, our contribution can be summarized in the following points.

1. We propose a new approach towards constructing ensembles of homogeneous image segmentation networks trained in different ways that are combined in a sequential manner to improve the segmentation outcome. We demonstrate that this approach is more effective than voting-based ensembles.
2. We exploit the simple BC-based skin segmentation for training multiple CNNs that are later treated as an ensemble.
3. We demonstrate that the CNNs trained from grayscale and color images are focused on different features, thus they can be effectively combined within the proposed ensemble scheme.

2 Proposed Approach

The proposed method consists in preparing a set of diverse deep CNN models, whose outcomes are fed to the second-level CNN that integrates the first-level decisions. In Sect. 2.1, we outline the exploited baseline techniques, namely the Skinny CNN [28] that serves as a base model in our ensemble and BC-based skin segmentation [19] which we employ to increase the diversity of the trained models. The details of our approach to constructing skin segmentation ensembles is presented in Sect. 2.2.

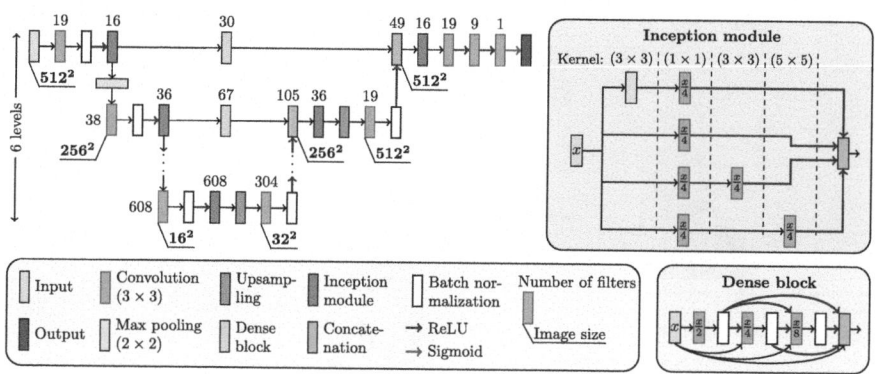

Fig. 1. Architecture of the exploited Skinny network for human skin segmentation.

2.1   Baseline Techniques

We have selected the Skinny network [28] to construct our CNN ensemble, as it is easy to train while offering the state-of-the-art performance (Fig. 1). It is a lightweight U-Net architecture [23] composed of six levels and $7.5 \cdot 10^6$ learnable parameters. We used Skinny both to obtain the base classifiers, as well as for aggregating the decisions to obtain the final segmentation result. By default, it processes color images with red, green and blue (RGB) channels, but it can be straightforwardly applied to processing data of different modality, including single-channel grayscale images.

To employ the BC classifier for skin segmentation [19], at first we compute the histograms for the skin (C_s) and non-skin (C_{ns}) classes and we obtain the probability of observing a given color value (v) in the C_x class as: $P(v|C_x) = C_x(v)/N_x$, where $C_x(v)$ is the number of pixels in the x-th class having v color and N_x is the number of pixels in that class. The probability that a pixel of v color presents skin is then obtained as: $P(C_s|v) = \frac{P(v|C_s)P(C_s)}{P(v|C_s)P(C_s)+P(v|C_{ns})P(C_{ns})}$. The *a priori* probabilities are commonly set to $P(C_s) = P(C_{ns}) = 0.5$.

2.2   Ensembles for Skin Segmentation

The proposed ensemble is composed of two levels—at the first one, we prepare a set of Skinny CNNs trained to focus on different features, which are combined using another Skinny CNN trained to aggregate the first-level responses into the final segmentation map. At first, we prepared two different base models trained from color and grayscale images, thus obtaining Skinny-RGB and Skinny-GS models, respectively. To further diversify the base models, we exploit the BC-based skin segmentation, as depicted in Fig. 2. The BC splits every color image into skin (P_B^S) and non-skin (P_B^{NS}) regions, and we train two Skinny models from these two exclusive training sets (Skinny-P_B^S and Skinny-P_B^{NS}, respectively). During training, the Skinny is presented with the whole color image, but the loss function is computed only in the areas indicated as skin and non-skin by the BC

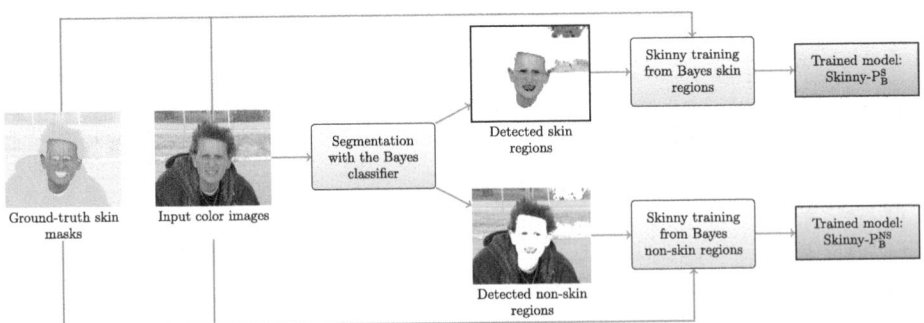

Fig. 2. The BC-based skin segmentation employed to split the input images into skin and non-skin regions, from which two different Skinny models are trained.

for Skinny-P_B^S and Skinny-P_B^{NS}, respectively. In this way, Skinny-P_B^S is trained to eliminate the BC's false positives, while Skinny-P_B^{NS} is focused on refining false negative pixels. As the color feature is already exploited by the BC, we may expect that these two models are more focused on extracting other types of features. As the datasets obtained after BC-based stratification are imbalanced (especially Skinny-P_B^{NS}), for the loss function we couple the binary cross entropy with the Dice coefficient during training.

The final ensemble (Ensemble-S) is outlined in Fig. 3. Three Skinny models trained from the grayscale images and from the regions indicated as skin and non-skin by the BC are employed to retrieve three skin-presence probability maps. These maps are stacked together to form an input (three-channel image) that is presented to the second-level Skinny which renders the final segmentation outcome. During our research, we have considered different variants of ensembling a variety of Skinny models which are discussed later in Sect. 3.

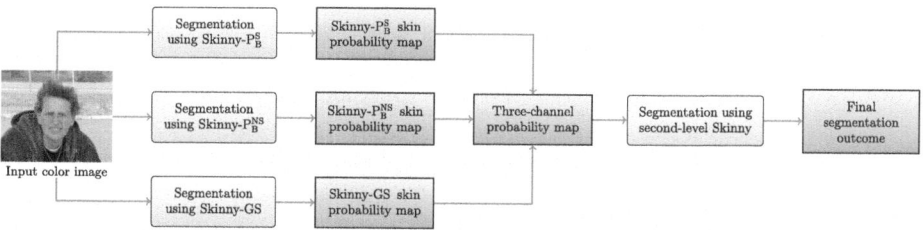

Fig. 3. A flowchart presenting skin segmentation using Ensemble-S. The probability maps retrieved with three Skinny models trained from the grayscale image and from two exclusive regions in the color image, indicated as skin and non-skin by the BC, are aggregated using the second-level Skinny model to obtain the final segmentation outcome.

3   Experimental Validation

We have validated the proposed approach using the ECU skin dataset [19] that contains 4000 color images which we have randomly split into training, validation

Table 1. The investigated base models for skin segmentation.

Method name	Description
BC	Bayesian classifier trained over the RGB color space
Skinny-RGB	Skinny network trained over the RGB color space (a baseline Skinny model)
Skinny-GS	Skinny trained from the grayscale images (single-channel)
Skinny-P_B^S	Skinny trained over the regions indicated as skin by the BC
Skinny-P_B^{NS}	Skinny trained over the regions indicated as non-skin by BC

Table 2. The constructed ensembles investigated in our experimental study that aggregate the grayscale channel (GS) and skin probability maps extracted using different Skinny models.

Method name	Ensembling technique	Input channel				
		GS	Skinny-RGB	Skinny-GS	Skinny-P_B^S	Skinny-P_B^{NS}
Ensemble-S	Skinny			✓	✓	✓
Ensemble-V	Voting			✓	✓	✓
Ensemble-S^{RGB}	Skinny		✓		✓	✓
Ensemble-S^{-S}	Skinny			✓		✓
Ensemble-S$_A$	Skinny	✓			✓	✓
Ensemble-S$_B$	Skinny	✓	✓	✓		
Ensemble-S$_B^{-GS}$	Skinny				✓	✓
Ensemble-BC+S	BC selection				✓	✓

Table 3. Quantitative scores obtained with the base classifiers and with the investigated ensembles for the ECU test set. The best scores in each group are boldfaced.

Method	F-score	Precision	Recall
BC	0.7394	0.7504	0.7288
Skinny-RGB	0.8751	0.8767	0.8752
Skinny-GS	0.8020	0.8146	0.7899
Skinny-P_B^S	0.8690	0.8690	0.8690
Skinny-P_B^{NS}	**0.8789**	**0.8797**	**0.8781**
Ensemble-S	**0.8965**	0.9001	**0.8930**
Ensemble-V	0.8718	**0.9142**	0.8331
Ensemble-S^{RGB}	0.8912	0.8963	0.8860
Ensemble-S^{-S}	0.8913	0.8931	0.8894
Ensemble-S$_A$	0.8893	0.8883	0.8903
Ensemble-S$_B$	0.8917	0.8933	0.8902
Ensemble-S$_B^{-GS}$	0.8864	0.8900	0.8883
Ensemble-BC+S	0.8757	0.8749	0.8764

and test sets with 1750, 250 and 2000 images, respectively. To decrease the computational burden, we downsampled all the images preserving the aspect ratio, so that they do not exceed the size of 256×256 pixels. The Skinny network was trained in 200 epochs with the learning rate set to 10^{-3}. For the quantitative evaluation, we report the precision, recall and F-score metrics, and we also render the precision-recall curves. The investigated base models are enlisted in Table 1 and the constructed ensembles are shown in Table 2. Skinny requires around 50 ms when run using NVIDIA RTX 2080Ti (11 GB VRAM) GPU to process a single image.

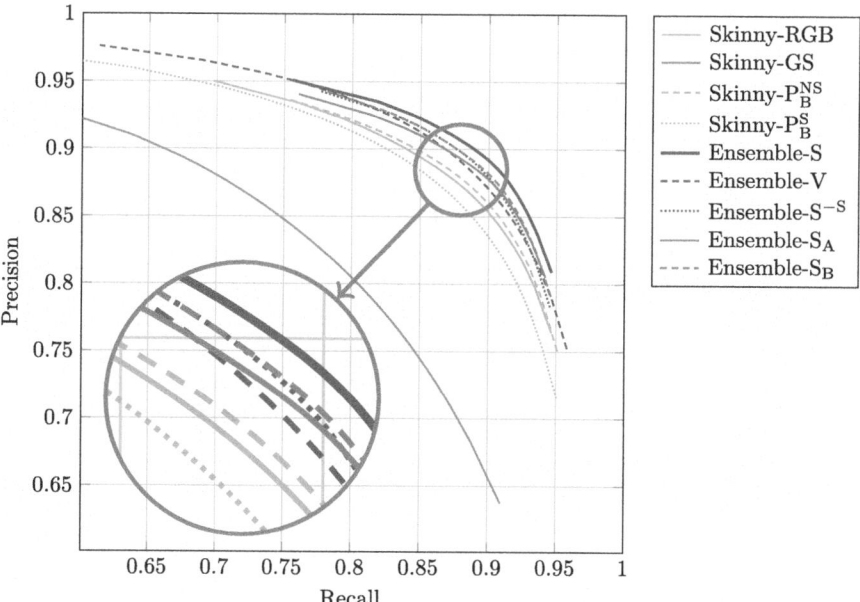

Fig. 4. Precision-recall curves for the selected base models and ensemble classifiers for the ECU test set.

The obtained quantitative scores are reported in Table 3 and the precision-recall curves for the selected models are presented in Fig. 4. All the Skinny models, including Skinny-GS that does not exploit color information, render better scores than the BC, and the best result is retrieved with the model trained from the BC's non-skin regions (Skinny-P_B^{NS}). In order to justify our choice of the ensembling technique, we have considered several different variants of combining multiple segmentation outcomes (Table 2). In Ensemble-V, we combine the outcomes retrieved with the same base models relying on the majority voting instead of employing the second-level Skinny model. Although the voting increases the precision compared with Ensemble-S, the obtained recall is much worse, resulting in the F-score lower than for the base models. We also substituted the Skinny-GS model with Skinny-RGB in Ensemble-S^{RGB}, and we tried excluding the weakest Skinny-P_B^S base model (Ensemble-S^{-S})—in both cases, the results were worse than for Ensemble-S. In Ensemble-S_A, we used the grayscale image instead of the Skinny-GS outcome and in Ensemble-S_B, we excluded the models trained from the data stratified using the BC (also without using the grayscale channel in Ensemble-S_B^{GS}). Finally, in Ensemble-BC+S, we apply a two-branch approach, in which we select either Skinny-P_B^S or Skinny-P_B^{NS} model for the pixels classified as skin or non-skin by the BC, respectively. All the ensembles that exploit Skinny at the second level are better than the best base model, with Ensemble-S rendering the highest F-score. Although the quantitative differences between Ensemble-S and other ensembles are not large,

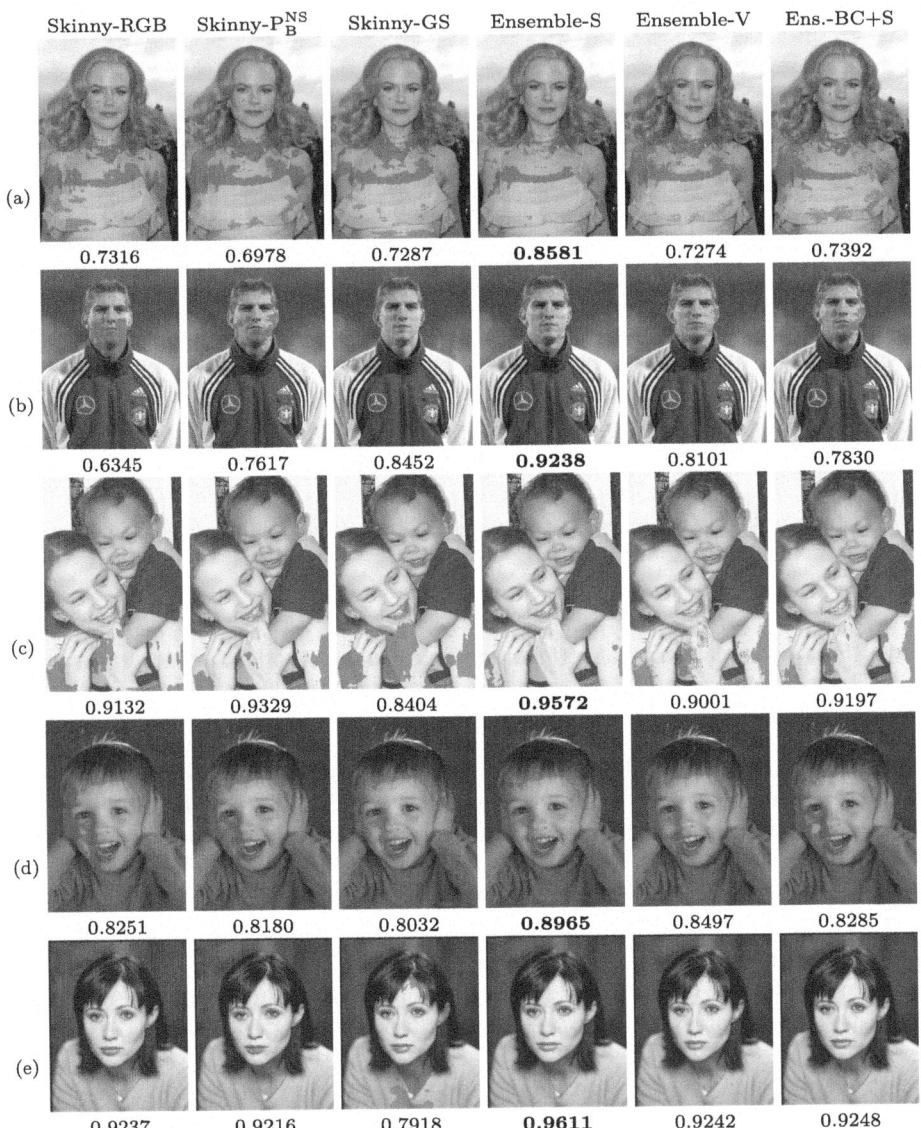

Fig. 5. Examples of skin segmentation performed using different techniques. False positives are indicated with red color and false negatives with blue color, and F-score is provided under each outcome (the best score for each example is boldfaced). (Color figure online)

they are all statistically significant according the the two-tailed Wilcoxon test ($p < 0.05$). Also, the precision-recall curve for Ensemble-S dominates all the remaining techniques.

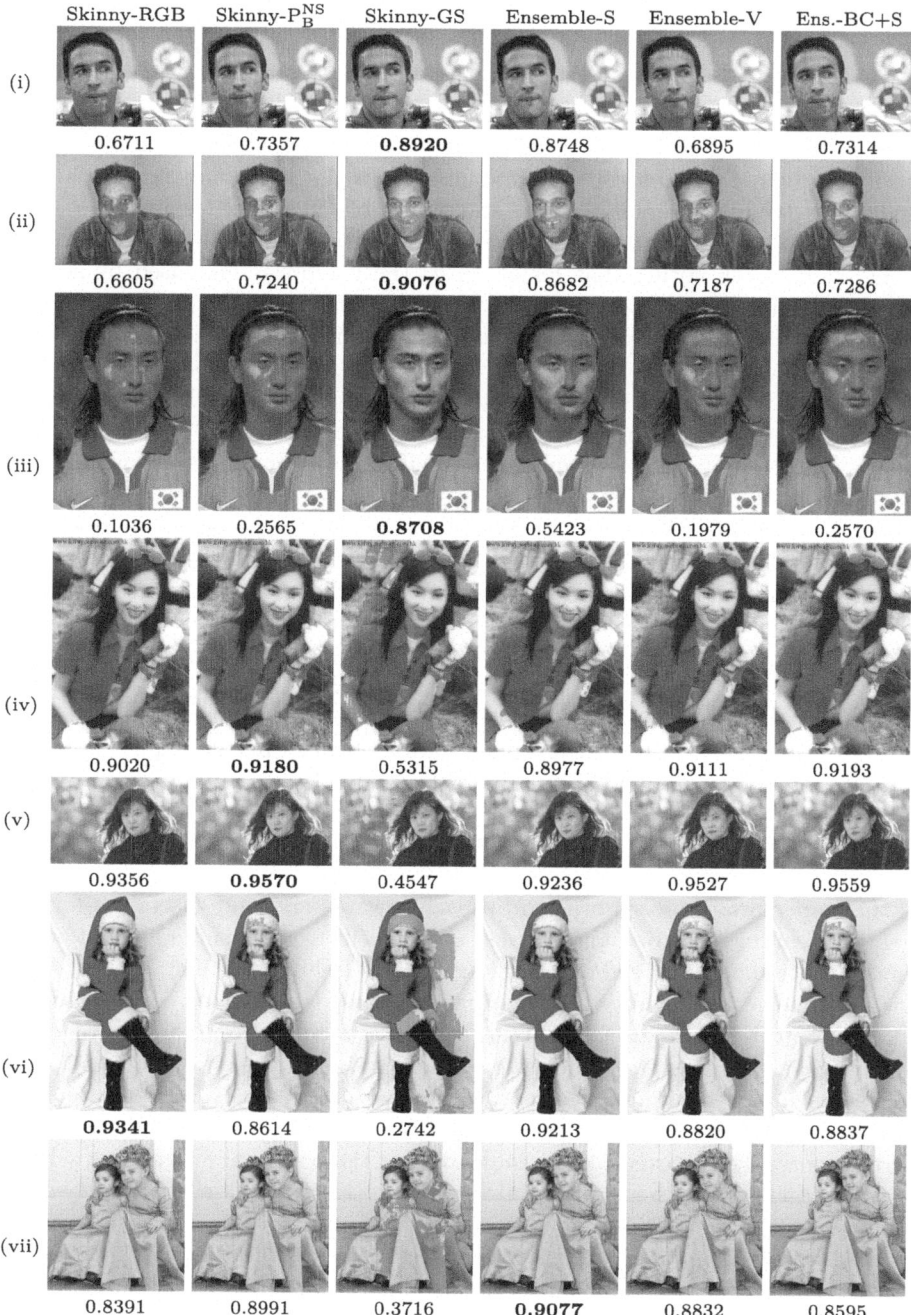

Fig. 6. Examples of skin segmentation for more challenging cases performed using different techniques. False positives are indicated with red color and false negatives with blue color, and F-score is provided under each outcome (the best score for each example is boldfaced). (Color figure online)

In Figs. 5 and 6, we present examples of the segmentation outcome that allow us to analyze the results qualitatively. We showcase the results obtained using the vanilla Skinny-RGB model along with Skinny-P_B^{NS} and Skinny-GS base models, and we also include the outcomes retrieved using three ensembles of different kind: the Skinny-based Ensemble-S, the voting-based Ensemble-V, and the two-branch Ensemble-BC+S. In Fig. 5, both the RGB and grayscale base models render rather good results, and Ensemble-S allows for improving the outcome in all the cases, reducing both false positives (red) and false negatives (blue). It is worth noting that for (a, b, d) the F-score is increased by around 0.1 compared with the base models, and it is also definitely more improved compared with the two remaining ensembles. In Fig. 6, we present several more challenging cases, in which either the color models (i, ii, iii) or the Skinny-GS model (iv, v, vi, vii) fail to segment the skin regions accurately. Although the proposed Ensemble-S surpasses all the base models only in (vii), it is clear that in the remaining cases it manages to effectively combine the base models, rendering the scores that are close to the best one, better than the voting and two-branch schemes. Even in the example (iii), where the color-based models fail completely in detecting the skin, Ensemble-S manages to reduce the false negatives significantly (which does not happen for the two remaining ensembles). Also, false negatives are substantially reduced for (i) and (ii) relying on the outcome retrieved with Skinny-GS. In cases (vi) and (vii), Skinny-GS renders high false positive errors, which are successfully compensated by the ensemble models—as the remaining base models segment the skin regions mostly correctly here, the voting and two-branch schemes are also much better than Skinny-GS, however they are still outperformed with Ensemble-S.

4   Conclusions and Future Work

In this paper, we proposed a new approach towards creating ensembles for segmenting skin regions from color images. We benefit from learning a set of homogeneous base models (based on the Skinny CNN architecture [28]), whose diversity is achieved by guiding the training to obtain models that are focused on different features. Throughout our experimental study, we demonstrated quantitatively and qualitatively that the models trained from grayscale and color images, including training data stratification relying on the BC, are complementary and therefore may be effectively combined relying on ensemble learning. Furthermore, the reported results indicate that the proposed ensembling scheme which extracts the contextual spatial features of the initial skin probability maps using a second-level CNN outperforms the pixel-wise voting that is already a well-established approach for semantic segmentation [17].

The reported results are encouraging and in our future work we want to further extend the proposed approach. Our study was limited to creating two-level CNN ensembles—we expect that multi-level structures, possibly composed of lighter models, may improve the segmentation outcome even further, while making it possible to control the balance between the processing time and segmentation quality. While ensembling neural networks is quite convenient for creating resource-frugal solutions [11], in our approach, the segmentation outcome can be gradually refined when passed through subsequent levels of the sequentially-connected CNNs, so by choosing arbitrarily the number of processing levels employed during segmentation, the time vs. quality trade-off can be managed effectively. Also, it may be worth considering to fine-tune the connected CNNs, thus resulting in a multi-branch architecture. Last, but not least, we believe the developed approach can also be applied to other semantic segmentation tasks [5], in particular those that can benefit from heterogeneous image fusion.

Acknowledgements. This work was supported by the National Science Centre, Poland, under Research Grant 2022/47/B/ST6/03009.

References

1. Arsalan, M., Kim, D., Owais, M., Park, K.: OR-Skip-Net: outer residual skip network for skin segmentation in non-ideal situations. Expert Syst. Appl. **141**, 112922 (2020)
2. Baldissera, D., Nanni, L., Brahnam, S., Lumini, A.: Postprocessing for skin detection. J. Imaging **7**(6), 95 (2021)
3. Chen, Y., Hu, K., Ruan, S.: Statistical skin color detection method without color transformation for real-time surveillance systems. Eng. Appl. Artif. Intell. **25**(7), 1331–1337 (2012)
4. Ding, S., Liu, Z., Lei, Z.: A color attention mechanism based on yes color space for skin segmentation. J. Real-Time Image Proc. **20**(3), 53 (2023)
5. Hao, S., Zhou, Y., Guo, Y.: A brief survey on semantic segmentation with deep learning. Neurocomputing **406**, 302–321 (2020)
6. He, Y., et al.: Semi-supervised skin detection by network with mutual guidance. In: Proc. IEEE International Conference on Computer Vision (ICCV), pp. 2111–2120 (2019)
7. Jones, M., Rehg, J.: Statistical color models with application to skin detection. Int. J. Comput. Vis. **46**, 81–96 (2002)
8. Kawulok, M., Kawulok, J., Nalepa, J.: Spatial-based skin detection using discriminative skin-presence features. Pattern Recogn. Lett. **41**, 3–13 (2014)
9. Kim, Y., Hwang, I., Cho, N.: Convolutional neural networks and training strategies for skin detection. In: IEEE International Conference on Image Processing (ICIP), pp. 3919–3923 (2017)
10. Lee, J., Kuo, Y., Chung, P., Chen, E.: Naked image detection based on adaptive and extensible skin color model. Pattern Recogn. **40**, 2261–2270 (2007)
11. Liang, F., Shen, C., Yu, W., Wu, F.: Towards optimal power control via ensembling deep neural networks. IEEE Trans. Commun. **68**(3), 1760–1776 (2019)
12. Lumini, A., Nanni, L., Codogno, A., Berno, F.: Learning morphological operators for skin detection. arXiv preprint arXiv:1908.03630 (2019)
13. Lumini, A., Nanni, L.: Fair comparison of skin detection approaches on publicly available datasets. Expert Syst. Appl. **160**, 113677 (2020)
14. Ma, C., Shih, H.: Human skin segmentation using fully convolutional neural networks. In: Proc. IEEE Global Conference on Consumer Electronics (GCCE), pp. 168–170 (2018)

15. Muhammad, K., et al.: Vision-based semantic segmentation in scene understanding for autonomous driving: Recent achievements, challenges, and outlooks. IEEE Trans. Intell. Transp. Syst. **23**(12), 22694–22715 (2022)

16. Naji, S., Jalab, H.A., Kareem, S.A.: A survey on skin detection in colored images. Artif. Intell. Rev. **52**(2), 1041–1087 (2019)

17. Nanni, L., Lumini, A., Fantozzi, C.: Exploring the potential of ensembles of deep learning networks for image segmentation. Information **14**(12), 657 (2023)

18. Paracchini, M., Marcon, M., Villa, F., Tubaro, S.: Deep skin detection on low resolution grayscale images. Pattern Recogn. Lett. **131**, 322–328 (2020)

19. Phung, S., Bouzerdoum, A., Chai, D.: Skin segmentation using color pixel classification: analysis and comparison. IEEE Trans. Pattern Anal. Mach. Intell. **27**(1), 148–154 (2005)

20. Poirier, G.: Human skin modelling and rendering. Master's thesis, University of Waterloo (2004)

21. Rahim, M.A., Miah, A.S.M., Sayeed, A., Shin, J.: Hand gesture recognition based on optimal segmentation in human-computer interaction. In: Proc. IEEE International Conference on Knowledge Innovation and Invention (ICKII), pp. 163–166 (2020)

22. Rodrigues, J., Puech, W., Bors, A.: Selective encryption of human skin in JPEG images. In: IEEE International Conference on Image Processing (ICIP), pp. 1981–1984 (2006)

23. Ronneberger, O., Fischer, P., Brox, T.: U-Net: Convolutional networks for biomedical image segmentation. CoRR abs/ arXiv: 1505.04597 (2015)

24. Sarkar, A., Abbott, A., Doerzaph, Z.: Universal skin detection without color information. In: Proc. IEEE/CVF Winter Conference on Applications of Computer Vision (WACV), pp. 20–28 (2017)

25. Senouci, B., Rouis, H., Han, D., Bourennanea, E.: A hardware skin-segmentation ip for vision based smart ADAS through an FPGA prototyping. In: Proc. International Conference on Ubiquitous and Future Networks (ICUFN), pp. 197–199 (2017)

26. Shifa, A., Imtiaz, M.B., Asghar, M.N., Fleury, M.: Skin detection and lightweight encryption for privacy protection in real-time surveillance applications. Image Vis. Comput. **94**, 103859 (2020)

27. Simonyan, K., Zisserman, A.: Very deep convolutional networks for large-scale image recognition. arXiv preprint arXiv:1409.1556 (2014)

28. Tarasiewicz, T., Nalepa, J., Kawulok, M.: Skinny: a lightweight U-Net for skin detection and segmentation. In: IEEE International Conference on Image Processing (ICIP), pp. 2386–2390 (2020)

29. Xu, H., Sarkar, A., Lynn Abbott, A.: Color invariant skin segmentation. In: 2022 IEEE/CVF Conference on Computer Vision and Pattern Recognition Workshops (CVPRW), pp. 2905–2914 (2022)

30. You, H., Lee, K., Oh, J., Lee, E.C.: Efficient and low color information dependency skin segmentation model. Mathematics **11**(9), 2057 (2023)

31. Zuo, H., Fan, H., Blasch, E., Ling, H.: Combining convolutional and recurrent neural networks for human skin detection. IEEE Signal Process. Lett. **24**(3), 289–293 (2017)

Risk Assessment for UAV Autonomous Landing in Urban Environments Using Semantic Segmentation

Jesús Alejandro Loera-Ponce[1], Diego A. Mercado-Ravell[1,2] ⓘ,
Israel Becerra[2](✉) ⓘ, and Luis Manuel Valentin-Coronado[3] ⓘ

[1] Centro de Investigación en Matemáticas A.C., CIMAT, Zacatecas, ZAC 98160,
Mexico
[2] Investigadores CONAHCYT, Centro de Investigación en Matemáticas, A.C.,
Guanajuato, GTO 36023, Mexico
israelb@cimat.mx
[3] Investigadores CONAHCYT, Centro de Investigaciones en Óptica, A.C., CIO,
Aguascalientes, AGS 20200, Mexico

Abstract. We address the vision-based autonomous landing problem
in complex urban environments using deep neural networks for semantic
segmentation and risk assessment for Uncrewed Aerial Vehicles (UAVs).
We propose employing the SegFormer for the semantic segmentation of
the complex and unstructured urban environments. The assessment is
done when the video feed from an RGB camera on the UAV is segmented into typical urban classes and then mapped to a level of risk,
considering in general, potential damage to property, the drone itself
or endanger people. This approach yields valuable information that can
be leveraged when UAV missions need to land in urban spaces, especially when in emergency resulting from system failures or human errors.
The proposed strategy is validated through several case studies, demonstrating the huge potential of semantic segmentation-based strategies to
determine the safest landing areas for landing, which will help unleash
the full potential of UAVs on civil applications within urban areas.

Keywords: Autonomous Landing · Semantic Segmentation · UAVs ·
Risk Assessment · Emergency Landing

1 Introduction

The rapid development of Unmanned Aerial Vehicles (UAVs) has allowed an
increase in their use in all sorts of applications, especially civilian, ranging from
simple entertainment up to more serious applications like infrastructure inspection [19]. However, even with the significant progress on UAVs, their deployment
in populated areas has been hindered due to safety concerns. Most applications
are restricted to areas devoid of people due to the latent risk that flying objects
pose on them or on property in the event of malfunction or human errors, limiting

the full potential of UAVs within urban spaces. Accordingly, regulations require to present risk mitigation measures to local authorities to allow flight missions in civilian spaces. The European Union Aviation Safety Agency (EASA) [3] advises to minimize the time flying near people. The U.S. Federal Aviation Administration (FAA) [4], forbids to operate in sustained flight over "open-air assemblies of human beings" unless otherwise authorized by the Administrator. Following those rules, the Joint Authorities for Rulemaking on Unmanned Systems (JARUS) created a document of guidelines for aerial work operations called Specific Operations Risk Assessment (SORA) [11].

Existing rotary-wing UAVs offer limited Autonomous Landing (AL) capabilities, capping application development in urban populated areas. Although some sophisticated drones may be equipped with safety mechanisms for emergency situations, the majority of them are constrained by return-to-home protocols in the event of malfunction. This involves recording the take-off location and returning to it in a straight path if required. The device may also emit lights and sounds to alert unaware bystanders in the area. Nevertheless, such strategies do not warranty the safety of the landing missions, especially during a system failure or a battery shortage. More advanced drones could search for pre-known visual tags for landing, but this involves previously preparing the area, which is not practical for most applications.

Hence, AL in urban environments is a complex and open research problem, due to the enormous variety of scenarios present in cities, where the UAV operates within an unknown and unstructured environment under varying conditions, navigating alongside individuals, vehicles and other moving entities with unknown dynamics. When a dangerous situation occurs, the drone must find the best available space to land without causing any kind of damage, or the area where the risk of material damage and hurting people is minimized. This has proven to be very challenging due to the diverse and changing nature of urban populated areas. Proposing a solution that can be readily deployed, leveraging the existing components integrated into commercial UAVs, has the potential to greatly enhance their practical applications in urban settings.

In this regard, this work proposes the use of a monocular camera attached to the drone coupled with computer vision and state-of-the-art Deep Learning algorithms to analyze the scene and assess the risk of accidents that may produce significant material damage or hurt human beings during emergency landing situations. For this, it will be used a semantic segmentation network trained on aerial views of urban areas with labels at the pixel level of the most common classes for this context, such as people, cars, roads, vegetation, etc. The semantic segmentation model will provide pixel-level classification of RGB images, which in turn will be used to create a risk map assessing the probability of material damage or human accidents in the areas beneath the drone. This risk map can be used later to perform further analysis and help the UAV in the decision-making process to minimize the risk of accidents, making the system more reliable and resilient. No 3D information of terrain will be accounted for (solely considerations that objects generate a specific risk level because of verticality), and suppose air work space is clear so there is no risk of air collision.

This proposal aims to provide valuable insight into an Uncrewed Aircraft System (UAS) for a comprehensive understanding of the dynamic urban environment, particularly in areas where people and motor vehicles contribute to a multifaceted scenario. This work is organized as follows. Section 2 contains a discussion on related work on autonomous landing in general. Section 3 describes the different modules that comprise the proposed strategy. Experimental results are discussed and evaluated in Sect. 4. Finally, the conclusions and future work are presented in Sect. 5.

2   Related Works: Vision-Based Autonomous Landing

Since [13] helped pave the way to Deep Learning, advances in the field of Computer Vision (CV) and Machine Learning (ML) allowed the evolution of Autonomous Landing (AL) tasks from simple state machines using ground beacons and radio telemetry sensors such as GPS, to a great diversity of techniques using only monocular cameras most of the time.

Some of the first works reported in the literature were devoted to the detection of previously known visual markers [5], to avoid hazardous terrain [10] or to use visual feedback to provide additional awareness to UAVs. These methods were based on extracting and then tracking traditional features from images obtained from a monocular camera as done in [12]. However, vision was used mainly as an assistance system.

Furthermore, works based on people detection, such as [1,15], allowed the incorporation of a more safety-oriented approach to AL, achieved by detecting people to avoid human accidents. Using basic computational vision tools, it is possible to find areas to land in simple and controlled scenarios. However, those tools do not generalize enough the features like people's heads, do not account for changes in perspective and do not deal with occlusions and scene movement.

Subsequent works like [6] started focusing on improving human detection by obtaining Density Maps from their location on the images, generating counting and position data using classical tools as Probability Density Functions. But this approach still fails with perspective variations. To deal with this, [7] uses Convolutional Neural Networks (CNN) to create the Density Maps with improved generalization in a wide variety of environments and situations. Then, works such as [14,20,21] use Deep Neural Networks (DNN) to find landing zones free of people. In [21] a lightweight network is used to overestimate the Density Map of the crowd to prevent the drone from landing near any person. [14] uses the pitch and altitude information provided by the drone to feed the DNN to draw the Density Map on the heads plane preventing the system underestimating people. [6,7] use a Bayesian-Loss network to infer the Density Maps with high precision. Finally, the authors in [8] use the Density Maps to propose Safe Landing Zones by finding areas devoid of people and tracking them with Kalman Filters coupled with the Hungarian Algorithm.

More recently, the latest developments in Machine Learning (ML) have been integrated into AL schemes. [18] proposes implementations of state-of-the-art techniques and tools in the search for Safe Landing Zone (SLZ) for UAVs. The authors propose the use of different modules to create an annotated surroundings map, with landing sites and no-fly zones around people locations, and continuously check for paths towards SLZ. The approach considers finding people and different kinds of risks while also checking for a flat safe landing zone. Nevertheless, the design of the system is focused in a sparse and less dynamic scenarios. Finally, [2] uses a binocular-LiDAR to find flat areas to land with a network doing semantic segmentation simultaneously. This double function helps the system understand the morphology and semantic features of the terrain, selecting SLZ in complex environments with very high accuracy. This method is good to give terrain context to the UAV, but people and other urban agents are not accounted for.

Previous works are restricted to avoid accidents involving humans but do not account for potential material damage or indirect accidents involving people. Although some advances can be found in the recent literature, Autonomous Landing in dynamic urban unstructured environments continues to be a very complex open problem, where a lot of research effort is still required in order to have more reliable solutions. In this regard, the present work proposes a vision-based strategy for risk assessment in the context of autonomous landing in complex unstructured urban scenarios. The strategy consists in using a semantic segmentation DNN based on modern Visual Transformers to provide context to the UAV about the area below it. This is done by identifying classes of terrain and objects that can be used in the decision-making processes to avoid human accidents and material damage. Furthermore, the inferred image from semantic segmentation is then mapped to a heat map that assesses the potential risk of accidents involving people, directly or indirectly, as well as potential material damage to expensive infrastructure or obstacles such as cars or to the drone itself. The proposed strategy can be used as a framework for smart decision-making during emergency landing protocols. The risk map can be coupled with multi-objective optimization techniques to take decisions on where to land, prioritizing the zones with less risk, even if they are not ideal areas to land. The strategy is validated in various case studies, showcasing the huge potential of this proposal to assess risk confidently for autonomous landing in urban complex unstructured scenarios.

3   General Approach

The proposed risk assessment approach for Autonomous Landing is composed of two stages: Semantic Segmentation and Clustering by Risk Level.

3.1   Semantic Segmentation

Semantic Segmentation (SS) is the task of classifying images into different categories at a pixel-wise level. This method was chosen because it produces better context and location information of objects. Historically, this task has greatly improved with the use of CNNs, notably with the introduction of the U-Net proposed in [17]. More recently, a technique developed for Natural Language Processing known as Transformer have generated great interest because it efficiently process vast and complex datasets. This model proposed by [22] infused into other fields because it learns the dependencies between distant positions of the input and generates context. These context-aware capabilities have allowed state-of-the-art models of SS to be very efficient and accurate. They also have a straightforward implementation that enables many downstream applications like the one presented in this paper. The metrics used to measure the performance of the Semantic Segmentation are mean Intersection over Union ($mIoU$) and Dice Coefficient (DSC, also known as $F1$ score) which are similarity coefficients between the predicted classes and the ground truth.

Model. The model used in this proposal is the SegFormer . This sophisticated segmentation architecture that integrates Transformers and Multilayer Perceptron (MLP) blocks was proposed by [23]. Its design offers the advantages of minimal parameters, fast training, and extensive computational capacity. The network utilizes an auto-encoder structure which incorporates four transformer blocks that can produce features at multiple scales which are then combined by an MLP. Its desing offers a series of models with different sizes, from B0 to B5, allowing the flexibility of choosing the model best suited for the application needs. The one selected for this proposal was B0, which is the smallest with $3.8M$ parametres, but the fastest version with $8.4G$ FLOPs to run in real-time on an embedded system onboard UAVs. The model was implemented using the HuggingFace library of the SegFormer [16].

Fig. 1. Samples of aerial images in the Semantic Drone Dataset, captured in urban environments, using a down-looking camera mounted on a UAV. These images are labeled with several common classes found in urban areas, including grass, pavement, people, cars, vegetation, etc.

Dataset. A key element for DNNs is the dataset. To take advantage of the Seg-Former, it was finetuned on the Semantic Drone Dataset (SDD) [9], a dataset designed to enhance the safety of autonomous drone flight and landing procedures through improved semantic comprehension of urban environments. It is publicly available and contains 400 high-resolution aerial images of 6000×4000

pixels ($24Mpx$) from bird's eye view, acquired at altitudes between 5 to 30 meters above ground. Figure 1 depicts some samples of this dataset. Annotations are provided for SS and include 23 of the most common classes from outdoor environments. Table 1 presents a comprehensive overview of the dataset classes alongside their respective colors used in the dataset. Additionally, the proposed risk classification is displayed along with its corresponding color, as it will be explained in Subsect. 3.2

To finetune the SegFormer, the SDD was divided 80% for training, 10% for validation, and 10% for testing. Then, Data Augmentation was used to artificially increase the number and diversity of the data. The transformations applied for the augmentation included only brightness and contrast change, rotations, flips and random crops, which are transformations likely to occur in footage obtained with UAVs.

Table 1. Classes used for semantic segmentation and their corresponding Risk Levels.

Semantic Segmentation			Risk Level		Semantic Segmentation			Risk Level	
Class	Label	RGB	Class	RGB	Class	Label	RGB	Class	RGB
0	background				12	water			
1	dirt				13	wall			
2	grass		0		14	window			
3	gravel				15	door		3	
4	ar-marker				16	bicycle			
5	paved-area		1		17	tree			
6	vegetation				18	bald-tree			
7	rocks				19	obstacle			
8	pool				20	dog			
9	roof		2		21	car		4	
10	fence				22	conflicting			
11	fence-pole				23	person		5	

3.2   Clustering by Risk Level

A large number of classes may provide unnecessary information and affect negatively its metrics. Besides, to assess the risk of an scene using all classes is more challenging, so it is useful to group classes with a similar risk profile.

It was decided to define 6 different levels of risk, based on possible human injuries (which establishes the highest risk), material damage, preservation of the integrity of the drone and the identification of ideal zones for landing, being 0 the most ideal area to land, and 5 the riskiest, as defined in Table 2. This parameter

Table 2. Risk level definition.

Risk Level	Definition
0	Ideal landing zones, including grass, dirt, gravel, etc.
1	Low level of material damage or damage to the UAV itself.
2	Moderate risk of losing or damaging the UAV, along with low risk of material damage.
3	Important material damage, the imminent risk of losing or critically damaging the drone, and the moderate risk of indirectly hurting people. It includes the classes water, tree, wall, among others.
4	This level comprises indirect risk of hurting people, direct risk of hurting fauna, and conflicting regions where there is uncertainty about the presence of people in the area.
5	The maximum risk and considers the direct risk of hurting people.

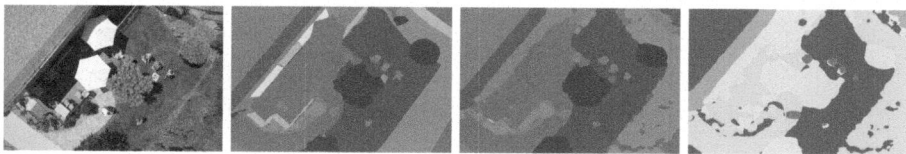

Fig. 2. The process of inference and risk assessment: first image shows the input image, the second is the ground truth reference of semantic segmentation, the third picture shows the prediction inferred from model (colored for visual interpretation), and last image presents the mapping to risk levels, where red regions are the riskiest and blue the safest. Reference for colors in Table 1 (Color figure online).

was selected to reflect the intrinsic risk class proposed by SORA but it could be easily adjusted for different applications. Depending on the risk associated, the segmented classes are classified as observed in Table 1.

In this work, conversion from semantic segmentation classes to risk levels is performed after inference, so the model can learn the characteristics of the same kind of classes and not lose generalization. Figure 2 shows an example of the different stages of the risk assessment strategy.

4 Experimental Validation

4.1 Experimental Setup

Development, training the model and testing were carried out with Python virtual environments on two desktop computers running on Ubuntu 20.04 equipped with NVIDIA graphic cards. An NVIDIA Jetson AGX Xavies (32 GB) was used only to obtain the minimum frame rate at which the SS model and the risk assessment proposal will run. The model was developed using Pytorch libraries and the SegFormer implementation by the HuggingFace community. At the HuggingFace Hub, NVIDA offers a MiT-B0 SegFormer encoder already pretrained which was finetuned on the SDD.

4.2   Experimental Results

Different training parameters were tested to obtain the best results of segmentation of the SegFormer on the SDD. The model produced the best results at 30 epochs. After integrating the SS model to the risk assessment pipeline, the system produced a minimum frame rate of 14 FPS at the Jetson Xavier, showing its viability to integrate the proposal to the flight system of certain UAVs. In terms of semantic segmentation, the results given by the model trained match the reported in the literature.

Table 3. Metrics obtained from experiments.

Experiment	$Acc.$	$Mean\ IoU$	$F1\ Score$	$Bal\ Acc.$
Class Segmentation	0.8831	0.4138	0.4888	0.5754
Risk Levels	0.8976	0.5811	0.6725	0.7110

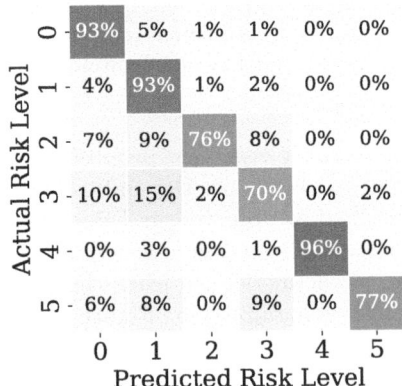

Fig. 3. Confusion matrix of the testing set on the six risk levels.

Evaluation on the test set produced the results shown in Table 3. It can be seen that grouping categories by risk level helps to improve the metrics obtained, caused in part for clustering neighboring classes. Figure 3 shows the confusion matrix when the prediction is transformed to risk levels. Risk levels 0 and 1, which could be considered desirable locations to land, had a correct labeling above 90%. The same for level 4. However, in the case of risk levels 2, 3 and 5, accuracy falls below 80%, something not so desirable, especially for the most critical level 5. Most classes and by extension, risk levels, were normally predicted to be of the classes with lower risk for up to 15%. Those risk levels contain the most common and extensive classes in the datasets, i.e. grass, dirt, paved-area, etc. The reason to this is that normally in the field of semantic segmentation, "background" classes as those mentioned earlier, compose the major part of the images due to its nature, technically unbalancing the dataset, in an unavoidable way. Even more, those classes encircle all the others, this being the main reason why all other classes have a considerable percentage being predicted as those. Thus, it is preferable to have more false positives on the high risk level classes than false negatives. In this regard, one strategy is to dilate the high levels areas, so the resulting prediction could be safer for critical classes.

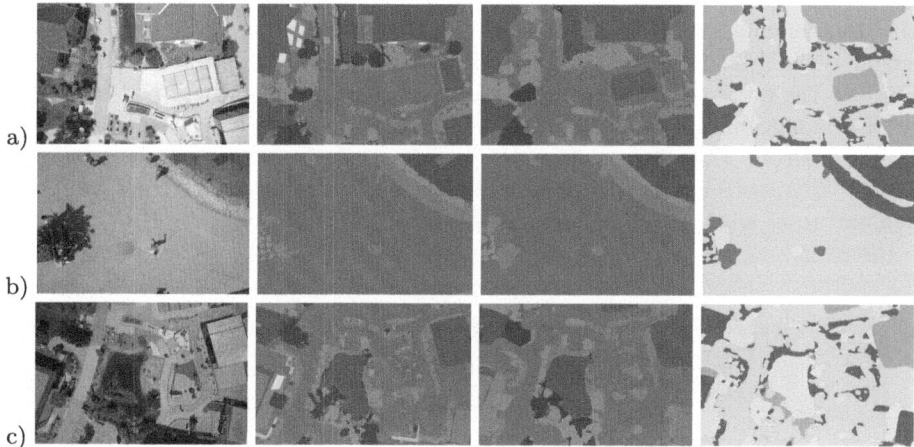

Fig. 4. Experimental results obtained for 3 different case studies within the test set. The first column presents the original image, the ground-truth annotated image is depicted in the second column, and the third column contains the inferred semantic segmentation. The last column shows the risk levels, where the red color represents the highest risk, passing through orange, yellow, green, cyan, to blue, which represents the safest areas. Reference for colors in Table 1 (Color figure online).

Figure 4 shows a variety of examples from the test dataset. One of the most notorious details from those is that aerial views from high altitude, the person class is not very detailed, sometimes do not even appearing in the prediction, as can be shown in rows *a* and *c*. This is caused by the small scale of these classes on the image and, as previously mentioned, their low count against low-risk classes. This might seem problematic, but in a real scenario it would be expected that while approaching land, the model would start identifying people as seen on *b*. Again, this is a problem of generalization and could be addressed by training the model with a larger dataset containing images taken from higher altitudes. In contrast, images closer to the ground have fewer classes and the predictions are more accurate. This is important because in a landing attempt, the UAV will descend start receiving more confident information to adjust the last maneuvers. Future work can consider the altitude to produce a confidence index so that the UAV will take the appropriate action based on it. In general, this work show the huge potential of SS-based techniques to provide useful information to UAS to determine more confidently the best options to land in case of an emergency.

Figure 5 show examples of inference from the trained model, but from different scenarios not contained on the original dataset. In this case, the uncontrolled and different scenarios of the pictures produce a variety of results. It can be seen that for test *b* the results are quite good and very useful: cyclist are clearly found by the model and details of surroundings are well defined. Results could be this good because this picture is similar to the ones at the SDD. By contrast, the other shows important misclassifications that results in an incorrect risk level map. In the case at row *a*, one car in the middle is identified as an AR-Marker, which puts it at very low risk. It also confuses an air exhaust as a person, although this does not cause any risk for the operation. These behaviors showcase the great complexity of the task of autonomous landing in real unstructured urban scenarios. Finally, example *c* presents objects not considered in the Semantic Drone Datasets, such as the train and its tracks. It can be seen that the model tries to categorize some features, as the train and its track as paved area and roof, which do not have the risk level that normally would be assigned to them.

Fig. 5. Experimental results obtained for different case studies using images with different aerial views not included in the training set. The first column presents the input image; then the inferred semantic segmentation map is depicted in the second column. Risk levels for the third column and a composition of the original image masked with the risk level map in the last column to visualize directly the mapping done by the model. Reference for colors in Table 1.

Those extra examples demonstrate that the model and the categorization of the classes is a viable tool for UAVs when they require to land autonomously in their surroundings. Some issues with generalization could be solved mainly by training in a more diverse dataset.

Based on the proposed method, there might be some approaches to land a UAS in an urban area, e.g., choosing the region with the lowest risk or lowest accumulated risk over a given time, or using Pareto optimality between the candidate spots. However, the proposal offers very useful insight to start with.

5   Conclusion

In this article, a semantic segmentation-based risk assessment strategy for autonomous landing in complex unstructured urban scenarios was proposed. First, a state-of-the-art visual transformer network was chosen and trained for input of semantic segmentation images from a UAS. Then, six different risk levels are obtained from the semantic segmentation model output, which gives more certainty to find high risk areas to avoid in case of an autonomous landing.

The method presented in this work was evaluated along different case studies and showcases the huge potential of using semantic segmentation and risk grouping to provide important contextual information to the UAV about the ground risk in its surroundings, useful for decision making, particularly in emergency landing situations. The work presented here adds robustness to the system in case of an emergency landing is needed, improving its SORA score and opening the way for applications in complex urban environments of being authorized.

Future work includes expanding the training dataset in order to capture more diversity of urban scenarios, camera angles, class variability, and more height. Also, it is of interest to use the risk assessment information to choose a landing spot using some formal decision process that accounts for uncertainty and other variables. Finally, it is planned to validate each module of the proposal in real settings, running each step in simulations and embedded in a UAV.

Acknowledgments. This work was supported by the the Mexican National Council of Humanities, Science and Technology (CONAHCYT), and by the U.S. Office of Naval Research (ONR) under award N62909-24-1-2001 with Arturo Ayon as technical representative.

Disclosure of Interests. The authors have no competing interests to declare.

References

1. Chan, A.B., Liang, Z.S.J., Vasconcelos, N.: Privacy preserving crowd monitoring: counting people without people models or tracking. In: 2008 IEEE Conference on Computer Vision and Pattern Recognition (CVPR) (2008)
2. Chen, L., Xiao, Y., Yuan, X., Zhang, Y., Zhu, J.: Robust autonomous landing of UAVs in non-cooperative environments based on comprehensive terrain understanding. Sci. Chin. Inf. Sci. (2022)

3. European Union Aviation Safety Agency: Unmanned Aircraft Systems (UAS) (2019)
4. Federal Aviation Administration: Unmanned Aircraft Systems (UAS). U.S. Department of Transportation (2021)
5. Garcia-Pardo, P.J., Sukhatme, G.S., Montgomery, J.F.: Towards vision-based safe landing for an autonomous helicopter. Rob. Autonom. Syst. (2002)
6. Gonzalez-Trejo, J., Mercado-Ravell, D.: Dense crowds detection and surveillance with drones using density maps. In: 2020 International Conference on Unmanned Aircraft Systems (ICUAS) (2020)
7. Gonzalez-Trejo, J.A., Mercado-Ravell, D.A.: Lightweight density map architecture for UAVs safe landing in crowded areas. Intell. Rob. Syst. (2021)
8. González-Trejo, J., Mercado-Ravell, D., Becerra, I., Murrieta-Cid, R.: On the visual-based safe landing of UAVs in populated areas: a crucial aspect for urban deployment. IEEE Rob. Autom. Lett. (2021)
9. Graz University of Technology: Semantic Drone Dataset (2019). http://dronedataset.icg.tugraz.at/
10. Johnson, A., Montgomery, J., Matthies, L.: Vision guided landing of an autonomous helicopter in hazardous terrain. In: Proceedings of the 2005 IEEE International Conference on Robotics and Automation (ICRA) (2005)
11. Joint Authorities for Rulemaking of Unmanned Systems: Specific Operations Risk Assessment (SORA) (2019)
12. Kendoul, F., Fantoni, I., Lozano, R.: Adaptive vision-based controller for small rotorcraft UAVs control and guidance. IFAC Proceedings Volumes (2008)
13. Krizhevsky, A., Sutskever, I., Hinton, G.E.: ImageNet classification with deep convolutional neural networks. In: Advances in Neural Information Processing Systems (2012). https://doi.org/10.1145/3065386
14. Liu, W., Lis, K., Salzmann, M., Fua, P.: Geometric and physical constraints for drone-based head plane crowd density estimation. In: 2019 IEEE/RSJ International Conference on Intelligent Robots and Systems (IROS) (2019)
15. Rabaud, V., Belongie, S.: Counting crowded moving objects. In: 2006 IEEE Conference on Computer Vision and Pattern Recognition (CVPR) (2006)
16. Rogge, N.: Fine-tune a semantic segmentation model with a custom dataset (2022). https://huggingface.co/blog/fine-tune-segformer
17. Ronneberger, O., Fischer, P., Brox, T.: U-Net: Convolutional networks for biomedical image segmentation. In: International Conference on Medical Image Computing and Computer-Assisted Intervention (2015)
18. Symeonidis, C., Kakaletsis, E., Mademlis, I., Nikolaidis, N., Pitas, I.: Vision-based UAV safe landing exploiting lightweight deep neural networks. In: Proceedings of the 2021 4th International Conference on Image and Graphics Processing (2021)
19. Teixeira, K., Miguel, G., Silva, H.S., Madeiro, F.: A survey on applications of unmanned aerial vehicles using machine learning. IEEE Access (2023)
20. Tzelepi, M., Tefas, A.: Human crowd detection for drone flight safety using convolutional neural networks. In: 2017 25th European Signal Processing Conference (EUSIPCO) (2017)
21. Tzelepi, M., Tefas, A.: Graph embedded convolutional neural networks in human crowd detection for drone flight safety. IEEE Trans. Emerg. Top. Comput. Intell. (2019)
22. Vaswani, A., et al.: Attention is all you need. Adv. Neural Inf. Process. Syst. (2017)
23. Xie, E., Wang, W., Yu, Z., Anandkumar, A., Alvarez, J.M., Luo, P.: SegFormer: simple and efficient design for semantic segmentation with transformers. Adv. Neural Inf. Process. Syst. (2021)

Visual SLAM in Underground Environments: Preliminary Results

Bliman Federico$^{(\boxtimes)}$ ⓘ, Monzon Pablo, and Llofriu Martin

Facultad de Ingenieria, UdelaR, San Juan, Uruguay
`federico.bliman@fing.edu.uy`

Abstract. Subterranean Simultaneous Localization and Mapping (Sub-T SLAM) is an active and challenging research area that is generally tackled with high-end hardware and multiple sensors. In this paper, we focus on RGB-only visual SLAM in underground environments, specifically in loop closure detection under opposite viewpoints. We propose a solution using Visual Transformers (ViT) features and simple feature matching that shows promising results.

Keywords: Underground SLAM · Computer Vision · ViT

1 Introduction

Autonomous robots are widely used for multiple applications nowadays, in part thanks to significant advances in SLAM over the last few decades. SLAM is crucial for robots to navigate safely by understanding their surroundings. However, some scenarios, such as underground environments, present special challenges where SLAM solutions are still not robust. This paper explores loop closure for Subterranean Simultaneous Localization and Mapping (Sub-T SLAM) using a Visual-Inertial (VI) setup, one of the most common and simple SLAM setups. Loop closure, an important task in SLAM, involves recognizing previously visited places to correct drift, merge maps, and optimize existing ones. This requires algorithms for encoding images or shapes and databases optimized for similarity metric search.

Sub-T SLAM presents extra challenges, such as the absence of GPS, poor lighting, poor signal propagation, and uniform and repeated shapes and textures. Driven by the DARPA challenge [6], multiple solutions have been developed for these environments using LiDAR, cameras, and multiple robots [5,11,12,16,21]. Our tests with state-of-the-art VI SLAM solutions yielded regular results in odometry but poor results in loop closure, motivating our work in loop closure in Sub-T SLAM, specifically in sewage pipes.

This paper is organized as follows: Sect. 2 reviews previous work on SLAM, emphasizing Sub-T and VI SLAM, and the evolution from traditional techniques to AI-based approaches. Section 3 introduces our novel approach for loop closure detection. In Sect. 4, we present our initial results. Finally, we outline our future research directions.

L. Correia et al. (Eds.): IBERAMIA 2024, LNCS 15277, pp. 209–220, 2025.
https://doi.org/10.1007/978-3-031-80366-6_18

2   Related Work

SLAM research began in the late 80 s [25] and gained traction with the availability of computational power in the early 2000s [19]. VI SLAM is usually classified based on image matching techniques: feature-based or dense methods. Feature-based methods, such as PTAM [17] and ORB-SLAM 3 [3], extract sparse features for matching and ego-motion estimation. Dense methods, like DTAM [15,20], use dense image pixels for motion estimation under photometric consistency. Both methods use factor-graphs to model robot poses, movements, and landmarks [7].

AI advancements rapidly transformed SLAM. AI-based feature extraction and matching, like SuperPoint [9] and SuperGlue [23], and CNN-based odometry and depth prediction [1,28], now dominate the field. Visual Transformers (ViTs) [10] are increasingly replacing CNNs.

2.1   Loop Closure in Sub-T SLAM Using RGB Cameras

Loop closure techniques vary based on available sensors and processing capabilities. We focus on RGB cameras, which are widely accessible and offer a spectrum of image analysis algorithms. However, underground navigation with RGB cameras faces challenges such as the absence of natural light, reliance on artificial lighting, and highly repetitive patterns in tunnels. Despite these challenges, cameras are valuable sensors for Sub-T SLAM. Our goal is to develop robust loop closure techniques using RGB camera data, exploring strategies to overcome limitations posed by artificial lighting and repetitive patterns. Integrating depth information can further enhance accuracy and reliability.

2.2   Traditional Techniques

Feature extraction is the cornerstone of traditional computer vision-based loop closure. Techniques such as SIFT (Scale-Invariant Feature Transform) [18], SURF (Speeded Up Robust Features) [2], and ORB (Oriented FAST and Rotated BRIEF) [22] are commonly used. These algorithms detect distinct points in images and describe them in a way that is invariant to scale, rotation, and, to some extent, lighting conditions. The matching process involves comparing these features between different images to identify common points, thereby facilitating the recognition of a previously visited location. These techniques are enhanced by the use of Bag of Words (BoW) [13], which involves creating a 'vocabulary' of visual features extracted from a set of training images. When a new image is encountered, its features are described in terms of this vocabulary, enabling the system to recognize places by comparing the distribution of these visual words.

2.3   The Use of AI in Loop Closure

The advent of Artificial Intelligence in SLAM has revolutionized the approach to loop closure. AI, particularly deep learning techniques like CNNs, has introduced new capabilities for feature extraction, matching, and scene recognition, overcoming some limitations of traditional computer vision methods.

CNNs in Feature Extraction and Matching. CNNs have emerged as powerful tools for feature extraction in images. Unlike traditional algorithms, CNNs learn feature descriptors directly from the data, making them more adaptable to a variety of environments. In the context of loop closure, CNNs can be trained to identify and describe key points in images, facilitating robust matching even in challenging conditions.

SuperPoint and SuperGlue. Significant advancements in AI-based loop closure include the development of SuperPoint [9] and SuperGlue [23]. SuperPoint is a deep learning model designed for keypoint detection and description. It automatically learns to identify points of interest in images, through an iterative process that starts with synthetic shapes, and describes them in a distinctive manner. SuperGlue, on the other hand, is a novel approach for keypoint matching. It uses a graph neural network to predict correspondences between sets of keypoints across different images. The combination of SuperPoint and SuperGlue offers a powerful solution for identifying loop closures with high accuracy.

Leveraging Intermediate Layers of CNNs for Image Descriptions. An emerging area in AI-driven loop closure is the use of intermediate layers of CNNs for generating image descriptions. Specifically, layers like the last convolutional layer or the first fully connected layer are being explored for their potential in capturing rich, abstract representations of images.

The Last Convolutional Layer. The last convolutional layer in a CNN is known for encapsulating high-level features of the input image. These features are the result of successive convolutions and non-linear transformations, and they carry significant spatial information about the image. By extracting feature maps from this layer, we can obtain a comprehensive description of the image, capturing both the local features and the overall spatial layout. This is particularly beneficial for loop closure, as it allows for a more semantic comparison between images, aiding in the recognition of previously visited locations even under varying conditions. These features are not especially robust to viewpoint changes.

The First Fully Connected Layer. The first fully connected (FC) layer in a CNN acts as a high-level feature aggregator. After the convolutional layers have extracted spatial features, the first FC layer combines these features, often leading to a more abstract and compact representation of the image. Utilizing this layer for image description in loop closure can be advantageous as it distills the essence of the image into a format that is conducive to efficient matching and recognition and can cope with very different viewpoints of the same scene.

Advantages in Complex Environments. Utilizing these intermediate layers of CNNs for image description in loop closure is particularly advantageous in complex environments, such as underground settings. These layers capture a blend of local and global features, making the image descriptions more robust to variations in lighting, texture, and perspective. This robustness is critical in environments where traditional feature extraction methods may fail due to repetitive patterns or poor lighting conditions.

Challenges and Considerations. While leveraging intermediate CNN layers for image descriptions shows promise, it also introduces challenges. One significant challenge is the computational cost associated with processing and comparing high-dimensional feature maps. Additionally, the choice of layer and the method of feature extraction need to be carefully considered to ensure that the resulting image descriptions are optimally suited for loop closure.

In summary, the exploration of intermediate layers of CNNs for image descriptions opens new possibilities for enhancing loop closure in SLAM. As research in this area progresses, it could lead to more robust and efficient loop closure methods, particularly in challenging navigation environments.

3   Novel Approach for Visual Loop Closure in Sewage Environments

Our research proposes a novel approach for visual loop closure in challenging subterranean environments like sewages. Recognizing the unique aspect of navigating sewages in opposite directions, our method uses the first fully connected (FC) layer of CNNs for feature extraction, focusing on spatial orientation.

3.1   Methodology

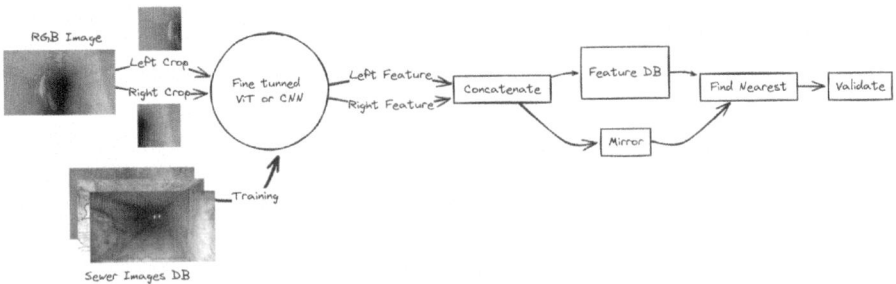

Fig. 1. Loop closure detection pipeline using VGG19 pretrained network or DINO Fine-Tuned ViT

We divide captured images into left and right halves and extract features using the first FC layer of the CNN. Interchanging these features during the matching process allows us to recognize loop closures more effectively when the direction of travel is reversed (Fig. 1).

Our experiments leverage pretrained networks like VGG19 [24], showing promising results, see Fig. 2. Fine-tuning networks with sewage-specific images adapts intermediate representations to capture unique patterns and features. Our initial attempts to use a CNN trained for sewage defect detection with a labeled dataset [14] did not significantly improve accuracy over the pretrained VGG19.

ViTs represent a significant advancement in image matching. They use the transformer's self-attention mechanism to process image data, capturing global context more effectively. ViTs are more computationally intensive than CNNs, but can extract richer features than CNNs in particular scenarios like ours.

Fig. 2. Loop closure guess using VGG19 pretrained network. The same spot seen from opposite directions

3.2   Visual Transformers for Image Matching

ViTs represent a significant advancement in image matching, inspired by the success of the "Attention Is All You Need" paper [27], which introduced the transformer architecture primarily for language processing tasks. Unlike traditional CNNs that process images through localized filters, ViTs utilize the transformer's self-attention mechanism to process image data. This approach allows ViTs to dynamically focus on different parts of an image, assessing relationships and features across the entire image space, which can be especially beneficial for tasks like image matching.

ViTs work by first converting images into sequences of flattened patches and then processing these sequences similarly to tokens in a text sequence in natural language processing. This method not only captures the global context of images more effectively but also enhances the model's ability to generalize from one task to another, thereby improving performance on a variety of computer vision tasks, including image matching. The adaptability and efficacy of ViTs demonstrate their potential as a robust alternative to traditional CNNs in the realm of computer vision.

As ViTs are more computationally and data-intensive than CNNs, a lot of research has been done on how to improve the training process. Solutions like [26] use distillation, taking advantage of the token architecture of attention to improve training performance. A similar idea uses DINO [4] for self-supervised training for image feature extraction. DINO, short for self-distillation with no labels, enables the ViT to further refine its feature extraction capabilities without the need for labeled data. By seeking consistency between the outputs of differently augmented views of the same image, DINO helps the model to focus on the most informative parts of the image and hence extract significant features.

3.3   Fine-Tuning ViT for Sewages

We apply self-supervised fine-tuning methods on a pretrained ViT, initially trained with the DINO framework on the ImageNet dataset [8]. This adapts a general-purpose visual model to the nuanced environments of sewage systems, enhancing its capability to identify and match relevant features effectively.

The ViT model, pretrained on the ImageNet dataset, leverages a broad range of visual features from diverse images. This provides a robust base for subsequent specialization.

The fine-tuning process requires some prior discussion regarding the nature of the use we are going to give the ViT model. In our basic test set, where we go forward and backward in the same pipe, our first objective is to detect loop closures from opposite points of view. We are doing a directed matching by searching nearest neighbors between images, previously split into right and left, and matching right to left and vice versa so that it effectively looks for matching images in opposite directions. In this use case, it may be reasonable to train on the test set if we are doing an offline SLAM. For online SLAM, that is not possible so we have to fine-tune with similar images. We will briefly show in the result section both cases.

4   Results

We present the first experiments to validate our approach and advance towards refining the proposed technique. We worked with 3 different datasets and several training configurations. One of the datasets will be used for testing, given that it is the one with more controlled conditions, and from which we can infer a pseudo ground truth. From it, we will introduce an accuracy metric that we will explain below, to have quantitative results. The objective of the experiments is finding loop-closure candidates when traversing a sewage in opposite directions.

4.1   Datasets

Our experiments are based on 3 Datasets with different characteristics. Two of them were recorded by our team in the same sewage in Montevideo, and the third one is a subset of sewage images from a larger dataset [14].

- Ducto_Buceo_1: 7129 images, 1920×1080 pixels. This dataset was recorded in one of Montevideo's sewages and will be used as the test set. It consists of 7129 images extracted from the RGB video channel. It starts from the main entrance and follows to a secondary pipe. After entering the secondary pipe, the camera looks up to a first duct access chamber and then traverses about 200 m to a second duct access chamber, and goes back to the first duct access chamber. The segment between the two access chambers was traversed at almost constant speed back and forth, so it can easily be determined which images correspond to the same section in each of the passes. This will be used as ground truth and from there we will introduce the accuracy metric we used.
- Ducto_Buceo_2: 33372 images, 1920×1080 pixels. This dataset was recorded in the same sewage, mainly in the same sections. The video has more movement and does not follow a uniform movement pattern and hence will only be used for training purposes.
- Defects_dataset: 40k images, 352×288 pixels. This dataset is a random subset of 40k images from a publicly available dataset presented in [14] with the objective of training CNNs for automatic defect labeling. The quality of the images is much lower than in the previous datasets (Fig. 3).

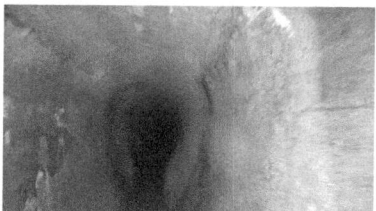 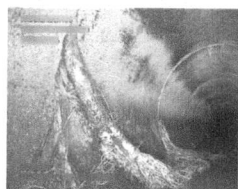

Fig. 3. Sample images from datasets 1 and 2 (left) and dataset 3 (right). Images from our test environment were recorded in higher definition than the Defects_dataset. Although it can be seen that they are similar environments and present the same kind of shapes and textures.

4.2   Algorithm

For finding potential loop closure candidates for each position (image), we will use a ViT to extract features from each image and match them to the most similar image using cosine distance. The features will be computed using a ViT trained using the DINO method, and we will take the class token as the image descriptor. Furthermore, we will split each image in two halves, right and left, and create a feature vector by concatenating both, see Fig. 1. The basic DINO configuration uses the vanilla ViT [10], which inputs 224×224 RGB images and returns an embedding of size 768. So each of our image descriptors will be of length $1586 = 768 * 2$, $D(x) = [D(x_L), D(x_R)]$, and the mirrored embedding $D_M(x) = [D(x_R), D(x_L)]$, being x_L and x_R the left and right crops of image x. In this first stage, we are doing offline processing, so we first calculate the embedding of all images and then we match them.

4.3   Accuracy Metric

As we previously explained, the first dataset that we introduced was recorded in very controlled conditions and hence we can use it as a pseudo ground truth. We will assume that both passes of the segments between the access chambers were traversed at uniform speed. The first duct access is recorded at position p_1, $T = t_1$. From there it walks looking forward to the second access at position p_2, $T = t_2$, and goes back to the first location at $T = t_3$. $t_3 - t_2 = t_2 - t_1$, so we can assume that the speed is almost constant through $[t_1, t_3]$ and $t_x + t_y = t_1 + t_3$, being x an image from the forward pass and y its corresponding image from the backward pass. From this relationship, we will introduce the accuracy metric for these experiments as $acc = \frac{\sum_{t_x=t_1}^{t_3}(t_1+t_3-A)<t_x+t_{M(x)}<(t_1+t_3+A)}{t_3-t_1}$, being A a tolerance parameter. This metric tells us how many of the images were matched within a tolerance of its corresponding section of the opposite pass.

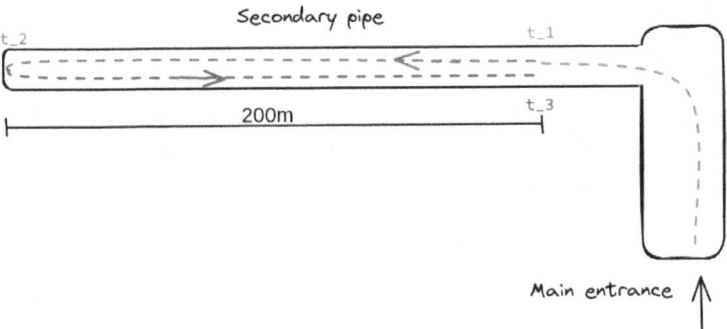

Fig. 4. Test sewage diagram. Ducto_Buceo_1 and Ducto_Buceo_2 were recorded in this location

4.4   Results

We compared the results of the image matching under different fine-tuning conditions always using DINO and the ViT small pretrained on ImageNet as initial conditions. For the purpose of this paper, we will present the results of training on dataset 2 and testing on dataset 1, which we believe is the most representative setup. All the training was done with the default DINO parameters for 10 epochs.

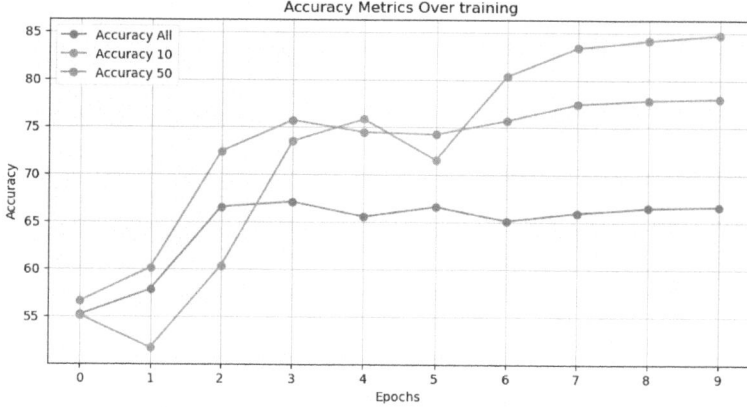

Fig. 5. Accuracy of loop closure matching as a function of training for Dataset 2

Figure 5 shows the accuracy as a function of training. We plot the general accuracy, and also the top 50% and 10%, taking only into account the images whose matching distances were in the top 50% and 10% respectively, see Fig. 7. The system ends up achieving good matching for the top 10%, so setting the correct image threshold would give us quite confident loop closure matches. We can also see the evolution of training in Fig. 6, where we end up with most images matching their corresponding image in the opposite direction or either themselves. Points around the diagonal $x = y$ correspond to images matching themselves or very close images, and points around the line $x = 8300 - y$ correspond to images matching the same section in the opposite pass. Ideally, we would expect most images in the second line, but as some images are highly symmetrical, it is reasonable that some images match themselves. The first 1700 images correspond to the entrance of the pipe and do not have matching images in the opposite view, see Fig. 4.

By looking at Fig. 7, we can also see how the ViT becomes more domain-specific. Most matches start around 0.9 (cosine distance), meaning all images are quite similar to each other as features come from a wider "language" (ImageNet pretrain). As the ViT becomes more sewage-specific, distances across images get bigger, but more precise. Our feature vectors occupy a larger space and hence become richer in describing sewage.

We ran training with other configurations and also learned that:

- Training with the test set may yield worse accuracy due to overfitting, because the right and left crops end up with almost identical features, matching themselves or nearby points instead of the opposite viewpoint as desired.
- Training with images of very different quality is not trivial in ViTs, even if the images are qualitatively similar; we couldn't achieve good results.
- For our specific application, training with half-images of sewage improves the performance of the matching.

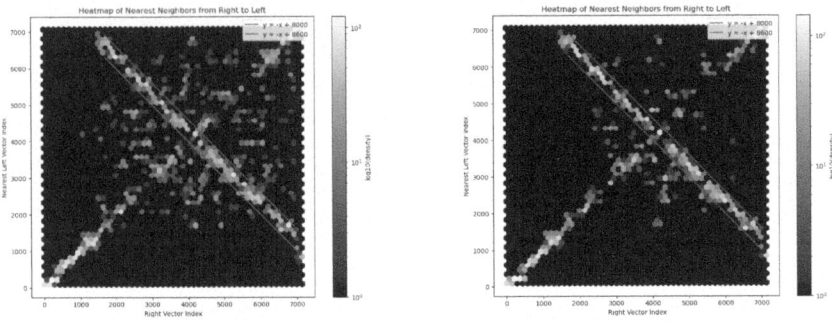

Fig. 6. Heatmap of matching images for epochs 1 and 10. Dataset 2

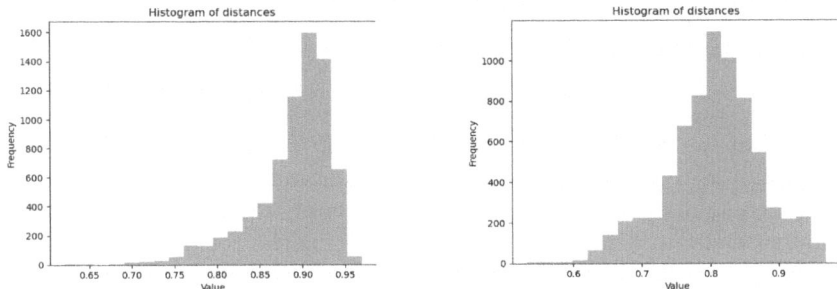

Fig. 7. Distance histogram of matching images for epochs 1 and 10. Dataset 2

– To further improve image matching and facilitate the calculation of transformations between images, the same technique can be employed by dividing each image into more sections and matching these sections symmetrically.

5   Conclusions

Our work demonstrates the potential of using Visual Transformers (ViT) for image matching and loop closure detection in challenging environments such as sewage systems. Our method, which leverages the robustness of ViTs and innovative feature extraction techniques, has shown promising results even under difficult conditions. The ability to accurately identify loop closures from opposite viewpoints represents a significant advancement in Sub-T SLAM. There is still work to be done to integrate this into a full SLAM pipeline, such as performing a geometric check and determining the relative position between matches. Additionally, in the ViT training area, there are still more architectures to explore and hyperparameters to tune. We believe that this technique can help deploy SLAM in specific environments such as sewages, paving the way for more robust and reliable autonomous navigation systems in subterranean settings.

References

1. Almalioglu, Y., Saputra, M.R.U., de Gusmao, P.P.B., Markham, A., Trigoni, N.: GANVO: unsupervised deep monocular visual odometry and depth estimation with generative adversarial networks. In: 2019 International Conference on Robotics and Automation (ICRA), pp. 5474–5480 (2019). arXiv:1809.05786

2. Bay, H., Ess, A., Tuytelaars, T., Van Gool, L.: Speeded-Up Robust Features (SURF). Comput. Vis. Image Underst. **110**(3), 346–359 (2008)

3. Campos, C., Elvira, R., Rodríguez, J.J.G., Montiel, J.M.M., Tard'os, J.D.: ORB-SLAM3: an accurate open-source library for visual, visual-inertial and multi-map SLAM. IEEE Trans. Rob. **37**(6), 1874–1890 (2021). arXiv:2007.11898

4. Caron, M., et al.: Emerging properties in self-supervised vision transformers. In: 2021 IEEE/CVF International Conference on Computer Vision (ICCV), pp. 9630–9640, Montreal, QC, Canada. IEEE (2021)

5. Chang, Y.: LAMP 2.0: a robust multi-robot slam system for operation in challenging large-scale underground environments. arXiv preprint arXiv:2205.13135 (2022)

6. DARPA: Subterranean challenge technical repository (2020)

7. Dellaert, F., Kaess, M.: Factor graphs for robot perception. Found. Trends Robot. **6**(1–2), 1–139 (2017)

8. Deng, J., Dong, W., Socher, R., Li, L.-J., Li, K., Fei-Fei, L.: ImageNet: a large-scale hierarchical image database. In: 2009 IEEE Conference on Computer Vision and Pattern Recognition, pp. 248–255. IEEE (2009)

9. DeTone, D., Malisiewicz, T., Rabinovich, A.: SuperPoint: self-supervised interest point detection and description. In: 2018 IEEE/CVF Conference on Computer Vision and Pattern Recognition Workshops (CVPRW), pp. 337–33712, Salt Lake City, UT, USA. IEEE (2018)

10. Dosovitskiy, A., et al.: An image is worth 16x16 words: transformers for image recognition at scale (2021)

11. Ebadi, K., et al.: Present and future of SLAM in extreme underground environments. arXiv:2208.01787 (2022)

12. Ebadi, K., et al.: LAMP: large-scale autonomous mapping and positioning for exploration of perceptually-degraded subterranean environments. arXiv:2003.01744 (2020)

13. Galvez-L'opez, D., Tardos, J.D.: Bags of binary words for fast place recognition in image sequences. IEEE Trans. Rob. **28**(5), 1188–1197 (2012)

14. Haurum, J.B., Moeslund, T.B.: Sewer-ML: a multi-label sewer defect classification dataset and benchmark. In: 2021 IEEE/CVF Conference on Computer Vision and Pattern Recognition (CVPR), pp. 13451–13462, Nashville, TN, USA. IEEE (2021)

15. Kerl, C., Sturm, J., Cremers, D.: Dense visual SLAM for RGB-D cameras. In: 2013 IEEE/RSJ International Conference on Intelligent Robots and Systems, pp. 2100–2106, Tokyo. IEEE (2013)

16. Khattak, S., Nguyen, H., Mascarich, F., Dang, T., Alexis, K.: Complementary multi–modal sensor fusion for resilient robot pose estimation in subterranean environments. In: 2020 International Conference on Unmanned Aircraft Systems (ICUAS), pp. 1024–1029 (2020)

17. Klein, G., Murray, D.: Parallel tracking and mapping for small AR workspaces. In: 2007 6th IEEE and ACM International Symposium on Mixed and Augmented Reality, pp. 1–10, Nara, Japan. IEEE (2007)

18. Lowe, D.: Object recognition from local scale-invariant features. In: Proceedings of the Seventh IEEE International Conference on Computer Vision, vol.2, pp. 1150–1157, Kerkyra, Greece. IEEE (1999)

19. Montemerlo, M., Thrun, S., Koller, D., Wegbreit, B.: FastSLAM: a factored solution to the simultaneous localization and mapping problem
20. Newcombe, R.A., Lovegrove, S.J., Davison, A.J.: DTAM: dense tracking and mapping in real-time. In: 2011 International Conference on Computer Vision, pp. 2320–2327, Barcelona, Spain. IEEE (2011)
21. Ramezani, M., et al.: Wildcat: online continuous-time 3D Lidar-inertial SLAM. arXiv preprint arXiv:2205.12595 (2022)
22. Rublee, E., Rabaud, V. Konolige, K., Bradski, G.: ORB: an efficient alternative to SIFT or SURF. In: 2011 International Conference on Computer Vision, pp. 2564–2571, Barcelona, Spain. IEEE (2011)
23. Sarlin, P.-E., DeTone, D., Malisiewicz, T., Rabinovich, A.: SuperGlue: learning feature matching with graph neural networks. arXiv preprint arXiv:1911.11763 (2020)
24. Simonyan, K., Zisserman, A.: Very deep convolutional networks for large-scale image recognition. arXiv preprint arXiv:1409.1556 (2014)
25. Smith, R.C., Cheeseman, P.: On the representation and estimation of spatial uncertainty. Int. J. Rob. Res. 5(4), 56–68 (1986)
26. Touvron, H., Cord, M., Douze, M., Massa, F., Sablayrolles, A., Jegou, H.: Training data-efficient image transformers and distillation through attention. In: Proceedings of the 38th International Conference on Machine Learning, pp. 10347–10357. PMLR (2021)
27. Vaswani, A., et al.: Attention is all you need. In: Advances in Neural Information Processing Systems, vol. 30. Curran Associates, Inc. (2017)
28. Wang, S., Clark, R., Wen, H., Trigoni, N.: DeepVO: towards end-to-end visual odometry with deep recurrent convolutional neural networks. In: 2017 IEEE International Conference on Robotics and Automation (ICRA), pp. 2043–2050 (2017). arXiv:1709.08429

Sound-Based Parakeets Detection System

Ernesto Rován, Pablo Monzón, and Facundo Benavides$^{(\boxtimes)}$ ⓘD

Universidad de la República, Montevideo, Uruguay
{ernesto.rovan,monzon,fbenavid}@fing.edu.uy

Abstract. In recent decades, the proliferation of parakeets has esca-
lated into a significant and urgent problem for crops in the Rio de La
Plata region. Their adaptability to different environments, high reproductive rates, and flexible diet have rapidly increased their population,
posing a serious threat to agricultural productivity. This work is part
of a crucial feasibility study for a parakeet detection system and subsequent deterrence based on real-time audio signals captured in the field.
Recent studies have unveiled a fascinating aspect of parakeet behaviour-
they possess a unique acoustic fingerprint. This allows them to encode
their identity through different calls, enabling them to recognise each
other and exchange information. This high level of acoustic complexity
makes their sounds distinguishable from other field noises, a key feature
that our proposed detection system aims to leverage. Our decision to
employ and train a convolutional neural network (CNN) in this research
was based on the recent breakthroughs of CNNs in classifying acoustic
events. The input for this model is the spectrogram of audio samples, and
it functions as a binary classifier, determining the presence or absence
of parakeets. The results of our initial tests have been highly promising,
bolstering our confidence in the potential of this system.

Keywords: Digital Audio Processing · Bioacustic Events · Deep
learning · Signal Processing

1 Introduction

The Argentine parakeet (*Myiopsitta Monachus*, Fig. 1), or monk parakeet, is a
species of psittaciform bird from the *Psittacidae* family native to South America,
with a strong presence in Uruguay, Argentina, Brazil, Paraguay, and Bolivia [1].
It primarily feeds on seeds from both wild and cultivated plants, such as thistle,
sorghum, corn, and rice. It also **consumes fruits** and flowers, as well as adult
insects and their larvae [2].

Monk parakeets are highly sociable birds, forming complex societies structured in levels and learning different calls for interaction. In the wild, they form
flocks ranging from a few individuals to large groups of separate colonies. Their
behaviour is described as *fission-fusion social dynamics*, where the size and composition of the community vary over time, and individuals change environments

© The Author(s), under exclusive license to Springer Nature Switzerland AG 2025
L. Correia et al. (Eds.): IBERAMIA 2024, LNCS 15277, pp. 221–233, 2025.
https://doi.org/10.1007/978-3-031-80366-6_19

Fig. 1. Argentine Parakeet perched in an eucalyptus tree. [1]

[3]. Monk parakeets build nests that, in their native state, correlate with the extent and abundance of eucalyptus trees, their preferred tree for nesting [7].

Many studies [4–7] show strong evidence of individual recognition among parakeets through their calls, indicating a high acoustic complexity in the signals they emit. This implies an *individual vocal signature* or *voice print* in each parakeet, identifying up to eleven different types of calls [4,8], such as contact call, alarm call, threat call, and other possible interactions.

In light of these factors, the monk parakeet has been a problem for crops for decades, particularly fruit trees (although not limited to these), impacting agricultural production. In fact, in Uruguay, it was declared a pest in 1947, and in 1981, the FAO estimated a loss of US\$ 6 million annually due to this problem, which today is probably higher [10]. In addition, traditional bird control methods have proven inefficient due to the parakeet's high adaptability to the environment. The current work aims to document a potential set of techniques to implement the bird detection and localisation stage so that the system can effectively dispatch drones to areas of interest, making the proposed solution feasible.

2   Sound-Based Detection System

Given the large size of the control area, equipment costs, processing time, and the potential frequency of the event, the best option seems to be a sound sensor system. Only a single audio source will be processed to simplify the problem and start by assessing the feasibility of detecting parakeets through audio. This leads to a binary classification problem, where continuous sensing involves processing audio samples to determine the parakeet's presence (True/1) or absence (False/0).

Considering [4–7], learning from their sounds seems feasible. Among the main techniques used are Spectrographic cross-correlation (SPCC) and Mel Frequency Cepstral Coefficients (MFCC); however, the analysis of the spectrogram image was considered the simplest and most efficient alternative, as it implicitly contains most of the information provided by the other indicators. The spectrogram is constructed by calculating the signal's frequency spectrum over several consecutive time windows, resulting in a three-dimensional graph (represented in the plane) that expresses the energy of each frequency component as a function of time. Given the power of modern image processing techniques, this method predominates significantly in modern sound classification systems [14,16,17]. Of the various studies on this subject, the BirdNET project (along with the eBird app) from the Cornell Lab of Ornithology is considered highly significant. It comprises a vast, user-generated database of bird recordings, capable of recognising any bird by capturing just a few seconds of a spectrogram using CNN [17–19].

However, there are limitations to consider: the computational cost and weight of the model to be used. The hardware (HW) must simultaneously process real-time data in the field from multiple microphones, which implies a processing and consumption limitation that must be considered. In addition, the system must be robust enough due to the environment's potential noise, which seems feasible with the spectrogram analysis. Considering these aspects, this work aims to conduct a proof of concept that, without relying on any pre-established HW, explores the potential for obtaining promising results with the least amount of resources possible.

3   The Data

For training, the parakeet's audio must include the various calls they emit and some variability in the number of emitting parakeets so that it works for a single bird, a small group, or a flock. On the other hand, the set representing ambient noise must consider all possible types of sounds in the field, including other birds or animals. A database was created with audio from various public sources extracted from the Internet and several recordings made at different locations across the country, recorded between November 2023 and March 2024, using the eBird app.

3.1   The Parakeets Spectrogram

The parakeets' spectrograms were studied to adjust their construction parameters to most clearly express the patterns inherent in the sounds emitted by the parakeets (their harmonics). See Figs. 2,3.

Based on these observations, considering the balance between measurement accuracy and the need for the temporal window to encompass sufficient acoustic information, and in view of some of the previously mentioned applications, it was decided to process audio in samples of five seconds in duration. This ensures that the spectrogram images used to train the neural network are consistent in their scales and construction parameters. For a given input image size, samples longer than five seconds lose important details about the main harmonics, as a larger

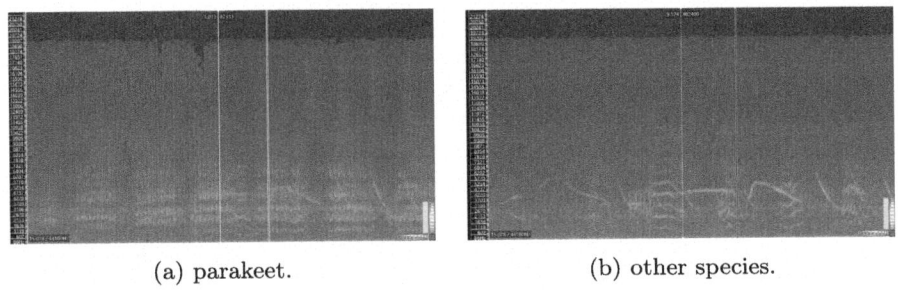

(a) parakeet.   (b) other species.

Fig. 2. Sonic Visualiser screenshots (Spectrogram: Window 512, 87.5%, dB), approx. two seconds duration.

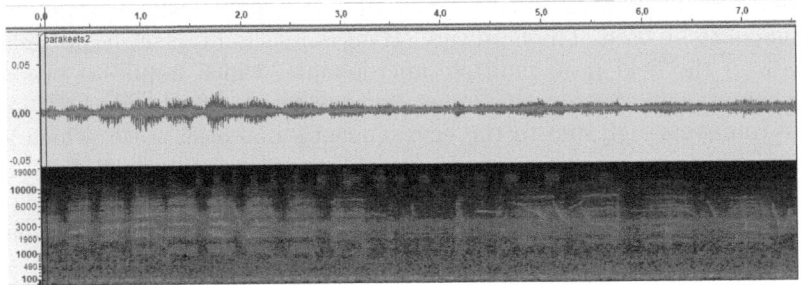

Fig. 3. Audacity capture with waveform and spectrogram.

time window is being covered with the same resolution. On the other hand, shortening the audio samples implies generating a greater number of images, increasing computational cost and processing time.

3.2 The Dataset

The dataset, with 30% set aside for validation, results in 511 samples (43.6 min of total recording) of the negative class and 309 samples (26.8 min) of the positive class for training. In total, including the validation data, there are 1171 recordings of 5 s each.

4   Classification Model: Convolutional Neural Network

4.1 Architecture

Based on the architecture suggested in [12], after numerous performance tests modifying various parameters, addressing the trade-off between the model's size and weight and its performance as a classifier, the model architecture is summarised in the diagram shown in Fig. 4.

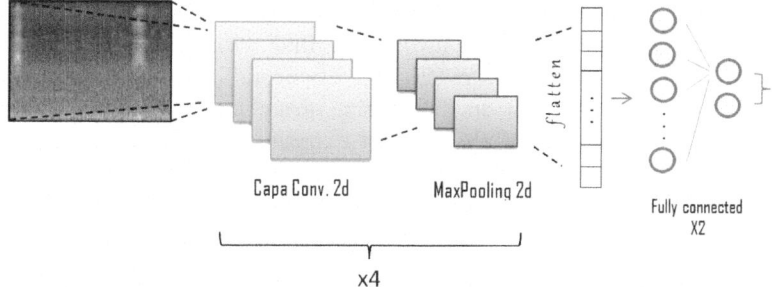

Fig. 4. Diagram of modelled CNN structure.

Note that this architecture, for an input size of 192×256, has approximately 700,000 trainable parameters and occupies 2.64 MB of memory, which is relatively low for a CNN. To limit the number of parameters, the most effective approach was to regulate the fully connected dense layer to maintain a certain depth in the convolutional layers to leverage their feature extraction capabilities.

4.2 Data Preprocessing

Next, the suitability of the image's input size was analyzed. Although this parameter has the highest impact on the computational cost, decreasing the image size deteriorates the spectrogram resolution. Initially, the dimensions were estimated at 196×256 pixels (preserving the original image proportions), which is sufficient for a good performance. Since the performance is very good, and this report is not made based on a given hardware, it was preferred that the use of resources be conservative. However, if computer power allows it, it is convenient to expand it to avoid losing details. In this work, an intermediate solution was sought.

Other spectrogram preprocessing techniques, such as variable changes or applying a band-pass filter, were also tested, but they did not improve performance. Once these points were determined, starting from the image dimensions mentioned above, several tests were performed by varying training and architecture parameters to maintain a limited size. The models with the best results are shown in Table 1.

Table 1. Best experiments.

Model	input_shape	# parameters	Memory	Acc. % (Train)	Acc. % (Val.)
Simple	192×256	691518	2.64 MB	85.75	86.65
HD	240×320	932158	3.56 MB	98.29	92.61
HD+DataAug	240×320	932158	3.56 MB	99.89	94.32
(TL + dense 10,2)	240×320	3247904	12.39 MB	100	98.01

For all experiments recorded in the table (except the last one), the architecture consists of four convolutional layers (the first with a kernel size of 32, the rest with 128) and two dense layers (20 and 2 neurons), Adam optimizer, binary cross-entropy as the loss function, batch size of eight (due to SW constraints), and twelve training epochs. While the first experiment with the base input size 192×256 is good, it is found that raising the input image dimensions somewhat produces a qualitative jump in the validation accuracy, and the model remains relatively small. After performing Data Augmentation on the training set, the model is retrained for 240×320 images. This measure increases the Accuracy in validation by two points, which is very positive considering that the model size remains the same. Figure 5 shows an example of inference for a spectrogram with the presence of a parakeet and the image output after the corresponding max pooling layer (the first one).

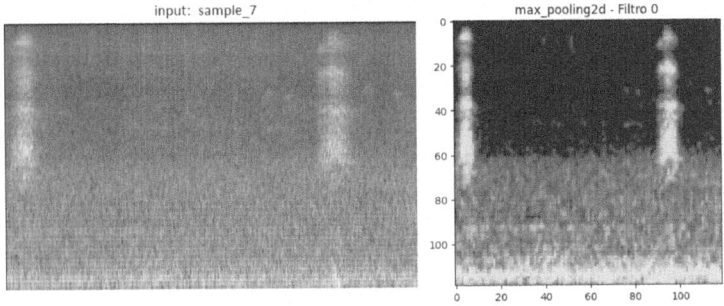

Fig. 5. Spectrogram (audio with parakeets sound) after first convolutional layer.

4.3 Data Augmentation

Data Augmentation is applied to 30% of the parakeet base (label = True) of the training set so that the classes are more balanced (while the validation set remains exactly the same). White Noise is added to 15% at two low power levels. Natural noise coming from nine background samples (field, rain, crickets, some bird sounds, etc.) is added to the other 15%.

The result was an increase in the sample of parakeets in the Train from 309 to 402, representing 44% of the total Train set, compared to 38% before the rise.

4.4 Transfer Learning

Finally, an experiment is performed to estimate the performance of using pre-trained models. As an example, at the suggestion of [12], the MobileNetV2 network was selected, which is Google's pre-trained CNN optimised for mobile devices, a small model that requires low computational power and memory. It is used to preprocess the input images to obtain their embeddings and then train a fully connected two-layer with the feature vectors extracted by that model. The result was a validation accuracy of 98%.

Using pre-trained models with spectrograms for audio classification would have been more accurate. However, strong evidence suggests that pre-trained networks in areas other than the spectrogram manage to transfer useful learned features, as discussed in [25]. That article proves that ImageNet-Pretrained standard deep CNN models constitute a solid basis for continuing training with spectrograms for audio classification. However, the model size would be expanded considerably.

4.5 Final Model

Although the best result was obtained using a pre-trained model, it was decided to keep "ModelHd+DataAug" (i.e. 240 × 320 images, trained with Data Augmentation) as the model to be used, weighting the performance-model size ratio. This work does not aim to obtain the best possible model but to outline the best lines as part of a feasibility study. It would be optimal to re-train a model of similar characteristics or the same model proposed with a database that takes into account the environment where the system is to be installed, including more and better recordings of parakeets. In particular, incorporating different types of calls and recording parakeets in the application field is considered of great interest when they come down to eat. In that aspect lies the main weakness of the model.

With the validation set, it was observed that the model tends towards false positives rather than false negatives, which seems appropriate for the purposes of this problem. It is better to make a mistake in sending the drone, even if there are no parakeets, than not to send it at all, and the parakeets continue to damage the crops.

5 Event Detector

Since the binary classifier expects five-second inputs to determine the presence or absence of parakeets, it does not seem feasible to compute the spectrogram and process it as an image in the neural network every five seconds permanently, especially considering that the sensing comes from a network of multiple microphones. Therefore, an event detector is needed as a previous instance of filtering, maintaining a low computational cost (and consumption). When a relevant acoustic event occurs, it sends five-second fragments to the classifier to determine whether it is a parakeet.

5.1 Sensing Dynamics: Sliding Window

Essentially, there are two requirements: to keep detection operations simple and to obtain an indicator that gives a positive result for all cases where there are

parakeets (no false negatives), even if it means detecting events without parakeets (false positives) but trying to keep them to a minimum. To this end, the
algorithm proposed consists of periodic calculation of the signal energy in the
last time window.

$$E_{W_n} = \sum_{W_n} x^2[n]$$

$$E_{W_n} > E_{W_{n-1}} * u \rightarrow Event! \tag{1}$$

For example, in (1), $x[n]$ is the signal in the time domain and u is a relative
threshold. Taking $u = 2$ implies that the signal energy must be doubled from
one window to another to capture the event.

5.2   Synthetic-Based Testing

Some tracks were taken (and others were created) to test and tune the event
detector. The objective is to prevent ambient noise from triggering the event.
The concept of a sliding window, and thus a moving average, seeks to adapt to
the variable acoustic conditions at a given time and place while also allowing for
gradual and 'smooth' changes, yet recognising sudden ones (Fig. 6).

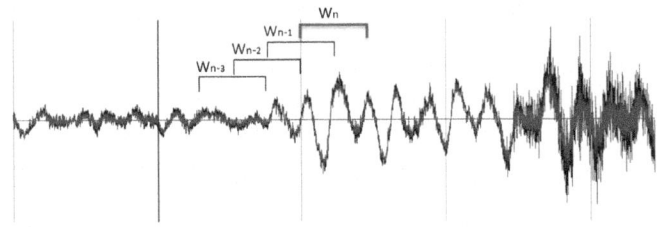

Fig. 6. Sliding window graphic.

5.3   Problematic Cases and Discussion

Although promising results are obtained for simple cases, it is not unusual for
an increase in noise (e.g., wind) to increase the signal energy considerably up to
peaks comparable to those produced with the presence of birds. The opposite
case also occurs, where the parakeet's sound is perfectly perceptible to the human
ear, but amid the ambient noise, it does not achieve a noticeable spike in the
energy graph.

Given the expected noise conditions, it seems that using only the signal
energy is not feasible, and exploiting the information derived from the waveform or frequency becomes unavoidable. This introduces a fundamental problem
since spectral analysis and derived indicators demand a high computational cost,
which must be avoided.

5.4 Band Pass Filter (BPF)

It is presented as the best alternative to limit the energy measurement of each time window to the signal components in a limited range of frequencies. Thus, a band-pass filter was incorporated (at the frequencies of the parakeets's calls) at the signal input, thereby mitigating the energy in the rest of the spectrum and making an increase in the range of interest more noticeable. Even if noise exists in that band, the relative weight of the noise energy in relation to the parakeet decreases enormously, making the previous algorithm based on energy thresholding feasible. Although the BPF was implemented via software in the tests, it should be implemented with hardware in practice.

Analysing the parakeets' spectrum, as shown in Fig. 7, the [2000, 7500]Hz band is initially selected to provide a flexible, non-exhaustive filter. This range tries to contemplate all the parakeet's calls, and it is considered satisfactory that other birds trigger the event, however, if necessary it can be narrowed down. The analysis (as in Sect. 3.1) was contrasted with the information provided in the cited research.

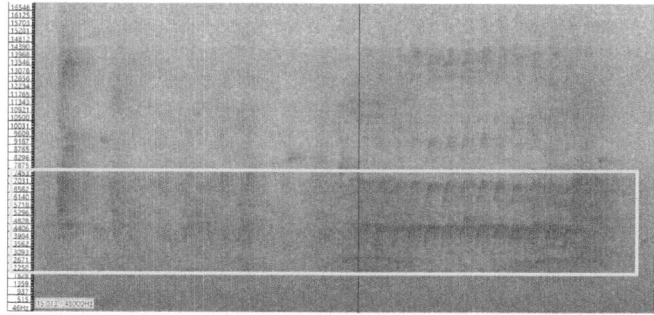

Fig. 7. Spectrum of parakeet, framed in the $[2 : 7.5]kHz$ band (Sonic Visualiser capture).

5.5 Detector Performance with Band-Pass Filter

The performance improves significantly, making the proposed algorithm feasible. It successfully detects all the events from a couple of tens of synthetic audios. Figure 8 shows a successful example of the filter, where the objective is to bypass the first energy increase section (due to the wind) and activate with the second and third peaks, produced in the frequency range corresponding to the parakeets' sounds.

Note that the orange curve represents the energy measurement corresponding to each instant for the last three seconds applying the band-pass filter, the blue curve represents the original energy without filter, and the red line represents the instants where there is event detection according to the Eq. 1 criterion. After several tests, the percentage threshold is set at 60%, although it is convenient to adapt it to the conditions where the system is installed.

The band-pass filter keeps the signal energy stable when no parakeets are present and increases it when they are. The only detected problem that still persists is the scenario where the noise is very low (close to zero), and any increase in noise in the frequency range of interest is sufficient to exceed 60% of the previous energy far, triggering the detector. However, as mentioned above, false positives are preferable to false negatives. If necessary, It could also be changed to a non-percentage threshold.

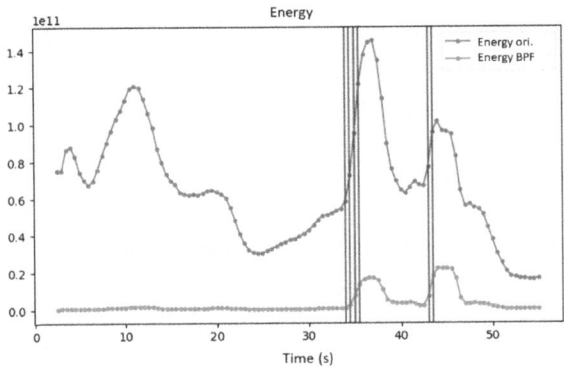

Fig. 8. Energy chart (with and without BPF) for each time window. Red lines indicate an event. (Color figure online)

6   Final System

The complete system would consist of the event detector, which runs constantly. When an event occurs, it sends the last 5 s to the classifier. The classifier converts the time window into an image from the spectrogram calculation, adapts its size, and feeds it into the trained neural network. If the classification is positive, the drone is sent, the detector is stopped for a prudent time to save energy, and a second detection is expected as confirmation. A simulation of the whole system was performed to test the entire pipeline so that it emulated the behaviour of the samples as if they were arriving in real-time.

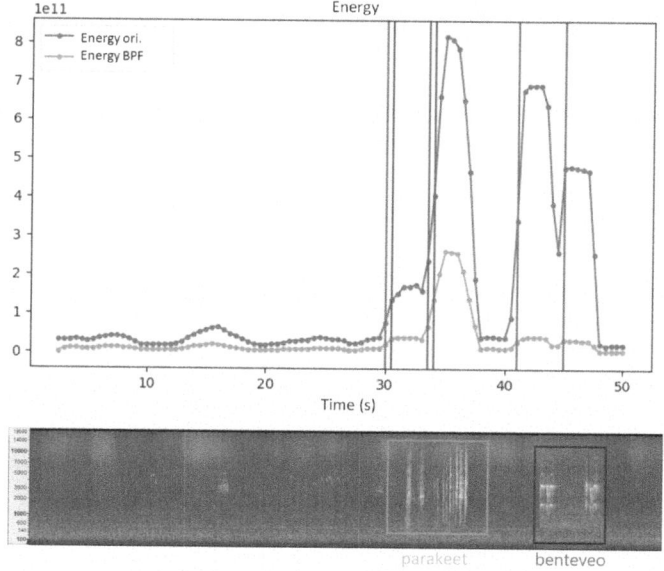

Fig. 9. Graphic result of the simulation.

An example is shown in Fig. 9, which corresponds to an audio file containing the sound of a parakeet and a *benteveo* bird, where both trigger the event detector but are classified differently. The vertical lines in the chart mark the time when events are detected; if the classification is positive, the line is green or red otherwise. Below the chart, the corresponding spectrogram is shown. Based on this, both classifications are correct.

7   Conclusions and Future Works

After a considerable survey of the literature on the detection of acoustic events in general and the sound emitted by parakeets in particular and the promising experimental results, we conclude that it is feasible to design a detection system, applying models already standard in the field for parakeets identification based on audio classification by spectrogram frequency. The main challenge is continuing to enhance the dataset. Besides, the level of information exchange and sound emission during foraging or ingestion is sufficiently intense for the model's sensitivity, being a potential complication that remains latent.

Moreover, the main future direction work consists of improving the classification model, especially by gathering more data with better balance and considering the environment in which the system will be inserted. The model's capabilities can be increased if information on the use of HW is available. It is not out of the question to try other models, particularly [26, 27], considering that the few-shot learning paradigm has recently gained popularity when applied to bioacoustic event detection. The event detector can also be robust based on new data, including setting a threshold learned from a more specific dataset.

Aknowledgments. We would like to thank the PEDECIBA Informática program for supporting this research.

References

1. SEOBirdLife. Cotorra argentina. https://seo.org/ave/cotorra-argentina/
2. Tala, C., Guzmán, P., González, S.: Cotorra argentina (Myiopsitta monachus), convidado de piedra en nuestras ciudades y un invasor potencial, aunque real, de sectores agrícolas (2005)
3. Honson, A., Avery, M., Wright, T.: The socioecology of Monk Parakeets: insights into parrot social complexity (2014)
4. Smeele, S., Senar, J., Aplin, L., McElreath, M.: Evidence for vocal signatures and voice-prints in wild parrot (2023)
5. Smith-Vidaurre, G., Pérez-Marrufo, V., Hobson, E., Salinas-Melgoza, A., Wright, T.: Individual identity information persists in learned calls of introduced parrot populations (2023)
6. Smith-Vidaurre, G., Pérez-Marrufo, V., Wright, T.: Individual vocal signatures show reduced complexity following invasion (2021)
7. Smith-Vidaurre, G., Araya-Salas, M., Wright, T.: Individual signatures outweigh social group identity in contact calls of a communally nesting parrot (2019)
8. Martella, M., Bucher, E.: Vocalizations of the Monk Parakeet (1990)
9. Mott, D.F.: Monk parakeet damage to crops in Uruguay and ITR control
10. Ministerio de Ganadería, Agricultura y Pesca. Cotorra (Myiopsitta monachus). https://www.gub.uy/ministerio-ganaderia-agricultura-pesca/politicas-y-gestion/cotorra
11. Viazzi, A., Blandón, J., Maciel Rios, A., Gil González, J.: A centered kernel alignment-based strategy for pest evolution tracing: Myiopsitta monachus case (2023)
12. Prosise, J.: Audio classification using convolutional neural networks (2021). https://github.com/jeffprosise/Deep-Learning/blob/master/Audio%20Classification%20(CNN).ipynb
13. Doshi, K.: Audio deep learning made simple: sound classification, step-by-step (2021). https://towardsdatascience.com/audio-deep-learning-made-simple-sound-classification-step-by-step-cebc936bbe5
14. Xiao, H., Liu, D.: AMResNet: an automatic recognition model of bird sounds in real environment (2022)
15. Kortas, M.: Sound-based bird classification (2020). https://towardsdatascience.com/sound-based-bird-classification-965d0ecacb2b
16. Mohanty, R., Kumar Mallik, B., Singh Solanki, S.: Automatic bird species recognition system using neural network based on spike (2020)
17. Kahl, S., Wood, C., Eibl, M., Klinck, H.: BirdNET: a deep learning solution for avian diversity monitoring (2021)
18. Hoffman, B., Van Horn, G.: From sound to images, part 1: a deep dive on spectrogram creation (2021). https://www.macaulaylibrary.org/2021/07/19/from-sound-to-images-part-1-a-deep-dive-on-spectrogram-creation/

19. Hoffman, B., Van Horn, G.: From sound to images, part 2: spectrogram image processing (2021). https://www.macaulaylibrary.org/2021/08/05/from-sound-to-images-part-2-spectrogram-image-processing/
20. Sawant, S., Arvind, C., Viral, J., Robin, V.: Spectrogram cross-correlation can be used to measure the complexity of bird vocalizations (2021)
21. Deruty, E.: Intuitive understanding of MFCCs (2022). https://medium.com/@derutycsl/intuitive-understanding-of-mfccs-836d36a1f779
22. Rovai, M.: TinyML made easy: sound classification (KWS) (2022). https://www.hackster.io/mjrobot/tinyml-made-easy-sound-classification-kws-2fb3ab
23. ARM: End-to-end tinyML audio classification with the Raspberry Pi RP2040 (2021). https://blog.tensorflow.org/2021/09/TinyML-Audio-for-everyone.html
24. Maayah, M., Abunada, A., Al-Janahi, K., Ejaz Ahmed, M., Qadir, J.: LimitAccess: on-device TinyML based robust speech recognition and age classification (2022)
25. Palanisamy, K., Singhania, D., Yao, A.: Rethinking CNN models for audio classification (2020)
26. Nolasco, I., et al.: Few-shot bioacoustic event detection at the DCASE 2023 challenge (2023)
27. Few-shot Bioacoustic Event Detection. https://dcase.community/challenge2022/task-few-shot-bioacoustic-event-detection

Knowledge Representation
and Reasoning

AI-Based Medical Education: Coping with Clinical Decisions in GLARE-Edu

Alessio Bottrighi[1,2] , Antonio Maconi[3(✉)] , Stefano Nera[1,2] ,
Luca Piovesan[1,2] , Erica Raina[1,2] , and Paolo Terenziani[1,2]

[1] Computer Science Institute, DISIT, University of Eastern Piedmont, Alessandria, Italy
{alessio.bottrighi,stefano.nera,luca.piovesan,erica.raina,
paolo.terenziani}@uniupo.it
[2] Integrated Laboratory of AI and Medical Informatics DAIRI + DISIT, Alessandria, Italy
[3] DAIRI, ASU SS. Antonio e Biagio e Cesare Arrigo, Alessandria, Italy
amaconi@ospedale.al.it

Abstract. Artificial Intelligence (AI) tools and methodologies can provide crucial contributions not only for medical decision support, but also for medical education. In particular, we operate in the AI in medicine context of Computer-Interpretable Clinical Guidelines, and, within a two-year national project, we are developing GLARE-Edu, a system specifically devoted to medical education. In this paper, we describe some of the main advances we achieved in the first year, concerning teaching and testing learners' capabilities about clinical decisions. Four facilities are proposed, to address different aspects of clinical decisions.

Keywords: AI for Education · Computer-Interpretable Clinical Guidelines · Knowledge Representation and Reasoning

1 Introduction

Medicine and healthcare are complex contexts, to which many efforts have been devoted by the AI community. In particular, many different AI approaches have been devised to support physicians' decisions (i.e., clinical decision support (CDS) systems). Since the end of the 90s, several AI approaches have based their decision support on clinical practice guidelines (CPGs). CPGs are "systematically developed statements to assist practitioner and patient decisions about appropriate health care in specific clinical circumstances" [1]. In other words, CPGs encode evidence-based best practices ("golden standards") to manage patients affected by a specific disease. Thousands of CPGs have been devised around the world (consider, e.g., http://www.g-i-n.net). AI systems devoted to CPGs usually provide an acquisition module to support the acquisition of medical textual CPGs in a computer format, to obtain Computer-Interpretable Guidelines (CIGs), as well as "execution" mechanisms connecting CIGs with patients' clinical data (e.g., Electronic Health Record) to suggest physicians how to act on patients, on the basis of the evidence in the CIGs. Some of them (the list is not exhaustive) are: Asbru, EON, De-GeL, GEM, GLARE, GLIF, GPROVE, GUIDE, PRODIGY, PROforma, SAGE (consider, e.g., the surveys in [2–4]).

Besides decision support, a fundamental and challenging task in this context is medical education, to teach medical students (or physicians, in continuous education) how to deal with patients using the "best" evidence-based procedures, and, in particular, how to take appropriate decisions about the patients. Though some CIG decision support system in the literature has also been applied in medical education (consider, e.g., [5–10]), such systems are born for decision support. And, indeed, the specificity of the educational task raises new challenges with respect to decision support. In this paper, we focus on one of such challenges, of crucial importance: how to support teaching and testing learners' capabilities to take appropriate clinical decisions. In traditional CIG systems, devoted to CDS, decision criteria are encoded in an internal format, and are automatically interpreted by the systems (considering patients' data) to propose the CIG-based decision to physicians. Dealing with decisions in the educational context arises additional intriguing problems, namely teaching and testing learners about complex aspects, such as *(i) which clinical parameters have to be considered to take a given decision, (ii) what are its possible alternative outputs (conclusions), (iii) what are the conditions that must be satisfied to reach such conclusions (i.e., what are the decision criteria) and, last but not least, (iv) how to apply knowledge in (i-iii) to take an appropriate decision regarding a specific patient.*

The starting point of our work is Guideline Acquisition, Representation and Execution (GLARE) [11], a CIG system (devoted to CDS) that we take as the basis to design and develop GLARE-Edu, a CIG-based system specifically devoted to medical education. GLARE-Edu is being developed within a two-year project, Personalized Training of Professional Competences with AI (PTPC-AI), that will be briefly introduced, together with GLARE, in Sect. 2of this paper. In Sect. 3 we describe four facilities that we are developing in GLARE-Edu to address tasks (i-iv) mentioned so far. Finally, in Sect. 4 we sketch conclusions and related works.

2   Preliminaries: GLARE, GLARE-Edu, and PTPC-AI Project

GLARE [11] is one of the *disease-independent* decision support system for **CIG**s in the Artificial Intelligence in Medicine (AIM) literature. We started to develop GLARE in 1996, in cooperation with Prof. Gianpaolo Molino of the Hospital San Giovanni Battista in Turin, one of the major hospitals in Italy. GLARE has been already applied to many different medical domains, including polytrauma, bladder cancer, heart failure, gastroesophageal reflux disease, ischemic stroke and melanoma. The kernel of GLARE, discussed in [11], consists of two main modules: GLARE's *acquisition module* provides an user-friendly support to acquire CPGs into a computer-interpretable format in GLARE's formalism; GLARE's *execution module* supports the application of an acquired CIG to a specific patient (through an automatic connection to the electronic clinical record of the patient), to provide "evidence-based" decision support to physicians.

The kernel of GLARE has been designed and realized within a long-term project leaded by hospital physicians. GLARE emerges from the CIG systems in the AI in Medicine literature for its extensive adoption of advanced AI techniques, to provide additional facilities, including cost-benefit and what-if analysis, CIG versioning and contextualization, temporal reasoning, model-based verification, conformance evaluation, treatment of comorbid patients [12–21]. In the following, for the sake of brevity we resume only the features of GLARE needed to make the paper self-contained.

In GLARE, as in most disease-independent CIG systems, CPGs are represented through the Task-Network Model (TNM). In TNM, a CIG is represented by a hierarchical oriented graph, in which nodes represent the actions in the CPG, and arcs represent the flow of control. Concerning nodes, GLARE has both *atomic* and *composite* action nodes. GLARE has a limited number of types of *atomic actions*, i.e., Work actions, Pharmacological prescriptions, Query actions, Decisions, and Conclusions, and *arcs* (sequence, concurrent, alternative and repetition), modeling control relations between action-nodes [11].

GLARE distinguishes between patient-status-based and cost-benefit decisions. **Patient-status-based decisions** (PSB-decisions for short) model cases in which alternative paths have to be selected, depending on (the data modeling the) clinical status of the patient. A typical case of PSB-decision is a diagnostic decision. Due to space limitations, this paper mainly focuses on PSB-decision only.

In GLARE, PSB-decisions are represented as a choice between two or more alternatives, to be taken based on Boolean conditions, to be evaluated considering the patient's clinical status. As long as a CIG system is used for decision, such an approach is effective. At execution time, the Boolean conditions are automatically evaluated by GLARE. Therefore, even complex Boolean formulae are not problematic. However, when switching from decision support to education, students have to learn decision criteria and how to apply them (to patients' data) on their own. In such a context, declarative Boolean expressions may not be the best option, since they may be very complex, and are "*flat*", in the sense that they do not promote a step-by-step evaluation of the different parameters involved in the decision.

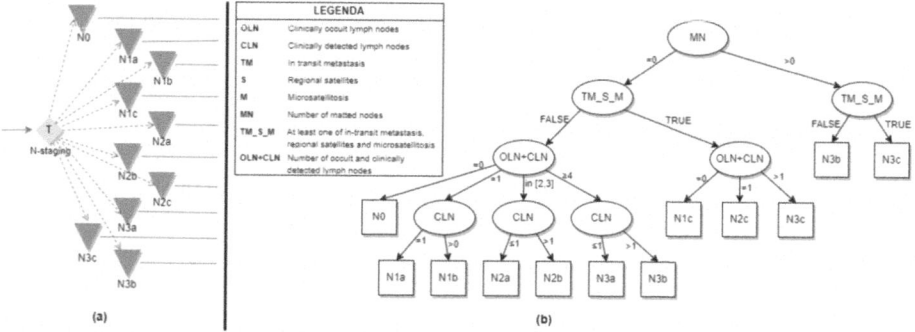

Fig. 1. (a) The PSB-decision node, with its conclusion nodes, for the N-staging in the melanoma CIG and (b) the corresponding decision tree in GLARE-Edu

Thus, we support the representation of PSB-decisions as decision trees. In GLARE-Edu, each node in a decision tree (excluding leaf nodes, that model conclusions) concerns a specific clinical parameter (expressed by a pair < data, attribute >), and the outgoing arcs contain atomic conditions on it, expressed as inequalities involving a parameter and a constant value (e.g., *37 ≥ < temperature, value > ≥ 38*, or, as another example, *< liver, dimension > = enlarged*). For example, Fig. 1(b) shows the decision tree used to model the N-staging in the melanoma CIG [22] and in Fig. 1(a) appears the corresponding PSB-decision node in the CIG, together with its conclusion nodes.

As discussed in the introduction, in the last years we have started to focus our attention not only on clinical decision support, but also on medical education, starting to develop a new version of GLARE specifically geared towards education, called GLARE-Edu [23]. We are now working hard on such a new research direction within the PTPC-AI project. The PTPC-AI project is part of a larger two-year project, *Learning Personalization with AI and of AI* (AI-LEAP), started on May 2, 2023 and involving three Italian Universities and many territorial partners, granted with 500,000 euros by Compagnia San Paolo and CDP. PTPC-AI involves, besides *University of Eastern Piedmont* (our University), *DAIRI (Integrated Activities Research Innovation Department)*, a department of the public hospital SS. Antonio e Biagio e Cesare Arrigo of Alessandria, Italy, and *Pop AI*, a non-profit association to foster the "culture of AI" to population. The main goal of the PTPC-AI project is to show that the adoption of AI techniques is useful to facilitate medical education. To achieve such a goal, the project will exploit advanced AI *methodologies* for *knowledge acquisition* and *representation*, and for *simulation* and *conformance evaluation*. It will consist of three main Workpackages:

(1) development of AI tools to simulate the application of CIGs to virtual patients (modelled as the evolution of their clinical findings), and to verify to what extent learners have acquired the knowledge in the CIG, and the ability to apply CIGs to specific patients,
(2) development of learning units, based on the above tools. In particular, using GLARE, we have already acquired the Italian CPG about melanoma as an application example, together with several case studies (i.e., clinical data of patients affected by melanoma).
(3) In the last six months of the project the new educational supports will be evaluated (see the discussion in the concluding Section).

3   Advances: Learning PSB-Decisions with GLARE-Edu

In the PTPC-AI project, we are considering many different aspects of medical education. This paper focuses on teaching and testing learners' competences about how to take PSB-decision. Of course, taking decision is one of the most important medical activities, so that we provide a set of facilities, that consider an increasing level of complexity. In such facilities, we assume that learners have already had the opportunity to learn the contents of a CIG and, specifically, the decision criteria in the PSB-decisions of the CIG, and propose support to test learners about: how to apply decision criteria (shown to the learner) to a specific patient (Sect. 3.1); how to take a given PSB-decision (decision criteria not shown to the student) for a specific patient (Sect. 3.2); the decision criteria to be used to take a given PSB-decision (Sect. 3.3); the decision tree to be used to take a given decision (Sect. 3.4). All the facilities take advantage of a basic engine which automatically applies a decision tree to a set of data, reaching the proper leaf node (conclusion). For the sake of precision and clarity, the facilities are described by high-level pseudocode. Only minimal additional explanations are added, for the sake of brevity.

3.1 Applying a (Shown) Decision Tree to a Patient

In such a facility, the goal is just to train learners to correctly apply a decision, modeled by a decision tree (which is shown to the learner). In the following, *DT-node* represents a node of the Decision Tree, and is initialized to the root node. *PD* represents the patient's data, and *Get-child(Node, Condition)* gives as result the child-node of *Node*, that is reached from *Node* following the arc with condition *Condition*.

Procedure ApplyDT(DT-node, PD)
IF DT-node is a conclusion-node THEN exit,
ELSE
Show to learner all the conditions associated to the arcs stemming from DT-node;
Ask the learner what is the condition that is satisfied by PD (let it be C_{learn});
Calculate (automatically) the correct condition satisfied by PD (let it be C_{CIG});
IF ($C_{learn} \neq C_{CIG}$) THEN show C_{CIG} to the learner;
ApplyDT(Get-child(DT-node, C_{CIG}), PD)

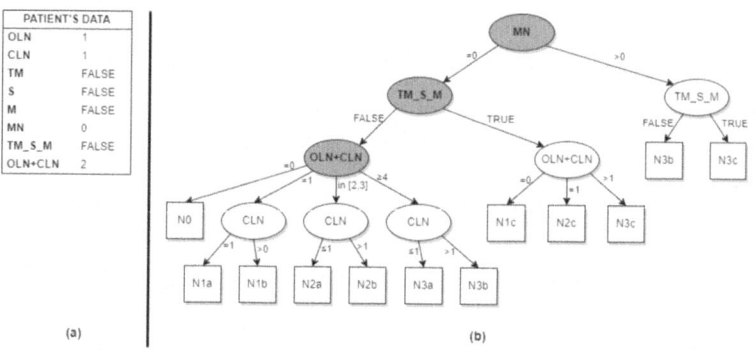

Fig. 2. (a) Example of patient's data and (b) situation at the third step of algorithm

For example, the decision tree in Fig. 1(b) and the patient's data in Fig. 2(a) are shown to learner. On the basis of such patient's data which make true some conditions along arcs, the learner has to select, step by step, the path leading from the root MN to the leaf N2a, i.e., MN = 0, TM_S_M = FALSE, OLN + CLN between 2 and 3, and CLN ≤ 1. The Fig. 2(b) shows the situation at the third step (choice concerning the OLN + CLN node).

3.2 Applying a Decision to a Patient

In such a facility, the goal is just to train learners to correctly apply a decision, modeled by a decision tree. The set of possible conclusions of the decision (*CD*) is shown to the learner, as well as the patient's data (*PD*), but not the decision tree. Notably, a given conclusion may be represented by multiple leaves of the decision tree (consider, e.g., the conclusion N3b in Fig. 1(b)), but only one leaf is the correct one based on patient's data. This fact is managed by the *Decide* procedure below.

Procedure Decide(DT, CD, PD)
Show CD and PD;
Ask the conclusion to take (let it be C_{learn});
Retrieve in DT the set of leaf-nodes corresponding to C_{learn} (let it be $NSet_{learn}$);
Find (automatically) what is the leaf (conclusion) node corresponding to the correct application of DT to PD (let it be N_{CIG}, and C_{CIG} the conclusion associated with it).
IF ($C_{learn} = C_{CIG}$) THEN return ("OK")
ELSE
    FOR EACH node $N_{learn} \in NSet_{learn}$ DO
        Find the closest common ancestor of N_{CIG} and N_{learn} in DT (let it be $N_{ancestor}$);
        Show the paths connecting N_{CIG} and N_{learn} to $N_{ancestor}$;
        Show the conditions associated to the arcs exiting $N_{ancestor}$ and leading to N_{CIG} and N_{learn} respectively

---

For example, the PSB-decision node with its conclusions in Fig. 1(a) and the patient's data in Fig. 2(a) above are shown to learner. We assume that s/he selects the conclusion N2c ($C_{learn} = N2c$). The correct choice, on the basis of the patient's data, is N2a (as explained in Sect. 3.1 above). The algorithm compares the two conclusions and shows to the learner the discrepancies, as it appears in Fig. 3: the closest common ancestor node has a marked border; the path from ancestor to the wrong leaf (selected by learner) is colored black and the one from ancestor to the correct leaf is colored gray.

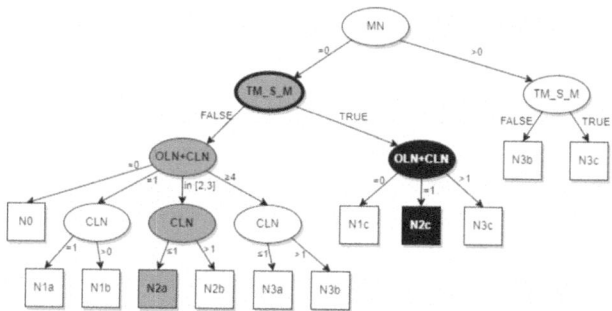

Fig. 3. Example of GUI shown to the learner containing discrepancies between two conclusions

3.3 Reconstructing the Decision Criteria

In such a facility, we show to the learner a decision node in a CIG, and we ask her to "reconstruct" such a decision, i.e., to specify (i) *what are the parameters needed to take such a decision*, (ii) *what are its possible conclusions*, and (iii) *what are the decision criteria associated with each one of the conclusions*. Notably, differently from the facility in Sect. 3.4 below, we ask for decision criteria, and not for the decision tree (i.e., we neglect the order in which atomic conditions are asked to discriminate). Please also notice that, in this facility, we focus on the decision "per se", independently of patients to which it should be applied.

In this facility, we exploit the fact that in GLARE-Edu decision trees (i) the condition to reach a given conclusion node is the conjunction of all the atomic conditions in the path from the root to the node, and (ii) (since the same conclusion may appear in different conclusion nodes in the tree) the overall condition to take a given conclusion is the disjunction of the conditions of each path to the conclusion (represented in the algorithm by $AndSet_{learn}$ and $AndSet_{CIG}$, which contain an element -i.e., a conjunction of atomic conditions- for each such disjunct).

In the algorithm, we consider the fact that we have a (disjunctive) set of conjuncts proposed by the learner, and a set proposed by the CIG, and we have to fix a correspondence between elements (i.e., conjuncts) of the two sets. We select the correspondence that minimizes the distance between conjuncts (i.e., that is more favorable for the learner), in terms of parameters included in the atomic conditions in each conjunction.

Procedure ReconstructDecision(CIG, Decision)

Ask which parameters are needed to take the decision (let $PSet_{learn}$ the set of such parameters, and let $PSet_{CIG}$ the set of parameters specified in Decision);

IF ($PSet_{learn} \neq PSet_{CIG}$) THEN show the differences;

Ask what are the conclusion of the decision (let $CSet_{learn}$ the set of such conclusions, and let $CSet_{CIG}$ the set of conclusions of Decision in the CIG);

IF ($CSet_{learn} \neq CSet_{CIG}$) THEN show the differences;

FOR EACH $C \in CSet_{CIG}$ DO

Ask each conjunction of atomic conditions that may lead to conclusion C (let $AndSet_{learn}$ the set of such conjunctions, let $AndSet_{CIG}$ the set of conjunctions in Decision);

Find the most favorable correspondence between pairs of conjuncts And_{learn} ($\in AndSet_{learn}$) and And_{CIG} ($\in AndSet_{CIG}$);

FOR EACH $And_{learn} \in AndSet_{learn}$ which has no correspondence DO

Show And_{learn} to the learner, with the message: "exceeding"

FOR EACH $And_{CIG} \in AndSet_{CIG}$ which has no correspondence DO

Show And_{CIG} to the learner, with the message: "missing"

FOR EACH pair <$And_{learn} \in AndSet_{learn}$, $And_{CIG} \in AndSet_{CIG}$> DO

Let $ParamSet_{learn}$ the set of parameters in And_{learn}, and $ParamSet_{CIG}$ the set of parameters in And_{CIG}

FOR EACH $Param_{learn}$ such that ($Param_{learn} \in ParamSet_{learn}$ AND $Param_{learn} \notin ParamSet_{CIG}$) DO

Show the atomic condition about $Param_{learn}$ with the message: "exceeding"

FOR EACH $Param_{CIG}$ such that ($Param_{CIG} \in ParamSet_{CIG}$ AND $Param_{CIG} \notin ParamSet_{learn}$) DO

Show the atomic condition about $Param_{CIG}$, with the message: "missing"

FOR EACH $Param_{learn}$ such that ($Param_{learn} \in ParamSet_{learn}$ AND $Param_{learn} \in ParamSet_{CIG}$) DO

Compare the corresponding atomic conditions, and show differences

---

For example, the PSB-decision node in Fig. 1(a) above is shown to learner without its conclusions. Let's assume that the parameters and the conclusions have already been determined, and the algorithm is asking to the learner the decision criteria associated with each one of the conclusions. In this example, we focus on the conclusion N3c. Supposing that the decision criteria identified by the learner (i.e., $AndSet_{learn}$) is ((MN = 0) AND (CLN = 1)) OR ((TM_S_M = FALSE) AND (MN > 0)), Fig. 4 shows the most favorable correspondence $AndSet_{learn}$ and the correct (on the basis of the decision tree shown in Fig. 1(b)) decision criteria $AndSet_{CIG}$.

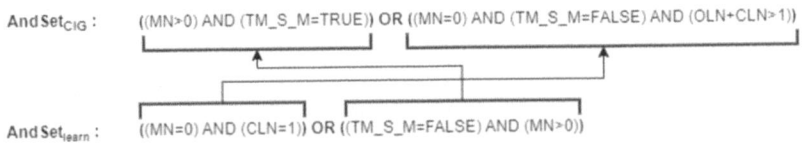

AndSet$_{CIG}$:   ((MN>0) AND (TM_S_M=TRUE)) OR ((MN=0) AND (TM_S_M=FALSE) AND (OLN+CLN>1))

AndSet$_{learn}$:   ((MN=0) AND (CLN=1)) OR ((TM_S_M=FALSE) AND (MN>0))

Fig. 4. The most favorable correspondence of conjuncts between the learner's decision criteria and the correct one in the CIG

In this example, there are not exceeding or missing conjuncts. The comparison between corresponding conjuncts is shown as in Fig. 5 below, and the discrepancies are shown to learner.

	CORRECT parameters, with		EXCEEDING parameters	MISSING parameters
	Correct atomic conditions	Wrong atomic conditions		
And$_{CIG}$: (MN>0) AND (TM_S_M=TRUE) And$_{learn}$: (TM_S_M=FALSE) AND (MN>0)	MN > 0	TM_S_M = FALSE	/	/
And$_{CIG}$: (MN=0) AND (TM_S_M=FALSE) AND (OLN+CLN>1) And$_{learn}$: (MN=0) AND (CLN=1)	MN = 0	/	CLN	TM_S_M OLN+CLN

Fig. 5. Discrepancies between parameters in the decision criteria of the CIG and in the one indicated by the learner

3.4 Reconstructing a Decision Tree

As in facility Sect. 3.3 above, we show to the learner a decision node in a CIG, and we ask her to "reconstruct" such a decision. However, differently from *ReconstructDecision* above, here we consider also the ordering in which the different parameters have to be considered (or, in other words, we ask to learners to reconstruct a decision tree). *DT* is the decision tree corresponding to *Decision*, and is asked "step by step" to the learner, as specified in the abstract procedure **ReconstructDT** below, whose core is the call to **RecursiveReconstruct**, which supports learners in the construction of a new decision tree, identical to DT. A graphical interface supports the learner in such a reconstruction, showing step-by-step the progresses in the construction of the learner's decision tree. Notably, though the final new decision tree is equal to DT, at each step the facility asks relevant questions to the learner, compares her answers to the DT, and signal differences (if any).

Procedure ReconstructDT(CIG, Decision)

Ask which parameters are needed to take the decision (let $PSet_{learn}$ the set of such parameters, and let $PSet_{CIG}$ the set of parameters specified in Decision);

IF ($PSet_{learn} \neq PSet_{CIG}$) THEN show the differences;

Ask to the learner what are the conclusions of the decision (let $CSet_{learn}$ the set of such conclusions, and let $CSet_{CIG}$ the set of conclusions of Decision in the CIG);

Let DT be the decision tree associated with Decision in the CIG, DTnode the root of DT, and N_{learn} the root of the decision tree to be constructed by the learner;

RecursiveReconstruct(DTnode, N_{learn}, $PSet_{CIG}$, $CSet_{CIG}$)

Procedure RecursiveReconstruct(DTnode, N_{learn}, $PSet_{CIG}$, $CSet_{CIG}$)

Ask what is the type of N_{learn}, compare to the type of DTnode, and (possibly) signal errors;

Set the type of N_{learn} to the type of DTnode;

IF N_{learn} is a conclusion-node THEN

    Ask what is the conclusion (in $CSet_{CIG}$) to be considered by N_{learn}, compare to the conclusion in DTnode, and (possibly) signal errors;

    Set the conclusion of N_{learn} to be the conclusion in DTnode

ELSE

    Ask what is the parameter (in $PSet_{CIG}$) to be considered by N_{learn}, compare to the parameter in DTnode, and (possibly) signal errors;

    Set the parameter of N_{learn} to be the parameter in DTnode (let it be P_{CIG});

    Ask how many arcs must stem from N_{learn}, compare to the number of arcs in DTnode, and (possibly) signal errors. Let $NumArc_{CIG}$ be the correct number;

    $CondSset_{learn} \leftarrow$ emptyset

    FOR i=1 TO $NumArc_{CIG}$ DO

        Ask to the learner an atomic condition regarding P_{CIG} and add it to $CondSet_{learn}$;

        Find the most favorable correspondence between conditions in $CondSet_{learn}$ and conditions on the arcs stemming from DTnode (let $CondSset_{CIG}$ be the set of such conditions).

    FOR EACH pair of corresponding conditions $<Cond_{learn} \in CondSet_{learn}, Cond_{CIG} \in CondSset_{CIG}>$ DO

        Check differences and show them (if any) to the learner;

        Create a new node $NNew_{learn}$, and a new arc $ANew_{learn}$, with condition $Cond_{CIG}$, connecting N_{learn} to $NNew_{learn}$;

        Let DTnodeChild be the child of DTnode corresponding to the arc with condition $Cond_{CIG}$;

        RecursiveReconstruct(DTnodeChild, $NNew_{learn}$, $PSet_{CIG}$, $CSet_{CIG}$)

For example, the PSB-decision node in Fig. 1(a) above is shown to learner without its conclusions. Let's assume that the parameters and the conclusions have already been asked to the learner, and the algorithm is asking her/him to reconstruct the decision tree. Specifically, in this example, the learner has already acquired the first two nodes and, now, we focus on the "OLN + CLN" node. In the Fig. 6(a), we can see the graphical interface shown to learner before her/his choice, in which gray nodes indicate those not yet filled by the learner. Now, the algorithm asks the type of the node (let's assume that the answer is internal node), the parameter (i.e., OLN + CLN) and the number of the arcs stemming from the node (i.e., 4). After that, for each arc, the learner has to indicate an atomic condition: let suppose that learner's answers are " ≥ 0 ", " $= 1$ ", "in [2, 3]" and " ≥ 4 ". Then, the algorithm finds the most favorable correspondence: e.g., the " ≥ 0 " condition corresponds to the correct " $= 0$ "; so, the algorithm shows differences to the learner and creates a new node "CLN" and a new arc with the correct condition " $= 0$ ", as shown in Fig. 6(b). After that, the next recursive step allows the conclusion node "N0" to be reconstructed, with the first "IF" in the algorithm.

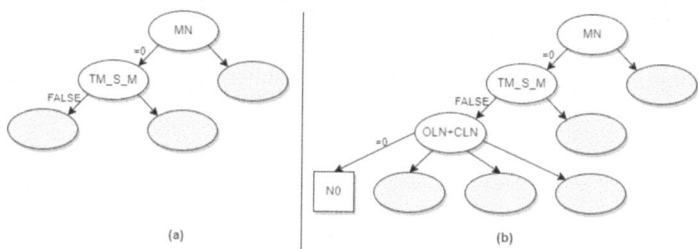

Fig. 6. Example of reconstruction of the decision tree: GUI shown to learner before (a) and after (b) her/his choice for the "OLN + CLN" node

4   Conclusions and Related Work

Within the PTPC-AI project, we are developing an innovative approach, GLARE-Edu, specifically devoted to medical education, based on CIG. In this paper, we focus on an essential aspect of the project: proposing facilities to facilitate learning and testing the competences needed to make appropriate decisions, and, specifically, addressing tasks (i – iv) discussed in the introduction.

The literature about CIG systems is extremely wide, and several CIG systems have been already applied (or have the potentialities to be applied) also for medical education. However, such systems are mainly designed for decision support, so that education-specific facilities such as the ones proposed in Sect. 3 of this paper have not been considered. The most closely related approaches are [8–10]. The paper [8] describes an educational approach for *arterial hypertension* based on CIGs. The paper [9] presents Shock-Instructor, a web-based e-learning tool that uses the CPGs about *seven types of shock*. Uribe-Ocampo et al. [10] have sketched a *serious game* called SIM-CIG, to simulate virtual patients in *antenatal health-care environment*, and evaluate the decision making of players. Our preliminary work in the area is described in [23], where we have planned an abstract roadmap to CIG-based medical education. None of such approaches, however, addresses the tasks described in Sect. 3 of this paper.

We have presented a summary of our goals in Sect. 2, sketching the PTPC-AI project. Notably, in this paper we have focused only on decision-making supports. However, GLARE-Edu also focuses on learning and testing the whole CIG process, including what *work* actions have to be performed on a patient, and when. In the rest of the project, we will overcome two major limitations of our current approach.

First, we will connect GLARE-Edu to a medical *ontology* (i.e., *SNOMED*), to provide learners with the possibility of choosing (during testing) clinical parameters, conclusions and clinical actions from the ontology (instead than from pre-defined menu's).

Second, the last part of the project we will be devoted to a systematic experimental evaluation of our approach. We will propose a six-month medical course to a set of physicians and medical students split into a class using GLARE-Edu, and a control class taking "traditional" medical education. A comparative evaluation of the learning level of the two classed will be performed, both in-itinere and at the end of the course.

Acknowledgments. This work is partially supported by Fondazione Compagnia di San Paolo and Fondazione CDP, Bando Intelligenza Artificiale 2, AI-LEAP project. Erica Raina is a PhD student enrolled in the National PhD program in Artificial Intelligence, XXXIX cycle, course on Health and life sciences, organized by Università Campus Bio-Medico di Roma.

References

1. Field, M.J., Lohr, K.N.: Clinical practice guidelines: Dir. New Program. National Academies Press (US) (1990)
2. Peleg, M., et al.: Comparing computer-interpretable guideline models: a case-study approach. J. Am. Med. Inform. Assoc. JAMIA. **10**, 52–68 (2003)
3. Bottrighi, A. et al.: Analysis of the GLARE and GPROVE approaches to clinical guidelines. In: Riaño, D., ten Teije, A., Miksch, S., Peleg, M. (eds.) Knowledge Representation for Health-Care. Data, Processes and Guidelines. KR4HC 2009. Lecture Notes in Computer Science(), vol. 5943. pp. 76–87. Springer (2010). https://doi.org/10.1007/978-3-642-11808-1_7
4. Peleg, M.: Computer-interpretable clinical guidelines: a methodological review. J. Biomed. Inform. **46**, 744–763 (2013)
5. Boxwala, A.A., Greenes, R.A., Deibel, S.R.: Architecture for a multipurpose guideline execution engine. Proc. AMIA Symp. 701–705 (1999)
6. Le, X.H., Luque, A.E., Wang, D.: Development of guideline-driven mobile applications for clinical education and decision support with customization to individual patient cases. Proc. AMIA 2012 (2012)
7. Le, X.H., Luque, A.E., Wang, D.: Assessing the usage of a guideline-driven interactive case simulation tool for insomnia screening and treatment in an HIV clinical education program. Stud. Health Technol. Inform. **192**, 323–327 (2013)
8. Real, F., Riaño, D., Alonso, J.R.: Training residents in the application of clinical guide-lines for differential diagnosis of the most frequent causes of arterial hypertension with decision tables. Proc. KR4HC 2014. pp. 147–159. Springer (2014)
9. Riaño, D., Real, F., Alonso, J.R.: Improving resident's skills in the management of circula-tory shock with a knowledge-based e-learning tool. Int. J. Med. Inf. **113**, 49–55 (2018)
10. Uribe-Ocampo, S., Torres, E.A., Luna, I.F., Florez-Arango, J.F., Smith, J.W.: SIM-CIG: a serious game to practice and improve clinical guidelines adoption based on computer-interpretable guidelines. Stud. Health Technol. Inform. **264**, 1997–1998 (2019)
11. Terenziani, P., Molino, G., Torchio, M.: A modular approach for representing and executing clinical guidelines. Artif. Intell. Med. **23**, 249–276 (2001)
12. Terenziani, P., Montani, S., Bottrighi, A., Torchio, M., Molino, G.: Supporting physicians in taking decisions in clinical guidelines: the GLARE "what if" facility. Proc. AMIA Symp. 772–776 (2002)

13. Terenziani, P., Carlini, C., Montani, S.: Towards a comprehensive treatment of temporal constraints in clinical guidelines. Proc. TIME (2002)
14. Giordano, L., Terenziani, P., Bottrighi, A., Montani, S., Donzella, L.: Model checking for clinical guidelines: an agent-based approach. AMIA Annu. Symp. Proc. AMIA Symp. **2006**, 289–293 (2006)
15. Leonardi, G., Bottrighi, A., Galliani, G., Terenziani, P., Messina, A., Della Corte, F.: Exceptions handling within GLARE clinical guideline framework. AMIA Annu. Symp. Proc. AMIA Symp. **2012**, 512–521 (2012)
16. Piovesan, L., Terenziani, P.: A mixed-initiative approach to the conciliation of clinical guidelines for comorbid patients. Proc. KR4HC 2015. **9485**, 95–108 (2015)
17. Piovesan, L., Anselma, L., Terenziani, P.: Temporal detection of guideline interactions. Proc. HEALTHINF 2015 (2015)
18. Bottrighi, A., Terenziani, P.: META-GLARE: A meta-system for defining your own computer interpretable guideline system-architecture and acquisition. Artif. Intell. Med. **72**, 22–41 (2016)
19. Spiotta, M., Terenziani, P., Dupré, D.T.: Temporal conformance analysis and explanation of clinical guidelines execution: an answer set programming approach. IEEE Trans. Knowl. Data Eng. **29**, 2567–2580 (2017)
20. Piovesan, L., Terenziani, P., Molino, G.: GLARE-SSCPM: An Intelligent System to Support the Treatment of Comorbid Patients. IEEE Intell. Syst. **33**, 37–46 (2018)
21. Andolina, A., Guazzone, M., Piovesan, L., Terenziani, P.: Temporal reasoning and query answering with preferences and probabilities for medical decision support. Expert Syst. Appl. **195** (2022)
22. Associazione Italiana Oncologia Medica: LG AIOM melanoma, https://www.aiom.it/wp-content/uploads/2020/10/2020_LG_AIOM_Melanoma.pdf. Accessed 20 Nov 2023
23. Bottrighi, A., Molino, G., Piovesan, L., Terenziani, P.: Towards an "operational" educational model in healthcare: exploiting computer-interpretable guidelines. Proc. HEALTHINF (2019)

Generating Contrastive Explanations from Gradual Semantics Rankings

Mariela Morveli-Espinoza$^{(\boxtimes)}$ (iD) and Juan Carlos Nieves (iD)

Department of Computing Science, Umeå University, Umeå, Sweden
morveli.espinoza@gmail.com, jcnieves@cs.umu.se

Abstract. Argumentation is a sub-field of Artificial Intelligence (AI) that gives a way for reasoning with inconsistent, uncertain, and incomplete knowledge. An Argumentation Framework (AF) is an argumentation approach made of a set of arguments and a relation between them called attack. Extension-based semantics and gradual semantics are two ways of reasoning in AFs. The former returns sets of consistent arguments and the latter assigns a value to each argument with the aim of assessing them. Explainable Artificial Intelligence aims to make clear the reasoning of AI systems to the humans (or to other systems) with which they interact. In literature, we can find some approaches for generating explanations for the results of extension-based semantics; however, there is no approach – to the best of our knowledge – for generating explanations for the results of gradual semantics. Thus, this work aims to generate explanations for these types of semantics. Specifically, we will generate contrastive explanations, which are useful and intuitive kind of explanation that explains decisions to lay people by imitating the way in which humans do it. Two types of contrastive explanations will be generated: Property-contrast explanations and Object-contrast ones. The former answers the question *why is an argument A in a position p in the resultant ranking rather than in position p'?* and the latter answers the question *why is an argument $\mathcal{E}$ in position p of the ranking whereas argument $\mathcal{E}'$ is in position p'?*. We use a scenario of decision making for illustrating our approach and make an analysts about when a constrastive explanation can be generated and when it cannot.

Keywords: Argumentation framework · bipolar argumentation frameworks · gradual semantics · contrastive explanations · uncertainty

1 Introduction

Explainability is one of the necessary ethical principles that must be respected in order to reach the trustworthy of AI systems [9,15]. In intelligent systems, it has gained attention in recent years due to their growing utilization in human-AI interaction applications such as recommendation or coaching systems in domains as e-health (e.g., [7]), UAVs (Unmanned Aerial Vehicle) (e.g., [8]), or smart environments (e.g., [14]). In these applications, the outcomes returned by the

© The Author(s), under exclusive license to Springer Nature Switzerland AG 2025
L. Correia et al. (Eds.): IBERAMIA 2024, LNCS 15277, pp. 250–261, 2025.
https://doi.org/10.1007/978-3-031-80366-6_21

systems can be negatively affected due to the lack of clarity and explicability about their dynamics, rationality, and reasoning. Thus, if these systems would be equipped with explicability abilities, then their understanding, reliability, and acceptance could be enhanced.

Formal argumentation provides intelligent systems with a way for reasoning with inconsistent, uncertain, and incomplete knowledge. The reasoning provided by argumentation techniques is considered transparent [16], which allows to generate explanations more easily than with other methods. An elemental Argumentation Framework (AF) is composed of a set of arguments and a binary relation encoding disagreements – called attacks – between arguments [6]. An extension of AFs, considers a positive relation called support, this extension is known as Bipolar AFs (BAFs) [3]. Hereafter, we will speak only about BAFs, because an AF is a BAF where its support relation is empty. An important notion in formal argumentation is the acceptability of arguments, which is assessed by applying argumentation semantics over a BAF. We can mainly distinguish extension-based semantics and gradual semantics. The former return sets of consistent arguments, that is, arguments that can be accepted together (these sets are known as extensions) and the latter are based on numerical evaluations (e.g. see [2,11,19]), and they aim to assign a numerical value to the arguments in order to rank them and establish their acceptability level. This work aims to generate explanations for gradual semantics results. This means that arguments in a BAF have numerical values associated. This type of BAFs are known as weighted BAFs (wBAFs) [1]. The idea is that a gradual semantics takes as input a wBAF whose arguments have initial associated values – known as base score values – and returns another values – known as strength values – which are calculated based on the relations of the arguments with their attacks and/or supports. Based on these strength values, a ranking is generated. The idea of our proposal is to take this ranking as input and return an explanation. Figure 1 shows a general view of our approach.

In recent years, formal argumentation and explainable AI (XAI) are closely related since formal argumentation has been applied to AI systems for providing explainability. In [5], the authors survey three forms of explanations based on AFs. These forms are mainly based on extension-based semantics. However, no form of explanation for gradual semantics is mentioned. Therefore, there is a need for methods that return explanations based on the output of gradual semantics in order to better understand the reasons for rankings. Thus, the research question that is addressed in this paper is: *How to characterize and generate explanations in the settings of gradual semantics?*

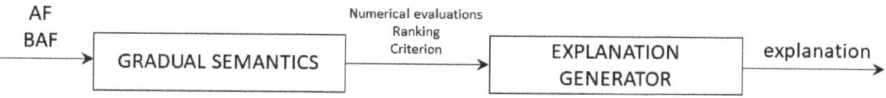

Fig. 1. General view of the proposed approach.

Regarding the generated explanations, a *contrastive explanation* is a type of explanation that is commonly employed by people and can bring benefits to the explanations exchange process. *Contrastive questions* provide an insight into the questioner's mental model, allowing to have a better understanding of what they do not know and contrastive explanations usually are more straightforward, more feasible, and less demanding for both the questioner and explainer [12]. In [17], the authors distinguish three types of contrastive questions:

- **(P-contrast)** Why does object o have property p, rather than property p'?
- **(O-contrast)** Why does object o have property p, while object o' has property p'?
- **(T-contrast)** Why does object o have property p at time t, but property p' at time t'?

In the case of wBAFs, it would be interesting to give explanations that compare the positions of arguments in the ranking in order the users better understand the reasoning of the system. For this work, we will consider P-contrast and O-contrast questions since we are not dealing with dynamic wBAFs where the factor time can be included. We will consider arguments as objects and positions in the ranking as properties. In this work, constrastive explanations are a set of beliefs. We illustrate how the approach works by generating explanations for a decision making scenario. We also analyse when an explanation can be generated and when it cannot.

The remainder of this paper is structured as follows. The next section gives a brief overview on wBAFs and gradual semantics. In Sect. 3, we present the approach for generating contrastive explanations. The use case based on the decision making scenario is presented in Sect. 4. We study the main properties of the proposed approach in Sect. 5. Finally, Sect. 6 is devoted to conclusions and future work.

2   Background

In this section, we will introduce the background concepts on which be base our proposal. We present what is a weighted BAF and a gradual semantics.

Definition 1. *(wBAF) [1] A weighted bipolar argumentation framework is a tuple of the form* $\text{wBAF} = \langle \text{ARG}, \mathcal{R}_{att}, \mathcal{R}_{sup}, \tau \rangle$ *where* ARG *is a finite set of arguments,* $\mathcal{R}_{att} \subseteq \text{ARG} \times \text{ARG}$ *is a binary relation that denotes attacks between two arguments,* $\mathcal{R}_{sup} \subseteq \text{ARG} \times \text{ARG}$ *is a binary relation that denotes support between two arguments, and* $\tau : \text{ARG} \to \mathbb{R}$ *is a function that assigns a base score value to each argument.*

For any $A, B \in \text{ARG}$, *the notation* $(A, B) \in \mathcal{R}_{att}$ *means that* A *attacks* B *and the notation* $(A, B) \in \mathcal{R}_{sup}$ *means that* A *supports* B.

The way of evaluating the arguments for generating a ranking is known as gradual semantics.

Definition 2. *(Gradual semantics)* *[1] Given a* wBAF $= \langle \text{ARG}, \mathcal{R}_{att}, \mathcal{R}_{sup}, \tau \rangle$, *a gradual semantics is a function* $\sigma : 2^{\text{ARG}} \longrightarrow \mathbb{R}$ *that assigns to each argument a real number.*

From the strengths returned by the gradual semantics, a ranking can be constructed. For the construction of the ranking, more than one criterion can be used. For example, a simple criterion could be to rank from the lowest to the greatest strength value. This means that more than one ranking can be constructed from the same output of a gradual semantics.

3 Generating Contrastive Explanations

In this section, we present how to generate contrastive explanations from a ranking –produced applying a gradual semantics – and a criteria used for such ranking.

Generally, gradual semantics measure arguments to be ranked; however, in [13], the authors employ gradual semantics for ranking extensions, that is sets of consistent arguments. This means that depending on the modelled problem by applying formal argumentation, other elements of a wBAF can be ranked. In order to generalise, we will call "component" (denoted by comp) to the elements of the ranking. Besides, we denote by COMPS to the set of all the components of a ranking and by $\text{RANK}_{\sigma}^{\text{crit}}$ to the rank constructed from the strength values obtained by applying the gradual semantics σ under a criterion crit. Notice that more than one component can share the same position in a $\text{RANK}_{\sigma}^{\text{crit}}$ because they may have the same strength value.

Now, let us present the contrastive questions. Given a component comp that appears in a ranking $\text{RANK}_{\sigma}^{\text{crit}}$, these can be expressed in the following way:

– **P-contrast**: WHY(comp, p(comp, $\text{RANK}_{\sigma}^{\text{crit}}$), pos$_{\sigma}$) (*Why is a component* comp *in position* p(comp, $\text{RANK}_{\sigma}^{\text{crit}}$) *rather than in position* pos$_{\sigma}$ *?*)
– **O-contrast**: WHY(comp, p(comp, $\text{RANK}_{\sigma}^{\text{crit}}$), comp', p(comp', $\text{RANK}_{\sigma}^{\text{crit}}$)) (*Why is a component* comp *in position* p(comp, $\text{RANK}_{\sigma}^{\text{crit}}$) *whereas a component* comp' *is in position* p(comp', $\text{RANK}_{\sigma}^{\text{crit}}$)*?*)

where p is a function that returns the position of a given component comp in a ranking $\text{RANK}_{\sigma}^{\text{crit}}$ and pos$_{\sigma} \in \mathbb{N}$ is a contrasted position in the same ranking such that p(comp, $\text{RANK}_{\sigma}^{\text{crit}}$) $\neq$ pos$_{\sigma}$.

For constructing the explanations, we base on some information. We consider the resultant strength values of a gradual semantics, a ranking constructed from such strength values, and a criterion for constructing the ranking. We call these information the frame for the contrastive explanations.

Definition 3. *(Contrastive Explanation Frame) A contrastive explanation frame is a tuple* CEF $= \langle \text{COMPS}, \sigma, \text{crit}, \text{RANK}_{\sigma}^{\text{crit}} \rangle$ *where (i)* COMPS *is a set of components, (ii)* σ *returns the resultant strength values of all components, (iii)* crit *is a criterion of ordering, and (iv)* $\text{RANK}_{\sigma}^{\text{crit}}$ *is the ranking constructed considering the strength values from* σ *under the criterion* crit.

The resultant contrastive explanations can be seen as sequences of observations that constitute explicative beliefs.

Definition 4. (Contrastive Explanation) *Given a* CEF $= \langle$COMPS$, \sigma,$ crit$,$ RANK$_\sigma^{\text{crit}}\rangle$, *a constrastive explanation* EXP *is a finite set of explicative beliefs, which are ground predicates of the form* $better\_under($crit$,$ comp$, x,$ comp$', y)$ *where* comp *and* comp$'$ *appear in* RANK$_\sigma^{\text{crit}}$, $x = \sigma($comp$)$, *and* $y = \sigma($comp$')$.

Now, let us present how the contrastive explanations are generated.

3.1 For the P-Contrastive Question

For the P-contrastive question WHY$($comp$, p($comp$,$ RANK$_\sigma^{\text{crit}}),$ pos$_\sigma)$, we consider two cases:

(i) When $p($comp$,$ RANK$_\sigma^{\text{crit}}) >$ pos$_\sigma$
(ii) When $p($comp$,$ RANK$_\sigma^{\text{crit}}) <$ pos$_\sigma$.

The former case aims to explain why a component comp is in a worse position than the contrasted one and the latter aims to explain why a component comp is in a better position than the contrasted one.

Algorithm 1 shows how to construct the explanation for case (i). It takes as input a a contrastive explanation frame and returns a contrastive explanation EXP. It begins by obtaining the set of components that have better positions than comp until the position pos$_\sigma$ (line 2). Since there could be more than one better-positioned components, the explanation can have more than one belief and we have to return an explanation that is the union of all the explicative beliefs (line 4).

Algorithm 1. Generation of the explanation for the P-contrast question WHY$($comp$, p($comp$,$ RANK$_\sigma^{\text{crit}}),$ pos$_\sigma)$: Case (i)

Input: CEF $= \langle$COMPS$, \sigma,$ crit$,$ RANK$_\sigma^{\text{crit}}\rangle$
Output: EXP
  1: EXP $:= \emptyset$
  2: comp_prev $:= \{$comp$'$ *appears in* RANK$_\sigma^{\text{crit}} \mid p_\sigma($comp$') < p_\sigma($comp$)$ and $p_\sigma($comp$') \geq$ pos$_\sigma\}$
  3: **for all** comp$' \in$ comp_prev **do**
  4:    EXP $:=$ EXP $\cup\ better\_under($crit$,$ comp$', \sigma($comp$'),$ comp$, \sigma($comp$))$
  5: **end for**

Algorithm 2 shows how to generate the explanation for case (ii). Since the required explanation is for knowing why component comp is better located than the contrasted one, we first obtain the component(s) that are in the contrasted position by applying function COMPONENTS_IN $: \mathbb{N} \longrightarrow 2^{\text{COMPS}'}$, where COMPS$' \subseteq$ COMPS (line 1). Then, the explanation is constructed (line 4).

Algorithm 2. Generation of the explanation for the P-contrast question $\text{WHY}(\text{comp}, p(\text{comp}, \text{RANK}_\sigma^{\text{crit}}), \text{pos}_\sigma)$: Case (ii)

Input: $\text{CEF} = \langle \text{COMPS}, \sigma, \text{crit}, \text{RANK}_\sigma^{\text{crit}} \rangle$
Output: EXP
1: $\text{comps\_in} = \text{COMPONENTS\_IN}(\text{pos}_\sigma)$
2: $\text{EXP} := \emptyset$
3: **for all** $\text{comp}' \in \text{comps\_in}$ **do**
4:     $\text{EXP} := \text{EXP} \cup better\_under(\text{crit}, \text{comp}, \sigma(\text{comp}), \text{comp}', \sigma(\text{comp}'))$
5: **end for**

3.2 For the O-Contrastive Question

For the O-contrastive question $\text{WHY}(\text{comp}, p(\text{comp}, \text{RANK}_\sigma^{\text{crit}}), \text{comp}', p(\text{comp}', \text{RANK}_\sigma^{\text{crit}}))$, we consider two cases:

1. When $p(\text{comp}, \text{RANK}_\sigma^{\text{crit}}) < p(\text{comp}', \text{RANK}_\sigma^{\text{crit}})$. This means that comp has a better position than comp'. In natural language the question would be: *Why does* comp *have a better position than* comp'*?*
2. When $p(\text{comp}, \text{RANK}_\sigma^{\text{crit}}) > p(\text{comp}', \text{RANK}_\sigma^{\text{crit}})$. This means that comp' has a better position than comp. In natural language the question would be: *Why does* comp *have a worse position than* comp'*?*

Unlike the explanations for the P-contrastive questions – where more than one explicative belief could be generated – in this case only one belief will be generated. Thus, given a $\text{CEF} = \langle \text{COMPS}, \sigma, \text{crit}, \text{RANK}_\sigma^{\text{crit}} \rangle$, the explanations for each case presented above are:

1. $\text{EXP}(\text{CEF}) = \{better\_under(\text{crit}, \text{comp}, \sigma(\text{comp}), \text{comp}', \sigma(\text{comp}'))\}$
2. $\text{EXP}(\text{CEF}) = \{better\_under(\text{crit}, \text{comp}', \sigma(\text{comp}'), \text{comp}, \sigma(\text{comp}))\}$

4 Use Case: Decision Making Scenario

In this section, we show how to generate constrastive explanations for an scenario of decision making. This case study was extracted from [13].

4.1 The Scenario

This scenario consider a factitious discussion between a group of medicine students (which can be represented by agents[1]). The discussion is about the diagnosis of a patient. In this context, arguments represent assertions about symptoms, a likely diagnosis, or facts about the patient; the attacks represent conflicts between arguments, and supports represent causality relations. Figure 2 shows

---

[1] An agent is a computer system. According to Wooldridge e Jennings [18], an agent can be characterized by using mentalistic notions, such as knowledge, belief, intention, and obligation.

the argumentation graph where nodes represent arguments and edges the attacks (or supports) between arguments. In the graph, arguments A and B represent two possible diagnosis namely measles and chickenpox, respectively, and the aim is to make a group decision about the diagnosis based on the opinions of the students. After applying the imprecise gradual semantics proposed in [13], we obtain a ranking of the extensions where the first ranked – following a given criterion – is extension $\{A, D\}$, the second one is $\{A, D, G\}$, the third one is $\{A\}$, the forth one is $\{B\}$, the fifth one is $\{A, G\}$, the sixth one $\{B, C\}$, the seventh one is $\{A, D, E, F, G\}$, and in the last position is extension $\{C, E, F, D, G\}$. We can notice that the the three first ranked extensions include argument A, which corresponds to diagnosis *"The patient has measles"* whereas B= *"The patient has chickenpox"* appears in the forth and sixth positions. Thus, after the results, it would be interesting to explain the users why extension $\{A, D, G\}$ is in the second position rather than in the first one or why an extension that includes argument B is not in the first position.

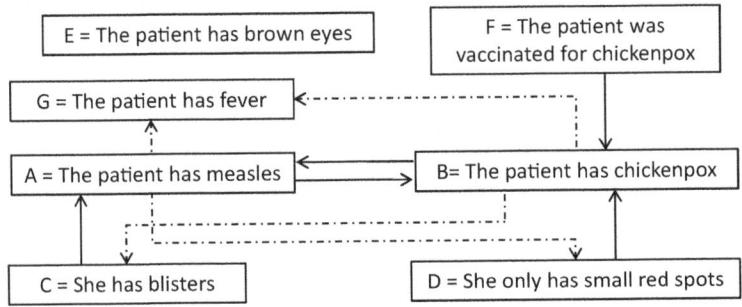

Fig. 2. Bipolar argumentation graph for the decision making about the diagnose of a patient. Solid lines represent attacks (conflicts) and traced lines represent supports (modelling causality relations).

In [13], the authors model this scenario by means of a BAF. For making a decision, they evaluate the BAF in two steps, first they apply an extension-based semantics and then they employ a gradual semantics for ranking the obtained extensions. Since they consider that each argument has associated more than one numerical value – which represent the opinion of an expert – they capture these group disagreeing opinions by means of imprecise probabilities. Specifically, they associate to each argument and to each extension an interval. This means that they rank intervals. The criteria they use for the ranking are the precision of the interval (denoted by PREC), the location of it (denoted by LOCA), and the combination of precision and location (denoted by COMB). Next, we show how each criterion is calculated for a better understanding of the scenario. Given an extension $\mathcal{E}$ and its associated interval $[\underline{P}(\mathcal{E}), \overline{P}(\mathcal{E})]$ where $\underline{P}(\mathcal{E})$ and $\overline{P}(\mathcal{E})$ are its lower and upper bounds, respectively, the calculation of each criterion is as follows:

$$\text{LOCA}(\mathcal{E}) = \left(\frac{\underline{P}(\mathcal{E}) + \overline{P}(\mathcal{E})}{2} \right)$$
$$\text{PREC}(\mathcal{E}) = 1 - (\overline{P}(\mathcal{E}) - \underline{P}(\mathcal{E}))$$
$$\text{COMB}(\mathcal{E}) = \text{LOCA}(\mathcal{E}) \times \text{PREC}(\mathcal{E})$$

Thus, we have that three possible rankings can be generated. Data of Table 1 was taken from [13]. In this table we have the set of components, that is extensions, the strength value returned after applying the gradual semantics and the numerical value of each criterion. Table 2 shows the rankings of extensions considering the three criteria. Notice that for criterion location there are two components – extensions – in position two and the same happens for criterion precision in position six.

Table 1. Values of the strength, and the criteria location, precision, and the combination of both. Extracted from [13].

Extension $\mathcal{E}$	σ	$\text{LOCA}(\mathcal{E})$	$\text{PREC}(\mathcal{E})$	$\text{COMB}(\mathcal{E})$
$\{C, E, F, D, G\}$	$[0, 0]$	0	1	0
$\{A, D, E, F, G\}$	$[0.03, 0.25]$	0.14	0.78	0.11
$\{A, D, G\}$	$[0.24, 0.45]$	0.34	0.79	0.27
$\{B, C\}$	$[0.1, 0.75]$	0.425	0.35	0.15
$\{B\}$	$[0.15, 0.7]$	0.43	0.45	0.19
$\{A\}$	$[0.25, 0.8]$	0.53	0.45	0.24
$\{A, D\}$	$[0.55, 0.7]$	0.35	0.85	0.53
$\{A, G\}$	$[0.1, 0.5]$	0.3	0.6	0.18

Table 2. Rankings considering location, precision, and the combination of both.

Order	LOCA	PREC	COMB
1	$\{A, D\}$	$\{C, E, F, D, G\}$	$\{A, D\}$
2	$\{A\}$	$\{A, D\}$	$\{A, D, G\}$
3	$\{B\}, \{B, C\}$	$\{A, D, G\}$	$\{A\}$
4	$\{A, D, G\}$	$\{A, D, E, F, G\}$	$\{B\}$
5	$\{A, G\}$	$\{A, G\}$	$\{A, G\}$
6	$\{A, D, E, F, G\}$	$\{A\}, \{B\}$	$\{B, C\}$
7	$\{C, E, F, D, G\}$	$\{B, C\}$	$\{A, D, E, F, G\}$
8			$\{C, E, F, D, G\}$

We can notice that the extension with the best location and combined value is $\{A, D\}$. $\{C, E, D, F, G\}$ has the best precision; however, the worst possible location. The second best precision is also for $\{A, D\}$. This extension indicates that the result of the decision*-making process – after aggregating the opinions of the students – is that the disease is measles and its strongest support is that the patient only has small red spots. On the other hand, notice that $\{C, E, F, D, G\}$ has the worst location and the worst combined value. This happens because there is no support relation between its elements, this means that all of them are independent, which impacts on the calculation. Besides, this extension does not include any of the diagnosis, so it could not support the decision-making. The notion of independence also impacts on the evaluation of $\{A, D, E, F, G\}$, note that its position in the ordering lists is almost the worst. Although this set includes A, which represents a diagnosis, it does not have any support relation with arguments E and F.

4.2 Explanations

In this subsection, we present three possible explanations for the rankings generated previously.

1. Question WHY($\{B\}, 4, 1$), which can be understood as *Why is not chickenpox the diagnosis?*. We consider the following frame for answering this question is CEF $= \langle$COMPS$, \sigma,$ COMB, RANK$_\sigma^{\text{COMB}}\rangle$, where COMPS and the values for σ are in Table 1 and the ranking RANK$_\sigma^{\text{COMB}}$ is taken from Table 2. Thus the generated explanation is:

   EXP(CEF) $= \{better\_under$(COMB$, \{A, D\}, [0.55, 0.7], \{B\}, [0.15, 0.7])$,
   $better\_under$(COMB$, \{A, D, G\}, [0.24, 0.45], \{B\}, [0.15, 0.7])$,
   $better\_under$(COMB$, \{A\}, [0.25, 0.8], \{B\}, [0.15, 0.7])\}$

   With this explanation, a user can notice that the three better ranked extensions include A, which correspond to the other diagnosis. Besides, the user will also know that these extensions are better than $\{B\}$ under the combination of precision and location. Finally, the user could analyse that the decision is not really chickenpox because measles is better supported by other arguments.

2. Question WHY($\{A, D\}, 1, 4$), which can be understood as *why $\{A, D\}$ – having less arguments than $\{A, D, G\}$ – is better ranked than it?*. We consider the following frame for answering this question is CEF $= \langle$COMPS$, \sigma,$ LOCA, RANK$_\sigma^{\text{LOCA}}\rangle$, where COMPS and the values of σ are in Table 1 and the ranking RANK$_\sigma^{\text{LOCA}}$ is taken from Table 2. Thus the generated explanation is:

   EXP(CEF) $= \{better\_under$(LOCA$, \{A, D\}, [0.55, 0.7], \{A, D, G\}, [0.24, 0.45])\}$

   With this explanation, a user can notice that $\{A, D\}$ is in position 1 because it is has a better location than $\{A, D, G\}$, which is in position 4. For a more informative explanations, it will necessary to know how the calculation of the intervals was done.

3. Question $\mathrm{WHY}(\{A, D\}, 2, \{B, C\}, 7)$, which can be understood as *why measles with an support is the diagnosis and not chickenpox with also one support?*. We consider the following frame for answering this question is $\mathrm{CEF} = \langle \mathrm{COMPS}, \sigma, \mathrm{PREC}, \mathrm{RANK}_\sigma^{\mathrm{PREC}} \rangle$, where COMPS and the value of σ are in Table 1 and the ranking $\mathrm{RANK}_\sigma^{\mathrm{PREC}}$ is taken from Table 2. Thus the generated explanation is:

$$\mathrm{EXP}(\mathrm{CEF}) = \{better\_under(\mathrm{PREC}, \{A, D\}, [0.55, 0.7], \{B, C\}, [0.1, 0.75])\}$$

With this explanation a user knows that the extension containing measles is more precise than the extension containing chickenpox, which indicates that the uncertainty in the information of the former is less than the uncertainty in the information of the latter. Thus, in this case the user may also have clues about what is happening with the information in extensions.

5   Properties of the Approach

In this section, we analyse when an explanation can be generated and when it cannot.

It is important to study the existence of contrastive explanations as defined in this work and in what circumstances a contrastive explanation does not exist. Let us recall that for generating a contrastive explanation we need a ranking constructed by using the numerical values returned by a gradual semantics and with base on a criterion. Thus, if such ranking exists, a contrastive explanation will indeed exist. However, there are three cases when we cannot generate a contrastive explanation:

1. Recall that the components of the ranking are elements of a wAF (or a wBAF), which can be arguments, extensions, attack relations, or any other element related to a wAF (or a wBAF). Thus, assume that the components are arguments and the evaluated wAF has only one argument. In this case, it is not possible to generate a ranking with more than one element, which is necessary for making the contrast of positions. The same happens when there is only one extension or any other component. This means that we cannot generate contrastive explanations from ranking that only have one component.
2. Following the previous idea, it could happen that there is more than one component; however, the numerical values returned by the gradual semantics are the same, which means that all the components are in the same position.
3. Finally, we can have more than one component and different numerical values; however, following the criterion all the components are in the same position.

Thus, we can say that a contrastive explanation cannot be generated when the ranking has only one position and it is necessary to have at least two positions for contrasting them.

Proposition 1. *(Non-existence) Let* $\texttt{CEF} = \langle \texttt{COMPS}, \sigma, \texttt{crit}, \texttt{RANK}_\sigma^{\texttt{crit}} \rangle$ *be a contrastive explanation framework. If* $|\texttt{RANK}_\sigma^{\texttt{crit}}| = 1$, *then* $\texttt{EXP}(\texttt{CEF}) = \emptyset$.

Proof. By reduction a*b absurbo*. Given a $\texttt{CEF} = \langle \texttt{COMPS}, \sigma, \texttt{crit}, \texttt{RANK}_\sigma^{\texttt{crit}} \rangle$ and a question is $\texttt{WHY}(\texttt{comp}, p(\texttt{comp}, \texttt{RANK}_\sigma^{\texttt{crit}}), \texttt{pos}_\sigma)^2$. Assume that $\texttt{EXP}(\texttt{CEF}) \neq \emptyset$. This means that exists an explanation that contrasts a position returned by function p and a contrasted position $\texttt{pos}_\sigma$. Given that $p(\texttt{comp}, \texttt{RANK}_\sigma^{\texttt{crit}}) \neq \texttt{pos}_\sigma$, it means that $|\texttt{RANK}_\sigma^{\texttt{crit}}|$ is at least 2, which contradicts the premise of the proposition.

On the other hand, when the ranking has more than one position, contrastive explanations can be generated. Next proposition asserts it.

Proposition 2. *(Existence)Let* $\texttt{CEF} = \langle \texttt{COMPS}, \sigma, \texttt{crit}, \texttt{RANK}_\sigma^{\texttt{crit}} \rangle$ *be a contrastive explanation framework. If* $|\texttt{RANK}_\sigma^{\texttt{crit}}| > 1$, *then* $\texttt{EXP}(\texttt{CEF}) \neq \emptyset$.

Proof. By reduction *ab absurbo*. Given a $\texttt{CEF} = \langle \texttt{COMPS}, \sigma, \texttt{crit}, \texttt{RANK}_\sigma^{\texttt{crit}} \rangle$ and a question is $\texttt{WHY}(\texttt{comp}, p(\texttt{comp}, \texttt{RANK}_\sigma^{\texttt{crit}}), \texttt{pos}_\sigma)$ case (i). Assume that $\texttt{EXP}(\texttt{CEF}) = \emptyset$ and $p(\texttt{comp}, \texttt{RANK}_\sigma^{\texttt{crit}}) = |\texttt{RANK}_\sigma^{\texttt{crit}}|$. Let us consider Algorithm (1), it means that $\texttt{comp\_prev} := \emptyset$. It means that $p(\texttt{comp}, \texttt{RANK}_\sigma^{\texttt{crit}}) = \texttt{pos}_\sigma$, which in turn means that $|\texttt{RANK}_\sigma^{\texttt{crit}}| = 1$, which contradicts the premise of the proposition.

6 Conclusions and Future Work

This work presented an approach for generating contrastive explanations for P-contrast and O-contrast questions in the settings of gradual semantics for BAFs. Based on the positions, three cases of explanations were identified, two for P-contrast question and one for O-constrast question. For constructing the explanations, we based on the ranking that can be constructed basing on the output of the gradual semantics and an ordering criterion. We presented a case study for a decision making scenario, which exemplifies the explanations and how they can be understood. Finally, some properties were studied.

Regarding related work, we only found one work that generates constrastive explanations in the context of wBAFS –or quantitative as they call it. In [10], the authors present an approach to explain the change of inference in Quantitative BAFS (QBAFs), that is, the reason a gradual semantics change its results. Three types of explanations are generated: sufficient ones, counterfactual ones, and necessary ones. In this case, explanations are made of arguments, which vary depending on the case and we only found one work that studies how to generate contrastive explanations in the context of extension-based semantics. In [4], the authors tackle the question *"Why is argument A acceptable and argument B not?"*.

As future work, we aim to study the T-contrast questions. We believe it has to do with dynamics in AFs due to the time component. We also would like to study other criteria for contrasting properties. Finally, we want to generate more precise and richer explanations for credal BAFs.

---

[2] We present the proofs for the P-Contrastive question due to the lack of space; however, these can be understood for the other questions because in all the cases the current position and the contrasted positions are taken into account.

References

1. Amgoud, L., Ben-Naim, J.: Evaluation of arguments in weighted bipolar graphs. Int. J. Approximate Reasoning **99**, 39–55 (2018)
2. Amgoud, L., Ben-Naim, J., Doder, D., Vesic, S.: Ranking arguments with compensation-based semantics. In: Fifteenth International Conference on the Principles of Knowledge Representation and Reasoning (2016)
3. Amgoud, L., Cayrol, C., Lagasquie-Schiex, M.C., Livet, P.: On bipolarity in argumentation frameworks. Int. J. Intell. Syst. **23**(10), 1062–1093 (2008)
4. Borg, A., Bex, F.: Contrastive explanations for argumentation-based conclusions. arXiv preprint arXiv:2107.03265 (2021)
5. Čyras, K., Rago, A., Albini, E., Baroni, P., Toni, F.: Argumentative xAI: a survey. arXiv preprint arXiv:2105.11266 (2021)
6. Dung, P.M.: On the acceptability of arguments and its fundamental role in non-monotonic reasoning, logic programming and n-person games. Artif. Intell. **77**(2), 321–357 (1995)
7. Guerrero, E., Nieves, J.C., Lindgren, H.: An activity-centric argumentation framework for assistive technology aimed at improving health. Argument Comput. **7**(1), 5–33 (2016). https://doi.org/10.3233/AAC-160004
8. Gunetti, P., Thompson, H., Dodd, T.: Autonomous mission management for UAVs using soar intelligent agents. Int. J. Syst. Sci. **44**(5), 831–852 (2013)
9. Jobin, A., Ienca, M., Vayena, E.: Artificial intelligence: the global landscape of ethics guidelines (2019)
10. Kampik, T., Čyras, K., Alarcón, J.R.: Change in quantitative bipolar argumentation: sufficient, necessary, and counterfactual explanations. Int. J. Approximate Reasoning **164**, 109066 (2024)
11. Matt, P.A., Toni, F.: A game-theoretic measure of argument strength for abstract argumentation. In: European Workshop on Logics in Artificial Intelligence, pp. 285–297. Springer (2008)
12. Miller, T.: Contrastive explanation: a structural-model approach. Knowl. Eng. Rev. **36**, e14 (2021)
13. Morveli-Espinoza, M., Nieves, J.C., Tacla, C.A.: Probabilistic causal bipolar abstract argumentation: an approach based on credal networks. Ann. Math. Artif. Intell. **91**, 1–20 (2023)
14. Nieves, J.C., Lindgren, H.: Deliberative argumentation for service provision in smart environments. In: European Conference on Multi-Agent Systems, pp. 388–397. Springer (2014)
15. Smuha, N.A.: The EU approach to ethics guidelines for trustworthy artificial intelligence. Comput. Law Rev. Int. **20**(4), 97–106 (2019)
16. Toni, F., Craven, R., Fan, X.: Transparent rational decisions by argumentation. PhilSci Archive (2013)
17. Van Bouwel, J., Weber, E.: Remote causes, bad explanations? J. Theory Soc. Behav. **32**(4), 437–449 (2002)
18. Wooldridge, M., Jennings, N.R.: Intelligent agents: theory and practice. Knowl. Eng. Rev. **10**(2), 115–152 (1995)
19. Yun, B., Vesic, S.: Gradual semantics for weighted bipolar SETAFs. In: European Conference on Symbolic and Quantitative Approaches with Uncertainty, pp. 201–214. Springer (2021)

Automatic Classification of Secondary and High School Students Dropout Risk via Knowledge Graphs and Machine Learning

Daniel Zapata-Medina[✉][iD], Albeiro Espinosa-Bedoya[iD],
and Jovani Alberto Jiménez-Builes[iD]

Artificial Intelligence in Education Research Group, Universidad Nacional de
Colombia, Medellín, Colombia
{dzapatame,aespinos,jajimen1}@unal.edu.co

Abstract. This study presents the implementation of an approach to classifying school dropout risk among secondary and high school students. By leveraging knowledge graph embeddings and machine learning algorithms, we aim to enhance classification accuracy and interpretability. We constructed a knowledge graph from a dataset of 1,830 students in a real scenario of a public educational institution; the records were composed of demographic and academic features. We trained the ComplEx model to generate embeddings. We subsequently used these embeddings in clustering and classification tasks via K-means, XGBoost, and Random Forest. The XGBoost classifier attained an F-score of 0.63 and an accuracy of 0.82, and the Random Forest classifier reached an F-score of 0.55 and an accuracy of 0.83. This integration of knowledge graph embeddings improved the model's performance concerning baseline prediction, providing more precise insights into the factors influencing dropout risk. This approach contributes to a deeper understanding of the most influential factors in student dropout for the educational institution under study, supporting efforts to detect early dropout risk and success in educational settings.

Keywords: Secondary and high school students · School dropout · Feature engineering · Knowledge graph embeddings · Clustering · Machine learning · XGBoost

1 Introduction

Today, data mining has opened up new avenues for understanding student data [11], with one of the most intriguing tasks being the early detection of dropout risk. Machine learning algorithms have been instrumental in this endeavor. The scientific community has grappled with this issue, exploring supervised, unsupervised, and semi-supervised learning approaches. However, there are still hurdles to overcome, such as the need for more interpretable models [2,13] and innovative ways to represent knowledge for training artificial intelligence algorithms [8].

© The Author(s), under exclusive license to Springer Nature Switzerland AG 2025
L. Correia et al. (Eds.): IBERAMIA 2024, LNCS 15277, pp. 262–271, 2025.
https://doi.org/10.1007/978-3-031-80366-6_22

Moreover, with the rise of learning analytics, predicting academic performance, a crucial factor in dropout risk, is also a focus [18].

The education system encompasses various scenarios, such as virtual, hybrid, and face-to-face education, across different educational levels (primary, secondary, high school, and higher education). Most studies have concentrated on university-level dropouts using supervised learning techniques [1]. This study uses data from face-to-face education at the secondary and high school levels. The approach presented is based on previous studies on knowledge representation using ontologies or knowledge graphs [4,6,10,19].

The analysis process of educational datasets requires considering various factors, including data quality, feature selection, and model validation, to ensure reliable results [22]. In this paper, we implement an approach to improve the automatic detection of dropout risk at secondary and high school levels. We use embedded knowledge generated by a graph in clustering and classification tasks. Embeddings enable representation learning, applying machine learning to knowledge graphs, and training a machine learning algorithm to predict dropout risk. Our research is not just an academic exercise; it has a practical application that could support student success and reduce dropout rates in secondary and high school educational settings. By harnessing graph-based representation learning, we can uncover some relationships between features and dropout risk, shedding light on how these factors influence machine learning algorithms' decisions. We will demonstrate the practical value of knowledge embeddings extracted from the graph in detecting dropout risk, thereby improving both the performance and interpretability of algorithms.

The structure of this paper is as follows: Sect. 2 provides a detailed explanation of our implemented approach for automatically detecting dropout risk, which is based on knowledge graph embeddings. Section 3 presents the experimental evaluation, including a description of the dataset, the experimental setup, and the results. Finally, Sect. 4 offers the main conclusions, discusses the limitations of our study, and outlines potential avenues for future research.

2   Methodology

The implemented approach utilized a methodology following the Ampligraph tutorial documentation [3], adapting it to assess the risk of school dropout. The process begins with representing knowledge in this area through a graph. We then integrate this graph into representation learning through clustering and classification tasks. During the knowledge graph (KG) construction process, we employed the RDF schema of the 3-tuples (triples) technique based on an educational dataset from a real scenario of a public educational institution at the secondary and high school levels. We used these triples to train the ComplEx embedding model [20] available in the AmpliGraph library [3]. We then evaluated the quality of the embeddings on a validation set. Next, we clustered the embeddings using the K-means algorithm. Then, we applied the embeddings as features in the classification task of the risk of school dropout by training two algorithms: gradient-boosting, XGBoost, and Random Forest (RF). Finally, we evaluated the algorithms on a test set.

2.1   Related Work

There remains a distrust towards machine learning algorithms due to their inability to explain their decisions [7]. Therefore, constructing or extracting knowledge graphs from datasets has been proposed. Other design pattern studies suggest feature engineering based on ontologies [6], which helps identify the most influential factors or predefine the inputs for machine learning algorithms. These approaches propose incorporating knowledge into training data based on representation learning [12]. These hybrid AI systems aim to produce more reliable, transparent, and reproducible models [6].

Current studies also propose, from an unsupervised learning perspective, clustering as a practical guide to exploring student categories and characteristics [9]. For instance, one study [17] used clustering techniques to predict university graduation status. Another study [2] described a method to extract topological features from the graph using distance measures (Euclidean and cosine) to calculate similarities between student data. The results showed that predictive performance significantly improved by combining initial data with the graph's topological features, as it captures structural correlations and provides deeper insights than isolated data analysis [2]. Another proposed model [8] used a graph neural network module based on learning analytics to learn local representations of academic performance in a virtual education setting. However, in face-to-face education, capturing interactions is more challenging due to the lack of availability and accessibility of such data.

In the educational context, identifying students at risk of dropping out is crucial for promoting academic success and overall well-being, thereby fostering social mobility. Meanwhile, machines face a bottleneck in knowledge representation. Therefore, we identified entities and relationships suitable for representation with RDF (Resource Description Framework) triples to implement representation learning from knowledge graphs. Vázquez and Miranda [21] describe research that concurs there is no single influential factor in dropout risk but a combination of several factors, especially from the family, social, and school environments where students interact. We employed sophisticated data analysis techniques and interdisciplinary expertise, and dataset characteristics are identified and related to dropout risk using machine learning algorithms and statistical models to identify associated patterns.

2.2   Knowledge Graph Construction

Experts in educational matters analyzed the results of statistical tests conducted on the educational dataset from a real scenario of a public educational institution, who inferred the relationship between certain factors in the decision to drop out in conjunction with previous theoretical studies in education and psychology [21]. We constructed the relationships using the RDF schema based on the information above, identifying entities (subjects and objects) and relationships (predicates) and structuring 3-tuples (triples) to represent knowledge through graphs [15]. The idea is that each student is an entity connected to their academic and demographic characteristics. The objective in this phase is to generate a new representation of the dataset where each point is a triple in the form: <subject, predicate, object>.

2.3   Knowledge Graph Embeddings

We first split the KG data into train and validation using a function of the Ampligraph library that extracts a test set that includes only entities and relationships in the training set. The implemented approach considers that the data points in the test set are two entities connected by some relationship (predicate). We implemented this using the AmpliGraph library, which employs multiple knowledge graph embedding models (TransE, ComplEx, DistMult, HolE). We chose to use the ComplEx model because, according to the authors' documentation, this model is frequently used to train embeddings that capture relevant relationships—the choice of hyperparameters involved using those mentioned in the ComplEx documentation applied to some benchmark datasets [16,20].

2.4   Evaluation of Embeddings

We used the performance evaluation function included in AmpliGraph to assess performance. To assess the subjective quality of the embeddings , we compared the cluster embeddings with native clusters, such as academic performance according to the Colombian Ministry of National Education: Low (1.0–2.9), Basic (3.0–3.9), High (4.0–4.4), and Superior (4.5–5.0). We considered the academic averages in high-intensity subjects (mathematics, language, natural sciences, social sciences, English) and low-intensity subjects (ethics, arts, religion, sports, technology, and entrepreneurship). We evaluate the quality of the clustering based on a predefined criterion. We used PCA to project the embeddings from 200D space (with k=100, ComplEx uses both real and imaginary numbers for its embeddings) to 2D space, representing each academic performance average, its category (Low, Basic, High, and Superior), and the clusters using K-means.

2.5   School Dropout Classification via Machine Learning

Researchers have frequently employed supervised learning to automatically classify students as dropouts or non-dropouts, specifically using decision trees [1]. In line with this, the algorithms implemented were XGBoost [5] and RF. The XGBoost trains an ensemble of decision trees that utilize the training data (with multiple features) to predict a target variable. In terms of regularization, defining the model's complexity by defining the trees and then learning the tree structure is necessary. Learning the tree structure presents a more significant challenge than the traditional optimization problem, as it must take more than just the gradient. So, learning all the trees is infeasible simultaneously; therefore, we use an additive strategy, adding one split at a time. This strategy may sometimes fail because it considers only one feature dimension at a time. At the same time, RF constructs multiple bagged random decision trees during training and generates outputs based on the average or majority ranking of the individual trees. In a real educational dataset, we extracted features from the knowledge embeddings for the averages of high- and low-intensity subjects. We then used these features to train the algorithms to classify students as dropouts or non-dropouts.

3   Experimental Evaluation

3.1   Dataset Description

The educational dataset comprised secondary and high school students at an educational institution. A total of 1,830 student records were collected, featuring 11 attributes, see Table 1. Of these, seven were demographic, and four were academic. Low-intensity subjects require a minimum of two hours of academic dedication (peace studies, arts, ethics, sports, religion, technology, entrepreneurship), while high-intensity subjects require three or more hours of dedication (natural sciences, social sciences, Spanish, English, mathematics). We split the dataset into training and testing sets for automatic classification. Table 2 describes the dataset distribution for both sets.

Table 1. Demographic and academic features of the initial dataset.

Feature		Description
Socio-economic	Siblings_School	Siblings enrolled at the same school (Yes-No)
	Social_Stratum	Socioeconomic level (Low-Middle-High)
	Sisbén_Level[a]	Level at the SISBEN III (Extreme_poverty,Moderate,Non-poverty)
Personal	Distance	Distance from the student's point of residence to the school's location
	Sex	Male-Female
	Age	Student's age
	Overage	Discrepancy between current and theoretical age[b]
Academic	Repetition	Repetition level
	GPA	Student's final average score obtained at the end of the school year
	PA_low_inten	Point Average in low-intensity subjects
	PA_high_inten	Point Average in high-intensity subjects

[a]SISBEN: System for the Identification of Potential Beneficiaries of Social Programs. Source: Authors.
https://www.mineducacion.gov.co/1780/articles-363305_recurso_1.pdf

We calculated three attributes (GPA, PA in low-intensity subjects, and PA in high-intensity subjects) from each student's final grades obtained at the end of the school year, including the target variable (dropout/non-dropout).

Table 2. Distribution of dataset for training and evaluation.

Class	Train	Test
Non-Dropout	854	427
Dropout	427	122

Source: Authors.

3.2   Experimental Setup

The experimental setup begins with constructing the graph based on the theoretical review [21] and insights from education experts to identify relationships between specific attributes and dropout risk [14]. After constructing the graph, the ComplEx embedding model was trained, with the hyperparameters configured based on examples from various datasets as documented in AmpliGraph. Next, we used the functions *MRR* (Mean Reciprocal Rank) and *Hits@*. The *MRR* function calculates the average position of the first relevant element in a ranking vector, and the *Hits@* function calculates how many elements of a ranking vector appear in the top n positions. However, these metrics indicate that the embeddings have learned a representation but do not define the embeddings' utility for subsequent model training. The Rand Index was used to measure the similarity between two clusterings. This metric considers all pairs of samples and counts the pairs assigned to the same or different clusters in the predicted and actual clusterings. The performance metrics F-score and accuracy (Acc) calculated per class were used to evaluate the classification of students at risk of dropping out, defining them by Eq. 1.

$$F - score = \frac{2 \times Precision \times Recall}{Precision + Recall}, Acc = \frac{TP + TN}{TP + TN + FP + FN} \quad (1)$$

3.3   Results

The knowledge graph was constructed from the previously identified entities (Fig. 1a) and relationships using the 3-tuple structure: subject, predicate, and object. Some of these identify student attributes, while others determine if they affect the decision to drop out: {'Student_416', 'hasDistance,' 'DistanceNear'}, {'Student_416', 'hasgender,' 'SexMale'}, {'Student_416', 'hasAverageHighHours,' 'Average_highhours3.21'}, {'DistanceNear,' 'affectsDesertionRiskBasedOnDistance,' 'Non-dropout'} (Fig. 1b).

The trained ComplEx model achieved a *MRR* of 0.56 and a *Hits@10* of 0.66, indicating that 66% of the time, the model classifies the correct entity within

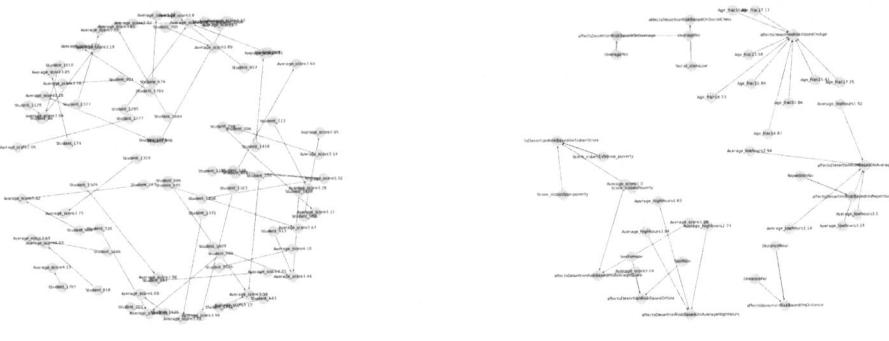

(a) Directed-graph from dataset.   (b) KG on a particular relation.

Fig. 1. Generated knowledge graph (KG). Source: Authors.

the top ten positions. To support these results, we visualized the embeddings in 2D (Fig. 2a) and displayed the clusters according to academic performance categories (Low, Basic, High, and Superior). We then plotted the embeddings in 2D but with the groups identified by K-Means (n_clusters=6), see Fig. 2b. These results are interpreted based on the Rand index, which measures the similarity between two clusterings, reaching 0.32, where 0 represents random labeling and 1 represents a perfect match.

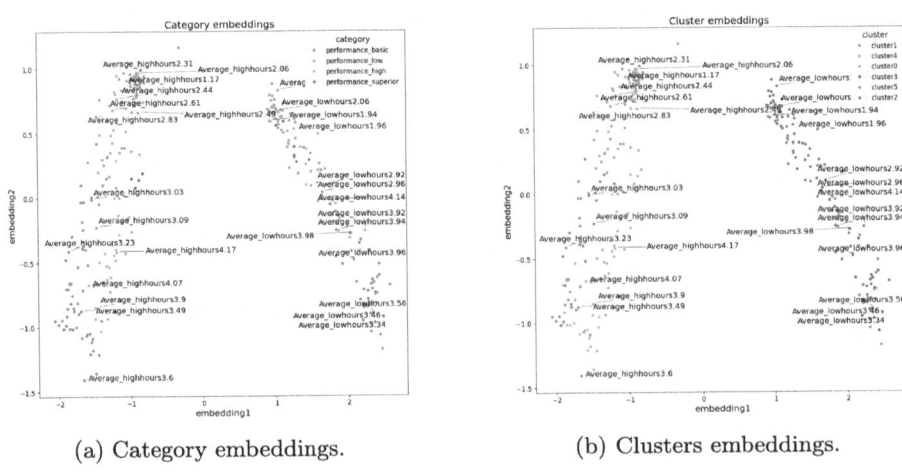

(a) Category embeddings.   (b) Clusters embeddings.

Fig. 2. Visualisation of the 2D embeddings. Source: Authors.

The results show that in the 200-dimensional embedding space, similar academic averages cluster closely together, which can be captured by a clustering algorithm [9]. Finally, using F-score and accuracy metrics, the best classification results with XGBoost on the test dataset were achieved with the parameters n_estimators=500 and max_depth=200, yielding an F-score of 0.63 and an accuracy of 0.82. The algorithms performed better using initial features compared to embeddings. XGBoost and RF achieved similar accuracy (0.82) and F-score (0.63) with booth features (see Table 3). When using embeddings, the performance of all algorithms tends to decrease. Random Forest with the parameters n_estimators=100 showed the highest accuracy with embeddings (0.83) but presented confusion in detecting the dropout class.

The embeddings did not capture information as effectively as the initial features, possibly due to their quality or the complexity of the graph models used. More advanced techniques or better hyperparameter tuning might be necessary to improve the embeddings and their data representation. The quality and quantity of data used to generate the embeddings could have impacted their performance. The embedding technique may not have been the most suitable for this dataset. Additionally, we used a single model knowledge graph embedding model (ComplEx) and did not consider another essential aspect for constructing the initial graph.

Table 3. Performance of machine learning algorithms in school dropout risk classification (XGBoost-XGB, Random Forest-RF).

Metrics	Initial Features		Embeddings	
	XGB	RF	XGB	RF
Accuracy	0.85	0.82	0.82	0.83
F1-Score	0.65	0.63	0.63	0.55

Source: Authors.

4   Conclusions, Limitations, and Future Work

The approach implemented in this article for automatically classifying dropout risk, utilizing educational data with demographic and academic features, knowledge graph embeddings, and machine learning algorithms, was successfully executed, achieving accuracy performance within reasonable ranges. The baseline accuracy for this problem is 77%, representing the frequency of the most common class (non-dropout), and with the incorporation of knowledge, the accuracy increased by 5%. Although the model exhibits some confusion in detecting the dropout class, it shows potential for further exploration. Additionally, the interpretation of the graph-based model reveals that the decision to drop out, according to the dataset from the educational institution under study, is related to academic averages in low and high-intensity subjects, as well as demographic factors such as age, grade repetition, overage, distance between school and home, SISBEN level (extreme poverty, moderate poverty, and poverty), and social stratum (low and middle).

The approach includes the graph construction process, validated by an education expert. However, the class imbalance remains complex, presenting an intriguing challenge for future research. Additionally, we should complement the performance evaluation of the embeddings by representing them and using clustering to validate whether they have learned an appropriate representation and captured implicit information from the graph. Another limitation was that conducting experiments with a limited and small dataset diminished the capacity to generalize and capture all relevant patterns.

Initial features were more effective for both models than embeddings, suggesting that the original data was highly relevant for prediction. XGBoost and Random Forest decreased performance with embeddings, indicating the need for better embedding techniques. Using the XGBoost algorithm requires systematic hyperparameter exploration through grid search using cross-validation to deepen the experiments and observe different results. This task is time-consuming, which could suggest exploring alternative machine learning methods with fewer parameters. We obtained the most robust results by considering the F-score.

Future research should aim to increase the number of samples for the dropout class and dataset, construct or extract new knowledge graphs based on techniques for measuring similarity, and ensure that the triplets are representative and capture meaningful relationships between entities. In addition, we should focus on improving embedding generation and exploring more advanced graph learning techniques. Indeed, we will evaluate state-of-the-art machine learning methods based on data and graph neural networks. We will also add more features to the model and tune its hyperparameters.

ComplEx has been a good starting point for this research; however, future exploration can include other models. Feature engineering, a crucial aspect in dropout prediction, will also be explored to extract relevant features. We will conduct experiments to include students' traceability or historical interactions over a time window, further emphasizing the importance of this aspect. Finally, semi-supervised or unsupervised learning will be further explored with other clustering techniques and evaluation metrics and visualizations such as t-SNE (t-distributed Stochastic Neighbor Embedding) to see if the embeddings reveal patterns that clustering does not capture, providing a more comprehensive view of the clustering performance.

References

1. Agrusti, F., Bonavolontà, G., Mezzini, M.: University dropout prediction through educational data mining techniques: a systematic review. J. E-Learn. Knowl. Soc. **15**, 161–182 (2019). https://doi.org/10.20368/1971-8829/1135017
2. Albreiki, B., Habuza, T., Zaki, N.: Extracting topological features to identify at-risk students using machine learning and graph convolutional network models. Int. J. Educ. Technol. Higher Educ. **20** (2023). https://doi.org/10.1186/s41239-023-00389-3
3. Costabello, L., et al.: AmpliGraph: a library for representation learning on knowledge graphs (2019). https://doi.org/10.5281/zenodo.2595043

4. El-Rady, A.A.: An ontological model to predict dropout students using machine learning techniques (2020). https://doi.org/10.1109/ICCAIS48893.2020.9096743
5. Friedman, J.H.: Greedy function approximation: a gradient boosting machine. Ann. Stat. **29** (2001). https://doi.org/10.1214/aos/1013203451
6. Ghidalia, S., Narsis, O.L., Bertaux, A., Nicolle, C.: Combining machine learning and ontology: a systematic literature review (2024)
7. Gunning, D., Aha, D.: Darpa's explainable artificial intelligence (xAI) program. AI Mag. **40**, 44–58 (2019). https://doi.org/10.1609/aimag.v40i2.2850
8. Huang, Q., Zeng, Y.: Improving academic performance predictions with dual graph neural networks. Complex Intell. Syst. (2024). https://doi.org/10.1007/s40747-024-01344-z
9. Iam-On, N., Boongoen, T.: Generating descriptive model for student dropout: a review of clustering approach. Hum.-centric Comput. Inf. Sci. **7** (2017). https://doi.org/10.1186/s13673-016-0083-0
10. Kanellopoulos, D., Kotsiantis, S.: Towards an ontology-based system for intelligent prediction of student dropouts in distance education. Int. J. Manag. Educ. **2** (2008). https://doi.org/10.1504/IJMIE.2008.018391
11. Lee, S., Chung, J.Y.: The machine learning-based dropout early warning system for improving the performance of dropout prediction. Appl. Sci. Switzerland **9** (2019). https://doi.org/10.3390/app9153093
12. Lin, Y., Han, X., Xie, R., Liu, Z., Sun, M.: Knowledge representation learning: a quantitative review. arXiv preprint arXiv:1812.10901 (2018)
13. Melo, E., Silva, I., Costa, D.G., Viegas, C.M., Barros, T.M.: On the use of explainable artificial intelligence to evaluate school dropout. Educ. Sci. **12** (2022). https://doi.org/10.3390/educsci12120845
14. Ministerio de Educación Nacional: School dropout in Colombia: analysis, determinants, and policies for welcome, well-being, and retention: technical note, p. 146 (2022) (in Spanish)
15. Nickel, M., Murphy, K., Tresp, V., Gabrilovich, E.: A review of relational machine learning for knowledge graphs (2016). https://doi.org/10.1109/JPROC.2015.2483592
16. Rossi, A., Barbosa, D., Firmani, D., Matinata, A., Merialdo, P.: Knowledge graph embedding for link prediction: a comparative analysis. ACM Trans. Knowl. Discov. Data **15** (2021). https://doi.org/10.1145/3424672
17. Sajjadi, S., Shapiro, B., Mckinlay, C., Sarkisyan, A., Shubin, C., Osoba, E.: Finding bottlenecks: predicting student attrition with unsupervised classifier, vol. 2018-Janua, pp. 1166–1172 (2018). https://doi.org/10.1109/IntelliSys.2017.8324279
18. Sansone, D.: Beyond early warning indicators: high school dropout and machine learning. Oxford Bull. Econ. Stat. **81**, 456–485 (2019). https://doi.org/10.1111/obes.12277
19. da Silva, E.M., Mutz, F.W., Ruy, F.B.: Uso de ontologias no suporte a aplicação de machine learning: um caso no domínio de evasão escolar, vol. 3346 (2022)
20. Trouillon, T., Welbl, J., Riedel, S., Ciaussier, E., Bouchard, G.: Complex Embeddings for Simple Link Prediction, vol. 5 (2016)
21. Vásquez, J., Miranda, J.: Student Desertion: What Is and How Can It Be Detected on Time?, pp. 263–283. Springer (2019)
22. Zapata-Medina, D., Espinosa-Bedoya, A., Jiménez-Builes, J.A.: Improving the automatic detection of dropout risk in middle and high school students: a comparative study of feature selection techniques. Mathematics **12**, 1776 (2024). https://doi.org/10.3390/math12121776

Machine Learning

Semi-supervised Hierarchical Bayesian Multi-label Classification

Jonathan Serrano-Pérez$^{(\boxtimes)}$ and L. Enrique Sucar

Instituto Nacional de Astrofísica, Óptica y Electrónica, Puebla, Mexico
{js.perez,esucar}@inaoep.mx

Abstract. In this work, a semi-supervised hierarchical Bayesian multi-label classifier (SSHBMC) is proposed. SSHBMC is a semi-supervised learning algorithm for hierarchical classification where the hierarchy is a directed acyclic graph and the labels can be associated to multiple paths of labels. SSHBMC builds pseudo-paths-of-labels for each unlabeled instance using information of its nearest labeled instances. A hierarchical Bayesian network classifier, that considers the data distribution while it models the hierarchy, is trained with the labeled and pseudo-labeled data, and later is used to classify new instances based on probabilistic inference. The method was tested in several datasets from functional genomics and compared against related methods, showing in most cases superior performance with statistical significance.

Keywords: Hierarchical classification · Semi-supervised learning · Bayesian networks

1 Introduction

In hierarchical classification the labels are arranged in a predefined structure, a hierarchy. The hierarchy contains relations among the labels such as *parent-child*; and instances can be associated to multiple paths of labels. Because of this characteristics, in hierarchical datasets, hand-labeling tend to be more difficult and time-consuming than in binary or multi-class datasets, leading to datasets with few labeled instances. Additionally, in hierarchical classification the data of a node is split from it to its children, so the deepest nodes of the hierarchy usually have few instances associated. Therefore, the problem of scarcity of data in hierarchical classification is serious. Scarce data occurs when hand-labeling data is time-consuming, expensive or difficult to label [4]; however, large amounts of unlabeled data can be obtained from different sources of information. Hence, suitable semi-supervised hierarchical classifiers that take advantage of labeled and unlabeled data are required.

In this work, the method *semi-supervised hierarchical Bayesian multi-label classifier* (SSHBMC) is proposed. It can handle any hierarchy of directed acyclic graph (DAG) type, and the instances can be associated to multiple paths of labels. SSHBMC extends SSHMC-BLI [14] by including the data distribution in

© The Author(s), under exclusive license to Springer Nature Switzerland AG 2025
L. Correia et al. (Eds.): IBERAMIA 2024, LNCS 15277, pp. 275–286, 2025.
https://doi.org/10.1007/978-3-031-80366-6_23

the classifier represented as a Bayesian network (BN), which also encodes the behaviour of the local classifiers; the BN improves the initial classification.of the local classifiers by enforcing the hierarchical probability constraint. SSHBMC builds pseudo-paths-of-labels using the paths of labels of the nearest labeled neighbors to each unlabeled instance, while considering if the unlabeled instance is similar to its labeled neighbors. Experiments on the gene ontology collection show that the performance of the proposed method is superior against related methods while taking into account different percentages of labeled and unlabeled data.

The main contributions of this manuscript are: (i) a semi-supervised classifier that can handle hierarchical problems where instances can be associated to multiple paths of labels, and hierarchies can be any DAG; (ii) the incorporation of a Bayesian network that represents the hierarchy and improves the initial classification of the local classifiers; and (iii) an experimental comparison of the proposed method on several real world datasets against related methods.

The document is organized as follow. Section 2 summarizes fundamentals of semi-supervised hierarchical classification. Section 3 reviews related work. Section 4 presents the proposed method. Section 5 presents the experiments and results. Finally, in Sect. 6, conclusions and some ideas for future work are given.

2   Fundamentals

2.1   Semi-supervised Learning

Semi-Supervised Learning (SSL) algorithms are able to handle labeled and unlabeled data to perform learning tasks [2]. These methods are appropriate in scenarios where labeled data is scarce and unlabeled data is available.

In SSL there are some assumptions on which most semi-supervised learning methods are based on, these methods intent to satisfy at least one of them [2]. They are: the *smoothness* assumption, which states that the labels for two input points close by in the input space should be the same label; the *low-density* assumption states that the decision boundary of a classifier should preferably pass through low-density regions; and, the *manifold* assumption, which is similar to the smoothness assumption, that is, data points on the same manifold must have the same label.

Semi supervised learning methods are commonly divided into two main groups, inductive and transductive [4]. The former produce a classification model that is used to predict the labels of instances that were not used in training, while the second is only focus on labeling the unlabeled instances. In this way, the proposed method belongs to the inductive group, because a classification model for predicting new data is generated.

2.2   Semi-supervised Hierarchical Classification

Formally, semi-supervised hierarchical classification can be defined as a tuple $SSHC=(HS,(X,Y),U)$, where: $HS=(L,E)$ is the hierarchy (a DAG in its general form), where L is the set of nodes/labels and E is the set of edges that link the nodes; (X, Y) is the labeled set, $X = \{x_1, x_2, ..., x_n\}$ contains n instances, where $x_i \in \mathbb{R}^d$, and $Y = \{y_1, y_2, ..., y_n\}$ contains the labels for X, where $y_i \in \{0, 1\}^{|L|}$, that is, each y_{ij} indicates whether the i-th instance is associated to the j-th label; and, $U = \{x_{n+1}, x_{n+2}, ..., x_{n+m}\}$ is the unlabeled set that contains m instances.

The task of *semi-supervised hierarchical classification* consists of assigning to a particular object, a subset of labels that comply the hierarchical constraint:

$$f_{SSHC} : \mathbb{R}^d \rightarrow \{0, 1\}^{|L|}. \tag{1}$$

Nevertheless, in hierarchical classification problems where the instances are associated to multiple paths of labels [6,15], the task is *modified* to assign to a particular instance, the probability of being associated to each node:

$$f_{SSHC} : \mathbb{R}^d \rightarrow [0, 1]^{|L|}. \tag{2}$$

But this prediction has to comply the *hierarchical probability constraint*, which states that the probability for an instance in the node l has to be equal or lower than the probabilities of all the parents of node l.

3   Related Work

First of all, two standard methods are presented. *Self-training for multi-label classification* (STML) which self-trains a binary classifier for each label using the whole unlabeled set; a prediction for a new instance is the union of the individual predictions of the binary classifiers. STML does not take into account the hierarchy neither in the training nor in the prediction phases, also predictions do not guarantee to comply the hierarchical constraint. *Self-training hierarchical classifier* (STHC) [14], in which each node of the hierarchy is self-trained, in the same way than STML. In the prediction phase, the individual predictions of the local classifiers are post-processed: if the probability of a node is greater than the probability of its parent with the lowest probability, its probability is reduced down to the probability of that parent; in this way, predictions comply the hierarchical probability constraint.

Metz and Freitas [8] proposed the first method for semi-supervised hierarchical classification. It is a hierarchical top-down classifier that can only handle hierarchies of tree type and predicts a single path of labels that always reaches a leaf node. For each node a decision tree is trained, which is self-trained following one of three possible strategies. However, reported results are not superior in a statistically significant way against a supervised hierarchical classifier.

Hierarchical multi-label classification using semi-supervised label powerset (HMC-SSLP) was proposed by Santos and Canuto [10]. It consists of training a hierarchical multi-label classifier with label powerset (HMC-LP) with labeled data, then it is used to pseudo-label a predefined proportion of the unlabeled data which is added to the training set for the next iteration, this process iterates until all unlabeled data is pseudo-labeled. HMC-LP combines all the classes of each example to generate a new hierarchy, nevertheless, examples of how to combine paths of different lengths are not shown. Also, Santos and Canuto [10] proposed hierarchical multi-label classifier using semi-supervised random K-labelsets (HMC-SSRAkEL), which, for each non leaf node, a RAkEL classifier is trained to predict its children. Later, a predefined proportion of the unlabeled data is pseudo-labeled with the top-down procedure, which is added to the training set for the next iteration, in the same way than HMC-SSPPL.

Hierarchical multi-label classification using semi-supervised binary relevance (HMC-SSBR) was proposed by Santos and Canuto [11]. From HMC-BR, BR classifiers are replaced by SSBR, a semi-supervised method for *multi-label classification*, where the unlabeled instances are pseudo-labeled with the prediction of a hierarchical top-down classifier, then the same steps than HMC-SSRAkEL to pseudo label the unlabeled instances are carried out. Nevertheless, HMC-SSBR and HMC-SSRAkEL lack of a way to select the instances with the most confident predictions, instead, they add the whole set of pseudo-labeled instances to the labeled set in each iteration.

Path cost-sensitive algorithm with expectation-maximization (PCEM) [16] was proposed for hierarchical text classification. As its name indicates, it is based on expectation-maximization procedure. First, a path cost-sensitive naive Bayes classifier (PCNB) [16] is trained with the labeled data, then it is used to pseudo-label the unlabeled instances. Second, the PCNB is trained with labeled and pseudo-labeled instances, and this process is iterated until the parameters of the PCNB converge. PCEM can not be applied to non-text domains, since it is designed using the bag-of-words representation.

Semi-supervised predictive clustering trees (SSL-PCT) was proposed by Levatic et al. [7], which is based on predictive clustering trees (PCT). PCT's consists of a hierarchically organized set of clusters, where the root cluster is recursively divided into smaller clusters as one goes deeper to the leaves. However, they stated that their results were *not so successful* on the functional genomics datasets, because the supervised classifier was rarely outperformed.

Serrano-Pérez and Sucar [12] proposed semi-supervised hierarchical classifier based on local information (SSHC-BLI), which is based on the smoothness assumption. SSHC-BLI got better performance than the supervised classifier, showing that making use of unlabeled data can help to improve the performance of a hierarchical classifier trained only on labeled data. Nevertheless, SSHC-BLI only works for hierarchies of tree type, and the instances have to be associated to a single path of labels. Later, SSHMC-BLI [14] was proposed for hierarchical problems where instances can be associated to multiple paths of labels and hierarchies can be any DAG.

Finally, *semi-supervised hierarchical Bayesian multi-label classifier* (SSHBMC) is proposed in this work, which is an extension of SSHMC-BLI [14], that is, SSHBMC takes into account the data distribution and the behaviour of the local classifier by modeling them into a Bayesian network. In this way, the method builds pseudo-paths-of-label for each unlabeled instance making use of the labels of its labeled neighbors, then only if the unlabeled instance is similar to its neighbors, it can be pseudo-labeled; later a hierarchical Bayesian network multi-label classifier is trained with the labeled and pseudo-labeled instances.

3.1   Evaluation Measures

The probabilities of being associated to each node of the hierarchy are the outputs of the proposed method. *Area under the average precision and recall curve*, $AU(\overline{PRC})$, also known as *average precision* (AP) [17], is the evaluations measure used to assess the methods:

$$AP = \sum_n (R_n - R_{n-1})P_n, \tag{3}$$

where P_n and R_n are the precision (Eq. 4) and recall (Eq. 5) at the n-th threshold, respectively.

$$P = \frac{\sum_{i=1}^{|L|} TP_i}{\sum_{i=1}^{|L|} TP_i + \sum_{i=1}^{|L|} FP_i}, \tag{4}$$

$$R = \frac{\sum_{i=1}^{|L|} TP_i}{\sum_{i=1}^{|L|} TP_i + \sum_{i=1}^{|L|} FN_i}, \tag{5}$$

where TP, FP and FN are true positives, false positives and false negatives, respectively.

4   Semi Supervised Hierarchical Bayesian Multi-label Classifier

SSHBMC follows the same assumption than SSHMC-BLI, that is, it is based on the smoothness assumption, where neighboring instances must share the same or similar paths of labels. In the same way, SSHBMC tries to build *pseudo-paths-of-labels* for each unlabeled instance using the paths of labels of its neighboring labeled instances; however, the pseudo-paths-of-labels in SSHBMC may be composed by multiple paths of labels. Furthermore, only if the unlabeled instance is similar to its labeled neighbors, it is pseudo-labeled, otherwise, it stays unlabeled; this process is iterated until all the pseudo-labels do not change.

Algorithm 1 shows the steps of SSHBMC. It is an iterative method that tries to build pseudo-paths-of-labels (line 10) to pseudo-label the unlabeled data using its nearest labeled neighbors (line 6), details of how pseudo-paths-of-labels are built can be found in Sect. 4.1. Also, the method considers the similarity of each unlabeled instances with its neighbors (line 7), if they are not *similar* the unlabeled instance stays unlabeled; *similarity of an instances with a set of instances* (SISI) [12] is the function used to estimate the similarity. Let p be an instance, let A be a set of instances, where $x_i \in A$, let k be the length of A, and let $d(X, Y)$ be the euclidean distance, then, SISI is defined in Eq. 6.

Algorithm 1. SSHBMC algorithm

Require: (X, Y): labeled data, U: unlabeled data, k: number of nearest labeled neighbors, THR: similarity threshold, $thLabel$: threshold to pseudo-label an instance, HS: hierarchy of DAG type, $maxIterations$: maximum number of iterations.

Ensure: f_{SSHC}: trained SSHBMC classifier

1: $T \leftarrow 1$
2: $LD \leftarrow X$
3: $CL \leftarrow Y$
4: **while** $True$ **do**
5:     **for each** $u_j \in U$ **do**
6:         $IND_j \leftarrow getNLN(k, u_j, LD)$ ▷ Nearest labeled neighbors
7:         **if** $SISI(u_j, IND_j) < THR$ **then**
8:             $PSL_j = \emptyset$
9:         **else**
10:             $PSL_j \leftarrow buildPPL(IND_j, LD, thLabel)$ ▷ Pseudo-path-of-labels for u_j
11:         **end if**
12:     **end for**
13:     **if** $(T > maxIterations)$ or $(PSL^T == PSL^{T-1})$ **then**
14:         **break** loop (while)
15:     **else**
16:         $CL \leftarrow Y \cup valid(PSL)$
17:         $LD \leftarrow X \cup U[valid(PSL)]$
18:     **end if**
19:     $T \leftarrow T + 1$
20: **end while**
21: $f_{SSHC} \leftarrow trainHBC(LD, CL, HS)$ ▷ Train the hierarchical Bayesian classifier

---

$$SISI(p, A) = \begin{cases} 1, & uavg(p, A) \leq lavg(A), \\ 0, & uavg(p, A) \geq n * lavg(A), \\ \frac{lavg(A) - uavg(p,A)}{(n-1)lavg(A)} + 1, & otherwise, \end{cases} \quad (6)$$

$$lavg(A) = \frac{\sum_{i=1}^{k} \sum_{j=i+1}^{k} d(x_i, x_j)}{\frac{k(k-1)}{2}}, \quad (7)$$

$$uavg(p, A) = \frac{\sum_{i=1}^{k} d(p, x_i)}{k}. \quad (8)$$

As it can be seen, SISI takes into account both the distances of the instance p with the set of instances A, $uavg$; and the distances among the set of intances A, $lavg$.

The loop finishes (line 13) either when the pseudo-paths-of-labels of the unlabeled data do not change from an iteration to other, or if the maximum number of iterations is reached. Finally, a hierarchical Bayesian network multi-label classifier is trained with the labeled and pseudo-labeled data (line 21), details can be found in Sect. 4.2.

4.1   Pseudo-Paths-of-Labels

Pseudo-paths-of-labels are built from the paths of labels of labeled instances in similar way than SSHC-BLI [12]. Let $Y = [y_1, ..., y_k]$ be the labels of k labeled instances close to the unlabeled instance x, where $y_i \in \{0, 1\}^{|L|}$. Then, psl is the new pseudo-path-of-label, which can be build from eq. 9 and a threshold, $thLabel$, to determine whether an instance is associated to the label.

$$psl_j = \begin{cases} 1 : ppsl_j \geq thLabel \\ 0 : ppsl_j < thLabel \end{cases}, \; \forall j \in \{1, ..., |L|\}, \tag{9}$$

$$0 \leq thLabel \leq 1$$

$$ppsl_j = \frac{\sum_{i=1}^{k} y_{i,j}}{k}, \; \forall j \in \{1, ..., |L|\}. \tag{10}$$

If psl is full of *zeros*, it means that x stays unlabeled.

4.2   Hierarchical Bayesian Network Classifier

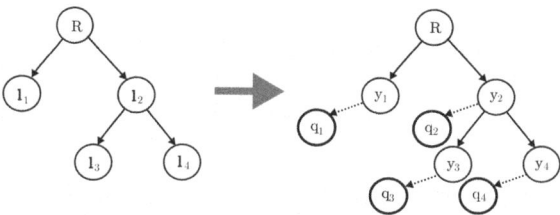

Fig. 1. The hierarchy on the left is transformed into a Bayesian network (right). The y_i nodes in in the BN correspond to the l_i nodes in the hierarchy, and the q_i nodes in the BN to the initial estimates from the local classifiers.

Initially, a hierarchical classifier based on local classifiers per node (LCN) could be trained, and the output of the local classifiers could be post-processed, for instance, following a top-down manner where the probabilities of each node are limited by the probabilities of their parents [6]. However, taking advantage of the hierarchy and the relations it represents, in this work a hierarchical Bayesian network classifier [1,13] is trained, where the hierarchy is modeled as a Bayesian network (BN) that also represents the data distribution. The BN receives the initial probability estimates for each label in the hierarchy, and then these are updated according to the hierarchical relations via probabilistic inference. Through this post-processing step, the BN guarantees the hierarchical probability constraint, and, in general, it improves the final classification results.

Two types of random nodes compose the Bayesian network as shown in Fig. 1. First, the data distribution is represented by y nodes, which are also in charge of maintaining the hierarchical constraint in the Bayesian network; for each node of the hierarchy, there is a y_i node in the Bayesian network. The parameters

Table 1. Conditional probability table of $P(y_i|pa(y_i))$. a is the number of instances associated to both y_i and its parents, $pa(y_i)$; b is the number of instances no associated to y_i but associated to all its parents, $pa(y_i)$. Laplace smoothing is applied only when $pa(y_i)=\mathbf{1}$.

$$pa(y_i)$$

		1	**$\neq\mathbf{1}$**
y_i	1	$\frac{a+1}{a+b+2}$	0
	0	$\frac{b+1}{a+b+2}$	1

for each node y_i are estimated by maximum likelihood from the training set, see Table 1, where $pa(y_i)$ is the set of parents of y_i given by the hierarchy. Note that if an instance is not associated to all the parents of $pa(y_i)$, then it will be no associated to y_i with probability one, $P(y_i=0|pa(y_i)\neq\mathbf{1})=1$, this is what maintains the hierarchical probability constraint in the Bayesian network.

q nodes are the second type of random nodes in the Bayesian network. Each y_i node has a child node, q_i, except the root node. q nodes model the behavior of the base classifier for predicting instances that were not used in training. Furthermore, q nodes will receive the predictions of the base classifier, which will be propagated in the Bayesian network.

Taking into consideration that the base classifier provides as prediction the probabilities of being associated to each i-th node, the distribution for each node q_i, $P(q_i|y_i)$, is modeled parametrically with Gaussian distributions [1]. That is, $P(q_i\in\mathbb{R}|y_i=0)\simeq N(\mu_0,\sigma_0^2)$ where μ_0 is the mean and σ_0^2 is the variance of the predictions of the base classifier in the instances no associated to y_i in the validation set; and the same for $P(q_i\in\mathbb{R}|y_i=1)\simeq N(\mu_1,\sigma_1^2)$ but considering the instances associated to y_i in the validation set.

On the other hand, local classifiers per node (LCN) are trained to feed the Bayesian network. LCNs require a *policy* that selects the positives and negatives instances for each binary classifier [5]. The *balanced bottom-up* policy is used, where for each node l, the positive instances are the instances associated to l, and the negatives are at most equal to the amount of positives, taking them first from its siblings, then from uncles and so on.

5   Experiments and Results

5.1   Datasets

The datasets belong to the Gene Ontology (GO) collection [15] which are from the field of functional genomics. Labels of the datasets are arranged in a hierarchy of DAG type, where some nodes have multiple parents. Also, instances can be associated to multiple paths of labels which can finish in an internal node.

Datasets were preprocessed in similar fashion than Ramírez-Corona et al. [9], nodes with less than 50 instances associated in the training set were removed. A description of GO datasets is shown in Table 2.

Table 2. Description of the GO datasets. *Train, validation* and *test* show the number of instances in each set; *Attr.* shows the number of attributes; *Nodes* shows the number of nodes/labels in the hierarchy; and *MD* is the maximum depth of the hierarchy.

Dataset	Train	Validation	Test	Attr.	Nodes	MD
cellcycle_GO	1625	848	1278	77	164	9
church_GO	1627	844	1278	31	164	9
derisi_GO	1605	842	1272	63	161	9
eisen_GO	1055	528	835	79	122	9
expr_GO	1636	849	1288	565	165	9
gasch1_GO	1631	846	1281	173	165	9
gasch2_GO	1636	849	1288	52	165	9
hom_GO	1661	867	1309	47034	166	9
pheno_GO	653	352	581	276	68	7
seq_GO	1692	876	1332	530	171	9
spo_GO	1597	837	1263	89	162	9
struc_GO	1659	859	1306	19628	169	9

First, training sets were split in a stratified way into labeled and unlabeled sets as follows: labeled {10, 30, 50, 70, 90}%; and unlabeled {90, 70, 50, 30, 10}%, complement with respect to labeled. The division of the training sets is carried out 3 times, so results are averages.

The validation set is used to tune the hyper-parameters of SSHBMC for each run. The parameters and values that they could take are: similarity threshold (THR): {0.3, 0.5, 0.7}; number of labeled neighbors (k): {3, 4, 5}. The configuration that maximizes the evaluation measure AP is the best. Later, the semi-supervised classifier is trained with the best configuration but only with the labeled and unlabeled sets; the validation set is not used for training.

5.2 Results

Results for some data sets[1] comparing the proposed classifier (SSHBMC) against SSHMC-BLI, standard methods (STML, STHC) and the supervised classifier (LCN) are shown in Fig. 2. The worse performance was obtained by STML, which does not consider the hierarchy for the training and prediction phases; also, it does not guarantee that predictions comply the hierarchical probability constraint. On the other hand, STHC, which considers the hierarchy in the prediction phase, shows better performance than STML, nevertheless its performance tend to be lower than the supervised classifier. Finally, the proposed method, SSHBMC, tends to outperform most of times the rest of methods, even

---

[1] Some graphs are omitted due to space restrictions, they can be found at: https://drive.google.com/drive/folders/12eadmNJ3R0TeWWZ2SSj3HpmtglfAxs01.

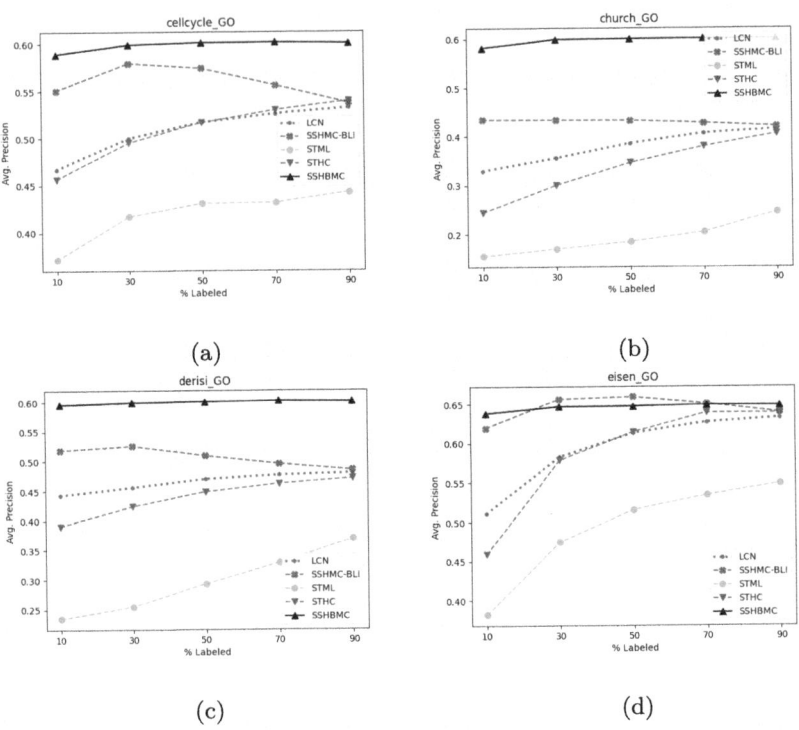

Fig. 2. Results in terms of average precision (AP) in the following datasets: a) cellcycle, b) church, c) derisi and d) eisen. The *x*-axis corresponds to the amount of labeled data, while its complement is the unlabeled data. (Best seen in color.) (Color figure online)

SSHMC-BLI; showing that making use of the Bayesian network is helping to improve the performance of the SSHMC-BLI classifier.

The Friedman test together with its post-hoc, Nemenyi test, were carried out to verify if there is statistical difference among the proposed methods and the rest of classifiers, as suggested by Demsar [3] when comparing multiple classifiers over multiple datasets.

First, let r_i^j be the rank of the j-th of l algorithms on the i-th of M datasets, then $R_j = \frac{1}{M} \sum_{i=1}^{M} r_i^j$ is the average rank of the j-th algorithm. So, the null hypothesis of the Friedman test states that all the algorithms are equivalent, therefore their average ranks (R_j) should be equal, against the alternative which states that they are not. Afterward, only if the null hypothesis was rejected, the Nemenyi test is used to compare all the classifiers against each other. Hence, the performance of two classifiers is significantly different if their average ranks differ by at least the *critical difference*.

The result of the Friedman test with $p = 0.05$ is that the null hypothesis can be rejected in favor of the alternative, in other words, the average ranks of the algorithms are not equal. Now the Nemenyi tests can be applied, Fig. 3

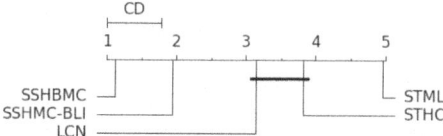

Fig. 3. Graphical representation of the Nemenyi test. Classifiers that are not significantly different, with $p = 0.05$, are connected. CD: critical difference. (Lower is better.)

shows the graphical representation of it. As it can be seen, the proposed method, SSHBMC, is significantly better from the rest of methods.

The Wilcoxon signed-rank test [3] was applied to SSHBMC and SSHMC-BLI, where the alternative hypothesis states that the distribution underlying z is greater than a distribution symmetric about zero, where z is the difference between the paired samples, in this case, the performances of SSHBC and SSHMC-BLI. Therefore, the result of the Wilcoxon signed rank test with $p = 0.05$ is that the null hypothesis can be rejected in favor of the alternative; in other words, the performance of SSHBMC is better than SSHMC-BLI.

6   Conclusions and Future Work

A semi-supervised hierarchical Bayesian multi-label classifier, SSHBMC, is proposed in this manuscript. First, SSHBMC builds pseudo-paths-of-labels for the unlabeled instances from the paths of labels of their labeled neighbors; the unlabeled instances only will be pseudo-labeled if they are similar to their labeled neighbors. Later, the relations of the hierarchy are represented in a Bayesian network, which also encodes the data distribution and the behaviour of the local classifiers; the BN improves the initial classification of the local classifiers by enforcing the hierarchical probability constraint.

Experiments on the Gene Ontology datasets were carried out. It was shown that making use of unlabeled data along with labeled helped to improve the performance of a hierarchical classifier trained only on labeled data. Additionally, taking into account the data distribution represented as a Bayesian network helped to improve the performance of the semi-supervised classifier.

As future work, the proposed method may be combined with convolutional neural networks so it could be applied to image classification problems where the labels are arranged in a hierarchy and labeled images are few.

Acknowledgements. J. Serrano-Pérez acknowledges the support from CONAHCYT scholarship number (CVU) 84075.

References

1. Barutçuoglu, Z., Schapire, R.E., Troyanskaya, O.G., DeCoro, C.: Bayesian aggregation for hierarchical classification, Princeton University, Technical report (2008)
2. Chapelle, O., Schlkopf, B., Zien, A.: Semi-Supervised Learning. The MIT Press, 1st edn. (2010)
3. Demšar, J.: Statistical comparisons of classifiers over multiple data sets. J. Mach. Learn. Res. **7**, 1–30 (2006)
4. van Engelen, J.E., Hoos, H.: A survey on semi-supervised learning. Mach. Learn. **109**, 373–440 (2019)
5. Fagni, T., Sebastiani, F.: On the selection of negative examples for hierarchical text categorization. In: Proceedings 3rd Lang Technology Conference (2007)
6. Giunchiglia, E., Lukasiewicz, T.: Coherent hierarchical multi-label classification networks. In: Larochelle, H., Ranzato, M., Hadsell, R., Balcan, M.F., Lin, H. (eds.) Advances in Neural Information Processing Systems, vol. 33, pp. 9662–9673. Curran Associates, Inc. (2020)
7. Levatić, J., Ceci, M., Kocev, D., Džeroski, S.: Semi-supervised predictive clustering trees for (hierarchical) multi-label classification. Int. J. Intell. Syst. **2024**(1), 5610291 (2024). https://doi.org/10.1155/2024/5610291
8. Metz, J., Freitas, A.A.: Extending hierarchical classification with semi-supervised learning. In: Proceedings of the UK Workshop on Computational Intelligence, pp. 1–6 (2009)
9. Ramírez-Corona, M., Sucar, L.E., Morales, E.F.: Hierarchical multilabel classification based on path evaluation. Int. J. Approximate Reasoning **68**, 179–193 (2016)
10. Santos, A., Canuto, A.: Applying semi-supervised learning in hierarchical multi-label classification. Expert Syst. Appl. **41**(14), 6075–6085 (2014)
11. Santos, A., Canuto, A.: Applying the self-training semi-supervised learning in hierarchical multi-label methods. In: 2014 International Joint Conference on Neural Networks (IJCNN), pp. 872–879 (2014)
12. Serrano-Pérez, J., Sucar, L.E.: Semi-supervised hierarchical classification based on local information. In: Bicharra Garcia, A.C., Ferro, M., Rodríguez Ribón, J.C. (eds.) Advances in Artificial Intelligence - IBERAMIA 2022, pp. 255–266. Springer, Cham (2022)
13. Serrano-Pérez, J., Sucar, L.E.: Hierarchical classification with Bayesian networks and chained classifiers. In: Proceedings of the Thirty-Second International Florida Artificial Intelligence Research Society Conference, Sarasota, Florida, USA, May 19-22 2019, pp. 488–493 (2019)
14. Serrano-Pérez, J., Sucar, L.E.: Semi-supervised hierarchical multi-label classifier based on local information. arXiv preprint arXiv:2405.00184 (2024)
15. Vens, C., Struyf, J., Schietgat, L., Džeroski, S., Blockeel, H.: Decision trees for hierarchical multi-label classification. Mach. Learn. **73**(2), 185 (2008)
16. Xiao, H., Liu, X., Song, Y.: Efficient path prediction for semi-supervised and weakly supervised hierarchical text classification. In: The World Wide Web Conference, pp. 3370–3376. WWW '19, Association for Computing Machinery, New York, NY, USA (2019)
17. Zhu, M.: Recall, Precision and Average Precision, vol. 2, no. 30. Department of Statistics and Actuarial Science, University of Waterloo, Waterloo (2004)

Advancing Photovoltaic Forecasting with Neural Networks: Integrating N-Beats and Sequential Models with Fourier Analysis

Gonzalo Surribas-Sayago[2] , Jose David Fernández-Rodríguez[1,2] ,
and Enrique Dominguez[1,2(✉)]

[1] ITIS Software, University of Málaga, Malaga, Spain
{josedavid,enriqued}@uma.es
[2] Department of Computer Science, University of Malaga, 29071 Malaga, Spain
surribasg@uma.es

Abstract. The rise of green hydrogen as a clean energy source underscores the need for advanced management of renewable energy inputs,
particularly photovoltaic (PV) power generation, which is known for
its variability and unpredictability. Effective integration and use of PV
power is essential for grid stability and sustainable green hydrogen production, which heavily depends on the availability of excess renewable
energy. This study introduces a hybrid deep learning model that combines the N-Beats architecture with a sequential model to enhance the
accuracy of PV power generation forecasts. The N-Beats model, known
for its ability to capture complex temporal relationships, is used alongside a sequential model that processes additional temporal dependencies.
Additionally, this study applies Fourier Analysis to identify significant
frequency components in the time variable, revealing key elements like
yearly, monthly, and daily cycles. The performance of the hybrid model is
compared against traditional algorithms like logistic regression, random
forests, support vector machines (SVM), and XGBoost, using metrics
such as accuracy, Mean Bias Error (MBE), Mean Absolute Percentage
Error (MAPE), and Root Mean Squared Error (RMSE). Our results suggest that the hybrid N-Beats model outperforms traditional algorithms
in terms of accuracy and stability, making it a promising option for
improving efficiency in green hydrogen production.

Keywords: photovoltaic energy · deep learning · N-Beats

1 Introduction

The transition towards renewable and clean energy sources, such as photovoltaic
(PV) power [12], is crucial to meet the growing demand for sustainable energy
and to reduce carbon emissions. PV power, as a distributed energy resource,
plays a key role in the production of green hydrogen [5], an energy carrier that

L. Correia et al. (Eds.): IBERAMIA 2024, LNCS 15277, pp. 287–297, 2025.
https://doi.org/10.1007/978-3-031-80366-6_24

can help decarbonize sectors like transportation and industry. However, PV production is inherently intermittent and unpredictable, posing challenges for planning and optimizing processes related to green hydrogen production.

Accurate forecasting of PV production is essential to address these challenges. By predicting how much PV power will be generated, it is possible to better plan and manage resources for green hydrogen production, optimizing the use of electricity for electrolysis processes [14]. A precise forecast facilitates the scheduling of hydrogen production, reducing reliance on non-renewable energy sources and improving overall sustainability.

Traditionally, several approaches have been used for time series forecasting [8], including physical models, persistence methods, and statistical models. Among these, models like SARIMA (Seasonal Autoregressive Integrated Moving Average) [2] and SARIMAX (which incorporates exogenous variables) have been employed to forecast PV production. However, these traditional methods often struggle with handling time series that exhibit high variability and non-linear behaviors.

In this context, deep learning-based models, such as N-Beats, have shown particular effectiveness in forecasting [1]. N-Beats is a block-based model that can capture complex trends and seasonality, making it an ideal choice for accurate forecasting. This paper explores the use of N-Beats networks, both in their pure form and as a hybrid model with sequential components, to improve the accuracy of PV forecasting. By addressing uncertainty and intermittency, these models offer innovative solutions to optimize green hydrogen production, facilitating better planning and resource management.

The performance of the hybrid model [4] is compared against traditional algorithms like logistic regression [13], random forests [3], support vector machines (SVM) [6], and XGBoost [9], providing a comprehensive assessment of the effectiveness of the N-Beats approach. The comparison allows us to identify which models offer the most accurate and reliable forecasting results for green hydrogen production, based on various metrics such as accuracy, Mean Bias Error (MBE), Mean Absolute Percentage Error (MAPE), and Root Mean Squared Error (RMSE).

2   N-Beats and Hybrid Model

N-Beats is a deep learning model designed specifically for time series forecasting. Developed by Microsoft Research, it is based on a block-based architecture, with each block containing a series of fully connected layers. A unique characteristic of N-Beats is its ability to capture both trends and seasonality in time series data. The model consists of two main stacks: the trend stack and the seasonality stack. The trend stack focuses on modeling long-term trends, while the seasonality stack addresses periodic fluctuations. The trend block uses a polynomial function to model the trend:

$$T(t) = \sum_{i=1}^{n} w_i \times t^i \tag{1}$$

where $T(t)$ represents the estimated trend at time t, w_i are the polynomial coefficients, and n is the degree of the polynomial. The seasonality block captures repetitive patterns over time using sinusoidal and cosinusoidal components:

$$S(t) = \sum_{i=1}^{m} a_i \times \sin(2\pi \times i \times t/\omega) + b_i \times \cos(2\pi \times i \times t/\omega) \qquad (2)$$

where $S(t)$ represents the estimated seasonality at time t, a_i and b_i are the coefficients for sinusoidal and cosinusoidal components, m is the number of frequency components, and ω is the period. The final forecast combines trend and seasonality:

$$Y(t) = T(t) + S(t) \qquad (3)$$

Our study introduces a hybrid version that combines the N-Beats model with a sequential network. This combination allows us to adapt to both short-term and long-term patterns in renewable energy analysis.

The N-BEATS architecture is described by Oreshkin et al. [10], in their first Figure. The first component in N-BEATS is the initial block (block input), where inputs are processed through a fully connected neural network with four layers (FC stack). The outputs bifurcate into two different FC blocks: one to compute the backcast coefficients (theta b) and another to compute the forecast coefficients (theta f). After calculating the coefficients, the input features are multiplied by theta b and theta f, generating two outputs: the backcast and the forecast.

These outputs are then passed to the next component, known as Stack Input. Here, the backcast and forecast are combined to form the inputs for the first block of the architecture (block 1). In this block, the backcast is used to correct the previous prediction and enhance future forecasting. The outputs of this block are then passed through multiple stacks of blocks, where each stack learns specific patterns in the time series.

Finally, the outputs from all the stacks are combined in the last component of the N-BEATS architecture. Here, the global forecast is generated, providing a comprehensive and refined view of the time series. This stacking approach allows the model to capture both local and global patterns in the data.

The hybrid model combines N-Beats with a sequential model to capture additional temporal dependencies and refine the forecast. This combination is effective in contexts with high variability or complex seasonality, such as photovoltaic production. Additionally, it is compared against traditional algorithms using standard performance metrics such as accuracy, Mean Absolute Error (MAE), and Root Mean Squared Error (RMSE).

In summary, the N-BEATS architecture utilizes a modular and stacked approach to capture complex patterns in time series data, combining input features, expansion coefficients, backcasts, and forecasts across multiple blocks and stacks. This approach provides flexibility and accuracy in time series prediction across various contexts.

3   Materials and Methods

This section describes the materials and methods used in the study, including the data, the hybrid N-Best model, data preparation, and the evaluation metrics used to assess the performance of the model. It details both the N-Best model and its hybrid sequential variant, providing a comprehensive view of the analytical techniques applied.

3.1   Data

The dataset utilized in this study, *Solar Power Generation Data*, was posted at Kaggle by Syed Afroz[1]. It encompasses photovoltaic power generation data and other relevant meteorological parameters recorded throughout the year 2017, comprising hourly data points across the full year. This totals 8,760 observations. Each data entry is timestamped and includes the following features:

- **Date-Hour (NMT)**: The exact date and hour of the observation.
- **Wind Speed** (*m/s*): The speed of the wind measured during each hour.
- **Sunshine** (*hours*): The duration of sunshine per hour.
- **Air Pressure** (*hPa*): Atmospheric pressure recorded at the time of measurement.
- **Radiation** (W/m^2): Solar radiation intensity.
- **Air Temperature** ($^\circ C$): The ambient air temperature.
- **Relative Air Humidity** (*%*): The percentage of relative humidity in the air.
- **System Production** (*kW*): The amount of electricity generated by the photovoltaic system.

The time span of the dataset covers from January 1, 2017, at 00:00 h to December 31, 2017, at 23:00 h, reflecting a complete year of operations minus the last hour. This comprehensive dataset allows for detailed analysis of patterns and impacts of environmental factors on photovoltaic power production. This dataset is primarily used to train and validate the forecasting models described in this study. The rich variety of meteorological data alongside the power output data enables the implementation of robust machine learning models to predict photovoltaic power production accurately. Figure 1 shows a visualization of the dataset.

3.2   Data Sparsity

The collected data of PV power output include night-time values, where there is a presence of many zeros due to the absence of solar radiation. These zero values are crucial for maintaining the integrity of the temporal data sequence, which is fundamental for forecasting models, especially when dealing with daily cycles of solar production. In the context of photovoltaic power forecasting using models

---

[1] https://www.kaggle.com/datasets/pythonafroz/solar-powe-generation-data.

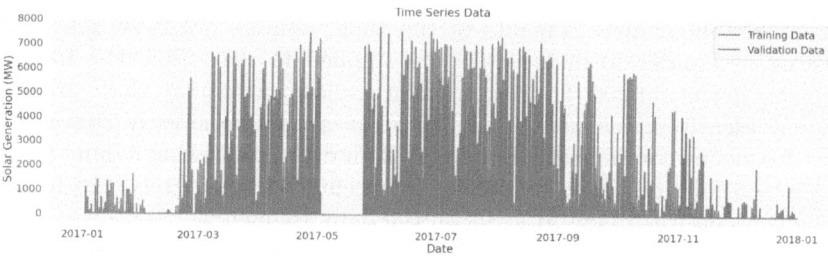

Fig. 1. A visualization of the dataset used to test our models (see Sect. 3.1. Blue samples are used for training, red samples for validation. (Color figure online)

like N-BEATS and its hybrid extension, N-BEATS-Sequential, the inclusion of zero values during nighttime is essential.

These models rely on understanding the complete pattern of solar production, including periods of non-production, to optimize the forecast accuracy throughout a 24-h cycle. Removing these zeros could lead to a misunderstanding of the daily production cycles and adversely affect the model's ability to predict zero output accurately during non-productive hours. Therefore, retaining these zero values in the dataset is imperative to ensure that the model can learn the true dynamics of solar power production, reflecting realistic operational conditions and enhancing prediction reliability.

An important factor in data quality is ensuring there are no missing or null values (NaNs). Upon analysis, this dataset was found to have no missing data in any of the relevant features. This absence of missing values indicates a high level of data quality, reducing the need for data imputation or additional data cleaning.

During the data analysis, it was observed that some data points had negative values in the *Radiation* column, indicating an inconsistency in the dataset. These negative radiation values were most likely due to measurement errors or data collection anomalies. Since negative radiation values are not physically meaningful, a data treatment process was applied to correct this issue.

The following steps were taken to address negative radiation values:

- Identify Negative Radiation: A filter was applied to identify all data points where the *Radiation* column had a negative value. Additionally, a subset of data points was identified where both *Radiation* and *SystemProduction* had negative values.
- Replace Negative Values with Zero: The rows identified with negative radiation were corrected by replacing the negative values with zero. This ensured that the dataset no longer contained physically implausible radiation values.

These data treatment steps were necessary to maintain data consistency and avoid erroneous inputs in the machine learning models. By correcting these anomalies, the dataset became more reliable for forecasting photovoltaic power generation.

To understand which elements of the time variable are most relevant for forecasting, a Fourier Transform [7] was applied to determine the frequency components in the dataset. Fourier Transform allows us to analyze the frequency of a signal, identifying elements that repeat at certain periods. In this analysis, the first frequency corresponds to a nearly 24-h cycle, indicating a daily pattern, while the second significant frequency represents approximately 11.5 h. Given this, the decision was made to focus on the daily frequency.

However, the original time variable had discontinuities that could confuse models like N-Beats. For example, the jump from February 28 to March 1 creates a discontinuity that can hinder the ability of the neural network to find suitable patterns. A solution was implemented by representing the time variable using sinusoidal and cosinusoidal functions, ensuring a cyclic behavior without discontinuities.

The following steps were taken to implement this solution:

- Sinusoidal and Cosinusoidal Representation: The time variable was converted into sine and cosine representations to create a continuous, cyclic pattern. This was done by considering the period of the signal and avoiding discontinuities.
- Implementation: New features were added to the dataset using these sinusoidal and cosinusoidal representations:
  - *dia11sin*: Represents the 11.32-h cycle with a sine function.
  - *dia11cos*: Represents the 11.32-h cycle with a cosine function.
  - *diasin*: Represents the daily cycle with a sine function.
  - *diacos*: Represents the daily cycle with a cosine function.

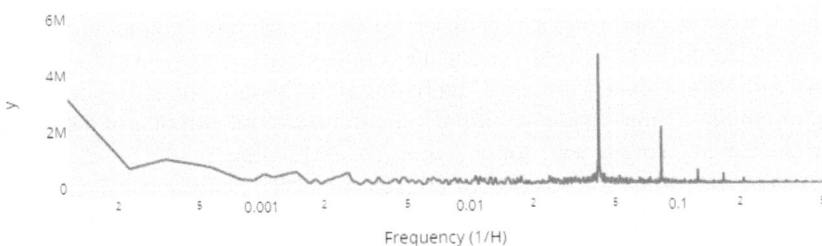

Fig. 2. Fourier Analysis to show the main frecuency components of the data.

This approach of using sinusoidal and cosinusoidal functions helps maintain the cyclic nature of the data, allowing the N-Beats model to detect and work with patterns more effectively. By ensuring a consistent representation of the day without discontinuities, the resulting dataset becomes more suitable for training and can improve the accuracy of forecasting. To understand which elements of the time variable are most relevant for forecasting, a Fourier Transform was applied to determine the frequency components in the dataset. This helped identify key cycles and patterns, such as a nearly 24-h cycle and an 11.5-h cycle. Based on this, new features were created using sinusoidal and cosinusoidal representations to ensure a cyclic behavior without discontinuities. In this respect, Fig. 2 illustrates the Fourier frequency analysis, showing the key cycles detected in the dataset.

To identify which elements of the time variable are most relevant for forecasting, a Fourier Transform was applied to determine the frequency components in the dataset. This analysis revealed key cycles and patterns, such as a nearly 24-h cycle and an approximately 11.5-h cycle. Based on this, new features were created using sinusoidal and cosinusoidal representations to ensure a cyclic behavior without discontinuities.

Feature scaling is a crucial step in data preprocessing, ensuring that variables with different scales do not lead to biased model behavior. This process also helps accelerate calculations in machine learning algorithms and improves convergence rates. In this study, we used normalization with *MinMaxScaler* to scale the dataset within a specific range, in this case, between –1 and 1. The normalization formula applied is:

$$x' = \frac{x - x_{\min}}{x_{\max} - x_{\min}}$$

where x is the original value, $x_{\min}$ is the minimum value, $x_{\max}$ is the maximum value, and x' is the normalized value. The *MinMaxScaler* allows us to set a custom feature range, which in this study is between –1 and 1. This range is commonly used in deep learning models to ensure that all features have a consistent scale, minimizing the risk of certain variables dominating the learning process due to their magnitude. Normalization with this scaler type is particularly useful in the context of deep learning models like N-Beats, which can be sensitive to differences in feature scales. By normalizing the dataset [11] to a consistent range, we help ensure a more stable learning process and improved model performance. For this study, the *MinMaxScaler* was applied to all relevant features in the dataset before training the models. This step helps maintain consistent data scales throughout the training process and contributes to more accurate and reliable model predictions.

3.3   Performance Evaluation Metrics

To evaluate the performance of the proposed hybrid deep learning model for photovoltaic (PV) power forecasting, we use two widely used forecasting accuracy metrics: Mean Absolute Error (MAE), which is defined in expression 4, and Root Mean Square Error (RMSE), which is defined in Eq. 5.

MAE represents the average of the absolute differences between the forecast values and the actual values. This metric offers a simple and direct measure of forecast accuracy, indicating the average magnitude of the errors:

$$\text{MAE} = \frac{1}{N} \sum_{t=1}^{N} \left| \hat{P}_t - P_t \right| \tag{4}$$

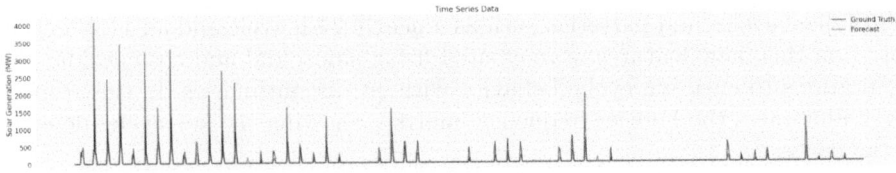

Fig. 3. Comparison between the ground truth and the forecast from our hybrid model.

$$\text{RMSE} = \sqrt{\frac{1}{N} \sum_{t=1}^{N} (\hat{P}_t - P_t)^2} \tag{5}$$

where N is the total number of observations, P_t is the actual value, and $\hat{P}_t$ is the forecast value of PV power output at time t. RMSE measures the square root of the average squared differences between the forecast and actual values. This metric penalizes larger errors more than MAE, offering a different perspective on the accuracy of the forecast.

These metrics are crucial for evaluating the performance of the deep learning model for PV power forecasting. While MAE provides a straightforward measure that is easy to interpret, RMSE highlights larger errors and gives a more nuanced view of the performance of the model.

4 Experiments and Results

The proposed hybrid N-BEATS model was trained using Google Colab with TPU (Tensor Processing Unit), which provided significantly enhanced processing capabilities compared to using GPU.

For this study, the proposed model was configured to process 24-h windows, meaning each input represents a complete day of data. Additionally, 10 additional features were used to reflect various parameters relevant to photovoltaic energy prediction. The prediction horizon for this model is one hour, indicating the goal is to predict photovoltaic energy production one hour ahead.

The model was compiled with a loss function based on Mean Absolute Error (MAE) and the Adam optimizer, known for its efficiency and stability in training deep learning models. Additionally, Root Mean Squared Error (RMSE) was used as an additional metric to evaluate model performance, providing a measure of mean squared error that helps identify significant deviations in prediction.

The training process for the N-Beats Hybrid model included several callbacks to optimize performance and prevent overfitting. Key callbacks were EarlyStopping and ReduceLROnPlateau, each serving a specific purpose. EarlyStopping stops training when the validation loss stops improving after a certain number of epochs, helping to prevent overfitting and allowing for restoration of the best weights. ReduceLROnPlateau reduces the learning rate if the validation loss does not improve over a specific period, adjusting training for better long-term performance. The training time for the hybrid N-BEATS model in the TPU environment was efficient considering the model's complexity, approximately 6.17 min in 37 epochs. In the hybrid model, the N-Beats component handled the 24-h windows, while the sequential part processed the 10 additional features. It was compiled similarly to the pure model, but with a lower learning rate for the Adam optimizer.

The use of these techniques and callbacks helped achieve a balance between performance and efficiency during the training process, allowing the model to converge in a stable and effective manner. With this configuration, the N-Beats Hybrid model proved suitable for photovoltaic data prediction, providing reliable and accurate results with a one-hour prediction horizon. The inclusion of RMSE as an additional metric provided a more comprehensive perspective on model performance, helping to identify areas for improvement and ensure precise predictions.

While the training time of the proposed model is longer compared to simpler approaches like XGBoost or ARIMA, the significant improvement in model accuracy justifies the use of advanced resources. The proposed model's ability to capture complex patterns in time series data provides more accurate predictions, making the additional training time and higher computational requirements a worthwhile investment. A comparison between the ground truth and the forecast from our model is shown in Fig. 3. Table 1 presents the metrics comparing the forecasts for our model and several other models applied to the same dataset by several data scientists from Kaggle: Aniket Kadam applied ARIMA[2], Mayowa Akinmolayan applied XGBoost[3], and Shekhar Chormal applied ExtraTrees[4]. As shown in the Table, our N-BEATS Hybrid model achieved the lowest MAE, indicating superior performance compared to other models.

Table 1. Model Comparison by MAE and RMSE. See text for details.

Model	MAE	RMSE
N-BEATS Hybrid (ours)	44.42	237.09
ARIMA	111.314	404.24
XGBoost	214.706	482.23
ExtraTrees	348.53	777.20

5   Conclusions

The performance of the N-BEATS Hybrid model developed in this study, shown in Table 1, underscores its effectiveness in predicting photovoltaic energy. These

[2] https://www.kaggle.com/code/aniketkadam702030/solar-power-machine-learnig.
[3] https://www.kaggle.com/code/mayowaakinmolayan/solar-power-generation-data-with-xgboost.
[4] https://www.kaggle.com/code/rajshekharchormale/solar-power-generation-prediction.

results are promising, considering the limitation of the available data to a single year, suggesting further potential for improvement with a more extensive and representative dataset. When compared to previous models, the N-BEATS Hybrid demonstrates significant superiority in accuracy and performance. It outperforms models like ARIMA, XGBoost, and ExtraTrees, highlighting its ability to capture complex patterns in photovoltaic energy production. The integration of N-BEATS with a sequential model proves to be an effective strategy for improving prediction accuracy and stability. This combination enables capturing both local and global patterns, contributing to better performance in predicting photovoltaic energy under diverse conditions. Fourier analysis and temporal variable transformation have proven to be valuable tactics for enhancing the ability of the model to learn temporal patterns. These techniques enrich the understanding of photovoltaic energy production dynamics and ultimately the accuracy of predictions. It is important to note that due to the nature of the data used, certain limitations were encountered. The dataset is confined to a single year, which may not be sufficient to fully capture annual seasonality and other long-term patterns. Additionally, using two months of low production for training may influence the final results and limit the ability of the model to capture longer seasonal patterns.

References

1. Ahmed, D.M., Hassan, M.M., Mstafa, R.J., et al.: A review on deep sequential models for forecasting time series data. Appli. Comput. Intell. Soft Comput. **2022** (2022)
2. Chaturvedi, S., Rajasekar, E., Natarajan, S., McCullen, N.: A comparative assessment of sarima, lstm rnn and fb prophet models to forecast total and peak monthly energy demand for India. Energy policy **168**, 113097 (2022)
3. Ibrahim, I.A., Khatib, T.: A novel hybrid model for hourly global solar radiation prediction using random forests technique and firefly algorithm. Energy Convers. Manage. **138**, 413–425 (2017)
4. Li, P., Zhou, K., Lu, X., Yang, S.: A hybrid deep learning model for short-term pv power forecasting. Appli. Energy **259**, 114216 (2020)
5. Liu, L., Zhai, R., Hu, Y.: Performance evaluation of wind-solar-hydrogen system for renewable energy generation and green hydrogen generation and storage: energy, exergy, economic, and enviroeconomic. Energy (Oxford) **276**, 127386 (2023)
6. Liu, Y., Zhou, Y., Chen, Y., Wang, D., Wang, Y., Zhu, Y.: Comparison of support vector machine and copula-based nonlinear quantile regression for estimating the daily diffuse solar radiation: A case study in china. Renewable Energy **146**, 1101–1112 (2020)
7. Marks, R.J., Gravagne, I.A., Davis, J.M.: A generalized fourier transform and convolution on time scales. J. Math. Anal. Appl. **340**(2), 901–919 (2008)
8. Montgomery, D.C., Jennings, C.L., Kulahci, M.: Introduction to time series analysis and forecasting. John Wiley & Sons (2015)
9. Obiora, C.N., Ali, A., Hasan, A.N.: Implementing extreme gradient boosting (xgboost) algorithm in predicting solar irradiance. In: 2021 IEEE PES/IAS PowerAfrica, pp. 1–5. IEEE (2021)

10. Oreshkin, B.N., Chapados, N., Carpov, D., Bengio, Y.: N-beats: neural basis expansion analysis for interpretable time series forecasting. In: International Conference on Learning Representations (ICLR) (2020)
11. Panigrahi, S., Behera, H.: Effect of normalization techniques on univariate time series forecasting using evolutionary higher order neural network. Inter. J. Eng. Adv. Technol. **3**(2), 280–285 (2013)
12. Wan, C., Zhao, J., Song, Y., Xu, Z., Lin, J., Hu, Z.: Photovoltaic and solar power forecasting for smart grid energy management. CSEE J. Power Energy Syst. **1**(4), 38–46 (2015)
13. Wang, G., Su, Y., Shu, L.: One-day-ahead daily power forecasting of photovoltaic systems based on partial functional linear regression models. Renewable Energy **96**, 469–478 (2016)
14. Zhao, H., Yuan, Z.: Progress and perspectives for solar-driven water electrolysis to produce green hydrogen. Adva. Energy Mater. **13**(16) (2023)

Efficiency of the Transformer Model in Time Series Forecasting: A Case Study in Wastewater Treatment Plants

Gonçalo Medeiros(✉) [iD], Francisco S. Marcondes[iD], Pedro Oliveira[iD], José Machado[iD], and Paulo Novais[iD]

ALGORITMI/LASI Centre, University of Minho, Braga, Portugal
pg50399@alunos.uminho.pt,
{francisco.marcondes,pedro.jose.oliveira}@algoritmi.uminho.pt,
{jmac,pjon}@di.uminho.pt

Abstract. Global energy demand has been growing over the last decades, having a significant impact on the environment. Alternatives that can mitigate these effects are crucial for the future of our planet. Wastewater Treatment Plants (WWTPs) are a vital infrastructure to manage residual waters, presenting an opportunity for energy production from the released biogas during the anaerobic digestion phase. The present study uses a multivariate recursive forecasting approach to evaluate the performance of Transformer-based models in forecasting electricity production in this context. Transformer-based candidate models were developed, and their hyperparameters were tuned using a grid search. The best Transformer candidate achieved the second-best Root Mean Square Error (RMSE) value of 359.4 kWh, outperforming the Gated Recurrent Unit (GRU) by 9%, although the Long Short-Term Memory (LSTM) model performed the best.

Keywords: Biodigestion · Deep Learning · Electricity Production · Energy Forecasting · Time Series Forecasting · Transformer · Wastewater Treatment Plants

1 Introduction

As incomes and population grow worldwide, more people can now afford to drive their vehicles, consume more goods, and use more services, resulting in a growing energy demand [4]. All this demand has an environmental cost given that most of the world's energy production doesn't come from clean sources but from fossil fuels (coal, oil, and gas) instead, the largest source of carbon dioxide (CO_2) emissions [18], producing many different types of pollution such as air, soil, and water. In this vein, Wastewater Treatment Plants (WWTPs) are an integral infrastructure for managing and treating residual water. There is an opportunity at these treatment plants to produce clean energy through the chemical reactions occurring in reactor tanks during the anaerobic digestion

L. Correia et al. (Eds.): IBERAMIA 2024, LNCS 15277, pp. 298–309, 2025.
https://doi.org/10.1007/978-3-031-80366-6_25

phase of the water treatment pipeline. Energy production from biogas during the anaerobic digestion phase of WWTPs has revealed promising economic gains and reduced environmental impact on the structures that use this method [1,13].

The incoming wastewater effluents of WWTPs are characterized by several indicators, such as flow rates, the concentration of pollutants, pH levels, temperature variations, energy consumption and production, and others, which are collected sequentially and continuously over time. The nature of this data collection leads to a time series problem. Analyzing the dependencies and trends of the data over time can help make predictions on future WWTP features of interest. Multiple Machine Learning (ML) models and statistical methods have been leveraged for this end [6,8,12,23], encompassing classical machine learning and deep architectures. The application of these techniques has led to an improvement in resource management and decreased operational costs.

This work aims to study the efficiency of the Transformer architecture applied to the time series forecasting problem present in a wastewater treatment plant, which has had good results in solving this task [10,22,25]; however, some recent findings question their efficiency [20,24]. To this end, candidate models were developed based on the Transformer architecture and tuned following a grid-search methodology. Long Short-Term Memory (LSTM) and Gated Recurrent Unit (GRU) models were developed in [15] for predicting electricity production using the same experimental setup. The results from this study serve as a baseline for comparing the effectiveness of the proposed candidates in this context.

The following section details a theoretical review of the Transformer model and its use; the third section covers the data collection, exploration, and preparation process and the model conception; the fourth and fifth sections specify the experiments and discuss the obtained results, respectively, followed by the final conclusions.

2   Review on the Transformer Architecture

The Transformer model, introduced in the paper "Attention is All You Need" [21], reshaping the Natural Language Processing (NLP) paradigm with its encoder-decoder model and self-attention mechanism. Self-attention typically refers to the scaled dot-product, a function that attributes different weights to different elements, determining how much focus each should receive during their processing. The function achieves this by taking the dot product of the query (Q) and key (K) vectors and then scaling it by the square root of the dimension of the key vectors. It is then passed through a *softmax* activation function, and the weights are multiplied by the value (V) matrix as given in Eq. 1 [5].

$$Attention(Q, K, V) = softmax(\frac{QK^{\top}}{\sqrt{d_k}})V \qquad (1)$$

Because of this mechanism, the model can take into account every element within a sequence and capture long-range dependencies and patterns in sequential data, presenting a promising opportunity to address the challenges of modeling complex processes in wastewater treatment.

Transformer-based architectures have performed well in NLP, achieving state-of-the-art in most tasks such as text generation and translation [3,17]. Transformer has also succeeded in capturing the temporal tendencies in time series problems, performing well in domains such as weather forecasting [7], finance [9], and energy consumption [11]. Applying this architecture to tackle the problem of predicting electricity production in a WWTP could yield promising results. In their study, Pereira et al. [16] developed a Transformer and an LSTM model for predicting the electrical conductivity of the influent water in the context of a water treatment plant. The work utilized a multivariate recursive multi-step forecasting approach to solve the time series problem. The study concluded that considering the top 5 candidate models for each, the Transformer performed better than the LSTM in this time series problem, with the best Transformer model performing 13.72% better regarding Root Mean Square Error (RMSE) than the best LSTM model.

In [19], Sapp et al. used a Temporal Fusion Transformer (TFT) to predict the biogas production rate in an anaerobic digestion tank of a WWTP. The data was collected over three years with a daily frequency. The study used measurements from two real-life digesters in a wastewater treatment plant with some missing values that were dealt by using linear interpolation. The three baseline models for comparing performance were k-Nearest Neighbors (KNN), Autoregressive Integrated Moving Average (ARIMA), and a modified Decoder-ANN. The TFT obtained a Mean Absolute Percentage Error (MAPE) of less than 8% and performed the best out of all tested models.

While multiple works indicate that Transformer-based models have performed well and thus effectively solve time series problems, some, such as [24], state the contrary. The authors propose various baseline models for the baseline comparison and conduct multiple experiments across popular time series datasets to compare with five Transformed-based models. The results show that the baseline models surprisingly outperform Transformer models in all considered cases, often by a big margin (20%–50%).

In a recent study [20], following up on [24], the efficiency of Transformers in solving time series problems is questioned against simpler linear models. They use new models of their own across some of the same datasets used in the original work, concluding that the best-performing models are oversimplified to the point where they are not considered Transformers but are more akin to shallow neural networks. The authors suggest that the scientific community may have prioritized the application of fashionable techniques over finding effective solutions.

In summary, this review suggests that there is a debate on the efficiency of the Transformer model applied to the time series problem, which this work aims to contribute to.

3   Materials and Methods

The data used in this study was collected by a wastewater treatment company. Between January 2020 and March 2023, 1186 observations were collected from a reactor tank used in the water treatment process, each entry containing multiple relevant features similar to the ones considered in [15], which are detailed in Table 1. A data analysis was conducted, and it was determined that there were no missing timesteps in the collected dataset, with each observation mapping to a specific day. However, there were seven missing values whose treatment is detailed in the following section.

Table 1. Features of the WWTP dataset, adapted from [15]

Feature	Description
date	Timestamp
Q	Volumetric water flow (m^3/day)
T	Temperature (žC)
%co2	Percentage of CO2 in the biogas
%ch4	Percentage of CH4 in the biogas
q_gas_prod	Biogas produced (m^3/day)
q_gas_cog	Biogas cogenerated (m^3/day)
q_gas_total	Partial pressure of biogas (bar)
p_gas_ch4	Partial pressure of methane (bar)
p_gas_co2	Partial pressure of carbondioxide (bar)
VS	Volatile solids (mg/L)
X_I	Inert particulate matter (mg/L)
Xe	Biodegradable matter (mg/L)
E	Produced electricity (kWh)

3.1 Data Preparation

An analysis was conducted on the regular features, applying a Kolmogorov-Smirnov test to determine if they followed a normal distribution. Since the p value obtained was less than 0.05, it is concluded that the features do not follow a normal distribution. As mentioned, the dataset had seven missing values in the target variable, electricity. Given that those seven missing values are dispersed throughout the curve and not sequential, linear interpolation was used to calculate and fill the missing target values. There were 78 rows with the values of *q_gas_cog* and *eletricity* equal to zero, a trend observed mostly from late November 2020 and early January 2021. Random values, above or below the standard deviation, were injected for the *q_gas_cog*. As for the target label, there is a direct correlation between energy production and biogas composition. Using this knowledge, the average efficiency of the dataset was calculated as the sum of the non-zero electricity values divided by the cogenerated gas, the value of the average energy efficiency obtained is 1.65 kWh/m^3. The energy efficiency value is then multiplied by the *q_gas_cog* for each entry in which the electricity value equals zero to obtain a valid electricity. After proceeding with this replacement strategy for the zero values, it is possible to observe that an increase in the production of electricity maps well with an increase in the value of the cogenerated biogas, confirming what was asserted.

A correlation analysis is required to get insights regarding how the other features influence the target value of electricity production. Given the previously presented features' non-normal distribution, a Spearman coefficient was taken to assess the relationship between the features and the target. The results obtained show that the only features with significant correlation to the target variable were *q_gas_cog*, *q_gas_prod*, and *T*. Considering this, all the unrelated features were removed, leaving the dataset with four final features, the three correlated ones plus the target variable.

The data was formatted through a sliding window approach, transforming the initial problem into a supervised learning one. The data was normalized to a range of 0 to 1, concluding the final step of data transformation. The data preparation process was similar to the one applied in [15], which serves as our baseline results to ensure that performance discrepancies pertain solely to model differences. Figure 1 presents the weekly averages of electricity production over three years (2020, 2021, and 2022), where it is observed that the values vary significantly. During the first year, 2020, the electricity production was lower at weeks 24 and 34. The same tendency is observed for the data collected in 2022, in which there are two minimum values at weeks 25 and 35. The following year, 2021, shows the lowest production of all years at weeks 16 and 45, while the maximum is attained during week 32. Nonetheless, the maximum production is reached during the year 2022 between the 12th and 18th weeks. Additionally, it's possible to observe that the electricity production stayed above 1250 every year during that same period. It is worth noting that the weekly average electricity production of 2021 and 2022 follows the same tendency and the same nominal value during the last 3 weeks, i.e., weeks 48 to 51.

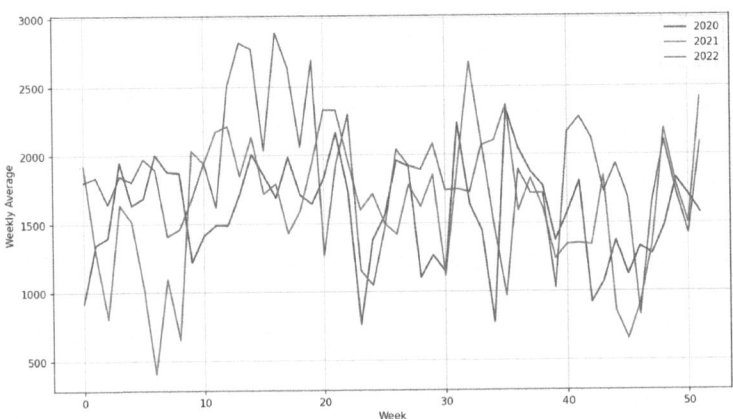

Fig. 1. Weekly averages of electricity production

3.2   Model Conception

It was necessary to conceive a model for the experiments conducted in this research. The developed candidates were based on the architecture illustrated in Fig. 2, inspired by the original Transformer paper [21]. It consists of N encoder blocks composed of the Multi-Head Attention layer with a dropout and a residual connection, followed by a one-dimensional convolutional layer with another dropout and residual connection. The encoder output is then fed to M feed-forward dense layers with dropouts. Finally, a global average pooling with the final linear output layer for the regression gives us the final prediction.

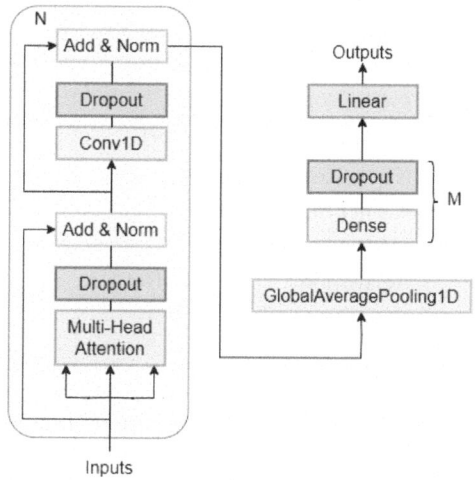

Fig. 2. Proposed Transformer-based Architecture

4   Experiments

Multiple experiments were conducted to determine the best-performing set of hyperparameters for the proposed model. These experiments include the model training and subsequent predictions, following a multivariate recursive forecasting approach to predict the next two values of produced electricity in the bioreactor, utilizing the first forecast as input to predict the second. A grid-search tuning methodology was applied to determine the best hyperparameters for the Transformer-based candidate models. The search space is described in Table 2.

The first two parameters, the timesteps and batch size, are related to the time series task, while the rest are exclusive to the model. For timesteps, the previous one, two, and three weeks are considered (values of 7, 14, and 21). Regarding the Transformer model, parameters such as attention head size, number of encoder blocks, and number of feed-forward layers are considered. Besides the mentioned parameters, the data format options *channels first* and *channels last* for the average pooling layer are also considered to evaluate how this affects the performance of the candidates. These different orderings correspond to input shapes *(batch, n_features, timesteps)* and *(batch, timesteps, n_features)* respectively. *Channels first* ordering doesn't have an effect on the timesteps, performing feature reduction instead, while *channels last* doesn't impact the number of features while reducing the timesteps [14]

Table 2. Hyperparameter Search Space

Parameters	Values
Timesteps	{7, 14, 21}
Batch Size	{10, 20}
Encoder Layers	{4, 8}
Neurons	{64, 128}
Num. Heads	{4, 8}
Kernel Size	{1, 2}
MLP Layers	{3, 4}
Activation	{tanh, ReLU}
Dropout	{0.0, 0.5}

The main evaluation metric considered for comparison of the candidate models is the RMSE [2], calculated as shown in Eq. 2, where y_i and $\hat{y}_i$ represent the actual values and the predicted values, respectively, and n is the total number of data points.

$$RMSE = \sqrt{\frac{\sum_{i=1}^{n}(y_i - \hat{y}_i)^2}{n}} \tag{2}$$

The RMSE is the standard deviation of the residuals, which measures how far away the predicted values are from the actual values. It provides a good understanding of the candidates' performances; the lower its value, the better the accuracy. A cross-validation technique ensures the candidate models perform well on new unseen data. Since this is a time series problem where the continuity of the data points is paramount to extracting the relationships between them, a *TimeSeriesSplit* is utilized, with a k value of 3, where each new fold is a superset of the former. No shuffling occurs between training sets to preserve the temporal component. An early stopping mechanism is used to prevent the model from learning noise instead of the relevant patterns, halting the training if the validation score does not improve after a set number of epochs. The experiments' reproducibility was considered, as the same seed value of 91195003 was used throughout all runs. It is relevant to note that to ensure the same experimental setup as the one detailed in the baseline study [15] and therefore guarantee reproducibility, this study imposes the same 3-fold cross-validation technique and the same initial seed besides the aforementioned data preparation. The RMSE is calculated for all the k folds, and the mean is taken. The candidates with the lower overall mean RMSE are considered the best. During the development of this work, Python version 3.10.12 and Tensorflow version 2.15 were used for the development, tuning, and evaluation of the proposed candidate models, along with auxiliary libraries such as Pandas, Numpy, Matplotlib, and sci-kit-learn. All hardware was provided and supported by Google's Collaboratory.

5 Results and Discussion

Having conducted all the defined experiments, the evaluation of the candidate models for predicting the energy production is shown in Table 3. This table highlights the 5 candidate models that obtained the lowest RMSE and their respective hyperparameters for each data formatting option. The best Transformer model achieved an RMSE value of 359.4 kWh, utilizing the previous 7 timesteps and a batch size of 20; the number of attention heads and encoding layers was 8. While the data formatting didn't impact the best model's performance, as the *channels first* and *channels last* best model have the same RMSE score, the second-best candidate uses *channels first* with an RMSE of 371.1 kWh, compared with *channels last's* second-best which achieved a score of 390.5 kWh, approximately a 5% improvement. The top 5 candidates' average RMSE of *channels first* and *channels last* is 381.6 kWh and 387.5 kWh, respectively, indicating a slight improvement from the former over the latter. Both data format options show some homogeneity regarding the number of neurons in the feed-forward part, as only 1 candidate in each used 64 instead of 128, and regarding the kernel size, where all but 1 candidate in each used size 2 instead of 1. A small number of timesteps seems to correlate with better performance, as only one of the candidates used 14 instead of 7, and no candidate used 21 timesteps. *Channels*

Table 3. Top 5 Transformer-based candidate models, for *channels first* and *channels last*

Timesteps	Batch	Enc Layers	Neurons	Num Heads	Kernel Size	MLP Layers	Activation	Dropout	RMSE
Channels first									
7	20	8	128	8	1	3	tanh	0.5	**359.4**
7	10	4	128	4	1	4	relu	0.0	371.1
7	20	4	128	8	2	4	relu	0.5	390.5
14	10	4	64	4	1	3	relu	0.0	391.3
7	10	4	128	4	1	4	relu	0.5	395.8
Channels last									
7	20	8	128	8	1	3	tanh	0.5	**359.4**
7	20	4	128	8	2	4	relu	0.5	390.5
7	10	4	128	4	1	4	relu	0.5	395.8
7	20	8	64	8	1	3	relu	0.0	395.8
7	10	8	128	8	1	3	tanh	0.5	396.2

first's best model used 8 encoding layers while the others only employed 4, and the preferred activation was *relu*. Most *channels last* candidates used 8 attention heads; the preferred dropout rate was 0.5. The remaining hyperparameters don't display the same homogeneity, alternating between options.

With the final Transformer candidates analyzed, the comparison is made with the results obtained in [15] for the same dataset, as can be seen in Table 4. The best Transformer does not outperform the best LSTM candidate, although it achieves the second-best result. However, 4 out of the 5 best Transformers outperform all the GRU candidates. The trends observed related to the number of timesteps, activation function, and dropout rate remain, with a majority value of 7, *relu* and 0.5, respectively. The main observed difference is in the number of neurons, as the Transformer candidates perform better with a higher neuron

Table 4. Top 5 candidate models for Transformer for both data formats, LSTM and GRU, from [15]

Timesteps	Batch	Enc Layers	Neurons	Num Heads	Kernel Size	MLP Layers	Activation	Dropout	RMSE	Data Format
Transformer candidate models										
7	20	8	128	8	1	3	tanh	0.5	**359.4**	first/last
7	10	4	128	4	1	4	relu	0.0	371.1	first
7	20	4	128	8	2	4	relu	0.5	390.5	first/last
14	10	4	64	4	1	3	relu	0.0	391.3	first
7	10	4	128	4	1	4	relu	0.5	395.8	first/last
LSTM candidate models										
7	5	–	128	–	–	–	relu	0.0	**347.9**	–
7	5	–	64	–	–	–	relu	0.5	362.8	–
7	5	–	64	–	–	–	relu	0.5	388.9	–
7	20	–	64	–	–	–	relu	0.5	391.1	–
7	10	–	32	–	–	–	relu	0.0	395.7	–
GRU candidate models										
7	5	–	32	–	–	–	relu	0.5	**394.8**	–
7	5	–	32	–	–	–	tanh	0.5	394.9	–
7	5	–	32	–	–	–	relu	0.5	399.5	–
7	10	–	32	–	–	–	relu	0.0	401.1	–
7	5	–	64	–	–	–	relu	0.5	401.5	–

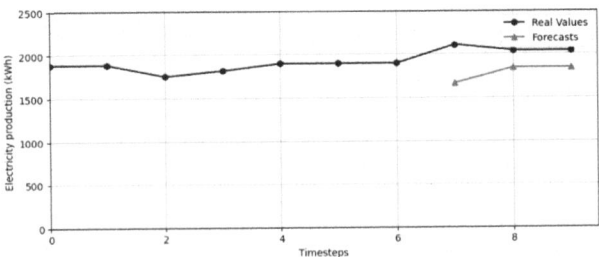

Fig. 3. Prediction for the next three timesteps

count in comparison to the lower number of neurons in the best LSTM/GRU candidates.

Figure 3 shows the prediction for the next three timesteps of the best-performing Transformer model. The first seven timesteps represented in the figure are used as input for the forecasts.

6   Conclusions

WWTPs are vital for treating residual waters before they reach main water bodies, helping control their pollution. These infrastructures have high energy costs, although there is an opportunity to produce energy from the released biogas during the anaerobic digestion phase. This can be helpful in managing energy consumption and reducing the environmental impact of these plants.

Throughout this study, several experiments were run with different candidate models based on the Transformer architecture to forecast biogas-based electricity production in the reactor tank of a wastewater plant. These models were tuned using a grid-search methodology, considering multiple parameters, including data formatting (*channels first* and *channels last*), and evaluated with appropriate metrics to determine the best-performing models.

The best candidate model achieved an RMSE of 359.4 kWh. Both data formats tied for the best model, although the top 5 candidates' average RMSE of *channels first* showed a slight improvement over *channels last*. In comparison with baseline LSTM/GRU candidate models developed in another study with the same dataset, the LSTM achieves the best performance while the Transformer ranks as the second-best candidate. This result shows that the Transformer-based model can be efficient compared to GRU models. However, it is noteworthy that for this dataset, other models, such as the LSTM, have the potential to outperform the Transformer, showing the non-conformity towards the efficiency of the model [24]. Therefore, for the future improvement of this work, different Transformer architectures can be explored, and a wider range of baseline models should be included to better measure the proposed candidates' efficiency. Additionally, including the standard deviation over each fold would provide a better understanding of the model's variability and robustness across the data. Furthermore, this study can be extended, comprising more datasets with different sizes, as it is of great interest to understand the impact of the data size on the models' performance.

Acknowledgements.. This work is financed by National Funds through the Portuguese funding agency, FCT - Fundação para a Ciência e a Tecnologia within project 2022.06822.PTDC (https://doi.org/10.54499/2022.06822.PTDC). The work of Pedro Oliveira was supported by the doctoral Grant PRT/BD/154311/2022 financed by the Portuguese Foundation for Science and Technology (FCT), and with funds from European Union, under MIT Portugal Program.

References

1. Capodaglio, A.G., Olsson, G.: Energy issues in sustainable urban wastewater management: Use, demand reduction and recovery in the urban water cycle. Sustainability **12**(1), 266 (2019). https://doi.org/10.3390/su12010266
2. Chai, T., Draxler, R.R.: Root mean square error (RMSE) or mean absolute error (MAE)?-Arguments against avoiding RMSE in the literature. Geoscientific Model Development **7**(3), 1247–1250 (2014). https://doi.org/10.5194/gmd-7-1247-2014
3. Devlin, J., Chang, M.W., Lee, K., Toutanova, K.: Bert: Pre-training of deep bidirectional transformers for language understanding. arXiv preprint arXiv:1810.04805 (2018)
4. EIA, U.: World Energy Outlook 2023 - DOE/EIA (2023). World Energy Outlook, 2023 (2023)
5. Han, K., et al.: A survey on vision transformer. IEEE Trans. Pattern Anal. Mach. Intell. **45**(1), 87–110 (2022). https://doi.org/10.1109/TPAMI.2022.3152247
6. Harrou, F., Cheng, T., Sun, Y., Leiknes, T., Ghaffour, N.: A data-driven soft sensor to forecast energy consumption in wastewater treatment plants: A case study. IEEE Sens. J. **21**(4), 4908–4917 (2020). https://doi.org/10.1109/JSEN.2020.3030584
7. Ji, J., He, J., Lei, M., Wang, M., Tang, W.: Spatio-temporal transformer network for weather forecasting. IEEE Trans. Big Data (2024). https://doi.org/10.1109/TBDATA.2024.3378061
8. Khalil, M., AlSayed, A., Liu, Y., Vanrolleghem, P.A.: Machine learning for modeling N2O emissions from wastewater treatment plants: Aligning model performance, complexity, and interpretability. Water Res. **245**, 120667 (2023). https://doi.org/10.1016/j.watres.2023.120667
9. Lezmi, E., Xu, J.: Time series forecasting with transformer models and application to asset management. Available at SSRN 4375798 (2023). https://doi.org/10.2139/ssrn.4375798
10. Liang, Y., et al.: Airformer: predicting nationwide air quality in China with transformers. In: Proceedings of the AAAI Conference on Artificial Intelligence, vol. 37, pp. 14329–14337 (2023). 10.1609/aaai.v37i12.26676
11. Nazir, A., Shaikh, A.K., Shah, A.S., Khalil, A.: Forecasting energy consumption demand of customers in smart grid using Temporal Fusion Transformer (TFT). Results Eng. **17**, 100888 (2023). https://doi.org/10.1016/j.rineng.2023.100888
12. Newhart, K.B., Marks, C.A., Rauch-Williams, T., Cath, T.Y., Hering, A.S.: Hybrid statistical-machine learning ammonia forecasting in continuous activated sludge treatment for improved process control. J. Water Process Eng. **37**, 101389 (2020). https://doi.org/10.1016/j.jwpe.2020.101389
13. Oladejo, J., Shi, K., Luo, X., Yang, G., Wu, T.: A review of sludge-to-energy recovery methods. Energies **12**(1), 60 (2018). https://doi.org/10.3390/en12010060
14. Oliveira, P., Fernandes, B., Aguiar, F., Pereira, M.A., Novais, P.: Evaluating unidimensional convolutional neural networks to forecast the influent ph of wastewater treatment plants. In: Yin, H., et al. (eds.) IDEAL 2021. LNCS, vol. 13113, pp. 446–457. Springer, Cham (2021). https://doi.org/10.1007/978-3-030-91608-4_44
15. Oliveira, P., Marcondes, F.S., Duarte, M.S., Durães, D., Martins, G., Novais, P.: Assessment of LSTM and GRU Models to predict the electricity production from biogas in a wastewater treatment plant. In: World Conference on Information Systems and Technologies. pp. 64–73. Springer (2024). https://doi.org/10.1007/978-3-031-60218-4_7

16. Pereira, J., Oliveira, P., Duarte, M.S., Martins, G., Novais, P.: Using deep learning models to predict the electrical conductivity of the influent in a wastewater treatment plant. In: International Conference on Intelligent Data Engineering and Automated Learning. pp. 130–141. Springer (2023). https://doi.org/10.1007/978-3-031-48232-8_13

17. Radford, A., Narasimhan, K., Salimans, T., Sutskever, I., et al.: Improving language understanding by generative pre-training (2018)

18. Ritchie, H., Rosado, P.: Electricity Mix. Our World in Data (2020), https://ourworldindata.org/electricity-mix

19. Sappl, J., Harders, M., Rauch, W.: Machine learning for quantile regression of biogas production rates in anaerobic digesters. Sci. Total Environ. **872**, 161923 (2023). https://doi.org/10.1016/j.scitotenv.2023.161923

20. Ughi, R., Lomurno, E., Matteucci, M.: Two steps forward and one behind: rethinking time series forecasting with deep learning. In: International Conference on Machine Learning, Optimization, and Data Science. pp. 463–478. Springer (2023). https://doi.org/10.1007/978-3-031-53969-5_34

21. Vaswani, A., et al.: Attention is all you need. Adv. Neural Inform. Process. Syst. **30** (2017)

22. Wen, Q., et al.: Transformers in time series: a survey. In: IJCAI International Joint Conference on Artificial Intelligence (2022). https://doi.org/10.24963/ijcai.2023/759

23. Wodecka, B., Drewnowski, J., Białek, A., Łazuka, E., Szulżyk-Cieplak, J.: Prediction of wastewater quality at a wastewater treatment plant inlet using a system based on machine learning methods. Processes **10**(1), 85 (2022). https://doi.org/10.3390/pr10010085

24. Zeng, A., Chen, M., Zhang, L., Xu, Q.: Are transformers effective for time series forecasting? In: Proceedings of the AAAI Conference on Artificial Intelligence, vol. 37, pp. 11121–11128 (2023). 10.1609/aaai.v37i9.26317

25. Zhang, Z., Meng, L., Gu, Y.: SageFormer: series-aware framework for long-term multivariate time series forecasting. IEEE Internet Things J. (2024). https://doi.org/10.1109/JIOT.2024.3363451

Bayesian Regularized Iterative Soft Thresholding Algorithm

Nicolas Cutrona$^{(\boxtimes)}$ ⓘD and Dominique Guillot ⓘD

University of Delaware, Newark, DE 19716, USA
{ncutrona,dguillot}@udel.edu

Abstract. Weighted Naive Bayes methods have recently been developed to alleviate the strong conditional independence assumption of traditional Naive Bayes classifiers. In particular, class-specific attribute weighted Naive Bayes (CAWNB) has been shown to yield excellent performance on many modern datasets. Such methods, however, are prone to over-fitting on small sample, large feature space data. In this work, we propose a Bayesian Regularized Iterative Shrinkage-Thresholding Algorithm (BARISTA), which includes both ℓ_1 and ℓ_2 regularization to mitigate this problem. As we show, estimating the parameters of BARISTA via maximum likelihood yields a convex objective that can be efficiently optimized using Iterative Shrinkage-Thresholding Algorithms (ISTA). We prove the resulting method has many attractive theoretical and numerical properties, including a guaranteed linear rate of convergence. Using several standard benchmark datasets, we demonstrate how BARISTA can yield a significant increase in performance compared to many state-of-the-art weighted Naive Bayes methods. We also show how the Fast Iterative-Shrinkage Thresholding Algorithm (FISTA) can be used to further accelerate convergence. (Our code and data are publicly available on this repository.)

Keywords: Naive Bayes · Convex Optimization · Soft Thresholding

1 Introduction

Naive Bayes (NB) is a popular white-box probabilistic classification algorithm that is used in a variety of machine learning applications. This method can produce accurate and interpretable results along with the ability to quantify the uncertainty of classifications. Given a set of m predictor attributes, $\{X_1, \ldots, X_m\}$, and a class attribute, C, Naive Bayes efficiently estimates the joint distribution of the data as

$$P(C, X_1, X_2, \ldots, X_m) = P(C) \cdot P(X_1|C) \cdot P(X_2|C) \cdot \ldots \cdot P(X_m|C), \quad (1)$$

Supplementary Information The online version contains supplementary material available at https://doi.org/10.1007/978-3-031-80366-6_26.

where $P(\cdot) \in [0,1]$ is a probability measure. Using this joint distribution, the Naive Bayes estimator for the class attribute of a sample $\mathbf{x} = (x_1, \ldots, x_m)$ is given by

$$\operatorname*{argmax}_{c \in \mathcal{C}} P(C = c | X = \mathbf{x}) = \frac{P(c) \prod\limits_{j=1}^{m} P(x_j | c)}{\sum\limits_{c \in \mathcal{C}} P(c) \prod\limits_{j=1}^{m} P(x_j | c)}, \tag{2}$$

where $P(c)$ is the prior probability, $P(x_j|c)$ is the likelihood of observing attribute x_j given the class, $P(c|\mathbf{x})$ is the posterior probability, and $\mathcal{C}$ denotes the set of possible attribute classes. Computing the posterior probability as in NB requires the assumption of conditional independence amongst the predictors. More formally, for any $i \neq j$, $X_i \perp X_j | C$, where the notation $\perp$ is used to denote independence. When this assumption does not hold, which is often the case for real world data, misclassifications become more prominent as the posterior probabilities can be severely biased [9]. This hindrance has motivated a variety of works aimed at offering techniques to mitigate the performance loss of NB, while preserving the simplicity and interpretability of the approach. Most notably, variations of attribute weighting have yielded the strongest results.

Attribute weighting techniques are used to boost the discriminative power of Naive Bayes to increase classification accuracy. The most effective weighting methods learn optimal weights through an iterative optimization procedure to minimize a given loss function. In the attribute weighted Naive Bayes (WANBIA) framework, a weight is assigned to every likelihood, irrespective of the class [13]. Zhang *et al.* extended this idea by introducing a more complex class-specific attribute weighting Naive Bayes (CAWNB) approach [6]. This procedure assigns a weight to each likelihood with respect to each class c and each attribute x_j to capture class dependencies in the data. More specifically, let $D = \{\mathbf{x}_1, \mathbf{x}_2, \ldots, \mathbf{x}_n\}$ be a sample of observations, where $\mathbf{x}_i = (x_{i1}, \ldots, x_{im}) \in \mathcal{C}^m$. The estimated posterior probability of class $c \in \mathcal{C}$ given $\mathbf{x} = (x_1, \ldots, x_m)$ is given by

$$\widehat{P}(c|\mathbf{x}) = \frac{\pi_c \prod\limits_{j=1}^{m} \theta_{c,j,\mathbf{x}}^{w_{c,j}}}{\sum\limits_{c'=1}^{l} \pi_{c'} \prod\limits_{j=1}^{m} \theta_{c',j,\mathbf{x}}^{w_{c',j}}}, \tag{3}$$

where $\pi = (\pi_1, \pi_2, \ldots, \pi_l)$ are the prior probabilities and where each π_c is the prior probability that a sample, $\mathbf{x}$, belongs to a class $c \in \mathcal{C}$. The $\theta_{c,j,\mathbf{x}}^{w_{c,j}}$ are the weighted likelihoods with respect to the jth attribute and class c, and l is the number of possible class values. Each likelihood is assigned a weight $w_{c,j}$ that is class-dependent [6]. The prior probabilities and the likelihoods are estimated from the data as follows

$$\pi_c = \frac{\sum\limits_{i=1}^{n} \delta(c_i, c) + \frac{1}{l}}{n+1}, \qquad \theta_{c,j,\mathbf{x}} = \frac{\sum\limits_{i=1}^{n} \delta(x_{ij}, x_j)\delta(c_i, c) + \frac{1}{n_j}}{\sum\limits_{i=1}^{n} \delta(c_i, c) + 1}, \tag{4}$$

where $\delta(\cdot, \cdot)$ is an indicator function that returns 1 if the inputs are the same and 0 otherwise. The $\frac{1}{l}$ and $\frac{1}{n_j}$ terms are smoothing constants to handle numerical instability during model learning, where n_j is the number of possible attribute values for the jth attribute.

The CAWNB framework, along with a multitude of other work regarding attribute weighting methods, use the MSE loss function for classification feedback to iteratively find an optimal set of weights. In this setting, the MSE function is:

$$L(\mathbf{W}) = \frac{1}{2} \sum_{\mathbf{x}_i \in D} \sum_{c \in \mathcal{C}} (P(c|\mathbf{x}_i) - \widehat{P}(c|\mathbf{x}_i))^2, \tag{5}$$

where $\mathbf{W} = (w_{c,j})$ is the matrix of class-specific attribute weights, and $P(c|\mathbf{x}_i)$ is the ground truth of sample $\mathbf{x}_i$, given by

$$P(c|\mathbf{x}_i) = \begin{cases} 1 & \text{if } c = c_i \\ 0 & \text{otherwise,} \end{cases} \tag{6}$$

where c_i denotes the class attribute of sample $\mathbf{x}_i$. Learning class-dependent weights significantly increases the complexity of this framework in comparison to simple attribute weighting methods. To handle over-fitting on small sample, large feature space data, Wang et al. propose to regularize the CAWNB framework by using the simpler WANBIA model as a constraint [9]. While their method does yield an increase in performance over CAWNB, Wang et al. state that ℓ_1 or ℓ_2 regularization would be a more robust approach, especially to handle noise and outliers in the data. Furthermore, they suggest that implementing these techniques would impose a severe computational burden.

In this paper, we show that incorporating ℓ_1 and ℓ_2 regularization within the CAWNB framework significantly increases performance and can be achieved with little impact on computational complexity. Moreover, in contrast to most previous work in the literature, we investigate the properties of the objective function to optimize and derive theoretical convergence guarantees of our algorithm. In particular, we first replace the MSE loss function (5) by a negative log-likelihood. With a careful analysis of the associated Hessian, we prove that the resulting objective function is convex, contrary to the MSE used in previous work. We then exhibit how the negative log-likelihood with added ℓ_1 and ℓ_2 penalties can be efficiently minimized by leveraging iterative shrinkage-thresholding algorithms (ISTA). We call the resulting approach BARISTA. While ISTA and FISTA are well known optimization techniques in the literature, verifying the assumptions of these methods to justify their use in BARISTA requires several non-trivial calculations, which constitutes the main contribution of our work. This analysis allows us to derive theoretical results to guarantee the linear convergence of BARISTA to its optimal value. We also illustrate the linear convergence of our algorithm in concrete experiments. Finally, we compare the performance of BARISTA with several state-of-the-art weighted Naive Bayes methods on eight classical real world datasets from the UCI machine learning repository that were previously used in the weighted Naive Bayes literature. Our results demonstrate how BARISTA can yield a significant increase in performance compared to competing methods, while maintaining computational efficiency and interpretability.

The rest of this paper is organized as follows. Section 2 describes prior work on attribute weighting techniques for Naive Bayes. The BARISTA method is formulated in Sect. 3, along with the necessary results to show convexity of the negative log-likelihood function. Section 4 details the theoretical convergence results of BARISTA. Performance results are presented in Sect. 5, and concluding remarks are made in Sect. 6.

2   Prior Work

Naive Bayes performs well on many machine learning and data mining tasks. However, the strong conditional independence assumption often hinders its classification capabilities in real world applications. There is thus interest in developing new robust Naive Bayes variations [13]. Attribute weighting is a popular strategy that has been shown to work well in practice. In this work, we focus on wrapper-based methods [9].

Wrapper-based methods [5,6,8,9,11,13–15] iteratively optimize over attribute weights to find an acceptable solution as opposed to non-iterative filter-based methods. Because wrapper-based methods are iterative in nature, they are computationally demanding but often yield stronger results [9]. Taheri et al. design an attribute weighted NB framework that learns optimal attribute weights by a local optimization quasi-secant method [8]. Zheng et al. also make use of attribute weighting by utilizing conjugate gradient descent and implementing an ℓ_1 penalty to regularize attribute weights [15]. In [13], Zaidi et al. give a detailed overview of different attribute weighting schemes for NB, as well as comparisons to other data mining methods. Works [8,13,15] represent simple attribute weighting for NB (WANBIA) [13]. Class-specific attribute weighted NB is a more discriminative version of WANBIA. CAWNB determines an optimal weight for an attribute with respect to each class to learn deeper patterns in the data, as opposed to disregarding class dependencies [9].

Jiang et al. argue that optimizing over class specific attribute weights should boost performance of wrapper-based methods [6]. While this has shown to be an accurate assessment, Wang et al. state that CAWNB methods introduce more computational overhead than previously discussed weighting methods. Wang et al. propose a regularized attribute weighted framework for NB (RNB) that is designed to handle the over-fitting issue that can occur using the CAWNB approach [9]. While the RNB method provides excellent performance on benchmark datasets, Wang et al. speculate that ℓ_1 or ℓ_2 regularization could further improve performance; however, they suggest implementing such a scheme would drastically increase the computational complexity of the learning algorithm.

3   Methodology

We now provide the theory and methodology of BARISTA. We begin by addressing the non-convexity of the MSE loss function. We then discuss how to optimize the penalized negative log-likelihood.

3.1   Convexity Analysis and Optimization

It is not difficult to show that, in general, the MSE (5) is not convex under the
CAWNB framework described in (3). To yield an easier optimization problem
and guarantee the uniqueness of its solution, we estimate the weights $w_{c,j}$ in
(3) via maximum likelihood. Let $\widehat{P}(c|\mathbf{x})$ be given as in (3). The negative log-
likelihood function associated to this problem is

$$f(\mathbf{W}) = -\sum_{i=1}^{n} \log\left(\widehat{P}(c_i|\mathbf{x}_i)\right) \tag{7}$$

$$= -\sum_{i=1}^{n}\left(\log\left(\pi_{c_i}\prod_{j=1}^{m}\theta_{c_i,j,\mathbf{x}_i}^{w_{c_i,j}}\right) - \log\left(\sum_{c'=1}^{l}\pi_{c'}\prod_{j=1}^{m}\theta_{c',j,\mathbf{x}_i}^{w_{c',j}}\right)\right),$$

where $\mathbf{W} = (w_{c,j}) \in \mathbb{R}^{l \times m}$ is the matrix of class specific attribute weights, m is
the number of attributes in the data, and the log given is the natural logarithm.

Theorem 1. *The negative log-likelihood function* (7) *is convex.*

Proof. In order to prove Theorem 1, we compute the Hessian matrix of f. The
calculations are non-trivial and are broken down into several lemmas. We direct
readers to the supplementary material[1] for the details.

Instead of directly optimizing (7), we append ℓ_1 and ℓ_2 regularization terms
to penalize the weights to form the following loss function

$$F(\mathbf{W}) = -\sum_{i=1}^{n} \log\left(\widehat{P}(c_i|\mathbf{x}_i)\right) + \rho_2\|\mathbf{W}\|_2^2 + \rho_1\|\mathbf{W}\|_1, \tag{8}$$

where $\rho_1, \rho_2 > 0$ are penalization constants. In general, ℓ_1 regularization can
be used for attribute selection by setting parameters associated with irrelevant
attributes to zero. The ℓ_2 regularization is used to achieve better numerical stability, decreased parameter variance, and a better conditioned design matrix. In
the same spirit as the Elastic Net used in the context of regression [16], using ℓ_1
and ℓ_2 regularization together allows BARISTA to produce accurate, and if necessary, sparse solutions. Because the ℓ_1 regularization term is non-differentiable,
conventional first order optimization techniques such as gradient descent cannot
be used in this framework. In Sect. 3.1, we proved that the negative log-likelihood
function is convex (Theorem 1). This allows us to handle the non-differentiable
ℓ_1 term using an iterative shrinkage-thresholding algorithm (ISTA), a special
case of proximal gradient descent.

Proximal gradient descent is an efficient first-order technique for solving problems of the form

$$\underset{\mathbf{x}\in\mathcal{X}}{\text{minimize }} F(\mathbf{x}) = g(\mathbf{x}) + h(\mathbf{x}), \tag{9}$$

---

[1] Please visit https://github.com/ncutrona/BARISTA to view the supplementary
material.

where $\mathcal{X}$ is a Hilbert space with inner product $\langle \cdot, \cdot \rangle$ and corresponding norm $||\cdot||$, $g : \mathcal{X} \to \mathbb{R}$ is a continuously differentiable convex function, and $h : \mathcal{X} \to \mathbb{R}$ is a lower semi-continuous, convex, and not necessarily smooth function [7]. The proximal operator associated to h, denoted by $\text{prox}_h : \mathcal{X} \to \mathcal{X}$, is given by

$$\text{prox}_h(\mathbf{x}) = \underset{\mathbf{y} \in \mathcal{X}}{\text{argmin}} \frac{1}{2} ||\mathbf{x} - \mathbf{y}||^2 + h(\mathbf{y}). \tag{10}$$

If h were differentiable, we could simply use the gradient descent update $\mathbf{x}^{(i+1)} = \mathbf{x}^{(i)} - \alpha \nabla F(\mathbf{x}^{(i)})$ to minimize F, where $\alpha > 0$ is a given step size. This update minimizes the following quadratic approximation of F at $\mathbf{x}^{(i)}$

$$\mathbf{x}^{(i+1)} = \underset{\mathbf{y}}{\text{argmin}} \, F(\mathbf{x}^{(i)}) + \nabla F(\mathbf{x}^{(i)})(\mathbf{y} - \mathbf{x}^{(i)}) + \frac{1}{2\alpha} ||\mathbf{x}^{(i)} - \mathbf{y}||^2. \tag{11}$$

When h is not differentiable, the proximal gradient descent proceeds in a similar manner by solving

$$\begin{aligned}
\mathbf{x}^{(i+1)} &= \underset{\mathbf{y}}{\text{argmin}} \, g(\mathbf{x}^{(i)}) + \nabla g(\mathbf{x}^{(i)})(\mathbf{y} - \mathbf{x}^{(i)}) + \frac{1}{2\alpha} ||\mathbf{x}^{(i)} - \mathbf{y}||^2 + h(\mathbf{y}) \\
&= \underset{\mathbf{y}}{\text{argmin}} \, \frac{1}{2\alpha} ||\mathbf{y} - (\mathbf{x}^{(i)} - \alpha \nabla g(\mathbf{x}^{(i)}))||^2 + h(\mathbf{y}) \\
&= \text{prox}_{\alpha h} \left(\mathbf{x}^{(i)} - \alpha \nabla g(\mathbf{x}^{(i)}) \right).
\end{aligned} \tag{12}$$

This technique is a cheap alternative compared to other optimization algorithms, especially when the proximal operator can be computed in closed form and can be efficiently evaluated. With respect to BARISTA, each term in (8) is separable, allowing us to express the problem in the form (9) with

$$g(\mathbf{W}) = -\sum_{i=1}^{n} \log \left(\widehat{P}(c_i | \mathbf{x}_i) \right) + \rho_2 ||\mathbf{W}||_2^2, \qquad h(\mathbf{W}) = \rho_1 ||\mathbf{W}||_1. \tag{13}$$

Using Theorem 1, we obtain

$$\nabla g(\mathbf{W}) = -\sum_{i=1}^{n} (v_{c_i,i} - v_i) + 2\rho_2 \mathbf{W}, \tag{14}$$

where

$$v_{c,i} = \sum_{k=1}^{m} \log \theta_{c,k,\mathbf{x}_i} \mathbf{e}_{c,k}, \qquad v_i = \sum_{c=1}^{l} \widehat{P}(c | \mathbf{x}_i) v_{c,i} \qquad (c = 1, \ldots, l). \tag{15}$$

The proximal operator of $\rho_1 ||\mathbf{W}||_1$ with $\rho_1 > 0$ is the entrywise soft thresholding operator $\eta_{\alpha \rho_1}(w_{i,j})$, given by

$$\left(\eta_{\alpha \rho_1}(\mathbf{W}) \right)_{c,j} = \text{sgn}(w_{c,j}) \left(|w_{c,j}| - \alpha \rho_1 \right)_+, \tag{16}$$

where $(x)_+ := \max(x, 0)$. The resulting proximal gradient update for the objective (8) is therefore

$$\mathbf{W}^{(i+1)} = \eta_{\alpha\rho_1}\left(\mathbf{W}^{(i)} - \alpha\nabla g(\mathbf{W}^{(i)})\right), \qquad (17)$$

with $\nabla g(\mathbf{W})$ as in (14). Since each update involves a soft-thresholding operation, the method is known as the iterative shrinkage-thresholding algorithm (ISTA) in the literature. See, e.g., [2] and [1, Chapter 10] for more details. We now discuss our implementation of BARISTA, which uses an accelerated version of ISTA to further improve convergence.

3.2 Bayesian Regularized Iterative Shrinkage-Thresholding Algorithm (BARISTA)

BARISTA implements both the ISTA approach given in (17) and the Fast Iterative Shrinkage-Thresholding Algorithm (FISTA) of Beck and Teboulle [2] to minimize (8). FISTA is an accelerated variation of the iterative shrinkage-thresholding algorithm. The method maintains the computational simplicity of ISTA, while providing a better global rate of convergence. Instead of using only the previous iterate $\mathbf{W}^{(k-1)}$ to find an updated set of parameters through the soft thresholding operator, a combination of the two previous iterates, $\mathbf{W}^{(k-2)}, \mathbf{W}^{(k-1)}$ is used. With either ISTA or FISTA, we use a backtracking line search to find an appropriate step size that decreases a quadratic approximation of (8):

$$Q_\alpha(x, y) = g(y) + \langle x - y, \nabla g(y)\rangle + \frac{1}{2\alpha}\|x - y\|^2 + h(x), \qquad (18)$$

where g, h are as in (13). FISTA, as applied to our framework, is given in Algorithm 1. The BARISTA algorithm is given in Algorithm 2. The initial choice of

---

Algorithm 1. FISTA with Backtracking Line Search

Step 0. Take $\alpha > 0$ and $\mathbf{W}^{(0)} \in \mathbb{R}^{l \times m}$. Set $\mathbf{Y}^{(1)} = \mathbf{W}^{(0)}$, $t_1 = 1$.

Step k. ($k \geq 1$) Find the smallest non-negative integer γ such that, for $\alpha^{(k)} = \alpha\left(\frac{1}{2}\right)^\gamma$,

$$F\left(\eta_{\alpha^{(k)}\rho_1}\left(\mathbf{Y}^{(k)}\right)\right) \leq Q_{\alpha^{(k)}}\left(\eta_{\alpha^{(k)}\rho_1}\left(\mathbf{Y}^{(k)}\right), \mathbf{Y}^{(k)}\right)$$

for Q as given in (18). Using $\alpha^{(k)}$, compute,

$$\mathbf{W}^{(k)} = \eta_{\alpha^{(k)}\rho_1}\left(\mathbf{Y}^{(k)}\right)$$

$$t^{(k+1)} = \frac{1 + \sqrt{1 + 4\left(t^{(k)}\right)^2}}{2},$$

$$\mathbf{Y}^{(k+1)} = \mathbf{W}^{(k)} + \left(\frac{t^{(k)} - 1}{t^{(k+1)}}\right)\left(\mathbf{W}^{(k)} - \mathbf{W}^{(k-1)}\right).$$

---

the weights is a matrix of 1's, which employs NB as the starting point. BARISTA computes the posterior distributions using the training data and finds $\mathbf{W}^*$ that minimizes (8) as in Algorithm 2. This solution is then used to compute the weighted posterior distributions of unseen data to make classifications.

Algorithm 2. BARISTA

---

input: Training Data $D = \{\mathbf{x}_1, \ldots, \mathbf{x}_n\}$, initial step size α, tolerance ϵ, ℓ_1 penalty ρ_1, ℓ_2 penalty ρ_2. Set initial iterate $\mathbf{W}_0 = \mathbf{1}$ with dimensions $l \times m$ and $\Delta = 2\epsilon$.
(1) Estimate the vector of priors $\pi \in \mathbb{R}^l$ from D as in (4).
(2) Estimate the likelihood matrices $\{\mathbf{\Theta}\}_{i=1}^n$ for each $\mathbf{x}_i \in D$ as in (4).
while $\Delta > \epsilon$ **do**
   (3) Derive the posterior probabilities $\widehat{P}(C|D)$ from (3).
   (4) Compute the model loss of using the current iterate as in (8).
   (5) Update $\mathbf{W}$ using ISTA or FISTA as in (17) or Algorithm 1.
   (6) Set $\Delta = \|\mathbf{W}^{(k)} - \mathbf{W}^{(k-1)}\|_1$.
end while
output: $\mathbf{W}^*$, where $\mathbf{W}^*$ is the solution found in by minimizing (8).

---

4 Convergence Analysis

We now examine the theoretical properties of our approach to minimize (8). We first show that the differentiable part of (8) has a Lipschitz continuous gradient.

Lemma 1. *Let $f(\mathbf{W})$ be as in Eq. (7) and $v_{c,i}$ as in Eq. (15). Then we have $\|\nabla^2 f(\mathbf{W})\|_2 \leq L$, where*

$$L := \sum_{i=1}^n \max_{c=1,\ldots,l} \|v_{c,i}\|_2^2. \tag{19}$$

As a consequence, the function $g(\mathbf{W})$ in Eq. (13) is $(L + 2\rho_2)$-smooth, i.e., its gradient is Lipschitz with constant $L + 2\rho_2$.

Proof. Please see supplementary material.

We next prove that the differentiable part of (8) is strongly convex.

Lemma 2. *Let $\lambda := \lambda_{min}(\nabla^2 f(\mathbf{W})) = \lambda_{min}(-\sum_{i=1}^n H_i)$ denote the smallest eigenvalue of the positive semidefinite matrix $\nabla^2 f(\mathbf{W})$, where $H_i = v_i v_i^T - \sum_{c=1}^l \widehat{P}(c|\mathbf{x}_i) v_{c,i} v_{c,i}^T$. Then the function $g(\mathbf{W})$ in Eq. (13) is σ-strongly convex with $0 < \lambda + 2\rho_2 \leq \sigma \leq L + 2\rho_2$, where L is given in (19).*

Proof. Please see supplementary material.

Our next result guarantees the convergence of the iterates (17) to minimize (8).

Theorem 2. (Convergence of ISTA in the *BARISTA* framework) *Let* $\mathbf{W}^{(k)}$ *denote the ISTA iterates (17) for minimizing Eq. (8), with either a fixed step size* $\alpha = 1/(L + 2\rho_2)$ *or a step size chosen via a backtracking line search analogous to the procedure described in Algorithm 1. Assume* $\mathbf{W}^*$ *is the optimal solution with optimal value* F^*. *Then*

1. For all $k \geq 1$, *we have* $F(\mathbf{W}^{(k)}) - F^* \leq \frac{c \cdot \|\mathbf{W}^{(0)} - \mathbf{W}^*\|_2^2}{2k}$.

2. For all $k \geq 0$, *we have* $\|\mathbf{W}^{(k+1)} - \mathbf{W}^*\|_2^2 \leq \left(1 - \frac{\sigma}{c}\right) \cdot \|\mathbf{W}^{(k)} - \mathbf{W}^*\|_2^2$,

where $c = L + 2\rho_2$ *in the constant step size case,* $c = \max(2(L + 2\rho_2), \frac{1}{\alpha})$ *if the backtracking line search is used,* L *is given in Eq. (19), and* σ *is the strong convexity constant of* $\nabla^2 g(\mathbf{W})$ *given in Lemma 2.*

Proof. See [1, Theorem 10.21] and [1, Theorem 10.29].

Remark 1. Notice that in Theorem 2(2), we always have $0 \leq 1 - \sigma/c < 1$ since $0 < \lambda + 2\rho_2 \leq \sigma \leq L + 2\rho_2$ (see Lemma 2). Moreover, the Hessian $\nabla^2 f(\mathbf{W})$ is a sum of n positive semidefinite matrices $-H_i$, where n denotes the sample size of the data. Since $\mathbf{W}$ is a $l \times m$ matrix, where l is the number of attribute classes and m is the number of features, we expect the Hessian to be positive definite when $n > l \times m$. In that case, $\lambda > 0$ and the ℓ_2 penalty in (8) is not necessary to guarantee linear convergence of the iterates.

Finally, a result analogous to Theorem 2(1) can be shown for FISTA, but with faster convergence.

Theorem 3 (Convergence of FISTA in the *BARISTA* Framework.) *Let* $\mathbf{W}^{(k)}$ *denote the FISTA iterates for minimizing Eq. (8), with either a fixed step size* $\alpha = 1/(L + 2\rho_2)$ *or a step size chosen via the backtracking line search procedure described in Algorithm 1. Assume* $\mathbf{W}^*$ *is the optimal solution with optimal value* F^*. *Then for all* $k \geq 0$

$$F(\mathbf{W}^{(k)}) - F^* \leq \frac{2c \cdot \|\mathbf{W}^{(0)} - \mathbf{W}^*\|_2^2}{(k+1)^2},$$

where $c = L + 2\rho_2$ *in the constant step size case,* $c = \max(2(L + 2\rho_2), \frac{1}{\alpha})$ *if the backtracking line search is used, and where* L *is given in Eq. (19).*

Proof. See [1, Theorem 10.34].

5   Performance Results

We now provide performance results to demonstrate the performance of *BARISTA* on real datasets. In Sect. 5.1, we measure the classification performance of *BARISTA* and give semilog plots to visualize convergence. In Sect. 5.2, we give details regarding experimental settings.

5.1   Benchmark Datasets

Table 1 provides the accuracy score of BARISTA on eight benchmark datasets (described in Sect. 5.2) in comparison to several Naive Bayes variations. Our goal here is to illustrate how the added ℓ_1 and ℓ_2 regularization of BARISTA can yield a significant accuracy improvement to weighted Naive Bayes methods. In selecting datasets for our experiments, we aimed to create a heterogeneous test bed for the new method. The datasets were chosen based on attributes such as the number of features and sample size, ensuring diversity in terms of size, complexity, and variability. This approach allows for a comprehensive evaluation of the method's performance across a wide range of scenarios, and avoids the bias of selecting datasets that favorably highlight our method's performance.

Table 1. Experimental accuracy of BARISTA and other competing methods.

Dataset	BARISTA	RNB [9]	CWANB [12]	CAWNB [6]	NB [10]	WANBIA [13]	FTAWNB [14]
breast-w	**97.86**	96.99	97.07	96.50	97.25	96.51	97.14
colic	**84.00**	83.42	82.83	83.07	81.20	83.72	81.75
heart-statlog	**88.15**	82.96	85.04	84.33	83.74	84.74	83.78
iris	**99.33**	97.33	94.60	94.67	94.33	94.33	95.53
kr-vs-kp	**95.40**	93.08	94.38	95.20	87.81	93.92	94.70
mushroom	99.90	**99.96**	99.84	**99.96**	98.03	99.90	99.85
segmentation	94.37	**95.84**	95.27	94.68	92.91	95.24	91.52
zoo	**100.00**	98.09	96.15	95.95	95.75	95.75	96.35
AVERAGE	94.88	93.46	93.15	93.04	91.38	93.01	92.58

On average, BARISTA performs nearly 1.5% better than the next best method and brings an improvement of more than 5% in accuracy compared to some of the

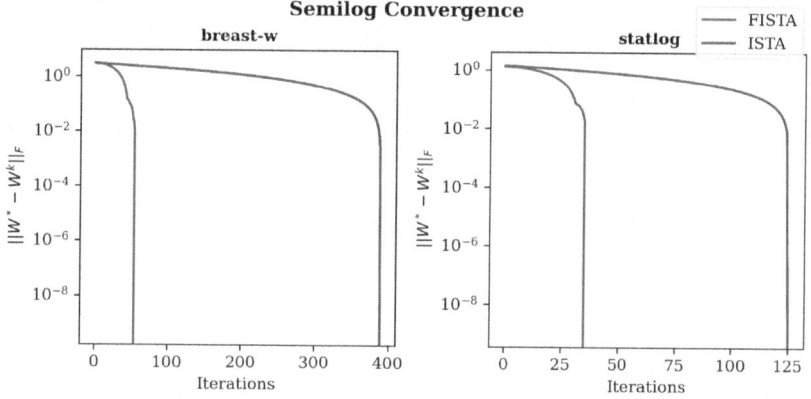

Fig. 1. Convergence rates of FISTA vs. ISTA for solving the BARISTA problem.

other competing methods, demonstrating the benefit of using both ℓ_1 and ℓ_2 regularization. Due to the attribute dependence in the selected datasets we see that BARISTA strongly outperforms vanilla Naive Bayes, demonstrating BARISTA's ability to mitigate performance loss in the presence of attribute dependencies in the data. Furthermore, comparing BARISTA to the CAWNB frameworks acts as an ablation study to evaluate performance gains when using ℓ_1 and ℓ_2 regularization.

Figure 1 illustrates the linear convergence rates of ISTA and the dramatically faster convergence of FISTA. Both techniques achieve super-linear convergence near a solution.

5.2   Experimental Settings

Each dataset listed in Table 1 can be found on the UCI machine learning repository [3]. With respect to pre-processing, missing values were replaced with the mean value or maximum frequency of the attribute. Numerical attributes were discretized using the MDL method [4]. To get an average accuracy score, a 5-fold cross validation procedure was employed. For each experiment, the initial step size was set to 0.1, and the tolerance, as defined in (Algorithm 2) by ϵ, was set to 10^{-6}. Finally, the hyper-parameters $\rho_1 \in [0.01, 0.03, 0.06, 0.09, 0.12]$ and $\rho_2 \in [0.00001, 0.0001, 0.001, 0.005, 0.01, 0.05]$, are reported based on the best performance during the 5-fold cross validation procedure. We report the best accuracy for our method based on these experiments in Table 1. As expected, smaller datasets obtain larger penalties to mitigate over-fitting (see Table 2).

Table 2. Characteristics of the datasets and optimal penalty parameters selected.

Dataset	breast-w	colic	heart-statlog	iris	kr-vs-kp	mushroom	segmentation	zoo
Instances	699	368	270	150	3169	8124	2310	101
Inst./Class	350	184	135	50	1598	4062	330	14
ℓ_1 Penalty	0.03	0.03	0.01	0.12	0.01	0.06	0.03	0.12
ℓ_2 Penalty	0.001	0.001	0.01	0.05	0.00001	0.005	0.01	0.05

Remark 2. Table 1 focuses on the accuracy metric, as is typically done in the weighted Naive Bayes literature (see, e.g., [6,9]). The above datasets do not present any major class imbalance, making accuracy a reasonable metric to compare performance. We note, however, that it would be useful for the community to report additional metrics in the future.

6   Conclusion

In this paper, an accelerated proximal gradient method was applied to a regularized class-specific attribute weighted Naive Bayes framework. A brief introduction to iterative shrinkage-thresholding algorithms, including ISTA and FISTA was given. The convexity of the negative log-likelihood loss function was proved and rigorous theoretical convergence guarantees for BARISTA were presented. The supplemental material contains the proofs of Theorem 1 and Lemmas 1 & 2, as well as plots analogous to Fig. 1 for all datasets considered. Performance results have also been shown, comparing the accuracy of BARISTA to other weighted Naive Bayes methods. The results indicate that BARISTA is very competitive, especially on small sample data where regularization can be a valuable tool. One should keep in mind, however, that no method will outperform all the others on every dataset. Methods based on Naive Bayes such as BARISTA are expected to perform well when the features are approximately conditionally independent. In the presence of strong dependence, one should consider techniques outside of Naive Bayes. Moving forward, it would be interesting to explore second order techniques that handle ℓ_1 regularization to further speed up convergence, as well as other approaches to alleviate the conditional independence assumption that hinders Naive Bayes classifiers in practice.

References

1. Beck, A.: First-order methods in optimization. SIAM (2017)
2. Beck, A., Teboulle, M.: A fast iterative shrinkage-thresholding algorithm for linear inverse problems. SIAM J. Imag. Sci. **2**(1), 183–202 (2009)
3. Dua, D., Graff, C.: UCI machine learning repository (2017)
4. Fayyad, U.M., Irani, K.B.: Multi-interval discretization of continuous-valued attributes for classification learning. In: International Joint Conference on Artificial Intelligence (1993)
5. Jiang, L., Kong, G., Li, C.: Wrapper framework for test-cost-sensitive feature selection. IEEE Trans. Syst. Man Cybernet. Syst. **51**(3), 1747–1756 (2021)
6. Jiang, L., Zhang, L., Yu, L., Wang, D.: Class-specific attribute weighted naive bayes. Pattern Recogn. **88**, 321–330 (2019)
7. Rolfs, B., Rajaratnam, B., Guillot, D., Wong, I., Maleki, A.: Iterative thresholding algorithm for sparse inverse covariance estimation. In: Pereira, F., Burges, C., Bottou, L., Weinberger, K. (eds.) Advances in Neural Information Processing Systems, vol. 25. Curran Associates, Inc. (2012)
8. Taheri, S., Yearwood, J., Mammadov, M., Seifollahi, S.: Attribute weighted naive bayes classifier using a local optimization. Neural Comput. Appl. **24**, 04 (2014)
9. Wang, S., Ren, J., Bai, R.: A regularized attribute weighting framework for naive bayes. IEEE Access **12** (2020)
10. Webb, A.R.: Statistical pattern recognition. John Wiley & Sons (2003)
11. Wu, J., Cai, Z.: Attribute weighting via differential evolution algorithm for attribute weighted naive bayes (wnb). J. Comput. Inform. Syst. **7**, 05 (2011)
12. Yu, L., Gan, S., Chen, Y., He, M.: Correlation-based weight adjusted naive bayes. IEEE Access **8**, 51377–51387 (2020)
13. Zaidi, N., Cerquides, J., Carman, M., Webb, G.: Alleviating naive bayes attribute independence assumption by attribute weighting. J. Mach. Learn. Res. **14**, 1947–1988 (2013)
14. Zhang, H., Jiang, L.: Fine tuning attribute weighted naive bayes. Neurocomputing **488**, 402–411 (2022)

15. Zheng, Z., Cai, Y., Yang, Y., Li, Y.: Sparse weighted naive bayes classifier for efficient classification of categorical data. In: 2018 IEEE Third International Conference on Data Science in Cyberspace (DSC), pp. 691–696, (2018)
16. Zou, H., Hastie, T.: Regularization and variable selection via the elastic net. J. Royal Stat. Soc. Ser. B (Stat. Methodol.) **67**(2), 301–320 (2005)

Performance Evaluation of Data Analysis Techniques in Dry Bean Seed Classification Using kNN and MLP

Victor Hugo Schneider Lopes[1] , Alessandro Bof de Oliveira[1]([⊠]) ,
Patricia Bof[1] , and Dante Augusto Couto Barone[2]

[1] Federal University of Pampa, Alegrete, RS 97546-550, Brazil
{victorlopes.aluno,alessandrooliveira}@unipampa.edu.br
[2] Federal University of Rio Grande do Sul, Porto Alegre, RS 91501-970, Brazil
barone@inf.ufrgs.br

Abstract. Seed classification, particularly for dry bean seeds, is crucial for ensuring high agricultural productivity and efficiency. This paper investigates the impact of data preprocessing techniques on the performance of machine learning models in the classification of dry bean seeds. Using a data set from the UCI Machine Learning Repository, derived from an experiment by Koklu and Ozkan [9], various pre-processing techniques, such as missing value imputation, outlier removal, and data normalization, were applied. The k-Nearest Neighbors (kNN) classification methods and Multi-Layer Perceptron (MLP) were used to evaluate the effectiveness of these preprocessing techniques. In addition, an enhanced MLP model with optimized parameters was proposed, including the learning rate and hidden layer configuration. The experimental results demonstrate the critical role of data normalization, with Z-Score normalization yielding the best performance improvements. The enhanced MLP model significantly outperformed the baseline model, highlighting the importance of robust preprocessing and careful model optimization. These findings underscore the necessity of comprehensive data preprocessing and fine-tuning of machine learning models to achieve high classification accuracy and efficiency.

Keywords: Machine Learning · Data Analysis · Beans Seed Classification

1 Introduction

The amount of data generated every day is growing exponentially, driven by mobile devices, sensors, social networks, and online transactions. This increase results in large volumes of data that challenge our ability to store, process, and extract useful knowledge. Data analysis, especially with machine learning (ML) techniques, has become essential for handling this amount of data. However, the effectiveness of these ML techniques depends on the quality of the data, which

L. Correia et al. (Eds.): IBERAMIA 2024, LNCS 15277, pp. 323–334, 2025.
https://doi.org/10.1007/978-3-031-80366-6_27

often contains errors, missing values, and discrepancies [2]. Therefore, about 80% of the data analysis work is dedicated to preprocessing, which includes handling missing data, detecting outliers, and normalizing data [7].

Recently, it has become more common for various fields of knowledge to use ML techniques to filter, organize, and extract knowledge from the datasets being worked on, playing a fundamental role in areas such as healthcare [2], economics [4], seismology [17], agriculture [10], among others. In agriculture, seed classification, such as for beans, is crucial for ensuring high productivity and automating this process with ML techniques can increase accuracy and efficiency, surpassing manual methods that are time-consuming and error-prone. Recently, Koklu and Ozkan [9] proposed multiple ML based methods to classify dry bean seeds, including K-Nearest Neighbor and Multi-Layer Perceptron. Although the results obtained in their work are sufficiently good, some adjustments can be made to ensure that the classification quality is as high as possible. Despite advances in seed classification, the impact of data preprocessing on the performance of ML techniques is still underexplored. This work proposes to investigate this impact, highlighting the importance of data quality for the effectiveness of ML models in bean seed classification.

2   Background

2.1   Seed Classification

Seed classification, especially the classification of bean seeds, is essential to ensure that the seeds used guarantee a high level of productivity [9]. This classification can be performed manually, but it is a time-consuming and error-prone process. Automating this process using ML-based techniques can help increase the efficiency and accuracy of seed classification [13].

Recently, numerous studies have explored the use of machine learning methods to classify various types of seeds. One such study by Koklu and Ozkan [9] aimed to classify dry bean seeds using multiple machine learning techniques, including decision trees, k-Nearest Neighbors (kNN), and Multi-Layer Perceptrons (MLP). This study demonstrated the potential of these methods to improve high classification accuracy, setting a benchmark for further research in seed classification.

Other research efforts have expanded the application of machine learning to different seed types. For instance, Xu et al. [18] utilized Convolutional Neural Networks (CNNs) to classify maize seeds. The CNN approach enhance the image recognition capabilities of deep learning, proving effective in distinguishing between various maize seed types based on their visual characteristics. Similarly, Zhao et al. [19] proposed a hybrid model combining hyperspectral imaging technology with machine learning techniques to classify wheat seeds. This approach integrated spectral data with advanced computational methods, enhancing the accuracy and reliability of seed classification. In [8], the authors used a three-step method, which included preprocessing, feature extraction, and ML-based classification, to classify 14 different types of rice seeds. Their approach demonstrated the effectiveness of a comprehensive pipeline that addresses data quality and feature relevance before applying classification algorithms. By first cleaning and normalizing the data, then extracting significant features, and finally employing machine learning techniques, they achieved high classification accuracy. These studies underscore the versatility and effectiveness of machine learning in agricultural applications, particularly in seed classification. By automating the classification process, these methods offer significant advantages over traditional manual techniques, which are often time-consuming and prone to errors. Machine learning models can process large datasets quickly and accurately, providing farmers and researchers with valuable insights to improve seed quality and agricultural productivity.

2.2   Data Preprocessing

The rapid growth in the volume of available data has created an environment where an immense quantity of information can be found; however, these data may be arranged in a manner that complicates their usability [6]. Such datasets often contain missing, discrepant, or duplicated entries, commonly referred to as original or raw data [12].

Missing data, a frequent issue across various application databases, can lead to biased results and significantly reduce the performance of machine learning algorithms that rely on such data [5]. This problem necessitates sophisticated techniques to handle these gaps and ensure the reliability of the data analysis process.

Moreover, databases might exhibit outliers, or anomalous values, which can serve dual roles; on one hand, they are invaluable for detecting credit card fraud, security breaches, and anomalies, as they represent rare and suspicious events [3]; on the other hand, outliers can negatively impact data analysis by distorting statistical measures and degrading the performance of machine learning techniques, potentially leading to erroneous conclusions and inaccurate classification models [15].

Additionally, datasets often include non-normalized data, where features exist on different scales; this discrepancy can cause features with smaller numerical ranges to be overshadowed by those with larger scales, thereby diminishing their importance [14].

3   Database and Methodology

3.1   Database

The study utilized the dataset "Dry Bean" from the UCI Machine Learning Repository (University of California, Irvine)[1], derived from the experiment conducted by Koklu and Ozkan [9]. This dataset consists of images of 13,611 dry

---

[1] https://archive.ics.uci.edu/dataset/602/dry+bean+dataset.

beans from seven different varieties determined by the Turkish Standards Institute, obtained from certified seed producers. The beans varieties include: Barbunya, Bombay, Cali, Dermason, Horoz, Seker, and Sira. These images were captured by a 2.2-megapixel RGB camera called Prosilica GT2000C, with a resolution of 2048 ×1088, a CMOS sensor, and an effective operating temperature range of –20 r̂C to +65 r̂C. Before capturing the images, the beans were placed on a dark background to facilitate the segmentation process. During the image capture process, the camera was positioned 15 cm above the samples in the upper part of the box, aiming to provide a homogeneous lighting environment. The lamps at the top of the box allowed for the elimination of environmental noise. The Fig. 1 shows an example of the seed image capture according to the described method.

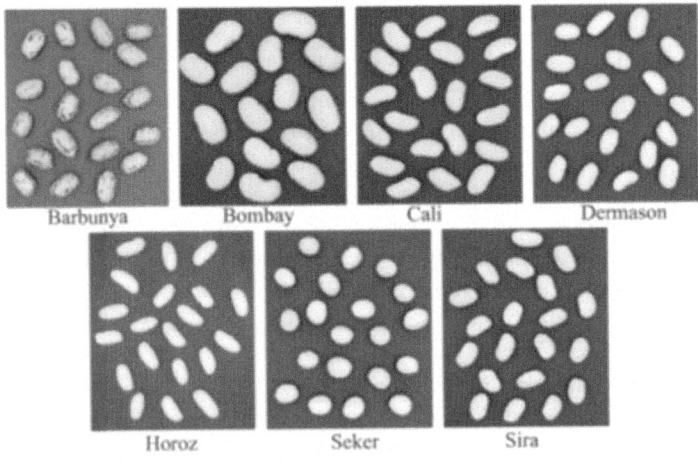

Fig. 1. Captured image of dry bean seeds. Source: Koklu and Ozkan [9]

Koklu and Ozkan [9] proceeded with the subsequent stage, entailing the elimination of shadows and background noise from the image. Subsequently, individual seeds were segregated through image segmentation, employing the Otsu global thresholding method, which binarizes the image, segregating it into background and foreground components. This process yields a binary representation of each seed, enabling the MATLAB[2] software to derive 16 distinct spatial characteristics. The resulting database is a text file with 13,611 rows and 17 columns, i.e., 16 features and a target class column. The explanation about the features can be viewed in [1]. The details of the methods used in image acquisition and feature extraction can be verified in the paper of Koklu and Ozkan [9].

---

[2] MATLAB is trademark of MathWorks Inc.

3.2   Classification Methods

In this work, the decision was made to employ the classification methods of kNN (k-Nearest Neighbors) and MLP (Multi-Layer Perceptron) neural network, as described in the study by Koklu and Ozkan [9]. This approach aims to provide a comprehensive and fair comparison of the impact of preprocessing techniques applied to the dataset. Thus, a machine learning classification model of the k-Nearest Neighbors type will be utilized, using the Euclidean distance as the distance calculation method and a value of k = 10. Additionally, a Multi-Layer Perceptron neural network model will be employed, following the structure proposed by Koklu and Ozkan [9], comprising 4 layers arranged as follows:

- Input Layer: 16 perceptrons ($x_1 - x_{16}$), representing the characteristics of the bean seeds as described in the previous section;
- Hidden Layers: According to [9], the optimal structure for this purpose consists of two hidden layers, the first with 12 perceptrons and the second with 3 perceptrons;
- Output Layer: 7 perceptrons ($O_1 - O_7$) representing the types of bean seeds, namely *Barbunya, Bombay, Cali, Dermason, Horoz, Seker*, and *Sira*.

The Table 1 presents the overall configuration of the MLP.

Table 1. Multi Layer Perceptron Parameters.

Parameter	Value
Hidden Layer Quantity	2
Hidden Layer Activation Function	Sigmoid
Output Layer Activation Function	Sigmoid
Learning Rate	0,3
Minimum Performance Gradient	$1e^{-5}$
Performance Goal	$1e^{-3}$
Maximum Number of Epochs to Train	500

For both classification methods, the cross-validation technique was used to ensure that the classification result accurately represents the quality of the method. This method involves dividing the dataset into subsets and using each subset as a test set while training with the remaining subsets, repeated until all subsets have been tested. This method increases the generalization of the tests, ensuring the quality of the results. In this work, the number of subsets used is 10, meaning each subset will have a size equal to 10% of the total dataset size. The kNN and MLP methods were implemented using Python language [16] and Scikit-learn library [11]. The code used to run the experiments is available on https://github.com/lopesvictor1/Beans-Classification.

3.3   Experiments

To evaluate the impact of preprocessing on the performance of classification methods, the following data preprocessing techniques will be tested:

- Missing Values Imputation by kNN (kNN-I);
- Missing Values Imputation by Linear Interpolation (LI);
- Outlier Removal by 3σ (3σ);
- Outlier Removal by Median Absolute Deviation (MAD);
- Data Normalization by min-max (min-max);
- Data Normalization by Z-$Score$ (Z-$Score$).

For conducting the experiments, different techniques will be combined, resulting in 16 different experiments, as outlined in Table 2. Each experiment is assigned a name (e.g., E1, E2, E3...) and involves the application of a missing values imputation technique (IVF), an outlier removal technique, a data normalization technique, and a classification method.

Table 2. Description of the proposed classification experiments using both kNN and MLP and preprocessed methods.

Experiment	MVI	Outliers	Normalization	Classification
Experiment 1 (E1)	kNN-I	3σ	min-max	kNN
Experiment 2 (E2)	kNN-I	3σ	min-max	MLP
Experiment 3 (E3)	kNN-I	3σ	Z-$Score$	kNN
Experiment 4 (E4)	kNN-I	3σ	Z-$Score$	MLP
Experiment 5 (E5)	kNN-I	MAD	min-max	kNN
Experiment 6 (E6)	kNN-I	MAD	min-max	MLP
Experiment 7 (E7)	kNN-I	MAD	Z-$Score$	kNN
Experiment 8 (E8)	kNN-I	MAD	Z-$Score$	MLP
Experiment 9 (E9)	IL	3σ	min-max	kNN
Experiment 10 (E10)	IL	3σ	min-max	MLP
Experiment 11 (E11)	IL	3σ	Z-$Score$	kNN
Experiment 12 (E12)	IL	3σ	Z-$Score$	MLP
Experiment 13 (E13)	IL	MAD	min-max	kNN
Experiment 14 (E14)	IL	MAD	min-max	MLP
Experiment 15 (E15)	IL	MAD	Z-$Score$	kNN
Experiment 16 (E16)	IL	MAD	Z-$Score$	MLP

3.4   Enhanced Multi Layer Perceptron

The proposed neural network stands out due to its unique parameters, particularly the initial learning rate and the hidden layer configuration. Another interesting parameter is the maximum number of epochs. Although 500 epochs are sufficient for a high learning rate, this limitation might interrupt the network's learning process before the error rate is sufficiently low. Moreover, the choice of the neural network's hidden layer topology is also noteworthy: the abrupt change from the first layer (containing 12 neurons) to the second layer (with only 3 neurons) can lead to information loss during the training process. Tests will also be conducted to improve the neural network proposed by Koklu and Ozkan [9], where the following parameters will be altered:

- Initial Learning Rate (ILR);
- Maximum Number of Epochs (MNE);
- Neurons in the First Layer (NFL);
- Neurons in the Second Layer (NSL).

For the experiments, different neural network sizes will be tested, generating a total of 16 different experiments, presented in Table 3, where each experiment is given a name (*e.g.*, A1, A2, A3...) and different combinations of neural network execution parameters.

Table 3. Description of the MLP configuration for the baseline (kolu and Ozkan proposed MLP) and our MLP proposed. The columns are Initial Learning Rate (ILR), Maximum Number of Epochs (MNE), Neurons in the First Layer (NFL), and Neurons in the Second Layer (NSL) respectively. Description of the MLP configuration for the baseline (kolu and Ozkan proposed MLP) and our MLP proposed. The columns are Initial Learning Rate (ILR), Maximum Number of Epochs (MNE), Neurons in the First Layer (NFL), and Neurons in the Second Layer (NSL) respectively.

Experiment	ILR	MNE	NFL	NSL
Koklu and Ozkan [9] (baseline)	0.3	500	12	3
Experiment A1	0.001	1500	10	5
Experiment A2	0.001	1500	10	10
Experiment A3	0.001	1500	10	15
Experiment A4	0.001	1500	10	20
Experiment A5	0.001	1500	15	5
Experiment A6	0.001	1500	15	10
Experiment A7	0.001	1500	15	15
Experiment A8	0.001	1500	15	20
Experiment A9	0.001	1500	20	5
Experiment A10	0.001	1500	20	10
Experiment A11	0.001	1500	20	15
Experiment A12	0.001	1500	20	20
Experiment A13	0.001	1500	25	5
Experiment A14	0.001	1500	25	10
Experiment A15	0.001	1500	25	15
Experiment A16	0.001	1500	25	20

4   Results and Discussion

4.1   Results

As presented in Table 3, 16 experiments were conducted, covering all possible combinations between the proposed preprocessing techniques and classifiers.

Since the dataset used does not contain missing data, 5% of missing data was inserted completely at random for all experiments. Each of the experiments was repeated 50 times to ensure consistent results and minimize the effects of variability. The results obtained from the proposed experiments will be compared with the results obtained from the repetition of Koklu and Ozkan's experiments [9]. Additionally, data normalization was performed before applying Koklu's methods, which resulted in numbers closer to those reported in their study. Throughout this chapter, the results that apply normalization before classification will be described with the suffix *min-max*, as this was the normalization method used.

Table 4. kNN Classification Results.

Experiment	Accuracy (%)	Precision (%)	Recall (%)	F1-Score (%)
Experiment 1	91.84	93.31	92.76	93.00
Experiment 3	92.07	93.54	93.04	93.26
Experiment 5	91.84	93.33	92.76	93.01
Experiment 7	92.10	93.56	93.07	93.29
Experiment 9	92.41	93.85	93.35	93.57
Experiment 11	92.60	94.04	93.60	93.79
Experiment 13	92.43	93.87	93.36	93.59
Experiment 15	92.59	94.03	93.57	93.78
Koklu KNN	71.96	73.26	71.63	72.09
Koklu KNN minmax	92.18	93.64	93.15	93.36

Table 4 presents the results of the experiments applying *k*NN classifiers. The results show a significant impact of data normalization on classification, especially when comparing the results of Koklu-KNN and Koklu-KNN-minmax. Despite this, it seems that other techniques have a relevant, yet minor, impact on the evaluation metrics.

Table 5. MLP Classification Results.

Experiment	Accuracy (%)	Precision (%)	Recall (%)	F1-Score (%)
Experiment 2	86.70	87.27	87.08	86.58
Experiment 4	90.41	91.97	91.40	91.51
Experiment 6	86.57	87.02	87.01	86.45
Experiment 8	90.33	91.80	91.31	91.38
Experiment 10	87.71	88.42	88.17	87.79
Experiment 12	90.89	92.40	91.93	92.00
Experiment 14	87.52	87.94	87.82	87.35
Experiment 16	90.91	92.38	91.97	92.02
Koklu MLP	26.05	3.72	14.29	5.91
Koklu MLP minmax	87.40	88.22	87.99	87.61

Table 5 reveals that MLP classifiers are highly sensitive to data normalization. The MLP showed inferior performance with unnormalized data but demonstrated significant improvements when applying various preprocessing techniques. Experiment 16, which also used Linear Interpolation, Median Absolute Deviation, and Z-Score, had the best performance, corroborating the effectiveness of this combination of techniques. We can note the greater efficiency when using the Z-Score data normalization technique compared to the min-max technique. Experiments 4, 8, 12, and 16 use this technique and show considerably higher evaluation metrics than experiments 2, 6, 10, and 14.

Based on the initial results, a new series of experiments was conducted focusing on adjusting neural network parameters, such as learning rate and hidden layer configuration. According to Table 3, 16 experiments were conducted, showing significant performance gains in evaluation metrics compared to the initial experiments and the experiments by Koklu and Ozkan [9].

Table 6 presents the results of the conducted experiments. Adjusting the parameters of the MLP model, including the initial learning rate and hidden layer configuration, led to notable improvements in performance. The initial learning rate of 0.3 used by Koklu and Ozkan was found to be suboptimal, causing abrupt adjustments during training. A more moderate learning rate, combined with a higher number of maximum iterations, allowed the network to learn more effectively and achieve lower error rates.

Table 6. Experimental results to baseline (Koklu and Ozkan proposed MLP method) and enhancements to the MLP proposals.

Experiment	Accuracy (%)	Precision (%)	Recall (%)	F1-Score (%)
Koklu and Ozkan [9] (baseline)	91.73	93.11	92.68	92.88
Experiment A1	92.95	94.11	93.86	93.97
Experiment A2	93.12	94.32	94.03	94.15
Experiment A3	93.07	94.27	94.01	94.12
Experiment A4	93.08	94.28	94.04	94.14
Experiment A5	93.07	94.29	94.03	94.14
Experiment A6	93.14	94.35	94.09	94.20
Experiment A7	93.15	94.37	94.09	94.21
Experiment A8	93.16	94.37	94.09	94.21
Experiment A9	93.13	94.39	94.11	94.23
Experiment A10	93.11	94.35	94.07	94.19
Experiment A11	93.14	94.37	94.10	94.22
Experiment A12	93.16	94.39	94.11	94.23
Experiment A13	93.05	94.31	94.00	94.13
Experiment A14	93.10	94.32	94.02	94.15
Experiment A15	**93.18**	**94.40**	**94.14**	**94.25**
Experiment A16	93.12	94.36	94.08	94.20

4.2   Discussion

The experiments applying k-Nearest Neighbors (kNN) classifiers revealed a significant impact of data normalization on classification performance. Specifically, the comparison between the original Koklu-KNN results and the normalized Koklu-KNN-minmax results demonstrated considerable improvements in accuracy, highlighting the importance of proper feature scaling. However, other preprocessing techniques, such as missing values imputation and outlier removal, had a less pronounced effect on the evaluation metrics. This suggests that while kNN classifiers benefit greatly from normalization, they are relatively robust to other types of data inconsistencies. The experiments revealed that MLP classifiers are highly sensitive to data normalization. The application of the Z-Score normalization technique resulted in superior performance compared to the min-max normalization. This suggests that for datasets with diverse feature scales, Z-Score normalization may provide a more balanced and effective approach to standardizing data. Moreover, the combination of Linear Interpolation, Median Absolute Deviation, and Z-Score normalization (Experiment 16) yielded the highest evaluation metrics, highlighting the effectiveness of this multifaceted preprocessing strategy. This combination addresses different aspects of data quality, such as missing values, outliers, and feature scaling, providing a comprehensive solution for enhancing classifier performance.

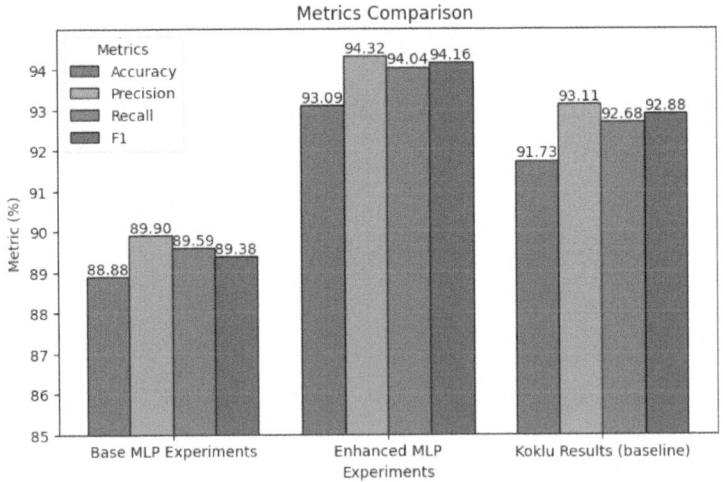

Fig. 2. Comparison between MLP classification results (Accuracy, Precision, Recall and F1-Score. First we have "Base MLP experients" with MLP configuration without preprocessing, after "Enhanced MLP experiments" with MLP with enhanced configuration and finally Koklu Results (baseline) with MLP configuration decribed [9].

On Fig. 2 we presents a comparison between the classification results using the initial results without applying data preprocessing, the results after enhancing the neural network parameters MLP and the baseline MLP as described in Koklu and Ozkan [9]. Despite the improvements observed with the application of data preprocessing techniques, the enhanced neural network MLP parameters further boosted performance metrics significantly. This comparison demonstrates the importance of the data preprocessing and optimization and tuning the MLP neural network parameters to achieve the highest classification accuracy and precision.

5   Conclusion

The experiments demonstrated the critical role of data preprocessing in enhancing the performance of machine learning models. Specifically, data normalization techniques, such as min-max and Z-Score, significantly impacted the effectiveness of both k-Nearest Neighbors (kNN) and MLP classifiers. The results indicated that MLP classifiers are particularly sensitive to data normalization, with the Z-Score method proving to be more effective than min-max. Experiment 16, which combined Linear Interpolation, Median Absolute Deviation, and Z-Score normalization, achieved the best performance, underscoring the synergy of these techniques. The findings highlight the necessity of thorough data preprocessing to maximize the accuracy and efficiency of ML models. Additionally, the enhanced MLP model, featuring optimized parameters such as initial learning rate and hidden layer configuration, showed considerable improvements over the initial experiments and the baseline model by Koklu and Ozkan [9]. This optimization led to significant gains in evaluation metrics, demonstrating the importance of fine-tuning neural network parameters to achieve superior classification results. In conclusion, this study confirms that meticulous data preprocessing and careful tuning of neural network parameters are crucial to improving classifier performance. Furthermore, we found an optimal neural network configuration for bean seed classification. Future work will explore further refinements in neural network architectures as well as their application to other types of seeds.

Disclosure of Interests.. The authors have no competing interests to declare that are relevant to the content of this article.

References

1. Dry Bean. UCI Machine Learning Repository (2020). https://doi.org/10.24432/C50S4B
2. Benhar, H., Idri, A., Fernández-Alemán, J.: Data preprocessing for heart disease classification: a systematic literature review. Comput. Methods Programs Biomed. **195**, 105635 (2020)
3. Blázquez-García, A., Conde, A., Mori, U., Lozano, J.A.: A review on outlier/anomaly detection in time series data. ACM Comput. Surv. (CSUR) **54**(3), 1–33 (2021)

4. Dharma, F., Shabrina, S., Noviana, A., Tahir, M., Hendrastuty, N., Wahyono, W.: Prediction of indonesian inflation rate using regression model based on genetic algorithms. Jurnal Online Informatika **5**(1), 45–52 (2020)
5. Emmanuel, T., Maupong, T., Mpoeleng, D., Semong, T., Mphago, B., Tabona, O.: A survey on missing data in machine learning. J. Big Data **8**(1), 1–37 (2021). https://doi.org/10.1186/s40537-021-00516-9
6. Hariri, R.H., Fredericks, E.M., Bowers, K.M.: Uncertainty in big data analytics: survey, opportunities, and challenges. J. Big Data **6**(1), 1–16 (2019). https://doi.org/10.1186/s40537-019-0206-3
7. Jamshed, H., Khan, S.A., Khurrum, M., Inayatullah, S., Athar, S.: Data preprocessing: a preliminary step for web data mining. 3C Tecnología: glosas de innovación aplicadas a la pyme **8**(1), 206–221 (2019)
8. Kiratiratanapruk, K., et al.: Development of paddy rice seed classification process using machine learning techniques for automatic grading machine. J. Sens. **2020**(1), 7041310 (2020)
9. Koklu, M., Ozkan, I.A.: Multiclass classification of dry beans using computer vision and machine learning techniques. Comput. Electron. Agric. **174**, 105507 (2020)
10. Macuácua, J.C., Centeno, J.A.S., Amisse, C.: Data mining approach for dry bean seeds classification. Smart Agric. Technol. **5**, 100240 (2023)
11. Pedregosa, F., et al.: Scikit-learn: machine learning in Python. J. Mach. Learn. Res. **12**, 2825–2830 (2011)
12. Ridzuan, F., Zainon, W.M.N.W.: A review on data cleansing methods for big data. Procedia Comput. Sci. **161**, 731–738 (2019)
13. Sarijaloo, F.B., Porta, M., Taslimi, B., Pardalos, P.M.: Yield performance estimation of corn hybrids using machine learning algorithms. Artif. Intell. Agric. **5**, 82–89 (2021)
14. Singh, D., Singh, B.: Investigating the impact of data normalization on classification performance. Appl. Soft Comput. **97**, 105524 (2020)
15. ur Rehman, A., Belhaouari, S.B.: Unsupervised outlier detection in multidimensional data. J. Big Data **8**(1), 1–27 (2021). https://doi.org/10.1186/s40537-021-00469-z
16. Van Rossum, G., Drake, F.L.: Python 3 Reference Manual. CreateSpace, Scotts Valley (2009)
17. Xie, Y., Ebad Sichani, M., Padgett, J.E., DesRoches, R.: The promise of implementing machine learning in earthquake engineering: a state-of-the-art review. Earthq. Spectra **36**(4), 1769–1801 (2020)
18. Xu, P., Tan, Q., Zhang, Y., Zha, X., Yang, S., Yang, R.: Research on maize seed classification and recognition based on machine vision and deep learning. Agriculture **12**(2), 232 (2022)
19. Zhao, X., Que, H., Sun, X., Zhu, Q., Huang, M.: Hybrid convolutional network based on hyperspectral imaging for wheat seed varieties classification. Infrared Phys. Technol. **125**, 104270 (2022)

Integrating Convolutional Neural Networks and Omics to Promote Precision Medicine in Atopic Dermatitis

Ana Duarte$^{(\boxtimes)}$ ⓘ and Orlando Belo ⓘ

Algoritmi R&D Centre / LASI, University of Minho, Campus of Gualtar,
4710-057 Braga, Portugal
id9618@alunos.uminho.pt, obelo@di.uminho.pt

Abstract. Advanced machine learning techniques, including convolutional neural networks, are widely used in the healthcare field. Despite their high efficiency, convolutional neural networks are primarily designed to analyze images and have limitations when processing non-image data such as gene expression data. For this reason, few studies apply convolutional neural networks to omics data. Gene expression data in particular play a central role in the development of precision medicine, which aims to provide medical care tailored to the individual needs of the patients. The few studies that have been conducted in this field focus mainly on cancer and other diseases that could benefit from such an approach, such as atopic dermatitis, are not the subject of investigation. Therefore, in this work we employed a convolutional neural network to transcriptomic data collected from the skin of atopic dermatitis patients. As far as we know, this is the first paper reporting the application of a convolutional neural network to gene expression data of atopic dermatitis. The data were converted into images using the DeepInsight method. Furthermore, we also conceived a convolutional neural network model to distinguish between lesions and non-lesions and to extract the most significant genes. With two independent datasets, we achieved accuracy values of 75% and 85.59%. These results are considerably high to support the hypothesis that the identified genes may be involved in the development of atopic dermatitis lesions.

Keywords: Convolutional Neural Networks · Deep learning · Atopic dermatitis · Transcriptome · Feature Selection

1 Introduction

Atopic dermatitis (AD), often referred to as atopic eczema, is a chronic and complex skin condition characterized by severe itching, dry skin, visible lesions and diverse immune responses [1, 2]. Despite the high heterogeneity of the disease, current medical treatment remains rooted in a one-size-fits-all approach, that lacks patient-centered therapies [3]. As a result, AD treatment generally follows a trial-and-error paradigm that is applied similarly to all patients, often yielding unsatisfactory results [4]. Embracing the principles of precision medicine, targeted therapies, including novel biologic treatments

© The Author(s), under exclusive license to Springer Nature Switzerland AG 2025
L. Correia et al. (Eds.): IBERAMIA 2024, LNCS 15277, pp. 335–343, 2025.
https://doi.org/10.1007/978-3-031-80366-6_28

such as dupilumab, offer a revolutionary opportunity to break this cycle and optimize patient outcomes [5]. Precision medicine, also called personalized medicine, combines the influence of genetics, lifestyle and environment to develop optimal treatment strategies tailored to each individual patient [6, 7]. Despite many successful studies in the management of specific diseases, namely rheumatoid arthritis, asthma, and psoriasis, precision medicine in AD is still poorly explored [5, 8]. Given the intrinsic nature of the disease and the multifaceted factors involved, precision medicine is of particular importance in AD management to avoid prolonged ineffective and costly treatments, as well as potentially adverse effects [9].

Rapid advances in medical and genetic technologies have led to a significant increase in biological data, particularly omics data related to DNA (genomics) and RNA (transcriptomics). Omics data quantify the entire population of certain biological components from tissues or cells. For example, transcriptomic data measure all the messenger RNA molecules present in a biological structure [10, 11]. The emergence of such large-scale data has opened up new avenues for the development of precision medicine, but it also brings significant new challenges. Conventional techniques are unsuitable for converting data from the existing repositories into useful knowledge due to the inherent complexity and non-linear patterns among features [12]. Machine learning (ML) techniques have gained increasing popularity in the field as a promising solution to overcome these challenges and revolutionize standard healthcare [13]. Some of the most commonly used ML models [14] include support vector machine (SVM) [15, 16], random forest (RF) [17, 18], and logistic regression (LR) [19, 20]. These traditional techniques have some limitations when dealing with high-dimensional data. Typically, these techniques require the execution of a prior dimensionality reduction step, which may exclude important features from the analysis. In contrast to these techniques, deep learning (DL) models can perform feature extraction automatically, demonstrating high performance when dealing with a large number of features [21, 22]. Convolutional neural network (CNN) is a popular DL algorithm that can be particularly effective in processing the complex nature of omics data [23]. However, since these models require images as inputs and omics datasets are generally characterized by thousands of genes and a small number of samples, only few investigations have been conducted applying CNN on this type of data [23, 24]. Therefore, the objective of this research is to convert omics data from AD patients into images and implement a CNN lesion classification model to extract the most relevant features. In this way, we expect to contribute to a more targeted management of AD by facilitating the discovery of potential molecular mechanisms involved in the formation of skin lesions.

Following this introduction, the paper is organized as follows. Section 2 addresses some of the literature that focuses on the application of DL to omics data. Section 3 explains the methodology used and Sect. 4 presents and analyzes the results obtained. Lastly, Sect. 5 highlights the main findings and discusses possible directions for future research.

2 Transcriptomic Applications

In the last decades, the advent of omics technologies has unlocked promising opportunities for biomedical research by providing researchers with access to vast amounts of biological data. For example, novel transcriptomics studies have been conducted in order to gain new insights and a better comprehension of certain diseases. Several of these investigations employ bioinformatics tools and ML/DL algorithms. Specifically, although exploration of CNN models is limited, there is an increasing number of publications on the topic based on gene expression data or using real spatial transcriptomics images.

In this context, Lyu and Haque [25] transformed RNA-seq data into 2D images and used a three-layer CNN model to classify 33 different tumor types. With this approach, the authors obtained an accuracy of 95.59% and identified potential biomarkers[1] from the most relevant genes. In another study, after converting transcriptomic data to images, Elbashir et al. [26] designed a CNN model with only two convolutional layers for breast cancer classification. The developed model demonstrated excellent results, achieving a 98.76% accuracy. Both of these experiments used sequential CNN models. However, some researchers opt to implement a parallel CNN architecture. For example, Sharma et al. [24] created the DeepInsight method that uses two parallel CNN with four convolutional layers each. DeepInsight is able to process several different types of non-image data such as RNA-Seq or text data. The authors compared the results of the new method with traditional ML models based on four different datasets, and observed that DeepInsight led to higher accuracy values. Other researchers also evaluated multiple CNN architectures to find the most suitable design. Mostavi et al. [27] constructed three alternative CNN architectures (1D-CNN, 2D-Vanilla-CNN, and 2D-Hybrid-CNN) to classify transcriptomic data from 33 cancer types and 1 non-cancer class. In the 1D model, the input data is represented as a vector subject to a one-dimensional kernel. The input data in 2D models corresponds to a matrix, and the hybrid model distinguished by the use of two parallel convolutional layers. All models achieved excellent accuracy values (93–95%), although the authors concluded that the 1D model has a more stable performance in the presence of noise. On the other hand, some papers detail the application of a CNN model to real spatial transcriptomics data such as Chang et al. [28] and Xu and McCord [29].

Most studies applying CNN models to omics data, namely transcriptomics, focus on cancer. However, these data could be extremely valuable in making precision medicine a reality for other diseases, such as AD. To our knowledge, this is the first study in the field centered on AD. Furthermore, because the number of genes in these problems is frequently much higher than the number of samples, we propose a prior selection of the differentially expressed genes (DEGs) between the conditions under analysis. Additionally, to properly test the implemented model, we also used two independent datasets containing RNA data from different experiments and sequenced by different platforms.

---

[1] Biological molecules that can be objectively measured to indicate what is occurring in a living organism at a given time.

3   Methodology

Before constructing our CNN model, we selected five different datasets (GSE65832, GSE121212, GSE224783, GSE160501, and GSE157194) from the Gene Expression Omnibus[2] (GEO) containing gene expression data of AD patients. GEO is a public repository that provides access to a wide range of biological data and is a valuable resource for biomedical research. All selected datasets were sequenced using high-throughput technologies and contain transcriptomic data collected from skin samples. Only the samples from AD patients were considered. An overview of the characteristics of each dataset is shown in Table 1. DEGs between lesional (AD-L) and non-lesional (AD-NL) skin were determined in R considering the datasets GSE65832, GSE121212 and GSE224783 and using the DESeq2 package. All genes with low counts were removed and the remaining DEGs were selected for posterior analysis.

Table 1. Summary of the datasets under analysis.

Use	Implementation of a CNN model			Independent test	
Dataset	GSE65832	GSE121212	GSE224783	GSE160501	GSE157194
Platform	GPL10999	GPL16791	GPL16791	GPL18573	GPL21290
Lesional	20	26	22	6	57
Non-lesional	20	27	11	6	54

Subsequently, we combined the gene expression data from the three datasets into a single matrix and filtered the columns to include only the identified DEGs. The resulting matrix was properly pre-processed using Python 3.6. First, we split the data into training (80%) and test (20%) using a stratified approach. Stratification ensures a similar distribution of AD-L and AD-NL samples in both sets. Second, the labels "non-lesional" and "lesional" were converted to 0 and 1, respectively. Finally, genes with low variance (less than 30%) were eliminated and the count values were normalized between 0 and 1 using the formula given in Eq. 1. In the equation, x' represents the normalized value and x the non-normalized value.

$$x' = \frac{x - \min(x)}{\max(x) - \min(x)} \tag{1}$$

After normalization, the data were converted into 2D images with 31×31 pixels using the DeepInsight method [24] and specifically the UMAP technique for dimensionality reduction. Instead of the CNN employed in DeepInsight, we developed a novel model using Keras running on TensorFlow. The architecture of our CNN model is illustrated in Fig. 1. Hyperparameters and optimizer were properly tuned to obtain the optimal values. Since the classification problem is binary, the output layer uses the sigmoid activation function, which returns a value between 0 and 1, and binary_crossentropy was

---

[2] https://www.ncbi.nlm.nih.gov/geo/

defined as the loss function. Considering 100 epochs, we compared the accuracy and loss of the training data with the validation data. The final model corresponded to the algorithm with the lower value of validation loss. After implementing the model, datasets GSE160501 and GSE157194 were used as independent tests to evaluate its performance. Table 1 summarizes the number of samples included in these two datasets. These datasets were subjected to the same pre-processing that was used to build the model. From the constructed model, we determined the importance of the features using a gradient-based method and selected the top 10 features. This step is particularly relevant to find out which genes might have a greater influence on the occurrence of AD lesions.

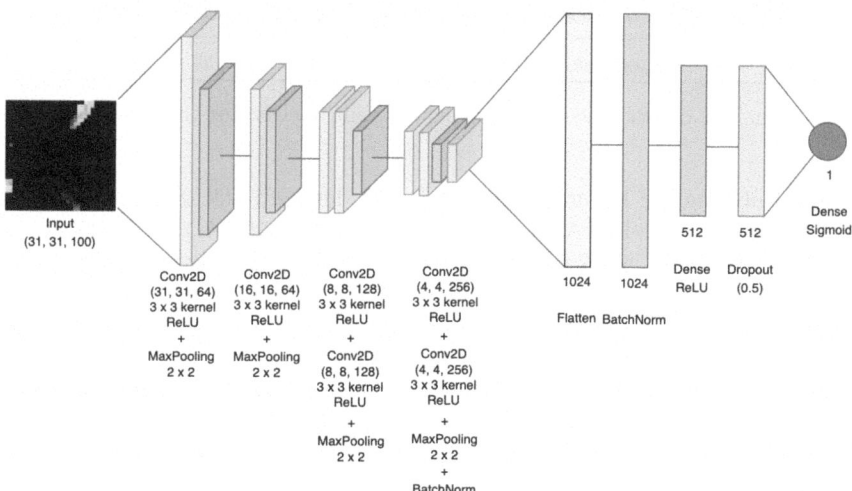

Fig. 1. Architecture of the CNN model.

4 Discussion and Results

Data preprocessing led to the identification of 929 DEGs that were used to create 100 images for training and 26 for validation, each with 31 × 31 pixels. Figure 2 shows some of the images obtained with the DeepInsight method. Hyperparameters optimization led to the selection of the Adam optimizer with a learning rate of 0.0015. Additionally, we set a momentum of 0.001 and 0.05 for the first and second batch normalization layers, respectively. We also applied a l2 regularization penalty (regularization factor = 0.04) in the dense layer of 512 neurons and established a batch size of 100 and 26 for training and validation, respectively. After implementing the model, the minimum validation loss was found at epoch 90. Both the training and validation datasets achieved 100% accuracy. Applying the same model to the independent tests, we obtained an accuracy of 75% for the dataset GSE160501 and 85.59% for the dataset GSE157194. In order to obtain more robust results, we have also analyzed the sensitivity and specificity values in the independent tests. For the GSE160501 dataset we obtained a sensitivity of 100% and a specificity of 40% and for the GSE157194 dataset 84% and 87%, respectively. Despite the limited number of samples for training, validation and independent test, these values are considerably high and demonstrate that the model is able to efficiently distinguish between lesional and non-lesional skin of AD patients.

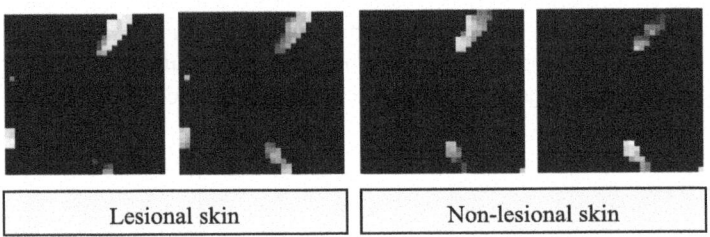

| Lesional skin | Non-lesional skin |

Fig. 2. Example of the generated images using the DeepInsight method.

When analyzing the features importance, we identified DNASE1L3, CARHSP1, FCMR, CFD, FABP7, VEGFA, CD274, CCR5, RASSF6, and CYP27B1 as the most important genes. These genes therefore appear to have a key role in the development of AD lesions and may be considered candidate biomarkers for AD. Further targeted research into these genes could expand the current understanding of the underlying biological pathways and help to develop novel approaches to control the disease. The association between some of the top genes found and AD has already been reported in the scientific literature, such as RASSF6 [30] and CCR5 [31]. These associations serve as an additional validation of the results obtained.

Despite the promising results, this research presents some shortcomings. The most important limitation is the lack of data, which affects the quality of the constructed model. Second, we did not compare the obtained results with classical ML models such as RF or SVM. Third, alternative strategies to convert the gene expression data into images were not considered. Finally, the initial reduction of the features in order to consider only the DEGs might exclude important genes from the analysis.

5   Conclusions and Further Research

Transcriptomic data are of fundamental importance to the development of precision medicine. The identification of the genes associated with the onset, progression or treatment of certain diseases enables a deeper understanding of the underlying biological pathways and contributes to patient-centered healthcare. ML techniques have been employed in this domain to unravel important genes. CNN is one of the most promising techniques. However, since the input data require an image format, the gene expression data must first be converted into a 2D matrix.

In our work, we used RNA-Seq data obtained from the skin of AD patients and implemented a classification model with a CNN. The high accuracy values in training and validation (100%) and independent testing (75% and 85.59%) indicate that our model is able to discriminate accurately lesions and non-lesions. The most significant features found are potential biomarker candidates for AD. Further in-depth investigation of the pathways associated with these features may aid biomedical research by providing new insights into the molecular mechanisms involved and suggesting novel therapeutic approaches that directly target these mechanisms. However, we need to take some limitations into account. The small number of samples, for example, is a critical point. Future research in the area should therefore analyze larger datasets in order to create a more robust model. We anticipate that the growing number of omics datasets in public repositories will help to overcome this shortcoming. In addition, alternative strategies for converting RNA-Seq data into images could be tested in the future to increase the performance of the CNN model.

Acknowledgements. This work has been supported by FCT – Fundação para a Ciência e Tecnologia within the R&D Units Project Scope: UIDB/00319/2020, and the PhD grant: 2022.12728.BD.

References

1. Naik, P.P.: Treatment-resistant atopic dermatitis: novel therapeutics, digital tools, and precision medicine. Asia Pac. Allergy. **12** (2022). https://doi.org/10.5415/apallergy.2022.12.e20
2. Katoh, N., et al.: Japanese guidelines for atopic dermatitis 2020. Allergol. Int. **69**, 356–369 (2020). https://doi.org/10.1016/j.alit.2020.02.006
3. Czarnowicki, T., He, H., Krueger, J.G., Guttman-Yassky, E.: Atopic dermatitis endotypes and implications for targeted therapeutics. J. Allergy Clin. Immunol. **143**, 1–11 (2019). https://doi.org/10.1016/j.jaci.2018.10.032
4. Flohr, C.: How we treat atopic dermatitis now and how that will change over the next 5 years. Br. J. Dermatol. **188**, 718–725 (2023). https://doi.org/10.1093/bjd/ljac116
5. De Bruin-Weller, M., et al.: Treat-to-target in atopic dermatitis: an international consensus on a set of core decision points for systemic therapies. Acta Derm. Venereol. **101** (2021). https://doi.org/10.2340/00015555-3751
6. Denny, J.C., Collins, F.S.: Precision medicine in 2030 - seven ways to transform healthcare. Cell **184**, 1415–1419 (2021). https://doi.org/10.1016/j.cell.2021.01.015
7. Manzari, M.T., Shamay, Y., Kiguchi, H., Rosen, N., Scaltriti, M., Heller, D.A.: Targeted drug delivery strategies for precision medicines. Nat. Rev. Mater. **6**, 351–370 (2021). https://doi.org/10.1038/s41578-020-00269-6
8. Muraro, A., et al.: Precision medicine in patients with allergic diseases: airway diseases and atopic dermatitis - PRACTALL document of the European academy of allergy and clinical immunology and the american academy of allergy, asthma & immunology. J. Allergy Clin. Immunol. **137**, 1347–1358 (2016). https://doi.org/10.1016/j.jaci.2016.03.010
9. Vestergaard, C., Skovsgaard, C., Johansen, C., Deleuran, M., Thyssen, J.P.: Treat-to-target in atopic dermatitis. Am. J. Clin. Dermatol. **25**, 91–98 (2024). https://doi.org/10.1007/s40257-023-00827-y
10. Vailati-Riboni, M., Palombo, V., Loor, J.J.: What Are Omics Sciences?. In: Ametaj, B. (eds.) Periparturient Diseases of Dairy Cows., pp. 1–7. Springer International Publishing (2017). https://doi.org/10.1007/978-3-319-43033-1_1

11. Poinsignon, T., Poulain, P., Gallopin, M., Lelandais, G.: Working with omics data: an inter-disciplinary challenge at the crossroads of biology and computer science. In: Neuromethods, pp. 313–330. Humana Press Inc. (2023). https://doi.org/10.1007/978-1-0716-3195-9_10

12. MacEachern, S.J., Forkert, N.D.: Machine learning for precision medicine. Genome **64**, 416–425 (2021). https://doi.org/10.1139/gen-2020-0131

13. Wilkinson, J., et al.: Time to reality check the promises of machine learning-powered precision medicine. Lancet Digit. Health. **2**, e677–e680 (2020). https://doi.org/10.1016/S2589-7500(20)30200-4

14. Monaco, A., et al.: A primer on machine learning techniques for genomic applications. Comput. Struct. Biotechnol. J. **19**, 4345–4359 (2021). https://doi.org/10.1016/j.csbj.2021.07.021

15. Zhu, H., et al.: Integration of genome-wide DNA methylation and transcription uncovered aberrant methylation-regulated genes and pathways in the peripheral blood mononuclear cells of systemic sclerosis. Int. J. Rheumatol. **2018** (2018). https://doi.org/10.1155/2018/7342472

16. Shen, Y., Wu, C., Liu, C., Wu, Y., Xiong, N.: Oriented feature selection SVM applied to cancer prediction in precision medicine. IEEE Access. **6**, 48510–48521 (2018). https://doi.org/10.1109/ACCESS.2018.2868098

17. Almlöf, J.C., et al.: Novel risk genes for systemic lupus erythematosus predicted by random forest classification. Sci. Rep. **7** (2017). https://doi.org/10.1038/S41598-017-06516-1

18. Pellegrino, E., et al.: Machine learning random forest for predicting oncosomatic variant NGS analysis. Sci. Rep. **11** (2021). https://doi.org/10.1038/s41598-021-01253-y

19. Queen, D., et al.: UV biomarker genes for classification and risk stratification of cutaneous actinic keratoses and squamous cell carcinoma subtypes. FASEB J. **34**, 13022–13032 (2020). https://doi.org/10.1096/FJ.202001412R

20. Liu, D., et al.: Integrative molecular and clinical modeling of clinical outcomes to PD1 blockade in patients with metastatic melanoma. Nat. Med. **25**, 1916–1927 (2019). https://doi.org/10.1038/s41591-019-0654-5

21. Sharma, A., Lysenko, A., Jia, S., Boroevich, K.A., Tsunoda, T.: Advances in AI and machine learning for predictive medicine. J. Hum. Genet. (2024). https://doi.org/10.1038/s10038-024-01231-y

22. Liu, J., Li, J., Wang, H., Yan, J.: Application of deep learning in genomics. Sci China Life Sci. **63**, 1860–1878 (2020). https://doi.org/10.1007/s11427-020-1804-5

23. Zompola, A., Korfiati, A., Theofilatos, K., Mavroudi, S.: Omics-CNN: a comprehensive pipeline for predictive analytics in quantitative omics using one-dimensional convolutional neural networks. Heliyon. **9** (2023). https://doi.org/10.1016/j.heliyon.2023.e21165

24. Sharma, A., Vans, E., Shigemizu, D., Boroevich, K.A., Tsunoda, T.: DeepInsight: a methodology to transform a non-image data to an image for convolution neural network architecture. Sci. Rep. **9** (2019). https://doi.org/10.1038/s41598-019-47765-6

25. Lyu, B., Haque, A.: Deep learning based tumor type classification using gene expression data. In: Proceedings of the 2018 ACM International Conference on Bioinformatics, Computational Biology, and Health Informatics, pp. 89–96. Association for Computing Machinery, Inc (2018). https://doi.org/10.1145/3233547.3233588

26. Elbashir, M.K., Ezz, M., Mohammed, M., Saloum, S.S.: Lightweight convolutional neural network for breast cancer classification using RNA-Seq gene expression data. IEEE Access. **7**, 185338–185348 (2019). https://doi.org/10.1109/ACCESS.2019.2960722

27. Mostavi, M., Chiu, Y.C., Huang, Y., Chen, Y.: Convolutional neural network models for cancer type prediction based on gene expression. BMC Med. Genomics. **13** (2020). https://doi.org/10.1186/s12920-020-0677-2

28. Chang, Y., et al.: Define and visualize pathological architectures of human tissues from spatially resolved transcriptomics using deep learning. Comput. Struct. Biotechnol. J. **20**, 4600–4617 (2022). https://doi.org/10.1016/j.csbj.2022.08.029

29. Xu, Y., McCord, R.P.: CoSTA: unsupervised convolutional neural network learning for spatial transcriptomics analysis. BMC Bioinf. **22** (2021). https://doi.org/10.1186/s12859-021-04314-1

30. Chen, G., Yan, J.: Integrated bioinformatics-based identification of potential diagnostic biomarkers associated with atopic dermatitis. Postepy Dermatol. Alergol. **39**, 1059–1068 (2022). https://doi.org/10.5114/ada.2022.114899

31. Zhou, B., et al.: Identification and validation of CCR5 linking keloid with atopic dermatitis through comprehensive bioinformatics analysis and machine learning. Front Immunol. **15** (2024). https://doi.org/10.3389/fimmu.2024.1309992

Multi Agent Systems

Reliability Analysis of Organization-Based Multiagent System Designs

Juan C. García-Ojeda[✉]

Facultad de Ingeniería, Programa de Ingeniería de Sistemas, Universidad de Cartagena,
Avenida del Consulado No. 48 – 152, 130015 Cartagena de Indias, Colombia
jcgarciao@unicartagena.edu.co

Abstract. Although several methodologies, processes, and frameworks are available for constructing sophisticated autonomous multiagent systems organizations, none of them provide techniques for the reliability analysis of multiagent systems designs. This is an important issue when designing a multiagent system because of the nature of the environments where it operates (dynamic, continuous, and partially accessible). Additionally, the multiagent system must be adaptive (self-organized) to adjust its behavior to cope with the dynamic appearance and disappearance of goals (tasks), their given guidelines, and the overall goal of the multiagent system. To address such an issue, we propose a novel approach for computing the reliability, in design time, of organization-based multiagent systems. This process consists of five steps. First, the multi-agent system is designed by adopting a modified version of the OMACS framework. Second, such a design is transformed into a P-graph model to take advantage of the combinatorial nature of the underlying structure. Third, algorithm SSG of the P-graph framework is used to generate all feasible assignment sets, which represents the different ways agents can play roles to achieve goals in the organization. Fourth, for each assignment set, a Markov chain is constructed, which captures the behavior of the system; finally, algorithm R_O is executed on each Markov chain to compute their steady states (either success or failure) for further analysis. The proposed approach is validated through the simulation of two organization-based multiagent systems from the robotics domain.

Keywords: Agent-oriented Software Engineering · Organization-based Multiagent Systems · Reliability · P-graph · Markov Chains

1 Introduction

Designing and implementing large, complex, and distributed systems by using autonomous or semi-autonomous agents that can reorganize themselves by cooperating with one another represents the future of software systems [1]. Trends in the field of autonomous agents and multiagent systems suggest that the explicit design and use of organization-based multiagent systems [2], which allow heterogeneous agents (either human or artificial entities) to rely on well-defined roles to accomplish either individual or system level goals [3, 4], is a promising approach to these new requirements [5].

© The Author(s), under exclusive license to Springer Nature Switzerland AG 2025
L. Correia et al. (Eds.): IBERAMIA 2024, LNCS 15277, pp. 347–359, 2025.
https://doi.org/10.1007/978-3-031-80366-6_29

In the literature, a set of methodologies [6], a selection of design processes [7], and a collection of frameworks [8, 9] are available to provide the basis for constructing sophisticated autonomous multi-agent organizations. Moreover, a set of metrics and methods has been suggested with the intention of providing useful information about key properties (e.g., complexity, flexibility, self-organization, performance, scalability, and cost) of these multi-agent organizations [2, 10, 11].

Nevertheless, in situations where the nature of the environment makes the organization susceptible to individual failures, these failures could significantly reduce the ability of the organization to accomplish its goals. The above-mentioned methodologies and frameworks, however, do not offer techniques for identifying the number of feasible configurations of agents that can be synthesized or designed, from a set of heterogeneous agents. This is an important issue when designing a multiagent system because of the nature of the environments in which it operates (dynamic, continuous, and partially accessible) [12]. The multiagent system must be adaptive (self-organized) to adjust its behavior to cope with the dynamic appearance and disappearance of goals (tasks), their given guidelines, and the overall goal of the multiagent system [1, 12, 13]. In what follows, an algorithmic method for assessing the design of organization-based multiagent systems supported by software tools at each step is discussed. The method is demonstrated by applying it to two robotic-based organization-based multiagent system designs.

2 Background

2.1 Organization Model for Adaptive Computational Systems: *OMACS*

OMACS Captures the knowledge required of a system's organizational structure and capabilities to allow it to organize and reorganize at runtime [1, 13]. Figure 1 shows a slightly modified version of *OMACS*. Specifically, an *OMACS* model is defined as the following tuple, $O = \langle G, R, A, C, possesses, requires, plays, assignment \rangle$. In *OMACS*, the organization, O, is composed of four entities: G, R, A, and C. G defines the goals of the organization (i.e., overall functions of the organization); R defines a set of roles (i.e., positions within an organization whose behavior is expected to achieve a particular goal or set of goals). A is a set of agents, which can be either human or artificial (hardware or software) entities that perceive their environment and can perform actions upon it. To perceive and to act, the agents possess a set of capabilities (C), which define the percepts/actions at their disposal. Capabilities can be soft (i.e., algorithms or plans) or hard (i.e., hardware-related actions).

Furthermore, a set of relations and functions are specified. The relation *achieves* : $R \times G$ defines the allocation of a goal to a role. This notation differs from the previous *OMACS* model [13], as no real value greater than zero is associated with it. This change is relevant since in any real organization, i.e., human-like, the achievement of a goal or a set of goals depends more on the agent (i.e., person) and their skills (i.e., capabilities) than on the role being played [14]. Also, the function *possesses* : $A \times C \rightarrow [0..1]$ defines the reliability of an agent´s capability. Unlike the previous *OMACS* model, where the *possesses* function captures how well an object performs its proper function, i.e., quality; in this work, the *possesses* function describes how well an object maintains its original level of quality over time through various conditions, i.e., reliability [15]. The function *requires* : $R \rightarrow \wp(C)$, where $\wp$ denotes the power set of C, defines the set of capabilities required to play a specific role in the organization. Consequently, the function *plays* : $A \times R \rightarrow [0..1]$, defines how reliable an agent is at playing a role. Mathematically, the function *plays* can be expressed as:

$$plays(a, r) = \prod_{\forall c \in requires(r)} possesses(a, c) \tag{1}$$

This is analogous to the definition of a series system. That is, "the reliability of a series system is equal to the product of the reliability of its component sub-systems" [16]. In addition, the function *assignment* : $A \times R \times G \rightarrow [0..1]$ is introduced (in the previous *OMACS* model, it is called *potential*). This function defines how well an agent can play a role to achieve a goal. This function is specified as

$$assignment(a, r, g) = \begin{cases} plays(a, r), \ if \ (r, g) \in achieves \\ 0, \ otherwise \end{cases} \tag{2}$$

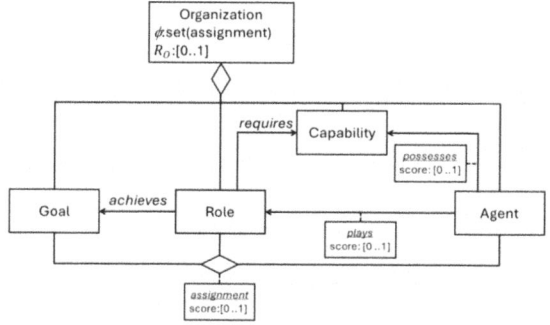

Fig. 1. Simplified and adapted OMACS metamodel

In this context, the assignment set, denoted by ϕ, represents the set of all current assignments in the organization; that is, $\phi : \wp(A \times R \times G)$. Thus, the reliability of the assignment depends on the reliability of playing a specific role in the organization. Recall that both roles and the goals are abstract elements of any organization [17]. Finally, to select the best set of assignments to maximize an organization's reliability in achieving its goals, this paper defines an organizational function, R_O, which is a function over the current assignment set, $R_O : \phi \rightarrow [0..1]$. Specifically, R_O replaces the *oaf* function, proposed in the previous *OMACS* model [1, 13]. In [11], the authors have shown that *oaf* function does not always capture the expected behavior of the organizational-based multiagent system design. Therefore, the design could lead the multiagent system to be unreliable at runtime [11, 15].

2.2 Process Graphs (P-graphs)

The P-graph methodology is a graph-based theoretical framework proposed for solving process-network synthesis problems (i.e., PNS) [18]. The methodology is grounded in two cornerstones: the P-graph (i.e., process graph) representation of a process network of interest and a set of five axioms for solution structures, specifically combinatorially feasible structures.

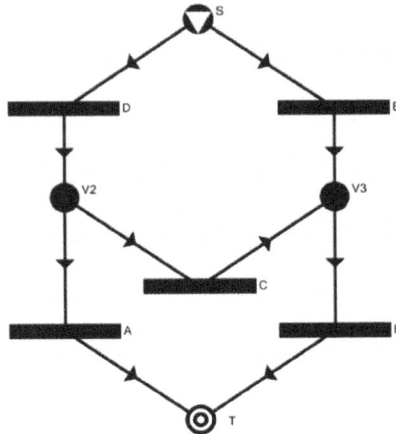

Fig. 2. Illustration of a P-graph, where $S,V2,V3$, and T are M-type nodes, and D,E,A,B and C are O-type nodes: represents raw material S or input el-ements; symbolizes intermediate materials $V2$ and $V3$, either as input or output elements to operat-ing units D,E,A,B and C; and represents prod-ucts, or output elements of the process.

A P-graph is a bipartite graph comprising M-type nodes denoting streams and O-type nodes denoting processes (see Fig. 2). Arcs signify relationships between streams and processes [18]. On the other hand, a set of axioms expresses necessary and sufficient properties to which a feasible structure should conform. The axioms are as follows [18]: (A1) every final product is represented in the graph, (A2) an M-type node has no input if and only if it represents a raw material, (A3) every O-type node represents an operating unit defined in the synthesis problem, (A4) every O-type node has at least one path leading to an M-type node representing a final product, and (A5) if an M-type node belongs to the graph, it must be an input to or output from at least one O-type node in the graph. These two cornerstones provide the mathematical foundations for the rigorous development of the component algorithms of the P-graph methodology, namely Maximal Structure Generation (MSG), Solution Structure Generation (SSG), and Accelerated Branch-and-Bound (ABB) [18]. The purpose of MSG is to generate a maximal structure (i.e., a non-redundant, rigorous superstructure) from a set of component process units. SSG aims to enumerate all structurally feasible process networks that can be formed from the available processes, with each solution structure being a subset of the maximal structure. Both MSG and SSG rely solely on structural information, i.e., the presence of physical connections between streams and processes. Finally, ABB's purpose is to provide rapid optimization once constraints of the process data are given. The algorithm ABB also allows near-optimal solutions to be generated in case they are of interest to the designer.

2.3 Markov Chains

In probability theory, a Markov chain is a stochastic model that describes a sequence of events where the probability of each event depends only on the state attained in the previous event (i.e., the next event depends only on the current state and not on the sequence of events that preceded it) [19].

Specifying a General Markov Chain. A Markov chain is formally described as follows. Let $\{X_n | n \in \mathbb{N} \wedge n > 0\}$ be a stochastic process, in discrete time, with finite or infinite discrete state space, S, such that for each $n \geq 1$, if A is an event depending on any subset of $\{X_n, X_{n-1}, X_{n-2}, ..., X_1\}$, then, for any states i and j in S,

$$P_{markov}(X_{n+1} = j | X_n = i \wedge A) = P_{markov}(X_{n+1} = j | X_n = i) \tag{3}$$

For any given states i and j.

$$P_{markov}(X_{n+1} = j | X_n = i) \text{ holds } \forall n \geq 1 \tag{4}$$

where, Eq. 4 is the Markov property. More generally, for each $n \geq 1$, and $m \geq 1$, if A (as defined in Eq. 3), then for any states i and j in S:

$$P_{markov}(X_{n+m} = j | X_n = i \wedge A) = P_{markov}(X_{n+m} = j | X_n = i) \tag{5}$$

$$p_{ij} = P_{markov}(X_{n+m} = j | X_n = i) \tag{6}$$

3 Motivational Example

To demonstrate the application of the proposed method for assessing the designs of organization-based multi-agent system, a survey is provided of a simplified Cooperative Robotic Search Team (CRST) system [2, 11, 20]. Essentially, we are to design a team of robots whose goal is to search for differ-ent areas of a given location on a map. The team should be able to search any area of the given loca-tion even when faced with failures of individual robots or specific ca-pabilities of those robots. This im-plies that the team must be able to: (1) assign areas based on individual team members reliability, (2) recognize when a robot is unable to perform its duties; and (3) reorganize the team to allow it to achieve its goals despite individual failures [2, 11, 20].

3.1 Overview of the CRST Organization

For illustration, it is assumed that four goals must be achieved by the CRST (see Fig. 3). In other words, $G = \{g_1, g_2, g_3, g_4\}$ where g_i for $1 \leq i \leq 4$ signifies "search area "i"." In the CRST, two roles are identified, i.e., $R = \{r_1, r_2\}$ where r_1 and r_2 represent the *Searcher* and *Patroller* roles, respectively. Role r_1 requires the *RangeFinder*, *GPS*, *Movement*, *Sonar* capabilities for achieving goals g_1, g_2, g_4, and g_4. Likewise, role r_2 requires the *GPS*, *Movement* and *Sonar* capabilities for achieving the same goals as those of role r_1. Moreover, for each role, r_i, and each

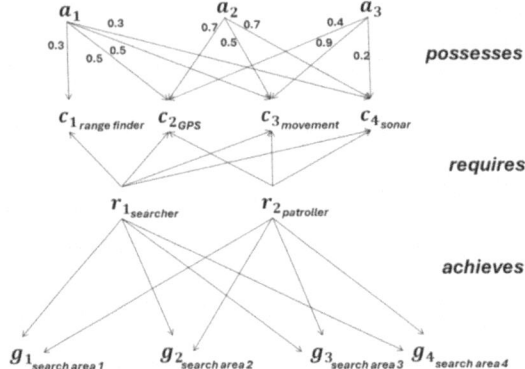

Fig. 3. View of the CRST Organization by adopting the modified version of the OMACS meta-model

goal, g_i, and an *achieve* relationship is established. This *achieve* relationship defines the allocation of a goal to a role. Both, the *requires* and *achieves* relations can be formally stated as: *requires* = $\{(r_1, \{c_1, c_2, c_3\}), (r_2, \{c_2, c_3, c_4\})\}$ and *achieves* = $\{(r_1, g_1), (r_1, g_2), (r_1, g_3), (r_1, g_4), (r_2, g_1), (r_2, g_2), (r_2, g_3), (r_2, g_4)\}$.

Additionally, four capabilities are specified, i.e., $C = \{c_1, c_2, c_3, c_4\}$. They are: *RangeFinder Sonar* (c_1), *GPS* (c_2), *Movement* (c_3), and *Sonar* (c_4). Capability c_1 captures information about all objects around agent a_i (in a $360°$ view). Capability c_2 allows agent a_i to move in any direction – north, south, east, or west (up, down, left, or right). Capability c_3 provides the ability to read the absolute position of agent a_i in the environment. Finally, capability c_4 renders it possible for agent a_i to measure the distance of the closest object directly in front of it. Also, three different agents are modeled, i.e., $A = \{a_1, a_2, a_3\}$. Specifically, agent a_1 possesses capabilities c_1, c_2, c_3, and c_4 while both agents a_2 and a_3 possess capabilities c_2, c_3, and c_4. The possesses relationship is formulated as follows: *possesses* = $\{(a_1, c_1, 0.3), (a_1, c_2, 0.5), (a_1, c_3, 0.5), (a_1, c_4, 0.3), (a_2, c_2, 0.7), (a_2, c_3, 0.5), (a_2, c_4, 0.7), (a_3, c_2, 0.4), (a_3, c_3, 0.9), (a_3, c_4, 0.2)\}$.

4   Methodology

Figure 4 depicts the proposed approach for assessing the reliability of multiagent systems. The proposed methodology consists of five steps: *Modeling, Reduction, Computing Feasible Assignment Sets, Computing Reliability*, and *Analyzing Results*.

In the first step, the designer sketches the model of an OMACS-based multiagent system (refer to Fig. 3). Next, such a model is transformed into a P-graph model by executing algorithm R_O (the second step). Algorithm R_O is a relaxed version of algorithm OMACS2PNS described in [11]. The term "relaxed" refers to a P-graph model where the constraints of the PNS problem are not crucial, as the network model of the PNS problem stands independently. Specifically, each agent of the organization is reduced into a raw material in the resulting P-graph model. Additionally, each goal of the system is reduced into an intermediate material. Moreover, each assignment of the systems, i.e., $A \times R \times G$,

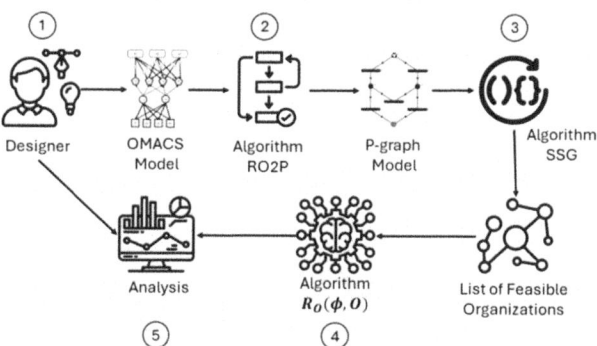

Fig. 4. Proposed methodology for assessing organization-based multiagent system design models.

is reduced into an operating unit; an operating unit *Goals* is also mapped in the resulting P-graph. Finally, the product *SystemsGoals* is added to the P-graph model (see Fig. 5). Subsequently, in step three, the algorithm SSG of the P-graph framework is executed [18]. As a result, a total of 50625 feasible organization assignment sets are revealed. This action was accomplished in 1.839 s, on a laptop computer (AMD Ryzen 3 5300U, Radeon Graphics @ 2.60 GHz, 8 GB RAM. For the sake of the hypothetical example, one of the feasible organization assignment sets comprises the following assignments $\{\{a_1, r_1, g_1\}, \{a_1, r_1, g_2\}, \{a_1, r_1, g_3\}, \{a_1, r_1, g_4\}, \{a_1, r_2, g_1\}, \{a_1, r_2, g_2\}, \{a_1, r_2, g_3\}, \{a_1, r_2, g_4\}, \{a_2, r_2, g_1\}, \{a_2, r_2, g_2\}, \{a_2, r_2, g_3\}, \{a_2, r_2, g_4\}, \{a_3, r_2, g_1\}, \{a_3, r_2, g_2\}, \{a_3, r_2, g_3\}, \{a_3, r_2, g_3\}, \{a_3, r_2, g_4\}\}$.

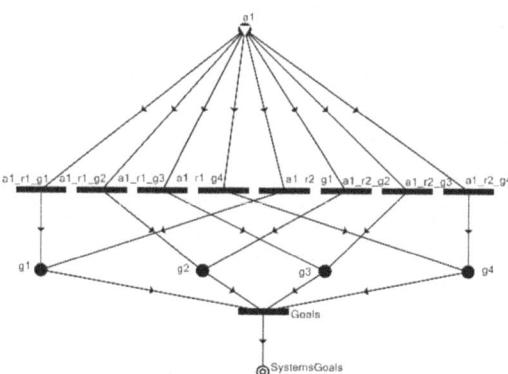

Fig. 5. Excerpt of the Maximal structure for the hypothetical relaxed example.

Afterwards, in step four, algorithm R_O is invoked on each feasible organization assignment set. The aim of algorithm R_O is to transform, first, each organization-based multiagent system assignment set into an absorbing Markov chain, P_{markov}, and, then, compute its steady state, $x_n^{(k)}$ [19]. In this contribution a multiagent system can reach two steady states: either the system achieves its goals (success) or not (fails). Therefore, the purpose of algorithm R_O is to evaluate whether an assignment set in ϕ leads an organization-based multiagent system into one of the absorbing states, i.e., either the agents' organization achieve all its goals (success state) or not (failure state). Figure 6 shows the steps required for algorithm R_O to assess every feasible assignment set in ϕ; as well as its complementary function *AMC-Spec*. More details of these two procedures can be found in [21]. Algorithm R_O comprises three major parts, the initialization, the recursion, and the calculation of the steady state, $x_n^{(k)}$. The initialization part (statement $st1$) specifies the sets for storing both the absorbing and transient states of P_{markov}. The recursion part (statement $st2$) specifies P_{markov} by describing its state space, S, based on the assignment set, ϕ. Finally, the calculation part (statements $st3 - st7$ and loop $lp1$) computes $x_n^{(k)}$[19].

Algorithm R_O operates as follows. The variable n is assigned the integer value 1. n keeps track of the total number of states in S. Subsequently, sets S_S, S_F, and S_T are initialized as empty sets, representing success state (i.e., the assignment set leads the organization to achieve its goals), failure state (i.e., the assignment set leads the organization to fail in the process of achieving its goals), and transient states (i.e., the assignment set leads the organization to a re-organization process), respectively [19]. Next, recursive procedure $AMC - Spec$ is invoked. The outcome of this recursive procedure is an absorbing markov chain, i.e., P_{markov} (see Fig. 7).

Based on the cardinality of A and G, procedure $AMC - Spec$ evaluates three different cases: (i) if $|A| = 1$ and $|G| > 1$, the transition probability is calculated as the system reliability of a simple series system. In other words, the reliability, $\hat{r}$, that one agent, in A, achieves the entire set of goals, in G, through a set of roles, in R, is equivalent to the product of the best assignment $\langle a_i, r_j, g_k \rangle$ for each $a_i \in A$, $r_j \in R$, and $g_k \in G$, where $|A| = 1$, $|R| \geq 1$, and $|G| > 1$. Thus, the basic equation for this case is:

$$\hat{r} = assignment(a_1, r_j, g_1) \times \ldots \times assignment(a_1, r_j, g_k) \tag{7}$$

(ii) if $|A| > 1$ and $|G| = 1$, the transition probability is calculated as the system reliability of a simple parallel system. Specifically, the reliability that more than one agent, in A, achieves one goal, in G, through a set of roles, in R, is equivalent to the product of the best assignment $\langle a_i, r_j, g_k \rangle$ for each $a_i \in A$, $r_j \in R$, and $g_k \in G$, where $|A| > 1$, $|R| \geq 1$, and $|G| = 1$. Thus, the basic equation for this case is:

$$\hat{r} = 1 - \begin{pmatrix} (1 - assignment(a_1, r_j, g_1)) \times \\ \cdots \times \\ (1 - assignment(a_i, r_j, g_1)) \end{pmatrix} \tag{8}$$

and the probability of failure is defined as $1 - \hat{r}$. Finally, if both $|A| > 1$ and $|G| > 1$, the system is considered neither series nor parallel. Therefore, the transition probability is calculated as the product of a finite set of mutually independent events, where each event can be either an agent achieving a given goal or failing to accomplish it. Note that the first case, $|A| = 1$ and $|G| > 1$, and the second case, $|A| > 1$ and $|G| = 1$, are the base cases of this sub-procedure; while the third case, $|A| > 1$ and $|G| > 1$, is the recursive case.

For the sake of the hypothetical example, recall the assignment set introduced previously in Sect. 3.1. The procedure $AMC - Spec$ runs as follows. Initially, with $|A| > 1$ and

Algorithm $R_O(\phi, O)$

Input:
ϕ subset of different assignments over of the organization O
O an OMACS model
Output:
r_O: reliability of ϕ
begin
st1: $n := 1; S_S := \emptyset; S_F := \emptyset; S_T := \emptyset;$
st2: $AMC - Spec(O, \phi, n - 1);$
st3: $S_S := S_S \cup \{(n, 1)\}; S_S := S_S \cup \{(n + 1, 0)\}; S_F := S_F \cup \{(n, 0)\}; S_F := S_F \cup \{(n + 1, 1)\};$
st4: $x[1:n + 2]; x[1] := 1;$
st5: $\delta := 1 - x[n + 1];$
st6: $\varepsilon := 10^{-6};$
lp1: **while** $\delta > \varepsilon$ **do**
   begin
   $v := x[n + 1];$
   $x := matrixProduct(S_T, S_S, S_F, x);$
   $\delta := \| x[n + 1] - v\|;$
   end
st7: $r_O := x[n + 1];$
end;

Fig. 6. Algorithm R_O written in Pidgin Algol

$|G| > 1$, the procedure constructs a matrix, M, describing the possible assignments that can be set up based upon the current sets A, R, and G, with the aim of identifying the best possible assignment set $\langle a_i, r_k, g_j \rangle$. As a result, each cell of the matrix M corresponds to the evaluation for each $a_i \in A$ and $g_j \in G$ to determine the best $r_k \in R$ by calculating the assignment reliability of a_i, r_k, and g_j (see Eq. 6). Thus, the matrix M is created

$$M = \begin{bmatrix} \langle\{a_1, r_1, g_1\}, 0.075\rangle & \langle\{a_1, r_1, g_2\}, 0.075\rangle & \langle\{a_1, r_1, g_3\}, 0.075\rangle & \langle\{a_1, r_1, g_4\}, 0.075\rangle \\ \langle\{a_2, r_2, g_1\}, 0.245\rangle & \langle\{a_2, r_2, g_2\}, 0.245\rangle & \langle\{a_2, r_2, g_3\}, 0.245\rangle & \langle\{a_2, r_2, g_4\}, 0.245\rangle \\ \langle\{a_3, r_2, g_1\}, 0.072\rangle & \langle\{a_3, r_2, g_2\}, 0.072\rangle & \langle\{a_3, r_2, g_3\}, 0.072\rangle & \langle\{a_3, r_2, g_4\}, 0.072\rangle \end{bmatrix}$$

Thereafter, the Hungarian method [22] is applied to matrix M as input. The purpose of the Hungarian method in our approach is to guarantee, at each step, the selection of the best assignment of a set of agents to a set of goals through a set of roles. Accordingly, an optimal assignment or minimum matching is obtained. Specifically, $\{\langle\{a_1, r_1, g_1\}, 0.075\rangle, \langle\{a_3, r_2, g_2\}, 0.072\rangle, \langle\{a_2, r_2, g_3\}, 0.245\rangle\}$. This minimum matching indicates that agent a_1 should play role r_1 in order to achieve goal g_1 with a probability of success, i.e., reliability, of 0.075; agent a_2 should play role r_2 in order to achieve goal g_3 with a reliability of 0.245; and, agent a_3 should play role r_2 in order to achieve goal g_2 with a reliability of 0.072. Subsequently, the procedure proceeds to create eight different state transitions, which in turn represent the combination of agents who are still operative, i.e., not broken and not undergoing repair, to achieve the available goals of the system. These are: $\{\{a_1, a_2, a_3\}, \{a_1, a_2\}, \{a_2, a_3\}, \{a_1, a_3\}, \{a_1\}, \{a_2\}, \{a_3\}, \varnothing\}$. For instance, Fig. 7 displays the initial recursive iterations of the $AMC - Spec$ procedure. From the initial state, i.e., state 0, where all agents are operable and all goals are

to be achieved, the probability of transitioning to state 1, where all agents are operative, and the goals assigned in the previous state have been achieved is equivalent to
$\hat{r} = assignment(a_1, r_1, g_1) \times assignment(a_3, r_2, g_2) \times assignment(a_2, r_2, g_3).$

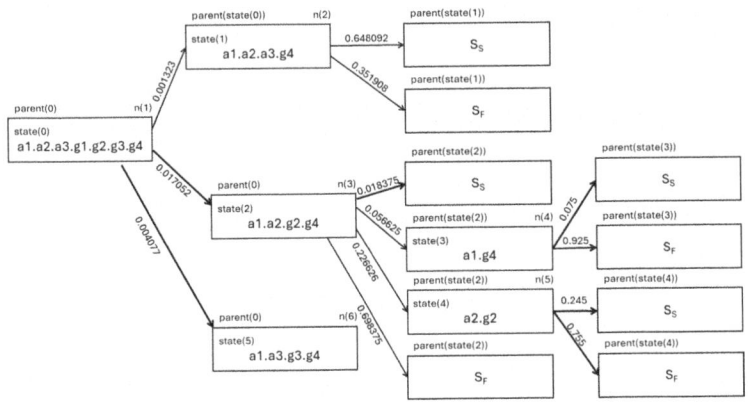

Fig. 7. Excerpt of the Markov Chain created by Procedure *AMC-Spec* for the assignment set ϕ, chosen for the sake of the hypothetical example.

Hence, $\hat{r} = 0.075 * 0.072 * 0.245$, which yields $\hat{r} = 0.001323$. From this state the system can reach each of the two absorbing states. Note that the set of available agents is $\{a_1, a_2, a_3\}$ and there is one goal available to achieve, i.e., $\{g_4\}$. Therefore, the $AMC - Spec$ chooses the best assignment, which is $\{\langle\{a_1, r_1, g_4\}, 0.075\rangle, \langle\{a_2, r_2, g_4\}, 0.245\rangle, \langle\{a_3, r_2, g_4\}, 0.072\rangle\}$. Hence, using Eq. 8, $\hat{r} = (1-((1-0.075)*(1-0.072)*(1-0.245)))$, which gives the transition probability from this state to the success state. Thus, $\hat{r} = 0.351908$. Additionally, the transition probability from this state to the failure state is $1 - \hat{r}$, which is 0.648092. Accordingly, the $AMC - Spec$ procedure continues until no more states are discovered for evaluation. Finally, after computing sets S_T, S_S, and S_F, algorithm R_O computes the steady state, $x_n^{(k)}$, of P_{markov} (steps $st3 - st7$ and loop $lp1$). Consequently, algorithm R_O adds two new elements to set S_s and S_F, respectively. These elements represent the absorbing transition probability of the given absorbing states. Subsequently, algorithm R_O creates the array x, which represents $x_n^{(k)}$. Afterward, the variable δ, representing the result after the $k + 1$ iteration, and ε, the accuracy of the result after $k + 1$, are initialized to $\delta = 0$ and $\varepsilon = 0.000001$.

Subsequently, algorithm R_O evaluates whether $\delta > \varepsilon$. If true, algorithm R_O computes the product, i, of array x and P_{markov}. It should be noted that P_{markov} is implicitly constructed in terms of sets S_T, S_S, and S_F. Table 1 shows the resulting steady state of P_{markov} after 3 iterations, i.e., $x_n^{(3)}$. State s_{13}, success state, represents the reliability of the organization-based multiagent systems in terms of the assignment set ϕ under evaluation (step $st7$). Thus, the reliability of the system is less than 1%, i.e., 0.6%. Hence, with a probability of 99.4%, the current system configuration is prone to failure.

Table 1. Probability distribution, after 3 iterations, given the initial probability vector, $x_n^{(0)}$, [1.0, 0.0, 0.0, 0.0, 0.0, 0.0, 0.0, 0.0, 0.0, 0.0, 0.0, 0.0, 0.0, 0.0, 0.0]

k	s_i	$s_{13}(success)$	$s_{14}(failure)$
1	...	0,000	0,648
2	...	0,004	0,986
3	...	0,006	0,994

5 Assessment of Organization Based Multi-Agent System Design by the Absorbing Markov Chain Model Method

To empirically evaluate the reliability of the different agent-based organization designs computed by algorithm R_O, we have developed a simulator that steps through the design of a CRST application. To measure system reliability, we follow a Bernoulli process [23]. For each assignment set, ϕ, a random system goal, $g_k \in G$, is selected. Subsequently, the reliability, $\hat{r}$, of the best available assignment, i.e., $assignment(a_i, r_j, g_k)$, is calculated. The best available assignment defines the reliability of an agent $a_i \in A$, for achieving a goal $g_k \in G$ while playing a role $r_k \in R$ (see Eq. 6). Afterwards, $assignment(a_i, r_j, g_k)$ is compared to a random variable X, which is uniformly distributed (Java SE Platform - JDK 21 pseudo-random number generator). If X is greater than $assignment(a_i, r_j, g_k)$, agent a_i is removed from A, it is assumed that a_i failed to achieve goal g_k; otherwise, g_k is removed from G. This process continues until either A or G is empty. If G is empty, 0 is returned; otherwise, 1 is returned. Note that, 0 represents a success and 1 represents a failure for each trial. A total of 5000 trials were conducted for the Bernoulli process. Finally, the resulting probability of success is calculated as an average of the trials to account for variations caused by the random generator used to simulate the success or failure of an assignment. Figure 8 (left) shows that the reliability of System #1 ranges between a maximum of 0.63% and a minimum of 0.055% with a median of 0.41%, a standard deviation of 0,24%, and a relative error of 2.45% (on average between the computed and simulated reliability values) for the first 100 assignment sets. As a result, System # 1 can be considered faulty overall. A second experiment was conducted where we modified the *possesses* function of the System #1 as follows: $possesses = \{(a_1, c_1, 0.98), (a_1, c_2, 0.95), (a_1, c_3, 0.95), (a_1, c_4, 0.93), (a_2, c_2, 0.97), (a_2, c_3, 0.95), (a_2, c_4, 0.97), (a_3, c_2, 0.94), (a_3, c_3, 0.99), (a_3, c_4, 0.92)\}$. Figure 8 (right) shows that the reliability of System #2 ranges between a maximum of 97,35 % and a minimum of 85,10 % with a median of 94 .04%, a standard deviation of 3,43%, and a relative error of less than 1% (in average). As a result, System # 2 can be considered reliable overall.

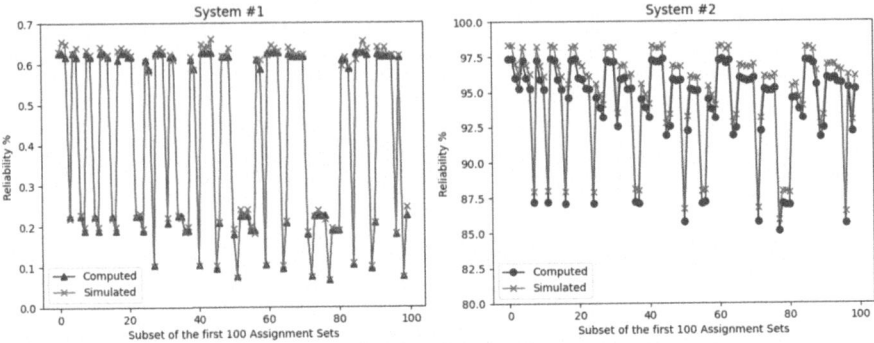

Fig. 8. Results for System #1 and System #2.

6 Conclusions and Future Work

In this paper, a novel approach is introduced to compute the reliability, during design time, of organization-based multiagent systems. The method comprises the following steps. First, the organization-based multi-agent system design is modeled by adopting a slightly modified version of the OMACS framework. Next, this design is transformed into a P-graph model to leverage the combinatorial properties of its structure. Subsequently, the algorithm SSG of the P-graph framework is used to generate all possible assignment sets, demonstrating the various ways agents can fulfill their roles to achieve their organizational goals. For each assignment set, a Markov chain is created to represent the system's behavior, and the algorithm R_O is applied to each Markov chain to determine its steady states, i.e., the probability of success or failure of the system. The proposed methoFigured is validated through simulations of two organization-based multiagent systems designed in the robotics field. The results demonstrate the effectiveness of the proposed methodology in assessing and enhancing the reliability of multiagent systems by systematically evaluating various configurations and optimizing assignments. As future work, we plan to evaluate how rules/policies would affect the reliability of the organization[20]; compare our approach with other similar methods [15]; elaborate on a method fully based on P-graph for reliability analysis [18]; develop a computational tool for analyzing OMACS-based organizational-based multiagent systems' reliability; and evaluate its applicability in real cases from different domains.

References

1. DeLoach, S.A., Oyenan, W.H., Matson, E.T.: A capabilities-based model for adaptive organizations. Auton. Agent. Multi-Agent Syst. **16**, 13–56 (2008). https://doi.org/10.1007/s10458-007-9019-4
2. Robby, DeLoach, S.A., Kolesnikov, V.A.: Using design metrics for predicting system flexibility. In: Baresi, L., Heckel, R. (eds.) Fundamental Approaches to Software Engineering. FASE 2006. Lecture Notes in Computer Science, vol. 3922. Springer, Berlin, Heidelberg (2006).https://doi.org/10.1007/11693017_15
3. Dignum, V., Dignum, F.: A logic of agent organizations. Log. J. IGPL **20**, 283–316 (2012)

4. Wooldridge, M., Jennings, N.R.: Intelligent agents: theory and practice. Knowl. Eng. Rev. **10**, 115–152 (1995)
5. Luck, M., McBurney, P., Shehory, O., Steven, W.: Agent technology: computing as interaction (a roadmap for agent based computing). AgentLink III, Liverpool, UK (2005)
6. Abdalla, R., Mishra, A.: Agent-oriented software engineering methodologies: analysis and future direction. Complexity. (2021)
7. Cossentino, M., Hilaire, V., Molesini, A., Seidita, V.: Handbook on Agent-Oriented Design Pro-cesses. An IEEE-FIPA standard compliant description approach. Springer-Verlag, Berlin (2014)
8. Cardoso, R.C., Ferrando, A.: A review of agent-based programming for multi-agent systems. Computers. **10**, 16 (2021). https://doi.org/10.3390/computers10020016
9. Savaglio, C., Ganzha, M., Paprzycki, M., Bădică, C., Ivanović, M., Fortino, G.: Agent-based Internet of Things: state-of-the-art and research challenges. Future Gener. Comput. Syst. **102**, 1038–1053 (2020). https://doi.org/10.1016/j.future.2019.09.016
10. Benaboud, R., Marir, T.: Flexibility measurement model of multi-agent systems. Multiagent Grid Syst. **16**, 309–341 (2020). https://doi.org/10.3233/MGS-200334
11. García-Ojeda, J.C., Bertok, B., Friedler, F., Argoti, A., Fan, L.T.: A preliminary study of the application of the P- graph methodology for organization-based multiagent system designs: assessment. Acta Polytech. Hung. **12**, 103–122 (2012)
12. Rădulescu, R., Mannion, P., Roijers, D.M., Nowé, A.: Multi-objective multi-agent decision making: a utility-based analysis and survey. Auton. Agent. Multi-Agent Syst. **34** (2020). https://doi.org/10.1007/s10458-019-09433-x
13. DeLoach, S.A.: OMACS: a framework for adaptive, complex systems. In: Handbook of Research on Multi-Agent Systems: Semantics and Dynamics of Organizational Models, pp. 76–104. IGI Global (2009)
14. Keog, K., Sonenberg, L.: Designing multi-agent system organisations for flexible runtime behaviour. Appl. Sci. **10**, 5335 (2020). https://doi.org/10.3390/app10155335
15. Feng, Q., et al.: An agent-based reliability and performance modeling approach for multistate complex human-machine systems with dynamic behavior. IEEE Access. **7**, 135300–135311 (2019)
16. Băjenescu, T.I., Bazu, M.I.: Component Reliability for Electronic Systems. Artech House Publishers, London, UK (2009)
17. Jiang, J., Dignum, V., Tan, Y.H.: An agent based inter-organizational collaboration framework: OperA+. In: Proceedings of the International Workshop on Coordination, Organizations, Institutions, and Norms in Agent System, pp. 58–74. Springer, Berlin/Heidelberg (2012)
18. Friedler, F., Pimentel Lozada, J., Orosz, Á.: P-graphs for process systems engineering: Mathematical models and algorithms. Springer Nature, Switzerland (2022)
19. Knill, O.: Probability Theory and Stochastic Processes with Applications. Overseas Press, NY, NY (2009)
20. Harmon, S.J., DeLoach, S.A., Robby, Caragea, D.: Leveraging organizational guidance poli-cies with learning to self-tune multiagent systems. In: Proceedings of the Second IEEE International Conference on Self-Adaptive and Self-Organizing Systems, pp. 223–232. IEEE Computer Society, Venice, Italy (2008)
21. García-Ojeda, J.C.: On Modeling Building-Evacuation-Route Planning and Organization-based Multiagent Systems by Resorting to the P-graph Framework (2016)
22. Kuhn, H.W.: The Hungarian method for the assignment problem. Nav. Res. Logist. Q. **2**, 83–97 (1955)
23. Breuer, L., Baum, D.: An introduction to queueing theory: and matrix-analytic methods. Springer (2005)

Emotions Identification in Exchanges of Messages Between Agents

Thiago Dantas, Giovani Farias$^{(\boxtimes)}$ ⃝, Cleo Billa⃝, Eder Gonçalves⃝,
and Diana Adamatti⃝

Federal University of Rio Grande – FURG, Center for Computational Sciences – C3,
Rio Grande, RS, Brazil
giovanifarias@gmail.com, {edergoncalves,dianaadamatti}@furg.br

Abstract. This paper presents the identification of emotions using Watson, a natural language processing tool, applied to multi-agent systems. In this approach, Watson acts as an intermediary between agents and identifies the emotions expressed in exchanged messages to assist in the agents' decision-making. The identification of emotions was based on the OCC model using tokens – words related to each emotion – which are useful for identifying and indicating the intensity of emotions. We built a prototype using the Jason platform to test communication between two agents, with Watson acting as an intermediary to identify emotions based on the OCC model. As an experiment, we present a study in Prisoner's Dilemma and how emotions change agents' decision-making.

Keywords: Emotions · Agents · Decision-Making · Modeling

1 Introduction

This work focuses on developing a system to identify emotions defined by the OCC model (an acronym for its creators Ortony, Clore, and Collins) [11] through *Natural Language Processing* (NLP). We aim to enable cognitive agents to identify emotions expressed in messages to assist in their decision-making. The developed project seeks to evaluate the emotions expressed among agents and to respond to their messages based on the captured emotions, achieving the desired outcome by generating the emotions present in the interactions between agents.

The major goal is to improve the understanding of interactions that occur between agents, as agents, when recognizing a certain emotion, perform their actions influenced by this emotion. Emotion can be defined as a complex state of feeling that results in physical and psychological changes that influence thought and behavior. In other words, emotions are feelings that directly alter the agent's decision-making process. One of the most used models to represent emotions is the OCC model. This model provides information that allows for an interpretation of a situation for an agent and determines to which emotion this interpretation leads. To interpret the interactions that take place between agents, a tool is used that can perform natural language processing and identify emotions

L. Correia et al. (Eds.): IBERAMIA 2024, LNCS 15277, pp. 360–371, 2025.
https://doi.org/10.1007/978-3-031-80366-6_30

using the parameters presented in the OCC model. The tool we used was IBM's cognitive computing system, called Watson[1].

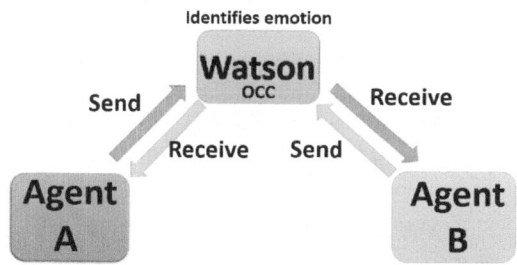

Fig. 1. Proposed model structure.

As shown in Fig. 1, Watson works as an intermediary that can understand the characteristics of the natural language used among agents, and from these it can identify emotions using the OCC model. Watson analyzes the sentences exchanged by the agents and identifies the emotions of each sentence. These emotions can then be used in the decision-making process.

This paper is structured as follows. Section 2 presents the OCC model and discusses natural language processing and Watson. Section 3 describes the techniques adopted in the implementation process. Following that, Sect. 4 presents the test case employed in this study and details the implementation of the OCC model with Watson. Finally, Sect. 5 presents the final considerations and future work resulting from this study.

2   Background

2.1   OCC Model

There are several proposals to simulate and identify emotions to enable their computational simulation [9]. In the context of *Artificial Intelligence* (AI), this research area encompasses various approaches, ranging from social and cognitive processes to biological processes [4]. It is a very complex topic and the subject of ongoing research, serving as a source of inspiration for new approaches to the development of intelligent agents [6].

One of the most widely used models is the OCC model, due to its simplicity and its ability to identify different emotions based on the contrast between positive and negative emotions. This approach is particularly useful in simulations where emotions play a critical role, such as in decision-making [9]. By classifying emotions into 22 different types, the OCC model simplifies the inherent complexity of emotion simulation, as shown in Table 1.

---

[1] https://www.ibm.com/Watson/br-pt/.

Table 1. The 22 emotions of OCC model, adapted from [11].

Emotion	Interpretation
Joy	(pleased about) a desirable event
Distress	(displeased about) an undesirable event
Happy-for	an event presumed to be desirable for someone else
Pity	an event presumed to be undesirable for someone else
Gloating	(pleased about) an event presumed to be undesirable for someone else
Resentment	(displeased about) an event presumed to be desirable for someone else
Hope	(pleased about) the prospect of a desirable event
Fear	(displeased about) the prospect of an undesirable event
Satisfaction	the confirmation of the prospect of a desirable event
Fear-confirmed	the confirmation of the prospect of an undesirable event
Relief	(pleased about) the disconfirmation of the prospect of an undesirable event
Disappointment	(displeased about) the disconfirmation of the prospect of a desirable event
Pride	(approving of) one's own praiseworthy action
Shame	one's own blameworthy action
Admiration	(approving of) someone else's praiseworthy action
Reproach	(disapproving of) someone else's blameworthy action
Gratification	one's own praiseworthy action and (being pleased about) the related desirable event
Remorse	one's own blameworthy action and (being displeased about) the related undesirable event
Gratitude	someone else's praiseworthy action and (being pleased about) the related desirable event
Anger	(disapproving of) someone else's blameworthy action and (being displeased about) the related undesirable event
Love	(liking) an appealing object
Hate	(disliking) an unappealing object

In addition to these classifications, each emotion has a specification divided into three parts, as shown in Fig. 2.

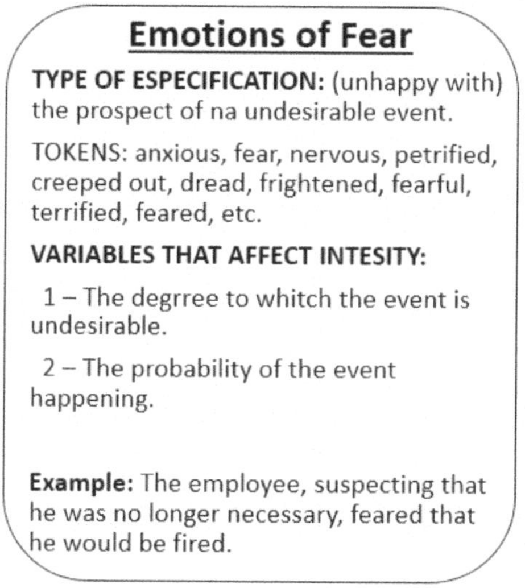

Emotions of Fear

TYPE OF ESPECIFICATION: (unhappy with) the prospect of na undesirable event.

TOKENS: anxious, fear, nervous, petrified, creeped out, dread, frightened, fearful, terrified, feared, etc.

VARIABLES THAT AFFECT INTESITY:

1 – The degrree to whitch the event is undesirable.

2 – The probability of the event happening.

Example: The employee, suspecting that he was no longer necessary, feared that he would be fired.

Fig. 2. Example of fear emotion, adapted from [11].

– **Type of specification**: describes conditions that cause a certain emotion.
– **Tokens**: a list of tokens that specify which words indicate a certain emotion. For example, "anxious", "dread", or "fearful" are types of fear.
– **Variables that affect intensity**: each emotion includes a list of variables that affect its intensity. These variables influence a single emotion; variables that affect multiple emotions are not considered. The greater the number of variables, the stronger the emotion.

2.2 Natural Language Processing and WATSON

Currently, the use of AI in various segments of the *Information Technology* (IT) field is notable. AI aims to enable machines and computers to imitate human actions to discover solutions to common problems [8]. AI employs learning methods to solve problems through acquired knowledge. For this, computers need to have certain capabilities, such as understanding natural language [10]. Communication between machines and humans occurs through natural language, which allows computers, based on statistical models and the analysis of linguistic behavior patterns, to translate the sentences provided by users into machine language [13].

Watson is one of the machine learning systems that use natural language to communicate with users through the simulation of human linguistic processing. It employs cognitive technology to simulate the processes carried out by the human mind, enabling it to learn from the information provided by users [3]. This system gained significant recognition and fame after its introduction in 2011, during the American question-and-answer program Jeopardy, a word recognition competition on American television. Watson became a champion because it could process natural language very efficiently and quickly, thanks to substantial investments in powerful hardware [3]. Watson comprises various services with different functions, including converting speech to text, converting documents, converting text to speech, optimizing document searches, and image recognition.

3   Proposed Model

After identifying the tokens for each of the 22 emotions present in the OCC model and understanding how the chosen natural language processing tool works, the next step is modeling how the system should behave. In the proposed model, Watson is solely responsible for identifying the tokens present in the exchanged messages, so it can be replaced by another tool capable of performing the same task. There are two possible types of messages to be interpreted by Watson:

- **Statement with emotion included**: these statements contain at least one token in their structure. For example:
  - I am resentful of him;
  - We are euphoric about the trip;
  - You are a bitter person.

  In these sentences, the tokens `resentful`, `euphoric`, and `bitter` are identifiable.
- **Statement with no emotion included**: these are usually monosyllabic direct responses. For example:
  - Yes;
  - No.

  However, they can also be longer sentences without emotional expression. For example:
  - I am going to the market;
  - My neighbor's son was born tonight;
  - He is building his house.

  None of these statements contains tokens that express emotion. Thus, Watson will not pass on an emotion from the issuing agent to the receiving agent.

The functioning of the system can be better visualized in Fig. 3, which presents an example.

The processing order for this example is:

1. Agent A sends a message to Agent B;
2. This message is processed by Watson, which looks for tokens. Upon finding `terrified`, it identifies the expressed emotion as fear;
3. Watson communicates to Agent B that Agent A is expressing fear;

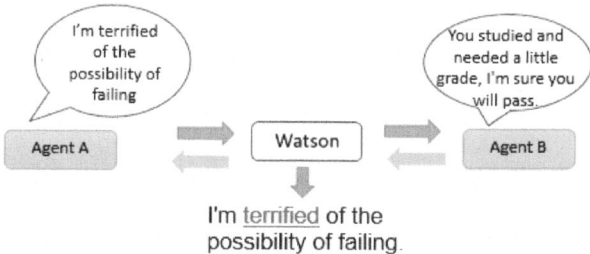

Fig. 3. Example of system operation.

4. Agent B responds to Agent A to calm him down;
5. Watson processes the message from Agent B ("You studied and needed a little grade, I'm sure you will pass.") and finds no tokens indicating that no emotion was expressed.

If an emotion is identified in a sentence exchanged between agents, a counter for that emotion is incremented. If the same emotion is identified in the next sentence, the counter is incremented by 1 again. However, if the emotion is not identified in the next sentence, the counter is reset to 0 because it represents short-term memory (appearing only once). If the emotion has been identified twice or more consecutively (becoming long-term), the counter is not reset, but divided by 2 instead. Long-term memory is not forgotten, but only weakened.

We integrated Watson with the Jason platform to implement cognitive agent systems. This platform was chosen because it supports the development of distributed applications and comprises autonomous entities based on the BDI (*Belief, Desire, Intention*) model [1], which is widely explored in the literature. Additionally, Jason provides an interpreter-based multi-agent system development environment for an extended version of the AgentSpeak(L) language. It implements the operational semantics of the AgentSpeak(L) language and offers a robust platform for developing multi-agent systems with numerous resources.

In addition, there is an integration between the Jason platform and one of the Watson modules to facilitate message exchanges between agents. To enable this integration, a Java-based process automation IDE called Selenium was used to send messages from Jason to Watson and vice versa. Figure 4 illustrates how these three systems are integrated.

4 Analysis: Experiment with Prisoner's Dilemma

The test case employed in this study revolves around a problem rooted in game theory known as the "Prisoner's Dilemma". This scenario, developed in 1950 by Merrill Flood and Melvin Dresher, serves as a standard technique for game analysis, particularly in discerning Nash equilibrium. This concept encapsulates a scenario wherein, within a game involving two or more players, no individual stands to benefit from unilaterally altering their strategy [12].

Fig. 4. Communication between Jason and Watson through Selenium.

For this validation, the rules were adjusted slightly, incorporating the notion that a suspect is more inclined to confess if the police officer can establish a connection and/or offer comfort during the conversation through emotional cues [5]. Originally, in the standard game analysis devised by Merrill Flood and Melvin Dresher, two prisoners may choose to report or remain silent regarding their crime partner's involvement. However, in this study, the suspect can choose to confess to the crime or remain silent without snitching anyone else.

These rules were based on a theory presented in the book: *Getting to the Truth: a practical, scientific approach to behavior analysis for professionals* [7], which presents in its studies that generating empathy leads the person, in this case the officer, to confess. That is, it is necessary to create a relationship of positive emotions, an empathetic connection, which is known in psychology as *rapport*[2], otherwise, the police officer will not obtain the suspect's confession to solve the case. There will come a point in the interrogation where the criminal's only desire will be to confess to the crime. However, if the officer continues to press the suspect, they will withhold the information and confess. Therefore, according to this theory, pressuring the suspect is not the best alternative. Thus, in the system, the officer will try to converse with the suspect, maintaining empathy in the conversation.

Table 2. Payoff matrix to Prisoner's Dilemma.

Prisoner 1	Prisoner 2	
	Confess	Remain in Silent
Confess	Prisoner 1 gets 5 years Prisoner 2 gets 5 years	Prisoner 1 is free Prisoner 2 gets 10 years
Remain in Silent	Prisoner 1 gets 10 years Prisoner 2 is free	Prisoner 1 gets 1 year Prisoner 2 gets 1 year

In this scenario, agents with short-term memory reduce the emotion value by 1 if that emotion is not repeated in the subsequent interaction. However, if the emotion persists, it transitions into long-term memory. Thus, instead of

---

[2] *Rapport* originated in psychology, used to designate the technique of generating an empathetic bond with another person, for communication without resistance.

subtracting 1, when the emotion is not reiterated, its intensity is halved. This ensures that while the intensity may diminish, the agent retains the memory, allowing it to be rekindled if identified in a future interaction, as shown in Fig. 5.

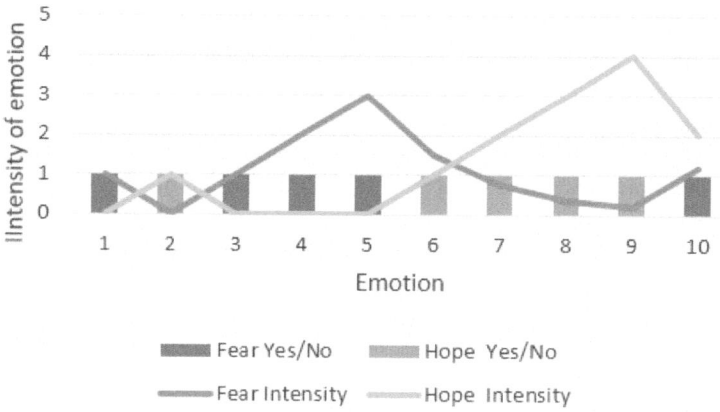

Fig. 5. Representation of memory using the emotions of hope and fear.

To justify these calculations, we considered that short-term memory can fade, while long-term memory remains intact, albeit diminishing in intensity over time. In the graph presented in Fig. 5, the bars indicate whether a particular emotion was identified during each interaction (0 absent, 1 present). In general, the graph illustrates 10 memory interactions. The blue bars denote the emotion of fear, and the gray bars signify hope, while the lines represent the variation in emotion intensity for the agent. To formalize the progressive decay of memory across interactions, Table 3 is presented.

Table 3. Data from Fig. 5, presented memory representation for one agent.

Emotion	Memory Representation for one Agent									
Fear	Yes/No	1	0	1	1	1	0	0	0	0	1
	Intensity	1	0	1	2	3	1.5	0.75	0.37	0.18	1.18
Hope	Yes/No	0	1	0	0	0	1	1	1	1	0
	Intensity	0	1	0	0	0	1	2	3	4	2

Table 3 represents the 10 interactions corresponding to the graph illustrated in Fig. 5, detailed in items 1 to 10:

1. The agent identifies the emotion of fear, and its intensity is incremented by 1, still remaining a short-term memory, as it has not been reinforced yet.

2. The agent identifies the emotion of hope, and its intensity is incremented by 1, still remaining a short-term memory, as it has not been reinforced yet. The intensity of the emotion of fear is reduced to zero because it has not occurred again, thus not being reinforced.

3. The agent identifies the emotion of fear, and its intensity is incremented by 1, still remaining a short-term memory, as it has not been reinforced yet. The intensity of the emotion of hope is reduced to zero because it has not occurred again, thus not being reinforced.

4. The agent identifies the emotion of fear and increments its value by 1, making it a long-term memory as it has been repeated, with its intensity equal to 2.

5. The agent identifies the emotion of fear and increments its value by 1, now having an intensity of 3.

6. The agent identifies the emotion of hope and increments its value by 1. The intensity of the emotion of fear is halved because it is a long-term memory, becoming 1.5.

7. The agent identifies the emotion of hope and increments its value by 1, becoming a long-term memory with an intensity of 2. Meanwhile, the intensity of the fear emotion of fear is halved because it is a long-term memory, becoming 0.75.

8. The agent identifies the emotion of hope and increments its value by 1, becoming 3. Meanwhile, the intensity of the emotion of fear is halved because it is a long-term memory, becoming 0.375.

9. The agent identifies the emotion of hope and increments its value by 1, becoming 4. Meanwhile, the intensity of the emotion of fear is halved because it is a long-term memory, becoming 0.1875.

10. The agent identifies the emotion of fear and increments its intensity by 1, becoming 1.1875. The emotion of hope is halved, becoming 2.

Thus, for this interaction, the predominant emotion is hope, as its final intensity is greater than that of fear. The preceding discussion explained the functioning of memory. Now, we will present some interactions involving Jason and one of the possible outcomes for the Prisoner's Dilemma as adapted in this work.

In the first dialogue, prisoner 1 confessed, as the police officer managed to generate a higher intensity of positive emotion than negative emotions. In the graph shown in Fig. 6, it is evident that hope was the most intense emotion. Despite some drops, it remained above the level of fear, and its intensity at the end of the dialogue was higher than that of fear.

Table 4 shows the intensity values of the emotions for each interaction in the dialogue presented in the graph of Fig. 6.

After the dialogue with prisoner 2, he declared himself innocent, as the police officer couldn't generate a higher intensity of positive emotion than negative emotions. In the end, prisoner 1 was released for confessing, while prisoner 2 received a sentence of 10 years for denying. This can be observed in the graph in Fig. 7.

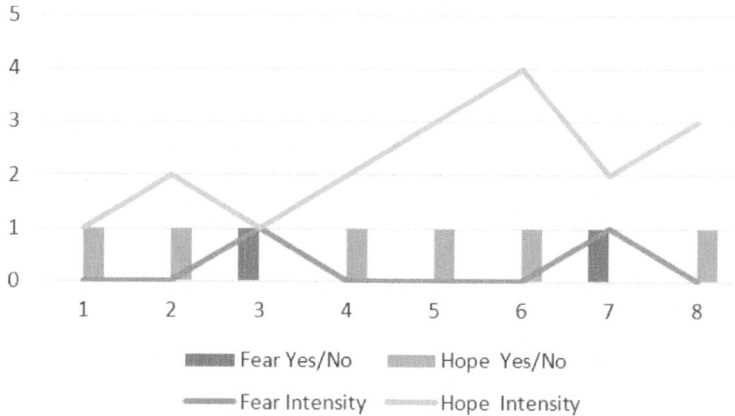

Fig. 6. Representation of the interaction between the police officer and prisoner 1.

Table 4. Data from Fig. 6, presented the dialogue 1 for prisoner 1.

Emotion	Dialogue 1 - Prisoner 1									
Fear	Yes/No	0	0	1	0	0	0	1	0
	Intensity	0	0	1	0	0	0	1	0
Hope	Yes/No	1	1	0	1	1	1	0	1
	Intensity	1	2	1	2	3	4	2	3

Table 5 shows the intensity values of the emotions for each interaction in the dialogue presented in the graph of Fig. 7.

Table 5. Data from Fig. 7, presented the dialogue 1 for prisoner 2.

Emotion	Dialogue 1 - Prisoner 2							
Fear	Yes/No	1	1	1	0	0	1	0	1
	Intensity	1	2	3	1.5	0.75	1.75	0.87	1.87
Hope	Yes/No	0	0	0	1	1	0	1	0
	Intensity	0	0	0	1	2	1	2	1

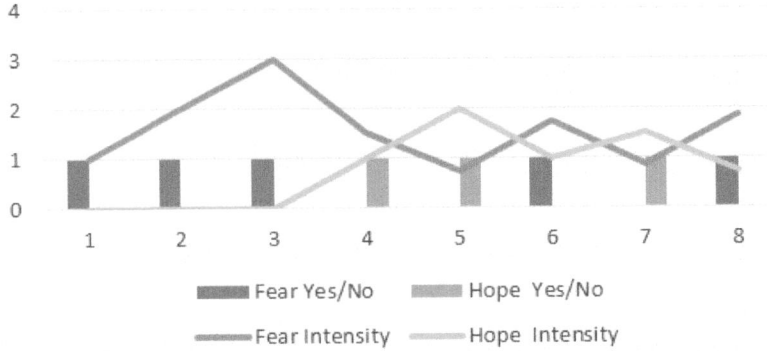

Fig. 7. Representation of the first dialogue between the police officer and prisoner 2.

5   Conclusions and Future Work

In this work, we proposed a system that uses Watson to identify emotions, through the OCC model, in the exchange of messages between agents, which would assist in the development of intelligent agents with the ability to understand and express emotions.

The OCC model was chosen as the basis for structuring emotions, as it tries to simplify the development of a model focusing on various cognitive aspects of emotions. Watson, on the other hand, was chosen because of its capacity to work with natural language processing. One contribution of this work was the integration of the OCC model, the Watson NLP tool, and the Jason platform. The project contributed with a practical approach to enable the study of emotions in multi-agent systems.

Through the conducted studies, it became evident that several tools are essential for defining the requirements of an architecture where emotions and memory play integral roles in agents' behavior. These include an artificial intelligence capable of identifying emotions in dialogue, utilizing a validated emotion model like OCC; a functional and adaptable multi-agent system platform, facilitating integration with other systems; and a set of rules derived from the functioning of human memory.

The results obtained indicate that the work successfully achieved its objective, defining an architecture that incorporates concepts of memory and emotions within multi-agent systems, coupled with conducting experiments to analyze its functionality. Notably, the distinctive feature of the presented architecture lies in its integration of human memory and emotional concepts into a multi-agent system, enabling the simulation of scenarios requiring these elements. For instance, the adaptation of the Prisoner's Dilemma in this work illustrates the theory that establishing rapport through positive emotions enhances the likelihood of a guilty suspect confessing.

In this way, we conclude that the use of emotions is promising, since the proposed system was able to identify emotions in the interaction between agents. We believe that this type of approach can be an innovative way to understand, develop, manage, and maintain complex systems that use emotions.

The future works outlined by this study include: (I) the implementation of personality traits in agents, potentially leveraging the OCEAN model (Big Five) [2] to enhance agents' behavioral diversity and realism; (II) the refinement of the calculation methodology employed for memory within the architecture. Additionally, integrating other memory concepts like explicit, implicit, transient, sensory, prospective, episodic, or semantic memory could enrich the model's fidelity. Developing equations to represent the functioning of each memory type would be crucial for achieving this goal; (III) conducting simulations involving human participants to evaluate the architecture's performance. Integration with an input and output system would enable users to interact with the system. This interaction could take two forms: presenting users with predefined options to choose from for message selection or allowing free-form text input for open-ended conversations. Using Watson's capabilities, the architecture could process text inputs from users and respond accordingly, facilitating human-system interaction.

References

1. Bratman, M.: Intention, plans, and practical reason, vol. 10. Harvard University Press Cambridge, MA (1987)
2. Digman, J.M.: Personality structure: emergence of the five-factor model. Annu. Rev. Psychol. **41**(1), 417–440 (1990)
3. Ferrucci, D.A.: Introduction to "this is watson." IBM J. Res. Developm. **56**(3.4), 1–1 (2012)
4. Gratch, J., Marsella, S.: Modeling emotions in the mission rehearsal exercise. In: Proceedings of the 10th Conference on Computer Generated Forces and Behavioral Representation, pp. 15–17 (2001)
5. Gudjonsson, G.H.: The psychology of interrogations and confessions: a handbook. John Wiley & Sons (2003)
6. Kintsch, W.: The representation of meaning in memory (PLE: Memory). Psychology Press (2014)
7. Langley, C.: Getting to the Truth: A Practical. Scientific Approach to Behaviour Analysis for Professionals, Emotional Intelligence Academy (2017)
8. Makridakis, S.: The forthcoming artificial intelligence (ai) revolution: its impact on society and firms. Futures **90**, 46–60 (2017)
9. Marsella, S., Gratch, J., Petta, P., et al.: Computational models of emotion. Blueprint Affective Comput. Sourcebook Manual **11**(1), 21–46 (2010)
10. Norvig, P., Russell, S.: Inteligência artificial: Tradução da 3a edição, vol. 1 (2014)
11. Ortony, A., Clore, G.L., Collins, A.: The cognitive structure of emotions. Cambridge University Press (1990)
12. Poundstone, W.: Prisoner's Dilemma/John Von Neumann, game theory and the puzzle of the bomb. Anchor (1993)
13. da Silva, B.C.D.: O estudo lingüístico-computacional da linguagem. Letras de Hoje **41**(2), 103–138 (2006)

Explaining Task Delegation Through Argumentation Debates with Votes

Jeferson José Baqueta[✉][iD] and Cesar A. Tacla[iD]

Programa de Pós-Graduação em Engenharia Elétrica e Informática Industrial,
Universidade Tecnológica Federal do Paraná (UTFPR), Curitiba, Brazil
{jjbaqueta,tacla}@utfpr.edu.br

Abstract. In multi-agent systems (MAS), the task delegation process involves assigning tasks or responsibilities among the agents to optimize the system's overall performance. Nevertheless, choosing which agent must receive a specific task or responsibility can be seen as an exploitation/exploration problem. In the Multi-armed Bandits (MAB) context, such a problem can be solved by deciding which agents (partners) should be exploited or explored concerning a certain task. However, for a system composed of several agents, it is important that such a choice be made based on a delegation model capable of providing mechanisms to explain the agents' decision-making process, making the partner selection more transparent and understandable. This feature is especially desirable when different partners are available to execute a task, resulting in a large set of combinations of partners and tasks. In this paper, we introduce a Multi-armed Bandit (MAB)-based delegation model that can optimize the partner selection process and explain the agents' choices. Our approach allows agents to explain their partner choices based on quantitative argumentation with votes (QuAD-V). We validate our model by simulating petroleum product distribution via pipelines, where agents represent temporary storage bases in a complex delegation chain. The results demonstrate the effectiveness of our model in optimizing delegation decisions while maintaining clear, understandable explanations for agents' decisions.

Keywords: Task delegation · Augmentation with votes · Explainability

1 Introduction

In a multi-agent system (MAS), task delegation is a fundamental mechanism adopted by agents to solve problems that involve teamwork [7,14]. Such a mechanism contributes to agents accomplishing private or collective objectives beyond their capabilities [11,19,28,30]. A critical stage of task delegation is partner selection, where the agents must choose partners capable of performing the tasks delegated to them. Generally, this process is oriented by the trust degree associated

Supported by CAPES and CNPq (process 409523/2021-6).

with each partner [8]. In particular, the trust establishment relies on observations about the partners' social behavior and the environmental conditions [23,24].

Nevertheless, traditional task delegation approaches often lack the transparency required for users to trust and understand the decision-making processes of the agents [1,8,14]. The capability to produce explanations that justify the agents' choices is essential for the development of complex systems composed of several agents [29]. In this situation, the adoption of an explainable delegation model allows the agents to provide understandable explanations about their decision-making process. This is crucial for the trust and acceptance of the system by users [26]. Furthermore, explanations help identify failures and improve the system [2,13,18]. Additionally, in fields such as medicine or finance, legal and ethical compliance requires that automated decisions be transparent and comprehensible [16].

In this context, we propose a novel delegation model based on the Multi-armed Bandit (MAB) framework, a reinforcement learning approach that aims to maximize the agents' cumulative rewards over time while balancing exploration and exploitation [12]. In order to incorporate explainability into the partner selection process, our model is capable of generating explanations about the agents' choices based on quantitative argumentation with votes (QuAD-V) [21]. This method allows an agent to determine if a partner should be chosen based on votes regarding the partner's behavior, which are produced by the partner's assessors as they interact with it. To validate the effectiveness of our model, we simulate a scenario involving the distribution of petroleum products via pipelines. In this scenario, the pipeline network is modeled as a series of delegation chains, where intermediate agents represent temporary storage bases that facilitate the transportation of products from one point to another within the network. Our results demonstrate that the proposed model is able to not only improve the system's performance but also provide comprehensible explanations for the agents' decisions regarding partner selection.

The rest of this paper is organized as follows. Section 2 presents the basic concepts adopted in this work. Section 3 presents our task delegation model, discussing the modeling details. Section 4 presents our experiments and the obtained results. The conclusions and future work are summarized in Sect. 5.

2   Background

2.1   Task Delegation

A task delegation scenario is defined by an agent (*delegator*) who wants to achieve its goal g but cannot execute the action a that leads to the accomplishment of g. Therefore, to achieve this goal, the delegator needs to delegate a task τ, which specifies the action a that must be performed, to a partner (*delegatee*) who is able to complete τ [7,8]. Nevertheless, when several partners are available to execute τ, each one with different capabilities and competencies, the delegator must select the most suitable partner among them to maximize its chances of achieving its goal. This decision process is called partner selection, as the delegator must choose its delegatee by considering factors such as competence, reputation, know-how, and the performance history of the partners.

2.2   Multi-Armed Bandits

The *multi-armed bandit* (MAB) is a machine learning paradigm that can be seen as a single-state reinforcement learning approach [12]. In the multi-armed bandit (MAB) problem, a single agent repeatedly chooses from a finite set of options, aiming to maximize its obtained reward. An option is termed an *arm*, while the act of choosing an arm is termed an *arm-pull* [3]. Pulling an arm i n times will yield the rewards $\mu_{i,1}, \mu_{i,2}, \ldots, \mu_{i,n}$, which are independent of each other and associated with an unknown probability distribution. The decision about which arm to pull is based on a MAB policy [3,9,15]. Such a policy aims to identify a sequence of arm pulls that maximizes the received rewards over time. One of the most popular MAB policies is the ϵ-*greedy* policy [25]. This policy, based on the value of a random variable ϵ, selects the arm that yields the highest expected reward so far or picks a random arm otherwise [1].

2.3   Social Evaluations

Social control mechanisms allow the agents to make their decisions based on the others' social behavior, offering to agents means to punish undesirable behaviors by themselves [20]. Trust and reputation models are considered good solutions concerning social control [8,10,14,17,23]. In particular, the trust an agent places in a partner (required to delegate a task to it) can be updated through social evaluations, such as social image, reputation, and references. The *social image* consists of evaluative beliefs about a partner's competencies. These beliefs are produced from direct experiences, expressing a personal opinion about the partner [20]. *Reputation* is a meta-belief created based on third-party opinions. Its difference regarding the social image is the lack of commitment to the truth, generalization, and loss of reference [20,23]. At last, *references* can be seen as a type of reputation where a partner can share social evaluations about itself. The partner stores such evaluations as it interacts with other agents, similar to job references [17].

2.4   QuAD-V Framework

The QuAD-V framework has been proposed by [21] as an extension of the quantitative argumentation debate framework (QuAD) [6]. Its main advantage is the possibility to solve a debate using a voting system, where a set of users vote for or against arguments. As presented in [21], a QuAD-V is a *6-tuple* (A, C, P, R, U, V), in which A is a finite set of answer arguments, C is a finite set of con arguments, P

is a finite set of pro arguments, $R \subseteq (C \cup P) \times (A \cup C \cup P)$ is an acyclic binary relation, U is a finite set of users, and $V : U \times (A \cup C \cup P) \rightarrow \{-, ?, +\}$ is a total function, such as $V(u, a)$ is the vote of user $u \in U$ on argument $a \in (A \cup C \cup P)$.

In the QuAD-V framework, the arguments can attack or support one another. The attackers and supporters of an argument are defined based on the con and pro arguments, respectively. Thus, for any argument $a \in (A \cup C \cup P)$, the set of *attackers* of a is $R^-(a) = \{b \in C | (b, a) \in R\}$ and the set of *supporters* of a is $R^+(a) = \{b \in P | (b, a) \in R\}$. Besides, each argument a has a vote base score $\tau_v : A \cup C \cup P \rightarrow \mathbb{I}$ (for scale $\mathbb{I} = [0, 1]$), which is computed according with the users voting, such as following:

$$\tau_v(a) = \begin{cases} 0.5 & if\ |U| = 0 \\ 0.5 + (0.5 * \frac{N^+(a) - N^-(a)}{|U|}) & if\ |U| \neq 0 \end{cases} \quad (1)$$

where, $N^+(a)$ is a counter that sums the positive votes for a, such that, for any argument $a \in (A \cup C \cup P), N^+(a) = |V^+(a)|$, and $V^+(a) = \{u \in U : V(u, a) = +\}$ is the set of users voting for a. Whereas, $N^-(a)$ is a counter that sums the negative votes for a, such that, for any argument $a \in (A \cup C \cup P), N^-(a) = |V^-(a)|$, and $V^-(a) = \{u \in U : V(u, a) = -\}$ is the set of users voting against a.

A final score is recursively computed for each argument based on their attack and support relationships. This process is performed from the leaf arguments up to the answer arguments. Specifically, the final score of an argument determines its degree of acceptability (strength), for example, 1 for accepted, 0.5 for neutral, and 0 for rejected. The Discontinuity-Free QuAD (DF-QuAD) [22] is an example of an algorithm that aggregates the base scores of arguments with the strength of their attackers and supporters to produce their acceptability degree.

An example of the *QuAD-V* framework is presented in Fig. 1, where the answer argument indicates whether a partner should be selected to execute a task or not. Note that the acceptability of the answer relies on a series of supporting and attacking arguments.

3   Delegation Model

We model task delegation as an exploitation/exploration problem. In this scenario, a delegator must decide whether to delegate a task to a known partner (exploitation), expecting a likely good outcome, or to an unknown partner (delegatee) in the hope of achieving better results (exploration). Therefore, the partner selection process is modeled as a MAB approach [27], where the delegator interacts with its partners (arms that can be pulled) to identify those most likely to maximize the expected rewards.

The expected reward for a partner determines its likelihood of being selected as a delegatee, which reflects the partner's probability of successfully completing a task while minimizing the delegator's frustration with the task outcomes. Such a likelihood is computed using the *QuAD-V* framework shown in Fig. 1. In particular, each partner maintains its own *QuAD-V* framework. After a task is

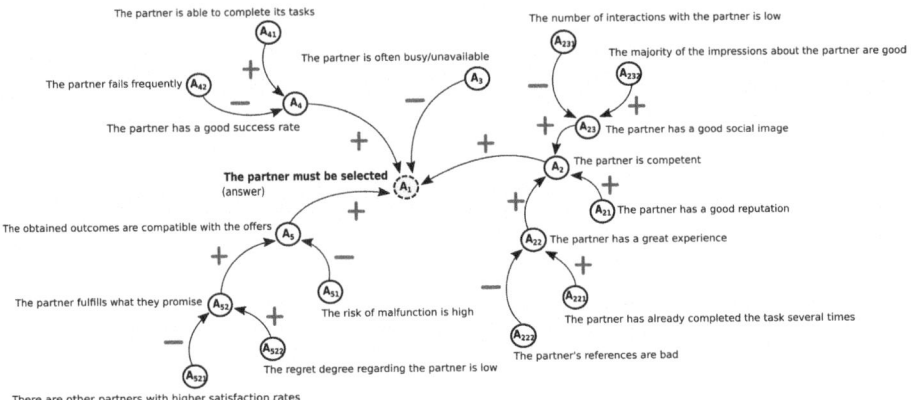

Fig. 1. Example of a QuAD-V Framework used by a delegator to determine whether a partner is suitable to perform a task based on collective voting.

completed, the delegator votes to express agreement or disagreement with the attackers and supporters of each argument within the partner's *QuAD-V* framework. The strength of the answer argument determines whether the delegator will choose this partner again for future tasks. The voting performed by a delegator (α) regarding a task (τ) executed by a delegatee (β) is defined based on four decision factors:

- Partner's competencies (argument A_2): $Comp(\alpha, \beta, \tau) \in [0,1]$ is an estimation made by α regarding β's abilities and experiences for a given task τ. This estimation is based on β's social image, reputation, and know-how, which are computed through impressions formed during interactions between α and β. An impression, represented as $\langle \alpha, \beta, \tau, t, S \rangle$, includes ratings by α on β's behavior related to τ's criteria (e.g., cost, quality, or time). Social image aggregates α's impressions of β, reputation is calculated by aggregating the impressions about β's behavior produced by other agents (shared evaluations), and know-how combines β's references. Weighted mean aggregation summarizes all these impressions into a single value [24], resulting in β's competence measure.
- Partner's availability (argument A_3): ($Availability(\beta, \tau) \in [0,1]$) indicates how often β was unavailable to execute a certain task τ in the past. This measure is calculated by dividing the number of times β denied requirements for executing τ by the total number of requirements received by β for executing τ.
- Partner's success rate (argument A_4): ($Succ_{Rate}(\alpha, \beta, \tau) \in [0,1]$) expresses how successful a partner β has been in performing a task (τ) delegated by α over time. The success of β depends on completing τ. Thus, a success rate of 0 for β implies that it has never completed τ, whereas a value of 1 indicates that β has completed τ every time it attempted the task. The success rate of

β is determined by dividing the number of times the partner has successfully completed the delegated task by the total number of attempts.

- Delegator's satisfaction (argument A_5): ($Sat_{Rate}(\alpha, \beta, \tau) \in [0, 1]$) represents α's satisfaction regarding the outcome obtained after β executes τ. This measure represents β's capability of meeting α's expectations regarding the execution of τ. The higher a delegator's satisfaction, the lower the regret with its choice of partner. The satisfaction measure is calculated based on the difference between the partner's promises and the task outcomes [5].

The value of a decision factor determines whether a delegator will vote for or against the argument associated with this factor. Specifically, a delegator will vote for an argument if its respective decision factor is equal to or greater than a threshold value ($\sigma \in [0, 1]$); otherwise, it will vote against the argument. For the attackers and supporters of such an argument, the delegator's voting will be performed using a rational voting pattern. According to [21], in a QuAD-V framework, the users must vote rationally to ensure the coherence of the voting process. For instance, a user is voting irrationally when agrees with some argument A_i, agrees with one of its attackers, but does not agree with any of its supporters. Therefore, to ensure the rationality of the voting process employed by the agents, we are assuming that agents are totally for or totally against a point of view. In the first case, when a delegator agrees with an argument A_i (voting for it), it also agrees with all its supporters and disagrees with all its attackers (voting against them). When a delegator disagrees with an argument A_i, it also agrees with all its attackers and disagrees with all its supporters.

4 Experiments

4.1 Case Study: Oil Pipelines

The problem discussed in this paper focuses on a multi-modal transportation system for oil-derived products [4]. The problem is modeled as a graph $G(B, A)$ where B is a finite set of bases b, and A is a finite set of arcs a. Each arc $a \in A$ connects a pair of bases (b_i, b_j), where $b_i, b_j \in B$ are respectively source and destination bases. Product transfers between bases occur through a set routes, each route being a sequence of arcs forming a valid path from a source base b_i and a destination base b_j (delegation chain). Our experiments were conducted using a graph structure composed of 10 agents (bases), as shown in Fig. 2. Note that the agents can play the role of root, inner, or terminator. A root agent can only send products to others, an inner agent can receive and send products, and a terminator can only receive products from others. For simplicity, we assume that the agents transport only one type of product, in this case, a quantity finite of oil.

4.2 Simulation Setup

The simulation involves transporting a quantity of product from a root agent (origin base) to a terminator agent (destination base). As shown in Fig. 2, each

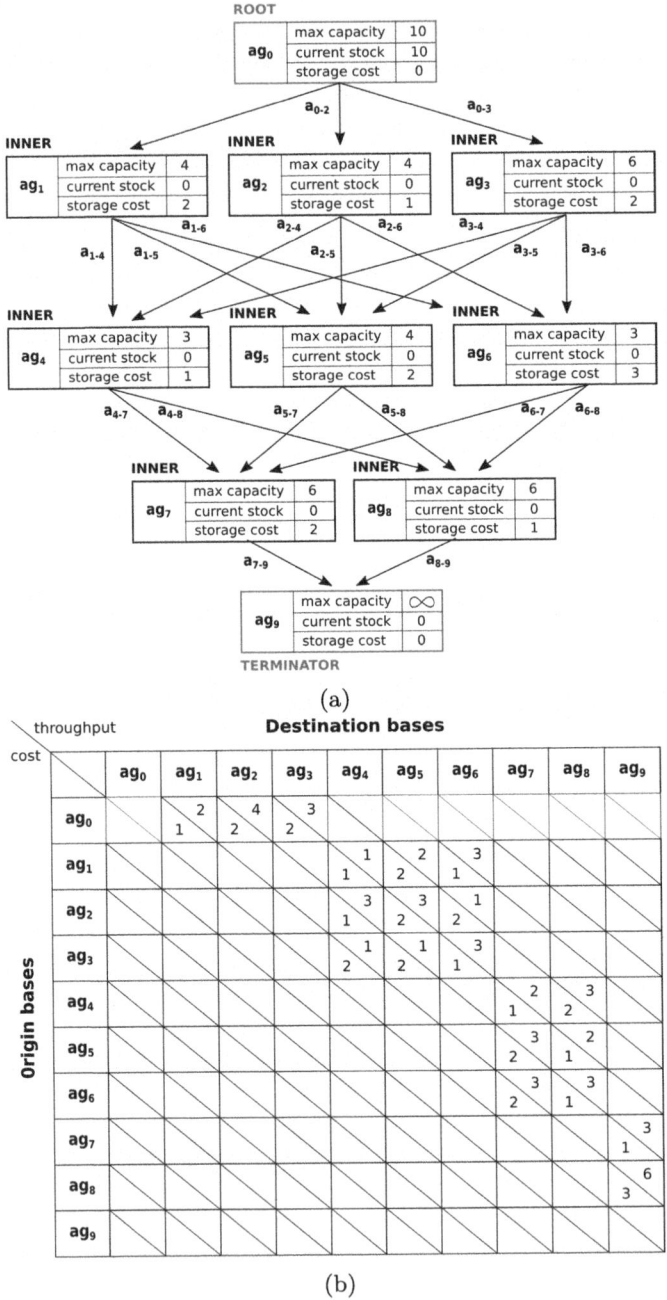

Fig. 2. Transportation bases: a) the initial states of the bases concerning their maximum capacity, current stock, and cost of storage per unit, and b) throughput and cost for each connection between two distinct origin and destination bases.

base can store a finite quantity of product (current stock). Note that moving a quantity of product between bases incurs a cost (operational costs), composed of storage and transportation costs, and takes time, which is calculated based on the time required to move the product considering the throughput of the arc that connects the bases. In this context, we assume that a task is defined by the action of transporting a quantity of product from one base to another and is assessed according to three criteria: the quantity of product to be transported, the cost of transporting and storing the product, and the time required to transport the product between the bases (from a delegator to a delegatee).

The delegation process begins when a delegator informs its partners about its intention to send a specific quantity of product. Upon receiving this notification, each partner responds with an offer (promise), which represents its expectations regarding the quantity of product that it will store, the operational cost, and the required time to execute the task. If the partner lacks available storage space or is currently executing a task for another agent, it declines the delegator task due to its unavailability. We remark that when a partner is chosen as a delegatee, it may not fulfill its promise (*e.g.*, taking longer than promised to store the product). This failure reduces the delegator's satisfaction with the task outcome, leading to regret about its partner choice[1].

After receiving offers from the partners, the delegator selects its delegatees based on their respective expected rewards. It is important to remark that when a delegatee completes its task, the delegator performs its vote and produces an impression considering the delegatee's performance (*i.e.*, calculating the success rate of the delegatee and the degree of satisfaction with the task performed by it). The delegator's impression is shared with the delegatee (reference) and sent to other delegators who may interact with the delegatee in the future (shared evaluation). This delegation process starts with the root and spreads through the network until it reaches the terminator agent, indicating the end of an iteration. During an iteration, a delegator can perform several delegation actions because its goal is to reduce the quantity of product stored in its stock to zero.

4.3   Results

MAB Policy. In our evaluation, partner selection is performed using a variation of ϵ-*greedy* algorithm, in which the parameter ϵ depends on the number of times a delegator α performed a task τ ($|(\alpha, \tau)|$) by mean a delegation action [25]:

$$\epsilon = \frac{1}{|(\alpha, \tau)| + 1} \tag{2}$$

where the probability of exploration becomes higher as ϵ approaches 1; otherwise, α tends to exploit its well-known options. Such an approach prevents α from making random choices in a situation where all possible options of partners

---

[1] The degree of regret of a delegator is defined as the difference between the satisfaction that could be obtained if the best possible combination of partners were selected and the satisfaction obtained with its current choice of partners.

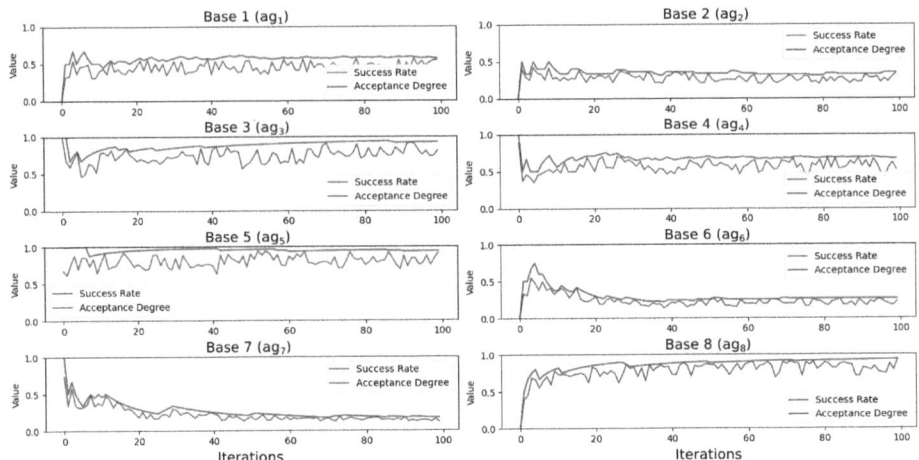

Fig. 3. Performance evaluation of agents (bases) acting as delegatees over 100 iterations, considering the success rate and acceptability degree.

have already been explored a considerable number of times, such as happens in strategies where ϵ receives a fixed value [1,25]. In the case of exploitation, α selects their partners by estimating their respective expected rewards concerning τ. The higher the expected reward associated with a partner, the higher the chance of it being chosen as an α's delegatee.

Simulation Outcomes and Discussions. Our simulation consisted of 100 iterations. In each iteration, the agents repeatedly delegate their tasks using the network presented in Fig. 2 as input (*i.e.*, respecting the original quantity of product stored for each agent in the network). However, the votes performed by delegators, as well as their impressions, are kept from one iteration to another, serving as a memory of their partners' behaviors.

In Fig. 3, the results obtained from the simulation are presented. Note that the success rate and the acceptance degree (*i.e.*, strength of the answer argument of a partner's QuAD-V framework) over iterations are shown for each inner agent (agents able to act as partners). Note that the curve of the success rate tends to follow the curve of the acceptance degree because, as expected, partners with higher success rates tend to be the best partners. At last, it is possible to notice that bases 3, 5, and 8 consistently displayed high performance (*i.e.*, high success rate with a high acceptance degree), indicating a good route for product transportation.

We emphasize that the success rate and the acceptance degree were obtained at the end of each iteration, indicating the general performance of a partner throughout each iteration. Nevertheless, we would like to point out that the strength of an answer argument is computed by the delegator during the partner selection process as a way to estimate the expected reward associated with a partner; however, at this point in the simulation, the obtained value depends on the number of votes tabulated at the moment of the strength computation.

Finally, the Fig. 4 shows the explanations produced to justify the behavior of one of the bases on the best route, in this case, base 8. This explanation was generated considering the agent's acceptance degree at the end of iteration 80, in which both the acceptance degree and success rate are high. We highlight that the process used to produce this explanation can be adopted to justify the choice of partner for any delegator. In particular, when a delegator selects a partner, it needs to estimate the partner's expected reward by computing the strength of the partner's answer argument. As discussed in Sect. 2, this process is performed recursively based on the attackers and supporters of arguments of the partner's QuAD-V framework (Fig. 1). Therefore, by using the strength of each argument in a partner's QuAD-V framework, the delegator can justify its choice of delegatee, indicating the reasons (arguments) that led it to select a particular partner as delegatee and reject others. Thus, in this experiment, the explanation for an argument is produced by selecting the attack and support arguments with a strength higher than a threshold factor ($\lambda \in [0,1]$). In our case, the value assigned to λ is 0.5.

The partner must be selected (argument accepted) :-
  (The partner has a good success rate :- The partner is able to complete its tasks)$\wedge$
  (The obtained outcomes are compatible with the offers :-
    (The partner fulfills what they promise :- The regret degree regarding the partner is low))$\wedge$
  (The partner is competent :-
    (The partner has a good reputation$\wedge$
      (The partner has a great experience :- The partner has already completed the task several times)$\wedge$
      (The partner has a good social image :- The majority of the impressions about the partner are good))

Fig. 4. Explanation produced through the delegators' voting concerning the behavior of agent 8 (base 8) at end of iteration 80.

5 Conclusions

In this paper, we present an easy way to generate explanations using the QuAD-V framework in a task delegation scenario. In addition to providing explanations, we also propose a delegation model based on a multi-armed bandit approach that employs the strength of arguments to produce a reward measure, which can be used to select good partners in partner selection applications. Therefore, we believe that the key contributions of this paper are twofold:

- The development of an MAB-based delegation model that optimizes partner selection;
- The integration of explainable AI techniques to provide transparent decision-making.

For future work, we intend to investigate how to use the QuAD-V framework to justify the selection of optimal routes in product transportation scenarios like the one presented in this work.

References

1. Afanador, J., Baptista, M.S., Oren, N.: Algorithms for recursive delegation. AI Commun. **32**(4), 303–317 (2019)
2. Alsaigh, R., Mehmood, R., Katib, I.: Ai explainability and governance in smart energy systems: a review. Front. Energy Res. **11**, 1071291 (2023)
3. Auer, P., Cesa-Bianchi, N., Fischer, P.: Finite-time analysis of the multiarmed bandit problem. Mach. Learn. **47**(2), 235–256 (2002)
4. Banaszewski, R.F., Arruda, L.V., Simão, J.M., Tacla, C.A., Barbosa-Póvoa, A.P., Relvas, S.: An application of a multi-agent auction-based protocol to the tactical planning of oil product transport in the Brazilian multimodal network. Comput. Chem. Eng. **59**, 17–32 (2013)
5. Baqueta, J., Tacla, C.A.: A task delegation model for delegation chains. In: C-MAS Workshop, pp. 1–7 (2023)
6. Baroni, P., Romano, M., Toni, F., Aurisicchio, M., Bertanza, G.: Automatic evaluation of design alternatives with quantitative argumentation. Argument Comput. **6**(1), 24–49 (2015)
7. Cantucci, F., Falcone, R., Castelfranchi, C.: A computational model for cognitive human-robot interaction: an approach based on theory of delegation. In: WOA, pp. 127–133 (2019)
8. Castelfranchi, C., Falcone, R.: Trust Theory: A Socio-Cognitive and Computational Model, vol. 18. John Wiley & Sons, Hoboken (2010)
9. Chapelle, O., Li, L.: An empirical evaluation of thompson sampling. Adv. Neural Inf. Process. Syst. **24** (2011)
10. Cho, J.H., Chan, K., Adali, S.: A survey on trust modeling. ACM Comput. Surv. (CSUR) **48**(2), 1–40 (2015)
11. Cui, Y., Idota, H., Ota, M.: Improving supply chain resilience with implementation of new system architecture. In: 2019 IEEE Social Implications of Technology (SIT) and Information Management (SITIM). pp. 1–6. IEEE (2019)
12. Drugan, M.M., Nowe, A.: Designing multi-objective multi-armed bandits algorithms: a study. In: The 2013 International Joint Conference on Neural Networks (IJCNN), pp. 1–8. IEEE (2013)
13. Endsley, M.R.: Supporting human-AI teams: transparency, explainability, and situation awareness. Comput. Hum. Behav. **140**, 107574 (2023)
14. Griffiths, N.: Task delegation using experience-based multi-dimensional trust. In: Proceedings of the Fourth International Joint Conference on Autonomous Agents and Multiagent Systems, pp. 489–496 (2005)
15. Hayes, C.F., et al.: A practical guide to multi-objective reinforcement learning and planning. Auton. Agent. Multi-Agent Syst. **36**(1), 1–59 (2022). https://doi.org/10.1007/s10458-022-09552-y
16. Herrmann, M., Wabro, A., Winkler, E.: Percentages and reasons: AI explainability and ultimate human responsibility within the medical field. Ethics Inf. Technol. **26**(2), 26 (2024)
17. Huynh, T.D., Jennings, N.R., Shadbolt, N.: Fire: an integrated trust and reputation model for open multi-agent systems. In: In Proceedings of the 16th European Conference on Artificial Intelligence (ECAI), pp. 18–22 (2004)
18. Karim, M.R., et al.: Explainable AI for bioinformatics: methods, tools and applications. Brief. Bioinf. **24**(5), bbad236 (2023)
19. Manavalan, E., Jayakrishna, K.: A review of internet of things (iot) embedded sustainable supply chain for industry 4.0 requirements. Comput. Ind. Eng. **127**, 925–953 (2019)

20. Pinyol, I., Sabater-Mir, J.: Computational trust and reputation models for open multi-agent systems: a review. Artif. Intell. Rev. **40**(1), 1–25 (2013)
21. Rago, A., Toni, F.: Quantitative argumentation debates with votes for opinion polling. In: An, B., Bazzan, A., Leite, J., Villata, S., van der Torre, L. (eds.) PRIMA 2017. LNCS (LNAI), vol. 10621, pp. 369–385. Springer, Cham (2017). https://doi.org/10.1007/978-3-319-69131-2_22
22. Rago, A., Toni, F., Aurisicchio, M., Baroni, P.: Discontinuity-free decision support with quantitative argumentation debates. In: Fifteenth International Conference on the Principles of Knowledge Representation and Reasoning (2016)
23. Sabater, J., Paolucci, M., Conte, R.: Repage: reputation and image among limited autonomous partners. J. Artif. Societ. Soc. Simul. **9**(2) (2006)
24. Sabater, J., Sierra, C.: Regret: reputation in gregarious societies. In: Proceedings of the Fifth International Conference on Autonomous Agents, pp. 194–195 (2001)
25. Sutton, R.S., Barto, A.G.: Reinforcement Learning: An Introduction. MIT press, Cambridge (2018)
26. Theis, S., Jentzsch, S., Deligiannaki, F., Berro, C., Raulf, A.P., Bruder, C.: Requirements for explainability and acceptance of artificial intelligence in collaborative work. In: International Conference on Human-Computer Interaction, pp. 355–380. Springer, Heidelberg (2023). https://doi.org/10.1007/978-3-031-35891-3_22
27. Turgay, E., Oner, D., Tekin, C.: Multi-objective contextual bandit problem with similarity information. In: International Conference on Artificial Intelligence and Statistics, pp. 1673–1681. PMLR (2018)
28. Xing, L.: Reliability in internet of things: current status and future perspectives. IEEE Internet Things J. **7**(8), 6704–6721 (2020)
29. Yao, B.M., Shah, A., Sun, L., Cho, J.H., Huang, L.: End-to-end multimodal fact-checking and explanation generation: a challenging dataset and models. In: Proceedings of the 46th International ACM SIGIR Conference on Research and Development in Information Retrieval, pp. 2733–2743 (2023)
30. Yliniemi, L., Agogino, A.K., Tumer, K.: Multirobot coordination for space exploration. AI Mag. **35**(4), 61–74 (2014)

Natural Language Processing

Te Ahorré Un Click: A Revised Definition of Clickbait and Detection in Spanish News

Gabriel Mordecki[1]([✉]), Guillermo Moncecchi[1], and Javier Couto[2]

[1] Universidad de la República, Montevideo, Uruguay
{gabriel.mordecki,gmonce}@fing.edu.uy
[2] PEDECIBA, Montevideo, Uruguay

Abstract. We revise the definition of clickbait, which lacks current consensus, and argue that the creation of a curiosity gap is the key concept that distinguishes clickbait from other related phenomena such as sensationalism and headlines that do not deliver what they promise or diverge from the article. Therefore, we propose a new definition: *clickbait is a technique for generating headlines and teasers that deliberately omit part of the information with the goal of raising the readers' curiosity, capturing their attention and enticing them to click*. We introduce a new approach to clickbait detection datasets creation, by refining the concept limits and annotations criteria, minimizing the subjectivity in the decision as much as possible. Following it, we created and release TA1C (for *Te Ahorré Un Click*, Spanish for *Saved You A Click*), the first open source dataset for clickbait detection in Spanish. It consists of 3,500 tweets coming from 18 well known media sources, manually annotated and reaching a 0.825 Fleiss' κ inter annotator agreement. We implement strong baselines that achieve 0.84 in F1-score.

Keywords: clickbait · clickbait definition · clickbait detection · corpus · news articles · social media · spanish · natural language processing

1 Introduction

Even though clickbait is a widely employed term, there remains a lack of consensus regarding its precise definition. Despite the different criteria about what clickbait is, there is some common ground: it differs from traditional headlines in their objective and style, leaving behind the goal of informing in favor of attracting attention [4,14,26,27]. Journalists use clickbait to lure, manipulate, bait the reader into clicking, as the name implies [4,6]. This phenomenon is also often linked to loss of journalistic quality [15,24,27] and the creation of disappointment among readers; and even when it does not, it is very hard to overlook because it captures our attention by creating curiosity.

L. Correia et al. (Eds.): IBERAMIA 2024, LNCS 15277, pp. 387–399, 2025.
https://doi.org/10.1007/978-3-031-80366-6_32

Clickbait started in low-reputation web-exclusive media that focused on political propaganda or soft-news, such as The Huffington Post, Buzzfeed and Upworthy [35], or even bad quality advertising such as in chumboxes[1]. However, it has recently gained prominence across all types of news and media.

Many authors define clickbait as a type of content [4,26] exclusively tying it to soft-news, yellow journalism and especially sensationalism, even to the point of considering them synonyms [17]. Also, several definitions make focus on that deceptive effect [5,11,17,30] created by the news failing to deliver what they promise, a notion created by examples like the first one in Fig. 1.

Fig. 1. Three examples of clickbait news. Left: *COVID-19 — Will vaccination against coronavirus be mandatory?* [in linked note:] *Meanwhile, in Peru, President Martín Vizcarra did not address the mandatory nature of vaccination last Thursday [...]* An example of clickbait where it disappoints by not answering the information-gap it created. Center: *What does the letter with which Evo Morales submitted his 'forced resignation' say?*, clickbait on hard news. Right: *Argentina's unofficial dollar: find out here its price today, Friday, February 26, 2021.* [in linked note:] *The so-called 'blue dollar' was trading at 145 pesos in the Argentine market [...]* A case where a clickbait news delivers exactly what it promises.

These two attributes were essential in the early stages, but they are no longer a requirement for a headline to be clickbait, especially since it was adopted by many of the most reputable and prestigious media sources. Currently, there are plenty of cases of hard news that are clearly clickbait, such as the second example in Fig. 1 about the words of a president overthrown by a coup d'état. Also, many teasers create curiosity and then answer exactly what they promise in the news text. Some of those news concern very concrete data like in the third news presented in the Fig. 1, along with the article's answer, and some answer in long articles with good journalistic quality like *Africa eradicates polio: How it achieved this historic milestone.*[2]

All of the available clickbait detection datasets that we have knowledge of are built assuming some of these characteristics or without following a concrete definition. In [28] Potthast et al. describe three distinct methods for creating them. The first approach, named the *reputation method* involves collecting teasers from media sources with established good reputations and categorizing all of them as

---

[1] https://en.wikipedia.org/wiki/Chumbox.
[2] All of the headline and tweet examples are from the TA1C dataset, except when noted, and have been translated to English by the authors.

non-clickbait whereas teasers from notorious clickbait-oriented media or websites are categorized as clickbait. The second approach is the *gatekeeper method* that relies on curated third-party forums dedicated to exposing clickbait in platforms like *Reddit* and *Twitter*. It consists of assigning all of them to the positive class, while teasers from news-focused forums that explicitly avoid clickbait are collected as the negative class. Lastly, the *importance method* involves the systematic sampling of teasers based on a predefined criteria of relevance and then manually label them.

These first two ways of creating datasets, however, lead to differences in the two classes in much more than their samples being clickbait or not. In the reputation based method, texts come from different media outlets that typically have different writing styles and do not cover the same topics. They even usually consider media that publish clickbait as part of the negative class. In a similar way, gatekeeper based methods create datasets that rely on the criteria of the selected intermediaries to choose which news to publish in their forums. Again, they typically differ in source and topics.

It is the third approach where a precise clickbait definition and annotation criteria becomes crucial. In this paper, we argue that the information gap theory [23] that explains why curiosity arises is not only relevant but also the fundamental difference between clickbait and other related phenomena such as sensationalism. Its essence is not about the content of the news, the language employed, or the creation of disappointment; rather, it is withholding information to provoke curiosity. We propose a new formal definition for the term in Sect. 3.

Based on this definition, we introduce a new approach to clickbait detection datasets creation. The Clickbait Challenge corpus construction [28] tackled the problem of having different perspectives of what clickbait is by relying on many annotations in a non-binary scale, their annotation task showed multiple clickbait definitions and a few examples, allowing each annotator to establish their own criteria on the boundaries of what is clickbait or not. Instead, we address these conflicting perspectives by defining clickbait as precisely as possible and refining the annotation criteria to minimize the subjectivity of the decision. Therefore, we reach a 0.825 Fleiss' κ inter-annotator agreement while reported agreements in the literature range from a Fleiss' κ of 0.36 up to 0.73 for binary annotation [8, 12, 22, 28], similar to those found in other subjective NLP tasks such as irony and humor detection.

As it usually happens in NLP, most of the publications on clickbait detection work with English-written news, and as far as we know there are no publicly available datasets in Spanish. We, therefore, create and publish TA1C: a dataset for clickbait detection in Spanish and implement baseline classification methods that reach an F1-score of 0.84 and average precision of 0.93 on the positive class.

2   Related Work

Clickbait detection works are heavily influenced by dataset construction. Some influential works in the area [1, 5, 11, 12, 30] report nearly perfect results. However, because of the way they created their datasets, machine learning models can learn to distinguish between the two classes based on features unrelated to clickbait, solving a significantly less challenging task instead.

As in the vast majority of problems in the NLP field, initial solutions were primarily traditional methods using features ranging from lexical (e.g., n-grams, word lists) [29] and syntactic features like amount of stopwords and length of syntactic dependencies [11] to tweet-related information (e.g., attachments, posting time) [29] to text-readability scores [5], forward-reference (a feature based in the seminal work by Blom and Hansen [6]) and distances between headline and article. Then, works transitioned to the prevalence of deep learning methods [1,25] and then to transformers architectures [34,36].

The Webis Clickbait Corpus, presented in [28] is the most important dataset on clickbait detection up to date. Many works use it, with the advantage that their results can be compared to each other. The best result was obtained by fine-tuning a RoBERTa model, reaching 0.74 of F1-Score in the positive class [17].

There are many works describing and analizing clickbait in Spanish in a qualitative way [2,4,14,15,27]. However, there are, to the best of our knowledge, no publicly available datasets in Spanish.

3   A Clickbait Definition

There is a general agreement that clickbait is a headline and/or teaser generation strategy that involves some sort of manipulation to bait the reader into clicking to easily increase traffic. Disagreements arise when it comes to determining which of the many methods that can be applied to increase the traffic are categorized as clickbait. It usually goes along with sensationalism because they are both answers to a usual question in the media: *How to easily increase traffic?*; however, they are two different responses to it. Unlike sensationalism, the core of the clickbait concept goes beyond what the news is about or what the article says: it only concerns how the news stories are presented.

We argue that the key feature that distinguishes clickbait from other forms of attracting attention is the explicit creation of a curiosity gap. As Loewenstein describes, curiosity is in one of its interpretations, ≪a form of cognitively induced deprivation that arises from the perception of a gap in knowledge or understanding≫ [23]. It is more intense when we know some about a topic, but not all about it [19], exactly what clickbait intends to generate. When curiosity arises, it has effects in our neurocognitive mechanisms, just as food-seeking or money does, creating a strong urge to act [3], even involuntarily [22]. It can lead us to irrationally seek information even when it is non-instrumental and non-relevant or even harmful, especially when we can get the answers immediately and with low cost [19,20,32].

Clickbait exploits these reward systems, making us susceptible to falling into an indulgent consumption of some news [31]. It can lead us to divert scarce resources such as time and attention to topics that don't actually interest us [32]. That is why it is so effective for the media to succeed in the attention economy, and generally so annoying for the general public.

The creation of a curiosity gap intended to attract attention is the key concept to distinguish clickbait from other headlines. Therefore, we propose a definition:

Clickbait is a technique for generating headlines and teasers that deliberately omit part of the information with the goal of raising the readers' curiosity, capturing their attention and enticing them to click.

Many works on clickbait cite the curiosity gap theory and some even agree that it is what makes clickbait work [16,22,31,33]. However, none of them creates a dataset using this definition.

3.1 Annotation Criteria

The proposed definition refers to the creation of an information gap as being deliberate. Since we cannot discern the author's intent, we must use an operational definition of clickbait to label the dataset: *teasers that omit part of the information not obvious by context, arising curiosity and enticing users to click.*

Even with this definition, some decisions are still left to individual interpretation. In the next paragraphs we describe some criteria, with the goal of better defining clickbait and reducing as much as possible the subjectivity in the annotation.

Take All the Context of the News Publication into Account. With this operational definition the problem of inferring the purpose of the writer disappears. However, a new one emerges: what is considered as *obvious by context?*

Given the limited space available to headlines, it is a common practice to omit obvious context and even words[3]. In a headline like *The massacre at Kabul airport results in at least 170 deaths*, and specially on Aug 27th, 2021, it is implied that readers will know what massacre they are talking about; the news is the update on casualties, not the event itself.

The boundaries between what is obvious and what is not can sometimes be blurry and subjective. The proposed criteria is to assume that anything within the common knowledge of an average reader of that publication is known, considering factors such as where, to whom, and when the news was released. In the previous example, the publication date is key, because the *the massacre at Kabul airport* took place the day before. When reading teasers from foreign media, the annotator must keep in mind that the target audience of the news will be familiar with the context, and assume it as known. Annotators can even seek additional information while tagging.

The headline *USA: A Donald Trump family scandal casts a shadow over the beginning of the Republican convention* is clearly clickbait: one wonders what the scandal is about. But in very similar case of *New York Governor Andrew Cuomo resigns amid the sexual harassment scandal* although it could be interpreted in the same way at first glance, it is actually referring to a widely known matter at

---

[3] It has even resulted in a specific jargon called Headlinese.

the time. None of the targeted public at the time wondered what scandal they were referring to, so in this case, it should not be considered clickbait.

There are instances when we require subtle contextual information to correctly judge if the omitted information is not so relevant in a news article and can be excluded for the sake of brevity, or it is so crucial that its absence would result in the creation of an information gap. The case of reports about professional athletes infected with COVID illustrates the point. In *Mexican golfer María Fassi tested positive for COVID-19* it is clear that the news is about who contracted it, especially knowing that she is a mexican athlete and it comes from mexican media. Contrarily, in *Juventud de las Piedras reports 17 COVID-19 positive cases* almost all of the team is sick, we should not expect the headline to list all of them, the relevant information is the number 17. These are clear cases of non-clickbait. In *A golfer tested positive for coronavirus during the PGA Tour tournament* some context is needed to judge: the tweet is from June 2020, it was the start of the pandemic and this represented the first case. We consider that the positive case is the news by itself, so it is not clickbait even if the name of the golfer is missing. Although *A Real Madrid player tested positive for coronavirus ahead of the match against Inter* seems the same, this news is from November, matches were played anyway, so the relevant part in this case is which Real Madrid's player will not be able to play: it is clickbait. Finally, in *NBA game postponed due to a positive COVID-19 case* whether it is news by itself or whether it is relevant to which player or game, really depends on the reader's interest in the NBA. As it is such a subjective case but the news could make sense by itself, the criteria is to tag it as non-clickbait.

Sensationalism is not Necessarily Clickbait. Clickbait is commonly mixed up with sensationalism, but it is not the same. Some news that could be labeled as clickbait in some other works, such as the headline *Leticia Brédice posted strong messages against her boyfriend: she accused him of being a 'thief' and 'bad person'* are tagged as negative in this dataset: there is no information gap.

In a similar fashion, if the news itself is what generates the curiosity rather than the presented information nor the writing style, it must not be considered clickbait. It is the case of *Video: Crocodile ate two sharks on an australian beach* where the headline details exactly what is found in the news, without omitting any crucial information.

Intense Adjectivation May Lead to Clickbait. There is a fine line when it comes to adjectivation: sometimes an excessive use of qualifications or an exaggeration is what creates the curiosity. The headline *How a space hurricane is, the spectacular phenomenon detected for the first time on Earth* is clickbait because it creates the information gap by not explaining at all why it is *spectacular*. It is very common in chronicles such as *Anna Delvey, the fake heiress who deceived New York's high society and whose luxurious lifestyle left substantial unpaid debts*. Here, the headline does not exactly omit information, it highlights specific details without the needed context to understand them, creating the curiosity to fill that new information gap. On the other hand, some strong adjectives can be understood in a context, like in *The bloody battle for the Sinaloa Cartel: Mayo, El Chapo's sons, and Caro Quintero's group compete for its control* where it is well known that *bloody* is not exaggerated in drug trafficking wars, therefore it is not clickbait.

Self-Responded Gaps are not Clickbait. Frequently, newspapers tweet the same news many times with different texts describing the same article. In some cases those texts answer the uncertainty created in the headline. They are negative because they close the information gap without the necessity of clicking.

Not All Questions are Clickbait. Framing headlines as questions is a typical way of creating it, but the presence of one does not always imply they are clickbait. There are examples such as when the question is already answered in the teaser, it is a cited question, it is addressed directly to the reader, or when it is clear by context that it will not be answered in the article like in *Football or rugby? Gaelic football, the fusion of two sports where Argentina excels.*

Some Direct Addresses to the Reader Create Clickbait. Talking directly to the reader is another common way to attract its attention. Many works include it as a characteristic of clickbait [2, 4, 27, 29]. Sometimes, it is indeed what creates an information gap, like in the case of *WhatsApp will stop working on several iPhone models: Will you need to switch to a new phone?* However, just the direct address to the reader does not imply clickbait, in *Over 15,000 Peruvians marched to reject the interim government. 'This Congress does not represent me' was one of the slogans of those who protested. We inform you,* that last appeal to the reader does not create an information gap because it does not add or promise any specific information at all. It is also the case of the question in *Do you remember what you were doing, where were you when Milton Wynants won the silver medal in Sydney 2000?*[4] where it is obvious that it can not be answered in the article.

Editorials can be Clickbait. There are works that exclude opinion articles because it is not possible to assess their truthfulness [15]. However, there are cases like *Why is the Government once again targeting the peace process?* where an editorial does omit information and arouses curiosity, they are clickbait.

4   The TA1C Dataset

Based on the previous definition and annotation guidelines, we created the TA1C dataset: a clickbait detection corpus representative of all Spanish-language mainstream media news, available at https://github.com/gmordecki/TA1C. It is composed of news collected from Twitter, where we prioritized reputable, national or

---

[4] Not in the dataset: https://twitter.com/ovacionuy/status/1307704183928377352.

international (not local), general (not focused on one topic) and popular[5] media. Also, with the objective of representing all of the Spanish language and its diversity, we chose newspapers from as many Spanish-speaking countries as possible, totaling 12, as well as international media that reach all of them. The complete list is available in Fig. 2, along with the proportion of clickbait we found in each.

4.1   TA1C Dataset Description

We downloaded a total of more than half a million tweets throughout a year between October 2020 and October 2021. Each tweet data was downloaded and is shared with the URL and clean HTML of the linked news, alongside the scraped headline, sub-headline, clean text (only the news body), images with their captions and embedded URLs (usually social media links).

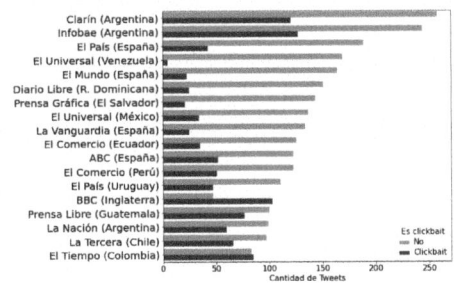

Fig. 2. Tweets in the TA1C dataset by media, divided by those labeled as Clickbait and not, sorted by amount of total tweets.

All retweets were discarded along with tweets without links or with links to other than the media main site and all of the cases where the news content crawling failed. For each newspaper we considered their main accounts and also secondary ones such as section-specific accounts. Some of them are specialized in soft-news and are more clickbait inclined, as Palau-Sampio describes [27].

The TA1C dataset contains 3500 randomly sampled tweets from those, guaranteeing that each of the 18 media sources contributes at least 150. The 3500 were annotated and split into a train set of 2100, and validation and test sets of 700 each. The splits were randomly made maintaining the same media and label balance as the complete dataset.

4.2   Annotation Methodology

Each tweet was labeled by three independent annotators: two of the authors plus one annotator with expertise in Communication Sciences. Initially, we labeled a sample of 100 news, then deliberated on the instances with differing annotations to ensure that they were distinctions of subjectivity rather than criteria discrepancies. We reached a Fleiss' κ inter annotator agreement of 0.825. The final annotation is decided by majority vote, but we publish all of the annotations.

---

[5] Based on Twitter followers, a Comscore ranking accessed in the press and the rank by 4 International Media & Newspapers.

5   Experiments and Results

We present some baselines on clickbait detection in Spanish using the TA1C datasets. All of the reported experiments used the training set to train and the validation set to define the best model, e.g. base model for transfer learning and hyper-parameters. Reported results are from a prediction in the test set.

The text from both the tweet and the news headline were used for training purposes. We did not employ the article's text, any images, or tweet data such as date and media. They are, although, part of the dataset and could be valuable for the task.

All of the experiments run on a preprocessed text where we substitute emojis, hashtags and account at signs for special tokens and remove URLs, special characters and multiple spaces. Then, we combined the headline with the text, keeping the longer one if one includes the other, or using the headline followed by the tweet if they are distinct.

5.1   Methods

LLMs - GPT4. LLMs classification with OpenAI's GPT-4 model using a *You are a helpful assistant. You are designed to understand text in English and Spanish and answer in English. Your answer is just one word and one number: yes or no, followed by the probability that the answer is yes* system prompt and a *Is the headline "[Headline]" clickbait? Answer:* user prompt. We report the GPT-4 metrics since it obtained better results than Cohere AI's LLMs.

Has Question. True if there is a question mark in the text.

Tf-idf + XGboost. Text vectorization with 1–3 grams TF-IDF features followed by an XGBoost classifier. We report the XGboost since it obtained better results than a linear model like LR or SVM.

fastText. Classification with the fastText library [18] using the supervised text version, with hyperparameter tuning via its *autotuneValidation* method.

Crafted features + XGBoost/LR. Mannualy crafted features extracted from the texts, adapted from the literature review [2,5,7,9,11,13,21,27] many of them including a translation process. Both XGBoost and Logistic Regression are reported since they got the best results among other trees and linear models respectively.

Finetuned Beto. Finetuning of Beto [10], a language model with BERT-base architecture trained in spanish text. Up to 10 epochs of train, best hyper-parameters: batch_size 32, dropout 0.2 and learning rate 6e-5.

5.2 Results

The results are available in Table 1. LLM's poor performance may reflect the inherent complexity of the problem, as well as the difficulty reconciling the different criteria for clickbait that certainly appear in their training texts. Tf-idf's and fastText bad results confirm that lexical information is not enough to discriminate clickbait. The Finetuned Beto results confirm the effectiveness of the transfer learning techniques for natural language detection problems, particularly when employed with a consistent dataset.

Table 1. Results obtained in clickbait detection on the TA1C dataset. Average Precision, F1-score, Precision and Recall are reported over the positive class.

Model	Avg Precision	F1-score	Precision	Recall	Accuracy
LLMs - GPT4	0.49	0.37	0.71	0.25	0.75
Has Question	0.27	0.41	0.93	0.26	0.78
Tf-idf + XGboost	0.74	0.61	0.85	0.48	0.82
fastText	0.76	0.64	0.77	0.55	0.82
Crafted features + LR	0.78	0.64	0.80	0.53	0.82
Crafted features + XGBoost	0.79	0.69	**0.81**	0.60	0.84
Finetuned Beto	**0.90**	**0.84**	0.80	**0.89**	**0.90**

6 Conclusion

Towards the goal of automatically detecting clickbait, we proposed a new definition of clickbait that differentiates it from other concepts like sensationalism and better fits the phenomenon after being adapted by mainstream and traditional media. We also stated precise criteria to determine whether a teaser qualifies as clickbait, in order to minimize subjectivity as much as possible.

We presented and released TA1C: the first publicly available dataset for clickbait detection in Spanish. It consists of 3,500 tweets from 18 media outlets representative of all of the Spanish language. It was manually annotated following the described criteria which helped us improve annotation agreement. We also implemented strong baselines on it. We hope that the proposed definition and criteria and the TA1C dataset can contribute to the progress of automatic clickbait classification and language resources in Spanish.

Disclosure of Interests. The authors have no competing interests to declare that are relevant to the content of this article.

References

1. Agrawal, A.: Clickbait detection using deep learning. In: 2016 2nd International Conference on Next Generation Computing Technologies (NGCT), pp. 268–272 (2016)
2. Araujo, A.B., Serrano-Puche, J., Jaso, M.F.N.: Uso del clickbait en los medios nativos digitales españoles. un análisis de el confidencial, el español, eldiario.es y ok diario. Dígitos. Revista de Comunicación Digital **1**(7), 185 (2021)
3. Aslan, S., Fastrich, G., Donnellan, E., Jones, D.J.W., Murayama, K.: People's naïve belief about curiosity and interest: a qualitative study. PLOS ONE **16**(9), 1–20 (2021)
4. Bazaco, Á., Redondo, M., Sánchez-García, P.: El clickbait, como estrategia del periodismo viral: concepto y metodología. Revista Latina de Comunicación Social (74), 94–115 (2019)
5. Biyani, P., Tsioutsiouliklis, K., Blackmer, J.: "8 amazing secrets for getting more clicks": detecting clickbaits in news streams using article informality. In: Proceedings of the Thirtieth AAAI Conference on Artificial Intelligence, AAAI 2016, pp. 94–100. AAAI Press (2016)
6. Blom, J.N., Hansen, K.R.: Click bait: forward-reference as lure in online news headlines. J. Pragmat. **76**, 87–100 (2015)
7. Brogly, C., Rubin, V.L.: Detecting clickbait: here's how to do it. Can. J. Inf. Libr. Sci. **42**(3), 154–175 (2018)
8. Broscoțeanu, D.M., Ionescu, R.T.: A novel contrastive learning method for clickbait detection on RoCliCo: a Romanian clickbait corpus of news articles. In: Proceedings of the 2023 Conference on Empirical Methods in Natural Language Processing (EMNLP). ACL (2023)
9. Cao, X., Le, T., Jason, Z.: Machine learning based detection of clickbait posts in social media (2017)
10. Cañete, J., Chaperon, G., Fuentes, R., Ho, J.H., Kang, H., Pérez, J.: Spanish pre-trained bert model and evaluation data. In: PML4DC at ICLR 2020 (2020)
11. Chakraborty, A., Paranjape, B., Kakarla, S., Ganguly, N.: Stop clickbait: detecting and preventing clickbaits in online news media. In: 2016 IEEE/ACM International Conference on Advances in Social Networks Analysis and Mining (ASONAM), pp. 9–16 (2016)
12. Chakraborty, A., Sarkar, R., Mrigen, A., Ganguly, N.: Tabloids in the era of social media? understanding the production and consumption of clickbaits in twitter. Proc. ACM Hum.-Comput. Interact. **1**(CSCW) (2017)
13. Coste, C.I., Bufnea, D., Niculescu, V.: A new language independent strategy for clickbait detection. In: 2020 International Conference on Software, Telecommunications and Computer Networks (SoftCOM), pp. 1–6 (2020)
14. García Orosa, B., Gallur Santorun, S., López García, X.: El uso del clickbait en cibermedios de los 28 países de la unión europea. Revista Latina de Comunicación Social (72), 1261–1277 (2017)
15. García Serrano, J., Romero-Rodríguez, L.M., Hernando Gómez, Á.: Análisis del "clickbaiting" en los titulares de la prensa española contemporánea/estudio de caso: Diario "el país" en facebook. Estudios sobre el Mensaje Periodístico **25**(1), 197–212 (2019)
16. Golman, R., Loewenstein, G., Molnar, A., Saccardo, S.: The demand for, and avoidance of, information. Manag. Sci. **68**(9), 6454–6476 (2022)

17. Indurthi, V., Syed, B., Gupta, M., Varma, V.: Predicting clickbait strength in online social media. In: Proceedings of the 28th International Conference on Computational Linguistics, pp. 4835–4846. International Committee on Computational Linguistics, Barcelona (2020)

18. Joulin, A., Grave, E., Bojanowski, P., Mikolov, T.: Bag of tricks for efficient text classification. In: Proceedings of the 15th Conference of the European Chapter of the Association for Computational Linguistics, vol. 2, Short Papers. pp. 427–431. Association for Computational Linguistics, Valencia (2017)

19. Kang, M.J., et al.: The wick in the candle of learning: epistemic curiosity activates reward circuitry and enhances memory. Psychol. Sci. **20**(8), 963–973 (2009). pMID: 19619181

20. Kruger, J., Evans, M.: The paradox of alypius and the pursuit of unwanted information. J. Exp. Soc. Psychol. **45**(6), 1173–1179 (2009)

21. Kuiken, J., Schuth, A., Spitters, M., Marx, M.: Effective headlines of newspaper articles in a digital environment. Digit. J. **5**(10), 1300–1314 (2017)

22. Li, X., Zhou, J., Xiang, H., Cao, J.: Attention grabbing through forward reference: an erp study on clickbait and top news stories. Int. J. Hum.–Comput. Interact. 1–16 (2022)

23. Loewenstein, G.: The psychology of curiosity: a review and reinterpretation. Psychol. Bull. **116**(1), 75–98 (1994)

24. Molyneux, L., Coddington, M.: Aggregation, clickbait and their effect on perceptions of journalistic credibility and quality. J. Pract. **14**(4), 429–446 (2020)

25. Omidvar, A., Jiang, H., An, A.: Using neural network for identifying clickbaits in online news media. In: Lossio-Ventura, J.A., Muñante, D., Alatrista-Salas, H. (eds.) SIMBig 2018. CCIS, vol. 898, pp. 220–232. Springer, Cham (2019). https://doi.org/10.1007/978-3-030-11680-4_22

26. Oxford English Dictionary: Clickbait definition (2023). https://www.oed.com/search/dictionary/?scope=Entries&q=clickbait. Accessed 26 Aug 2023

27. Palau-Sampio, D.: Reference press metamorphosis in the digital context: clickbait and tabloid strategies in elpais.com, vol. 29, pp. 63–79. Ediciones Universidad de Navarra, Pamplona (2016)

28. Potthast, M., et al.: Crowdsourcing a large corpus of clickbait on twitter. In: Bender, E., Derczynski, L., Isabelle, P. (eds.) 27th International Conference on Computational Linguistics (COLING 2018), pp. 1498–1507. The COLING 2018 Organizing Committee (2018)

29. Potthast, M., Köpsel, S., Stein, B., Hagen, M.: Clickbait detection. In: Ferro, N., Crestani, F., Moens, M.-F., Mothe, J., Silvestri, F., Di Nunzio, G.M., Hauff, C., Silvello, G. (eds.) ECIR 2016. LNCS, vol. 9626, pp. 810–817. Springer, Cham (2016). https://doi.org/10.1007/978-3-319-30671-1_72

30. Rony, M.M.U., Hassan, N., Yousuf, M.: Diving deep into clickbaits: who use them to what extents in which topics with what effects? In: Proceedings of the 2017 IEEE/ACM International Conference on Advances in Social Networks Analysis and Mining 2017, ASONAM 2017, pp. 232–239. Association for Computing Machinery, New York (2017)

31. Scott, K.: You won't believe what's in this paper! clickbait, relevance and the curiosity gap. J. Pragmat. **175**, 53–66 (2021)

32. Shin, D.D., Kim, S.I.: Homo curious: curious or interested? Educ. Psychol. Rev. **31**(4), 853–874 (2019)

33. Venneti, L., Alam, A.: How curiosity can be modeled for a clickbait detector (2018)

34. Wu, C., Wu, F., Qi, T., Huang, Y.: Clickbait detection with style-aware title modeling and co-attention. In: Sun, M., Li, S., Zhang, Y., Liu, Y., He, S., Rao, G. (eds.) CCL 2020. LNCS (LNAI), vol. 12522, pp. 430–443. Springer, Cham (2020). https://doi.org/10.1007/978-3-030-63031-7_31
35. Wu, T.: The raise of clickbait. In: The Attention Merchants, vol. 22, pp. 276–289. Knopf (2016). iSBN 9780804170048
36. Zhou, Y.: Clickbait detection in tweets using self-attentive network (2017)

Entrainment-Metrics: An Open-Source Toolkit for Quantifying Acoustic-Prosodic Entrainment in Spoken Dialogue

Erik Ernst[1], Ramiro H. Gálvez[2], and Agustín Gravano[2,3](✉)

[1] Universidad de Buenos Aires, Buenos Aires, Argentina
eernst@dc.uba.ar
[2] Universidad Torcuato Di Tella, Buenos Aires, Argentina
{rgalvez,agravano}@utdt.edu
[3] Consejo Nacional de Investigaciones Científicas y Técnicas, Buenos Aires, Argentina

Abstract. Speech entrainment, particularly acoustic-prosodic (a/p) entrainment, has gained significant interest in recent years for its implications on social interaction and dialogue systems. However, measuring a/p entrainment is challenging, mainly due to the scarce consensus on its characterization, and the lack of software libraries readily available to non-expert programmers. This paper presents Entrainment-Metrics, an open-source Python toolkit for quantifying a/p entrainment in spoken dialogue. It allows users to quantify different characterizations of a/p entrainment, measure its occurrence along different a/p features, and choose between different measurement strategies. The toolkit was designed for researchers and practitioners from a wide range of fields, with no need for a strong computational background, who are interested in studying a/p entrainment and its relation to other phenomena. This paper provides a concise introduction to using Entrainment-Metrics and presents an illustrative empirical study that highlights its capabilities.

Keywords: speech entrainment · acoustics · prosody · synchrony · convergence · proximity · Python library

1 Introduction

Speech entrainment, that is, the tendency of speakers to adjust their speech to match their interlocutors', has been shown to occur across multiple dimensions of speech. These include acoustic-prosodic (a/p) entrainment [18], linguistic entrainment [31], lexical entrainment [3], emotional entrainment [14], among others. Specifically, a/p entrainment, where speakers adapt to each other on their a/p features, has been shown to have significant implications for the dynamics of social interactions, correlating with factors such as task success [24,29], perceived competence and social attractiveness [1,18,25], and speaker engagement [7,11]. A/P entrainment has also been associated with the naturalness, effectiveness and rapport of dialogue systems [6,10,12], and it has garnered interest in

L. Correia et al. (Eds.): IBERAMIA 2024, LNCS 15277, pp. 400–411, 2025.
https://doi.org/10.1007/978-3-031-80366-6_33

recent years across different fields, such as the study of autism spectrum disorders [21,30], schizophrenia [16], tutoring systems [20,26], and second-language acquisition [17], among others.

Quantifying a/p entrainment in conversations faces challenges due to various factors. First, there is scarce consensus on its characterization, with the literature proposing multiple manifestations such as proximity, convergence, and synchrony [18]. Second, there are different strategies for its quantification, ranging from simple comparisons of the a/p feature values of adjacent *inter-pausal units* (IPUs) [18,28] to complex estimations using continuous functions. Last, empirical studies often require technically complex pipelines, posing a barrier for researchers who lack technical skills, thus leading to substantial experimentation and programming efforts in studies involving a/p entrainment.

In this paper, we present *Entrainment-Metrics*, a free, open-source Python toolkit for quantifying a/p entrainment in spoken dialogue. The toolkit is aimed at both researchers and practitioners with a strong background in the computational study of spoken dialogues, as well as researchers and practitioners without this background but whose research/work agenda involves studying a/p entrainment. Concretely, Entrainment-Metrics allows users to quantify different characterizations of a/p entrainment in dyadic conversations, to measure its occurrence along different a/p features, and to do so by means of two alternative measurement strategies. In this article, we introduce the toolkit, outline its functionalities, and, for illustrative purposes, present an empirical study that can be readily conducted using the toolkit—a task that would otherwise demand substantial programming efforts.

2   Theoretical Background

Various definitions of a/p entrainment exist, with one commonly used taxonomy outlined by Levitan (2011), which distinguishes three characterizations: 1) *Proximity*, where a/p features have similar mean values across conversation partners throughout the entire conversation; 2) *Convergence*, where a/p features become increasingly similar across conversation partners over time; and 3) *Synchrony*, where speakers adjust the values of their a/p features to align with those of their interlocutor. Even for a single characterization of entrainment, the literature lacks a standardized method for quantifying its presence. Generally speaking, the literature on a/p entrainment follows two alternative strategies. The first strategy, referred to as "utterance-based extraction" [7], treats individual utterances as the fundamental unit of analysis [11,13]. Studies following this approach typically conduct a three-step procedure: 1) pair adjacent IPUs; 2) measure the values of a specific a/p feature for these adjacent IPUs; and 3) use these values as input for a metric designed to capture a/p entrainment [18,27]. The second approach involves estimating the evolution of each speaker's a/p feature values over the course of a dialogue. Next, a/p entrainment metrics are computed by comparing the resulting time series. The present toolkit aligns with this latter approach.

Below, we briefly introduce two strategies proposed in the literature for evaluating the evolution of a speaker's a/p features during a conversation, both of which are implemented in our toolkit: 1) a discrete approach and 2) a continuous approach.

2.1   Discrete Approach

The *Time Aligned Moving Average* (TAMA) method [15] takes as input a series of IPUs, each with a start time, an end time, and the values of one or more a/p features of interest attached (e.g., the IPUs' mean energy values). Then, it divides the speaker's speech into overlapping frames of fixed length. The displacement of each frame is equal to a "time step" parameter (defined by the user), and the length of each frame is equal to a "length" parameter (also defined by the user, and typically half the time step). This method then computes the value of the a/p feature under analysis for each frame l as follows:

$$\mu_l = \sum_{i=1}^{N_l} \frac{\phi_i \cdot d_i}{\sum_{h=1}^{N_l} d_h} \tag{1}$$

where d_i stands for the duration of IPU i, ϕ_i stands for the value of the a/p feature ϕ in IPU i, and the summation is over all IPUs that overlap at least partially with frame l (assumed to be N_l) [7]. By replicating this process for all frames, the method generates a time series representing the evolution of the a/p feature ϕ throughout the dialogue ($U = \{\mu_1, \mu_2, ..., \mu_M\}$ - where M is the total number of frames). Having estimated the evolution of an a/p feature for all speakers in a conversation, metrics quantifying a/p entrainment can be calculated. In [23], two alternative synchrony metrics are proposed: the *signed synchrony* and *unsigned synchrony* measures. (For further detail on this approach, please refer to [23].)

2.2   Continuous Approach

An alternative continuous approach for estimating the evolution of a speaker's a/p features is presented in [9]. In this method, the value of an a/p feature at any given time point is predicted by first fitting a function using a machine learning-inspired approach. The authors use the speaker's IPUs as training data; each IPU is a pair (x_i, y_i), where y_i is the a/p feature value in IPU i, and x_i is its temporal midpoint. Subsequently, they estimate the evolution of a speaker's a/p feature by fitting a k-nearest neighbors (KNN) model to this training data.

After estimating the evolution of a specific a/p feature for each speaker in a conversation, metrics that capture various facets of entrainment can be computed. In the context of dyadic conversations, [9] adapted and empirically validated the three metrics originally introduced by [18] (proximity, convergence, and synchrony) to operate with the estimated continuous functions. (For further detail on this approach, please refer to [9].)

3   Entrainment-Metrics Toolkit Description

This section introduces the Entrainment-Metrics toolkit and its associated modules. The process of quantifying a/p entrainment using the toolkit is structured into three key steps: 1) the extraction of a/p features from IPUs, 2) the estimation of speakers' a/p evolution functions, and 3) the computation of a/p entrainment metrics based on these functions. (For a comprehensive list of available options and practical examples showcasing how to effectively use the toolkit, please refer to the corresponding GitHub repository and its accompanying documentation.[1])

3.1   Input Data

The amount of data required for using the toolkit is minimal—just the start and end times of each speaker's IPUs, along with the values of the a/p features under study for each IPU. Alternatively, when only the speech recordings and the IPU start/end times are available, the toolkit allows for the automatic extraction of a/p feature values using various strategies (discussed further below).[2]

3.2   Inter-Pausal Units

InterPausalUnit Construction. *InterPausalUnits* are modeled as objects requiring only a *start time* and an *end time* for their creation. They can also accept a dictionary, with a/p feature names as keys and their values as values. The toolkit offers a method for extracting IPUs from transcription files. The following code snippet illustrates how to instantiate an *InterPausalUnit* object.

```
1  from entrainment_metrics import InterPausalUnit
2
3  ipu = InterPausalUnit(start = 0.0, end = 4.0, feature_values = {'F0_MAX': 210.032})
```

Feature Extraction. If no estimates of the a/p features values are available, the toolkit provides different methods for a/p feature extraction. These methods require a separate WAV file for each speaker (recorded in good audio conditions), as well as the IPU start/end times, and make use of Praat [2], openSMILE [8], and the Allosaurus phoneme extractor for calculating speech rate [19].

3.3   Time-Aligned Moving Average

In this section, we present the primary methods implemented in the toolkit for obtaining entrainment metrics following the TAMA method [7,15].

---

[1] https://github.com/erikernst4/entrainment-metrics.
[2] If the IPU start/end times are unavailable, we recommend using a Voice Activity Detection model to obtain these times before using the Entrainment-Metrics toolkit.

Frame Construction. A *Frame* object can be instantiated with a *start time*, an *end time*, and a list of *InterPausalUnit* objects. The toolkit provides a method for extracting frames from transcription files. The following code snippet illustrates how to instantiate *Frame* objects; note that a *MissingFrame* may be initialized with only *start/end* times.

```
1  from entrainment_metrics.tama import Frame, MissingFrame
2
3  frame = Frame(
4      start=8.0, end=24.0, is_missing = False,
5      interpausal_units=[ipu1, ipu2, ipu3],
6  )
7
8  missing_frame = MissingFrame(start=0.0, end=0.4)
```

TimeSeries Construction. The TAMA module offers a method for calculating time series from a list of *Frame* objects. If the *InterPausalUnit* objects attached to the frames already have their a/p features values, then the method only needs the list of frames and the name of the target a/p feature. Otherwise, the method requires an extractor and a WAV file for feature extraction. Below, we provide a code snippet illustrating how to calculate a time series from a list of *Frame* objects when no a/p features values are pre-calculated.

```
1  from entrainment_metrics.tama import calculate_time_series
2  from typing import List
3
4  time_series_a: List[float] = calculate_time_series(
5      feature='F0_MAX', frames=a_list_of_frames,
6      audio_file='path/to/audio.wav',
7      extractor='praat', pitch_gender='F',
8  )
```

The method *calculate_sample_correlation* computes the sample cross-correlation between two time series and a given amount of lags. It returns a list of floating points where each index i corresponds to the sample cross-correlation at a given lag [15,23]. Additionally, *signed_synchrony* and *unsigned_synchrony* are methods for calculating signed/unsigned synchrony, and use the same inputs as the method for calculating the sample cross-correlation.

3.4   Continuous Approach

This section introduces the main methods used in the toolkit to estimate entrainment using the continuous approach outlined in Sect. 2.2.

TimeSeries Construction. To instantiate a *TimeSeries* object, the required parameters are: 1) a list of IPUs with their feature values pre-calculated, 2) a feature name, and 3) a method name. Currently, only the method *knn* is implemented, which uses the *KNNRegressor* from Scikit-learn [22]. However, the code is flexible and easy to extend for users to try other methods. The default value of k in the KNN estimation is equal to 7, but this value, as well as all other parameters, can be provided as inputs. Finally, during fitting, the toolkit allows discarding IPUs identified as outliers [9]—these outliers can be accessed as an instance attribute.

TimeSeries Usage. A *TimeSeries* object offers methods for predicting the values of a given set of timestamps, and also of a given time interval with certain granularity. It also has a *plot* method for visualizing the predicted evolution of the given feature (more on this in Sect. 5). The following code snippet illustrates how to instantiate a *TimeSeries* object and obtain predictions from it.

```
1   from entrainment_metrics.continuous import TimeSeries
2   import numpy as np
3
4   time_series: TimeSeries = TimeSeries(
5       interpausal_units=ipus,
6       feature='FEATURE_CALCULATED',
7       method='knn', k=8,
8   )
9
10  time_series_values: np.ndarray = time_series.predict_interval(
11      start=0.5, end=44.4, granularity=0.0001,
12  )
```

For metric calculation, the *calculate_metric* function expects two *TimeSeries* objects and a metric name (*synchrony, convergence,* or *proximity*).[3] The code snippet below illustrates how this function can be used to calculate synchrony.

```
1   from entrainment_metrics.continuous import calculate_metric
2
3   metric_result: float = calculate_metric(
4       'synchrony', time_series_a, time_series_b,
5   )
```

4   Evaluation

This section describes an empirical exercise that illustrates the capabilities of Entrainment-Metrics. We show how our toolkit may be used to measure a/p entrainment in a corpus of dyadic conversations, starting with only the audio recordings and time-aligned IPUs.

4.1   UBA Games Corpus

We analyze the UBA Games Corpus [4,5], which consists of recordings of participants playing the *Objects Games*, a collaborative computer game requiring verbal interaction. In these games, participants engage in task-oriented dialogues where one describes an object's position for the other to place it accordingly, simulating real-world scenarios. Participants alternate roles between Describer and Follower, earning points for accuracy without speech restrictions.

The corpus, collected at the University of Buenos Aires (UBA), includes two batches [4,5]. The first batch, from November to December 2012 at the Sensory Research Laboratory, involved 14 native Argentine Spanish speakers (seven

---

[3] Monte Carlo is used as the default integration method, and for synchrony, there is also an available slower method that uses the composite trapezoidal rule.

females and seven males) completing 14 Objects Game tasks per session, resulting in 6.4 h of dialogue. The second batch, recorded in April 2014 through a
collaboration between the Laboratory of Integrative Neuroscience and the Laboratory of Applied Artificial Intelligence, involved 20 native speakers (ten females
and ten males) completing 15 to 30 tasks per session, totaling 5.3 h of dialogue.
The entire corpus comprises 415 task recordings, all manually transcribed and
time-aligned at the IPU level.

4.2   Methods

For each of the 415 tasks in the corpus, we calculated convergence and synchrony
metrics for six acoustic-prosodic (a/p) features: pitch maximum (F0 Max), pitch
mean (F0 Mean), energy maximum (ENG Max), energy mean (ENG Mean),
voiced frames to total frames ratio (VF2TF), and speech rate (phonemes per
second). Our toolkit was used for feature extraction as well, as described in
Sect. 3.2.

The continuous approach from Sect. 2.2 was employed, with the number of
neighbors (k) set to 7. Potentially, 12 metrics could be measured for each task,
totaling 4,980 measurements. However, tasks were excluded when at least one
speaker had fewer than 12 IPUs with the target a/p feature. In such cases, the
estimated functions are nearly static by construction, while both convergence
and synchrony require some variability in a/p features during the dialogue. This
removal resulted in a total of 3,061 entrainment measurements.[4]

To assess the statistical significance of an entrainment measurement, we propose the following permutation strategy:

1. Calculate the a/p entrainment metric.
2. Randomly shuffle the a/p feature values across each speaker's IPUs to disrupt
   entrainment while preserving the IPU locations and mean feature values.
3. Recalculate the a/p entrainment metric using the shuffled IPUs, reflecting
   behavior without adaptation.
4. Repeat Steps 2 and 3 a total of 2,000 times to generate a distribution of the
   a/p metric, reflecting the expected behavior under no-entrainment conditions.
   We refer to this resulting distribution as the *null distribution*.
5. Compare the initial entrainment measurement from Step 1 with the null distribution. If the measurement is beyond the 97.5% or below the 2.5% threshold, consider the result statistically significant at a 5% level.[5]

Figure 1 illustrates the proposed test for assessing the significance of entrainment on F0 mean using two sample game tasks: one with a statistically significant entrainment measurement (left panel) and one without (right panel). Each
panel contains a histogram of the null distribution, dashed vertical lines at the
2.5% and 97.5% quantiles of the null distribution, and a vertical blue solid line
indicating the actual entrainment measurement.

---

[4] The results remain stable even when considering threshold values below 12.
[5] This is a *two-sided test*, considering both entrainment and disentrainment as speaker
adaptation instances [23].

4.3   Results

Table 1 presents the percentage of analyzed conversations that were classified as significant at a 5% significance level for each a/p feature. Deviations from this threshold indicate differences from random chance, with the magnitude of deviation quantifying the significance level. For example, an observed value of 10% suggests a significance twice as high as expected by chance alone. The results presented stem from having estimated 6,125,061 entrainment metrics.

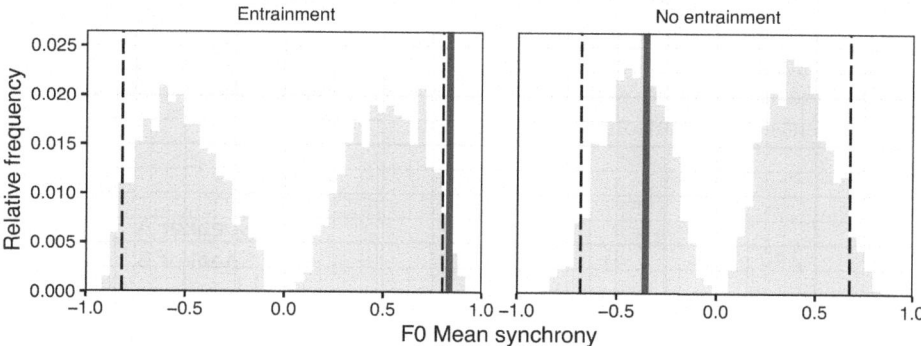

Fig. 1. Illustrative examples of the proposed methodology for assessing the statistical significance of entrainment metrics: Histograms of the *null distribution* for F0 mean in two sample game tasks. The dashed vertical lines are located at the 2.5% and 97.5% quantiles; the vertical blue solid lines indicate the actual entrainment measurement in each case, which is statistically significant (left) and non-significant (right).

Table 1. Percentage of conversations with significant convergence/synchrony for all analyzed a/p features at a 5% significance level.

A/P Feature	Convergence	Synchrony
ENG Max	9.02%	14.12%
ENG Mean	6.67%	9.80%
F0 Max	7.84%	7.45%
F0 Mean	7.84%	11.81%
VF2TF	5.49%	10.98%
Speech rate	3.91%	7.03%

Table 1 highlights several interesting patterns in the analyzed corpus. First, synchrony is more prevalent than convergence, as indicated by the higher proportion of significant synchrony estimates compared to convergence estimates for all features. Second, convergence is most pronounced in ENG Max, F0 Mean, and F0 Max, while the percentage of significant estimates for ENG Mean, VF2TF, and speech rate are close to chance levels. Third, the proportion of significant synchrony estimates exceeds chance for all features, with ENG Max, F0 Mean, VF2TF, and ENG Mean showing the highest proportions.

5   Visualization Capabilities

Entrainment-Metrics includes visualization tools. Users can plot the a/p feature values of individual IPUs and continuous approximations for multiple speakers. Figure 2 demonstrates this, with each horizontal segment representing an IPU's start and end times. The vertical axis shows the feature values, and each solid line represents the evolution of a *TimeSeries*.

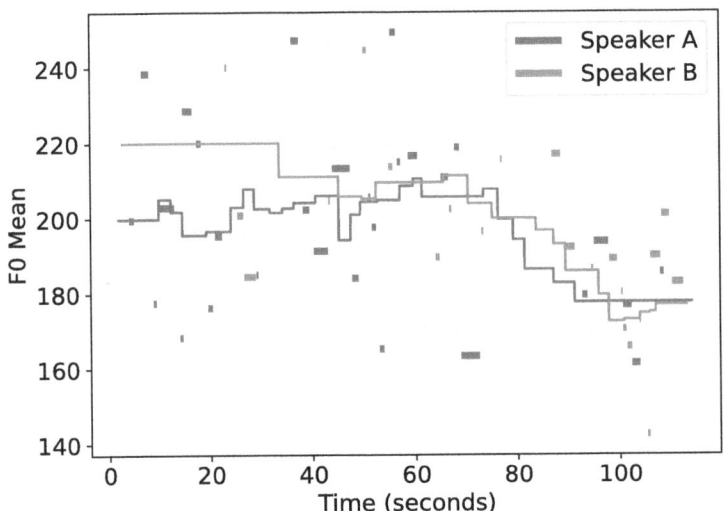

Fig. 2. Comparison of continuous approximations for multiple speakers: The blue and orange horizontal segments are measurements of the F0 mean of individual IPUs produced by speakers A and B. The lines are the corresponding continuous approximations computed using the described KNN approach. (Color figure online)

6   Conclusion

In this paper, we introduced *Entrainment-Metrics*, a free, open-source Python toolkit designed for quantifying a/p entrainment in spoken dialogue. In addition, we described an empirical exercise that showcases the potential uses of the proposed toolkit. By removing (or at least attenuating) many technical barriers, we expect this toolkit will prove valuable to researchers and practitioners interested in studying the intricate and subtle phenomenon of a/p entrainment.

Lastly, it is important to note that *Entrainment-Metrics* is an ongoing project. We anticipate incorporating new functionalities into the toolkit based on feedback from the research community.

Acknowledgments. *This research was supported by Universidad Torcuato Di Tella and CONICET. We thank the anonymous reviewers for their valuable comments and suggestions.*

References

1. Štefan Beňuš, Gravano, A., Levitan, R., Levitan, S.I., Willson, L., Hirschberg, J.: Entrainment, dominance and alliance in supreme court hearings. Knowl.-Based Syst. **71**, 3–14 (2014). https://doi.org/10.1016/j.knosys.2014.05.020. https://www.sciencedirect.com/science/article/pii/S0950705114002184

2. Boersma, P., Weenink, D.: Praat: doing phonetics by computer [computer program], version 6.3.17 (2023). Accessed 25 Sept 2023. http://www.praat.org/

3. Brennan, S.: Lexical entrainment in spontaneous dialog. In: Proceedings of ISSD (1996)

4. Brusco, P., Gravano, A.: Automatic offline annotation of turn-taking transitions in task-oriented dialogue. Comput. Speech Lang. **78**, 101462 (2023). https://doi.org/10.1016/j.csl.2022.101462. https://www.sciencedirect.com/science/article/pii/S0885230822000857

5. Brusco, P., Vidal, J., Štefan Beňuš, Gravano, A.: A cross-linguistic analysis of the temporal dynamics of turn-taking cues using machine learning as a descriptive tool. Speech Commun. **125**, 24–40 (2020). https://doi.org/10.1016/j.specom.2020.09.004. https://www.sciencedirect.com/science/article/pii/S0167639320302727

6. Cohn, M., Liang, K.H., Sarian, M., Zellou, G., Yu, Z.: Speech rate adjustments in conversations with an amazon alexa socialbot. Front. Commun. **6**, 671429 (2021)

7. De Looze, C., Scherer, S., Vaughan, B., Campbell, N.: Investigating automatic measurements of prosodic accommodation and its dynamics in social interaction. Speech Commun. **58**, 11–34 (2014). https://doi.org/10.1016/j.specom.2013.10.002. https://www.sciencedirect.com/science/article/pii/S0167639313001386

8. Eyben, F., Wöllmer, M., Schuller, B.: Opensmile: the munich versatile and fast open-source audio feature extractor. In: Proceedings of the 18th ACM International Conference on Multimedia, MM 2010, pp. 1459–1462. Association for Computing Machinery, New York (2010). https://doi.org/10.1145/1873951.1874246

9. Gálvez, R.H., Gauder, L., Luque, J., Gravano, A.: A unifying framework for modeling acoustic/prosodic entrainment: definition and evaluation on two large corpora. In: Proceedings of the 21th Annual Meeting of the Special Interest Group on Discourse and Dialogue, pp. 215–224. Association for Computational Linguistics, 1st virtual meeting (2020). https://aclanthology.org/2020.sigdial-1.27

10. Gratch, J., Lucas, G.: Rapport between humans and socially interactive agents. In: The Handbook on Socially Interactive Agents: 20 years of Research on Embodied Conversational Agents, Intelligent Virtual Agents, and Social Robotics, vol. 1: Methods, Behavior, Cognition, pp. 433–462. Association for Computing Machinery, New York (2021)

11. Gravano, A., Štefan Beňuš, Levitan, R., Hirschberg, J.: Backward mimicry and forward influence in prosodic contour choice in standard American English. In: Proceedings of Interspeech 2015, pp. 1839–1843 (2015). https://doi.org/10.21437/Interspeech.2015-67

12. Gálvez, R.H., Gravano, A., Štefan Beňuš, Levitan, R., Trnka, M., Hirschberg, J.: An empirical study of the effect of acoustic-prosodic entrainment on the perceived trustworthiness of conversational avatars. Speech Commun. **124**, 46–67 (2020). https://doi.org/10.1016/j.specom.2020.07.007. https://www.sciencedirect.com/science/article/pii/S0167639320302478

13. Heldner, M., Edlund, J., Hirschberg, J.B.: Pitch similarity in the vicinity of backchannels. In: Proceedings of Interspeech 2010 (2010). https://doi.org/10.7916/D8WS92R4

14. Kejriwal, J.: Relationship between speech entrainment and emotion. In: 2022 10th International Conference on Affective Computing and Intelligent Interaction Workshops and Demos (ACIIW), pp. 1–4 (2022). https://doi.org/10.1109/ACIIW57231.2022.10086027

15. Kousidis, S., Dorran, D., Mcdonnell, C., Coyle, E.: Time series analysis of acoustic feature convergence in human dialogues. In: SPECOM-2009, St. Petersburg, Russian Federation (2009)

16. Kruyt, J., Benus, S., Faget, C., Lançon, C., Champagne-Lavau, M.: Prosodic and lexical entrainment in adults with and without schizophrenia. In: Speech Prosody, Lisbonne, Italy (2022). https://hal.science/hal-03759753

17. Kuo, G.: Second language acoustic-prosodic entrainment in conversation and storytelling. J. Acoust. Soc. Am. **152**(4_Supplement), A288–A288 (2022). https://doi.org/10.1121/10.0016302

18. Levitan, R., Hirschberg, J.: Measuring acoustic-prosodic entrainment with respect to multiple levels and dimensions. In: Proceedings of Interspeech 2011, pp. 3081–3084 (2011). https://doi.org/10.21437/Interspeech.2011-771

19. Li, X., et al.: Universal phone recognition with a multilingual allophone system. In: ICASSP 2020 - 2020 IEEE International Conference on Acoustics, Speech and Signal Processing (ICASSP), pp. 8249–8253 (2020). https://doi.org/10.1109/ICASSP40776.2020.9054362

20. Molenaar, B., Soliño Fernández, B., Polimeno, A., Barakova, E., Chen, A.: Pitch it right: using prosodic entrainment to improve robot-assisted foreign language learning in school-aged children. Multimodal Technol. Interact. **5**(12), 76 (2021)

21. Ochi, K., et al.: Quantification of speech and synchrony in the conversation of adults with autism spectrum disorder. PLoS ONE **14**(12), e0225377 (2019)

22. Pedregosa, F., et al.: Scikit-learn: machine learning in python. J. Mach. Learn. Res. **12**(85), 2825–2830 (2011). http://jmlr.org/papers/v12/pedregosa11a.html

23. Pérez, J.M., Gálvez, R.H., Gravano, A.: Disentrainment may be a positive thing: a novel measure of unsigned acoustic-prosodic synchrony, and its relation to speaker engagement. In: Proceedings of Interspeech 2016, pp. 1270–1274 (2016). https://doi.org/10.21437/Interspeech.2016-587

24. Reitter, D., Moore, J.D.: Alignment and task success in spoken dialogue. J. Memory Lang. **76**, 29–46 (2014). https://doi.org/10.1016/j.jml.2014.05.008. https://www.sciencedirect.com/science/article/pii/S0749596X14000576

25. Street, R.: Speech convergence and speech evaluation in fact-finding interview. Human Commu. Res. **11**, 139–169 (2006). https://doi.org/10.1111/j.1468-2958.1984.tb00043.x

26. Thomason, J., Nguyen, H.V., Litman, D.: Prosodic entrainment and tutoring dialogue success. In: Lane, H.C., Yacef, K., Mostow, J., Pavlik, P. (eds.) AIED 2013. LNCS (LNAI), vol. 7926, pp. 750–753. Springer, Heidelberg (2013). https://doi.org/10.1007/978-3-642-39112-5_104

27. Weise, A., Levitan, S.I., Hirschberg, J., Levitan, R.: Individual differences in acoustic-prosodic entrainment in spoken dialogue. Speech Commun. **115**, 78–87 (2019). https://doi.org/10.1016/j.specom.2019.10.007. https://www.sciencedirect.com/science/article/pii/S016763931930086X

28. Weise, A., Silber-Varod, V., Lerner, A., Hirschberg, J., Levitan, R.: Entrainment in spoken hebrew dialogues. J. Phonet. **83**, 101005 (2020). https://doi.org/10.1016/j.wocn.2020.101005. https://www.sciencedirect.com/science/article/pii/S0095447020300966

29. Wynn, C.J., Barrett, T.S., Borrie, S.A.: Rhythm perception, speaking rate entrainment, and conversational quality: a mediated model. J. Speech Lang. Hear. Res. **65**(6), 2187–2203 (2022). https://doi.org/10.1044/2022_JSLHR-21-00293. https://pubs.asha.org/doi/abs/10.1044/2022_JSLHR-21-00293

30. Wynn, C.J., Borrie, S.A., Sellers, T.P.: Speech rate entrainment in children and adults with and without autism spectrum disorder. Am. J. Speech-Lang. Pathol. **27**(3), 965–974 (2018). https://doi.org/10.1044/2018_AJSLP-17-0134. https://pubs.asha.org/doi/abs/10.1044/2018_AJSLP-17-0134

31. Yu, M., Litman, D., Ma, S., Wu, J.: A neural network-based linguistic similarity measure for entrainment in conversations (2021)

Derivation Prompting: A Logic-Based Method for Improving Retrieval-Augmented Generation

Ignacio Sastre[✉], Guillermo Moncecchi, and Aiala Rosá

Instituto de Computación, Facultad de Ingeniería,Universidad de la República,
Montevideo, Uruguay
{isastre,gmonce,aialar}@fing.edu.uy

Abstract. The application of Large Language Models to Question Answering has shown great promise, but important challenges such as hallucinations and erroneous reasoning arise when using these models, particularly in knowledge-intensive, domain-specific tasks. To address these issues, we introduce Derivation Prompting, a novel prompting technique for the generation step of the Retrieval-Augmented Generation framework. Inspired by logic derivations, this method involves deriving conclusions from initial hypotheses through the systematic application of predefined rules. It constructs a derivation tree that is interpretable and adds control over the generation process. We applied this method in a specific case study, significantly reducing unacceptable answers compared to traditional RAG and long-context window methods.[1](Repo with all prompts: https://github.com/nsuruguay05/derivation-prompting)

Keywords: Large Language Models · Retrieval-Augmented Generation · Question Answering

1 Introduction

Question Answering (QA) has improved substantially with the advent of Large Language Models (LLMs). However, these models face important challenges, particularly in knowledge-intensive, domain-specific tasks, such as hallucinations and faulty reasoning [5,13]. The Retrieval-Augmented Generation (RAG) framework addresses these limitations by retrieving the most relevant document chunks from a trusted domain-specific document base, and grounding the LLM's generation on the retrieved information [2].

Substantial work has been done to improve the reasoning ability of LLMs [3]. Techniques like Chain-of-Thought (CoT) [15] have reliably improved performance across various tasks, including QA. However, these techniques do not explicitly define *how* the model should reason, as there are no restrictions on how the intermediate reasoning steps should be constructed.

In this work, we propose Derivation Prompting, an alternative approach for the generation step in the RAG framework inspired by logical derivations. In this

© The Author(s), under exclusive license to Springer Nature Switzerland AG 2025
L. Correia et al. (Eds.): IBERAMIA 2024, LNCS 15277, pp. 412–423, 2025.
https://doi.org/10.1007/978-3-031-80366-6_34

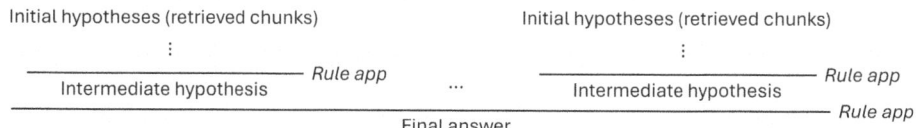

Fig. 1. Schematic illustration of a derivation tree constructed using derivation prompting.

method, a conclusion is inferred from initial hypotheses by applying well-defined rules to transform and/or combine these hypotheses. This novel approach offers some advantages over existing methods, mainly:

- **Interpretability:** The method not only generates a final answer but also produces a tree structure, referred to as a derivation (see Fig. 1). Each node in the derivation represents the application of an easily interpretable rule that transforms some of its children. This structure provides a straightforward way to identify errors the model could have made and to understand how it arrived at the final answer.
- **Controlled generation:** By generating the answer through the sequential application of predefined rules, this method provides a clearer reasoning path for the model to follow. This reduces hallucinations and faulty reasoning while ensuring that the generated answers remain grounded in the information from the documents.

The paper is structured as follows: Sect. 2 presents related work. Section 3 provides a detailed explanation of Derivation Prompting. Section 4 describes the case study conducted. Section 5 explains the evaluation method used. Section 6 presents the results and analysis. Finally, Sect. 7 contains the conclusions and outlines future work.

2   Related Work

Retrieval Augmented Generation
Derivation Prompting, as proposed in this work, is a prompting technique applied within the context of Retrieval-Augmented Generation (RAG). RAG augments LLMs with a retrieval component that recovers the most relevant information from an external knowledge base [4,8,11].

The naive RAG paradigm consists of two main steps: retrieval and generation. First, documents are segmented into smaller units, referred to as chunks, which are subsequently converted into vector representations. Upon a user query, during the retrieval step the query is converted to its vector representation, and similarity scores between the query and the indexed chunks are computed. The top k chunks are retrieved and used as context in the prompt for the generation step [2].

An alternative to using vector representations is to employ Cross-Encoder models that directly process each chunk with the query and return a similarity score [9,10]. While this approach tends to yield better results, it is significantly less compute-efficient, as it requires as many model inferences per query as number of chucks we have, compared to only one inference when using sentence embeddings.

Prompting Techniques for Enhancing Reasoning
Substantial work has been done to improve the reasoning ability of LLMs [3]. Chain-of-Thought (CoT) [15] involves prompting the model to generate a coherent series of intermediate reasoning steps that lead to the final answer. Few-Shot prompting [1] is applied and a chain-of-thought is added to each example. They show that sufficiently large LLMs can generate these reasoning chains, yielding promising results in arithmetic, commonsense, and symbolic reasoning tasks.

The Tree of Thoughts (ToT) framework [17] is an evolution of CoT that enables models to explore several different reasoning paths. In this approach, reasoning is conceptualized as searching through a tree, where each node represents a thought. This framework addresses some limitations of CoT, such as the inability to explore different continuations within the same reasoning chain or to backtrack when incorrect conclusions are reached.

While these methods significantly improve performance on various tasks, there is no control over how each thought is generated in the chain, as there is no systematic methodology the model has to follow. This lack of control can lead to erroneous reasoning and susceptibility to hallucinations, which are well-known problems when working with LLMs [5,13].

Logic and LLMs
Similar to Derivation Prompting, some works explore combining classical logic with prompting techniques to improve reasoning. The Logical Thoughts (LoT) prompting framework [18] uses logical equivalence, expressing premises in various logically equivalent forms to encourage the exploration of different solutions. This is achieved by incorporating a verification step for each thought, where an explanation is generated for both the thought as-is and its logical negation. The LLM is then tasked to decide between the two.

Symbolic CoT (SymbCoT) [16] is another proposed method that involves four LLM modules: (i) Translator: translating premises and the question to First-Order Logic formulas, (ii) Planner: dividing the original problem into smaller subproblems and developing a step-by-step plan, (iii) Solver: deriving the answer through a logical inference process and (iv) Verifier: validating the correctness of the translations and the Solver's output.

3   Derivation Prompting

This technique focuses on the generation step of the Retrieval-Augmented Generation (RAG) framework. It relies on the premise that the expected answer to a given query must be obtained by combining and/or transforming the most relevant information extracted from a document base, since our objective is to rely only on the information available in the documents and not on the information the model could have learned on its training phase.

The idea for this technique is inspired by how a derivation tree in propositional logic is constructed. In this context, a conclusion φ is derived from a set of premises or hypotheses $\Gamma = \{\delta_1, \ldots, \delta_n\}$. We denote $\Gamma \vdash \varphi$ if such a derivation exists. The class of derivations forms an inductively defined set characterized by a list of inference rules that explicitly state how to derive new conclusions from existing ones. These rules operate by systematically applying logical operations to the premises to construct a tree where each node represents an application of a rule, culminating in the conclusion at the root [14]. Figure 2 shows an example of a logic derivation.

$$\frac{p_1 \wedge p_2}{p_2}\, E\wedge \quad \frac{\dfrac{p_1 \wedge p_2}{p_1}\, E\wedge}{p_2 \wedge p_1}\, I\wedge$$

Fig. 2. Example of a derivation proving the statement $p_1 \wedge p_2 \vdash p_2 \wedge p_1$, where p_1 and p_2 are proposition symbols and $E\wedge$ and $I\wedge$ are the elimination and introduction rules respectively for $\wedge$, as defined in [14].

In a typical RAG framework, documents are divided into smaller units called chunks. Given a query, the n most relevant chunks are selected and used as context for generating the answer. Following the analogy with logic derivations, in Derivation Prompting, we consider these most relevant chunks as a set of hypotheses $\{h_1, \ldots, h_n\}$. The objective is to construct a derivation tree using predefined natural language rules, ultimately deriving a conclusion c, such that $h_1, \ldots, h_n \vdash c$, as depicted in Fig. 1.

In contrast to logic derivations, where we usually start from a candidate conclusion and seek to construct the proof, in this case, the conclusion is not known beforehand. Therefore, a query q is needed to guide the construction of the derivation tree, with the goal that the resulting conclusion serves as the answer to the query q.

For each step in the construction of the derivation, the task the LLM has to follow involves deciding which rule to apply, selecting the appropriate hypotheses, and generating the conclusion that arises from the application of the chosen rule. Although it may seem counter-intuitive to let the LLM decide which rule to apply, this is a key aspect for making this method viable due to the LLM's ability to disambiguate natural language in both the rule explanation and the hypotheses.

3.1   Rules

For Derivation Prompting, a set of derivation rules must be defined. These rules are specified in natural language and used by the language model to construct a derivation tree. We define a set of derivation rules that are convenient for our use case (Sect. 4). It is important to notice that these rules are specific to this problem, and any set of rules could be defined that best fits the type of combinations and/or transformations necessary in different use cases. Table 1 presents each rule with a description and Fig. 3 shows a toy example for each rule.

3.2 Algorithm

Algorithm 1 presents the pseudo-code for constructing a derivation. When looking at the algorithm in detail, it is important to notice that lines 3, 4, and 5 correspond to steps that the LLM should execute. The responsibilities of the

Table 1. List of defined rules with a brief description.

Name	Description
Extract	Given a hypothesis h, this rule extracts a specific part of h as a conclusion
Concat	Combines two independent hypotheses to generate the conclusion
Instantiate	Generates a conclusion by instantiating a generic hypothesis into a particular case
Compose	Combines two hypotheses that share a common element to generate a new conclusion
Refine	Given a hypothesis h, it slightly adapts it to better fit the question, without modifying the semantics or content of h
NoInfo	This rule is used when none of the hypotheses provide information to answer the question (or part of the question)

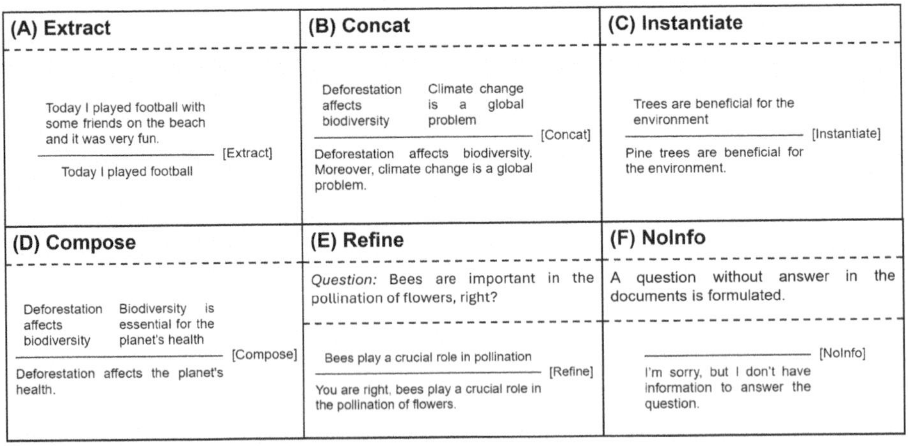

(A) Extract

Today I played football with some friends on the beach and it was very fun.
——————————— [Extract]
Today I played football

(B) Concat

Deforestation affects biodiversity   Climate change is a global problem
——————————— [Concat]
Deforestation affects biodiversity. Moreover, climate change is a global problem.

(C) Instantiate

Trees are beneficial for the environment
——————————— [Instantiate]
Pine trees are beneficial for the environment.

(D) Compose

Deforestation affects biodiversity   Biodiversity is essential for the planet's health
——————————— [Compose]
Deforestation affects the planet's health.

(E) Refine

Question: Bees are important in the pollination of flowers, right?
- - - - - - - - - - - - - - - -
Bees play a crucial role in pollination
——————————— [Refine]
You are right, bees play a crucial role in the pollination of flowers.

(F) NoInfo

A question without answer in the documents is formulated.
- - - - - - - - - - - - - - - -
——————————— [NoInfo]
I'm sorry, but I don't have information to answer the question.

Fig. 3. Toy examples of application for each rule. Examples (E) and (F) have information of the query for better understanding.

LLM are to decide which rule to apply and which hypotheses to use, as well as to construct the conclusion that arises from the application of the chosen rule. Additionally, the LLM is used to determine whether the conclusion serves as the final answer to the user's query.

Algorithm 1. Derivation Prompting pseudo-code

Require: $hypotheses\_list = \{h_1, \ldots, h_n\}$: List of hypotheses, q: Query
 1: $final\_answer \leftarrow$ **False**
 2: **while not** $final\_answer$ **do**
 3:     Decide which rule r to apply
 4:     Decide which hypotheses $\{h_i, \ldots, h_k\}$ to apply r to
 5:     $conclusion \leftarrow$ apply rule r over $\{h_i, \ldots, h_k\}$ and query q
 6:     **if** $conclusion$ is the final answer **then**
 7:         $final\_answer \leftarrow$ **True**
 8:     **else**
 9:         $hypotheses\_list.append(conclusion)$
10:     **end if**
11: **end while**
12: **return** $conclusion$

The rest of the algorithm is straightforward. If the conclusion is considered the final answer, the derivation is complete, and the last conclusion is used as the answer. If not, the conclusion is added to the list of hypotheses and can be used in subsequent rule applications (though it might potentially never be used). Optionally, each rule application can be stored with pointers to its arguments and conclusion to later reconstruct the derivation tree.

We explored different ways of implementing the aforementioned algorithm and considered two main alternatives:

1. **One-step prompt:** This approach isolates each rule application as an independent LLM call. Given the list of hypotheses in the middle of a derivation, the model is prompted to produce, in a single inference, the rule to apply, the hypotheses to use, the resulting conclusion, and whether it is the final answer. The algorithm is implemented similarly to the Algorithm 1, with lines 3, 4, and 5 replaced by this single call to the LLM, followed by parsing the result.
2. **Whole derivation prompt:** In contrast to the previous alternative, this approach allows the LLM to construct the entire derivation in one inference call, effectively emulating the execution of Algorithm 1. To achieve this, we applied a Few-Shot strategy [1], crafting six complete examples of manual executions of the algorithm to create different derivations using all the rules (Appendix A shows one of these examples). The model is then prompted to follow the same steps with a new query and initial hypotheses. The result is then parsed to obtain each rule execution and intermediate hypotheses.

In our experiments, the second approach yielded results as good as the first one but was much faster and computationally cheaper, as it replaces n LLM inferences with just one. Therefore, we decided to extensively use the whole derivation approach.

4   Case Study

We investigated this idea in the context of a specific use case: developing a platform for question answering in the domain of administrative information for the School of Engineering at Universidad de la República (UDELAR), for the Spanish language. Currently, the school operates the Orientation and Consultation Space (OCS), where students can ask questions via email or in person, and OCS staff provide answers. We explored the feasibility of building a tool to assist with this work by offering an automated system for students to obtain answers to their questions.

We gathered a small set of documents available on the school's webpage. Specifically, 17 websites were scraped and converted to markdown format using LangChain's `Html2TextTransformer` class.

For evaluation purposes, we constructed a QA dataset consisting of 135 real user queries. These queries were derived from past emails sent to the OCS over the last few years. Each student email was preprocessed using the Llama 2 7B model [12] to remove irrelevant information typical of email communication (e.g., greetings, apologies) and, most importantly, personal information such as names, identification numbers, and phone numbers.

In the context of this project, we explored several methods using LLMs which are explained below:

Retrieval Augmented Generation (RAG)
For the retrieval step, we explored using sentence embeddings generated with `intfloat/multilingual-e5-large` as well as `BAAI/bge-reranker-large` Cross-Encoder model. Given that the use case involves fewer than a hundred chunks, using a Cross-Encoder was feasible and, as expected, consistently yielded better results than using sentence embeddings.

For the generation step, we experimented with models from the Anthropic's Claude 3 family[1] (specifically, Haiku, which is faster but less capable, and Opus, which is the best performing and competitive with OpenAI's GPT-4). The k most relevant chunks to the user's query were added as context, and a prompt was crafted to explicitly instruct the model to use them for generating the answer.

Using Long Context Windows
Another approach we explored was leveraging the long context windows of closed models, specifically the Claude models, which support up to 200k tokens. In this method, we inserted all the full documents as context, thereby avoiding the retrieval step.

---

[1] https://www.anthropic.com/news/claude-3-family.

Derivation Prompting

Utilizing the retrieval method described in the RAG experiment, we explored the use of Derivation Prompting for the generation step, as detailed in Sect. 3. For this method, we used three initial hypotheses, corresponding to the three most relevant chunks obtained in the retrieval step. We did not explore using more hypotheses because of how the Few-Shot examples were designed, but this is an area for future work that we plan to investigate.

5   Evaluation

Evaluating Open-Domain Question Answering, especially when using LLMs, remains an open problem, and human evaluation still appears to have no substitute [6]. Nevertheless, it has been shown that state-of-the-art LLMs tend to exhibit a high degree of agreement with human evaluation when used as judges [19]. Therefore, we decided to follow this approach for evaluating each experiment separately. We are currently conducting human evaluation on the best performing experiments and results will be presented in future work.

We designed an evaluation prompt following the format defined for the Feedback Collection dataset [7], which encompasses four components:

1. **Instruction to evaluate:** The instruction for the task to evaluate. In our case, this is the particular question that the answer addresses.
2. **Response to Evaluate:** The response to the question that the LLM has to evaluate (with a score on a scale from 1 to 5).
3. **Reference Answer:** A reference answer that corresponds to a score of 5.
4. **Customized Score Rubric:** Specific criteria defined for our use case, specifying what the evaluator should focus on. This includes a description of the criteria and a detailed explanation for each possible score (1 to 5).

We defined the score rubric criteria as determining whether the generated answer is correct and truthful. This is clearly specified in each score description. Table 2 presents a brief explanation of each score.

Table 2. Score rubric criteria defined for each score, for evaluating generated answers.

Score	Explanation	Classification
1	Candidate contradicts reference; false information.	Unacceptable
2	Candidate has conflicts with reference; partially false information.	Unacceptable
3	Candidate does not contradict reference but does not provide any information either.	Acceptable
4	Candidate partially matches reference; correct but incomplete information.	Acceptable
5	Candidate completely matches reference; correct and complete information.	Acceptable

Additionally, we classified scores 1 and 2 as unacceptable and scores 3 to 5 as acceptable, thereby obtaining an aggregated metric for evaluation. Scores 1 and 2 correspond to answers that fully or partially contradict the reference answer and are therefore considered unacceptable. Scores 3 to 5 may have none of the information correct (but are not incorrect either, as no information is provided at all), part of the information correct, or be fully correct. In all these cases, the answers do not provide false or contradictory information and are considered acceptable.

6 Results

The evaluation was carried out using Claude Opus as the evaluator. Table 3 shows the percentage of acceptable answers and the distribution for each score from 1 to 5 for the best performing experiments. These experiments utilize Claude Opus and Claude Haiku, and where applicable, the Cross-Encoder for the retrieval step, with the number of chunks used as context set to $k = 3$.

Table 3. Percentage of acceptable answers, distribution of scores, average and standard deviation metrics for each experiment. CH is Claude Haiku and CO is Claude Opus.

Experiment	% Accep.	#1	#2	#3	#4	#5	Avg.	Std. Dev.
Long context - CH	65.2	35	12	61	22	5	2.63	1.14
RAG - CH	72.6	17	20	74	20	4	2.81	0.94
Derivation Prompting - CH	82.2	21	3	95	12	4	2.81	0.91
Long context - CO	76.3	17	15	73	25	5	2.90	0.97
RAG - CO	77.8	15	15	75	27	3	2.91	0.92
Derivation Prompting - CO	**89.6**	**10**	**4**	**92**	**25**	**4**	**3.07**	**0.79**

As can be observed in Table 3, Derivation Prompting with Claude Opus significantly reduces the number of unacceptable answers compared to the other experiments. However, it does not necessarily increase the number of answers with scores of 4 and 5. Many unacceptable answers from the other experiments receive a score of 3 in Derivation Prompting. There are two primary reasons for this: (1) The NoInfo rule has a more direct impact than simply prompting the model not to answer questions when the information is not available, as done in both the RAG and Long Context experiments; (2) Generating answers through the application of explicitly defined and constrained rules reduces the likelihood of hallucinations or misinterpretations of the context chunks and minimizes the potential for faulty reasoning. These results suggest that while there is minimal impact on recall (i.e., answering as much as possible), there is a significant improvement in precision (i.e., avoiding incorrect answers).

It is important to note that while Derivation Prompting with Claude Haiku also reduces unacceptable answers, it does have an impact on the number of answers with scores of 4 and 5, resulting in fewer such answers compared to RAG. This suggests that the size of the model is an important factor. Larger and more powerful models, such as Claude Opus, have a better understanding of the task of constructing the derivation and are more capable of applying the rules effectively, yielding better results.

Although unacceptable answers have been reduced, there are still some examples that scored 1 and 2. A significant advantage of Derivation Prompting is that the resulting derivation is interpretable, and it is easy to identify mistakes in the application of rules. This, when compared to simple RAG, is a notable advantage for users, as it often eliminates the need to verify answers directly from the source. Instead, users can follow the reasoning in the derivation and identify faulty steps. Figure 4 presents a real example of an incorrect derivation. In the application of the Refine rule, it is clear that the model has hallucinated facts not present in the hypotheses.

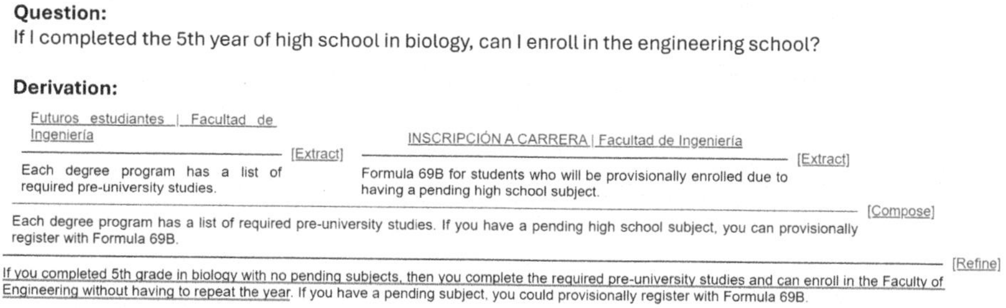

Question:
If I completed the 5th year of high school in biology, can I enroll in the engineering school?

Derivation:

Futuros estudiantes | Facultad de Ingeniería
——————————————— [Extract]
Each degree program has a list of required pre-university studies.

INSCRIPCIÓN A CARRERA | Facultad de Ingeniería
——————————————— [Extract]
Formula 69B for students who will be provisionally enrolled due to having a pending high school subject.

——————————————— [Compose]
Each degree program has a list of required pre-university studies. If you have a pending high school subject, you can provisionally register with Formula 69B.

——————————————— [Refine]
If you completed 5th grade in biology with no pending subjects, then you complete the required pre-university studies and can enroll in the Faculty of Engineering without having to repeat the year. If you have a pending subject, you could provisionally register with Formula 69B.

Fig. 4. Example of an incorrect derivation (translated from Spanish). In the application of the Refine rule, the model hallucinates that having completed 5th year of high school in biology fulfills the required pre-university studies (hallucination is underlined in red). (Color figure online)

7 Conclusions

In this paper we introduced Derivation Prompting, a new prompting technique inspired by logic derivations, to improve the generation step in the Retrieval-Augmented Generation framework for open-domain question answering. Our experiments showed that Derivation Prompting significantly reduces the occurrence of unacceptable answers compared to traditional RAG and long-context window approaches.

However, the performance of Derivation Prompting is influenced by the size and capability of the underlying LLM. While Claude Opus exhibited robust performance, smaller models like Claude Haiku showed a decrease in useful answers (though unacceptable answers were reduced), indicating the importance of model capacity when constructing effective derivations.

Future work will focus on refining Derivation Prompting by experimenting with different sets of rules and adjusting the number of initial hypotheses. We are also working on formalizing the underlying formal language behind the application of the rules, and using this to add further verification methods to ensure the correctness of the resulting derivation.

We believe that this method can be applied to additional use cases and may be generalized to non-RAG scenarios with different sets of rules. However, it is important to evaluate this method further to ensure its utility in such cases.

A   Few-Shot Example

Hypotheses: {*retrieved chunks removed for brevity, enumerated from 1 to 3*}
User question: What is the meaning of a credit and how many hours of weekly study should be dedicated to a 13-credit course?
Extract — 2 — A credit is a measure of the dedication required for a subject. It is assumed that if a subject has more credits, it requires more hours of dedication. — Not a final answer
New hypothesis: a. {*Previous conclusion is repeated, removed for brevity*}
Extract — 2 — A credit approximately equals one hour of weekly study, throughout an entire semester. — Not a final answer
New hypothesis: b. {*Previous conclusion is repeated, removed for brevity*}
Instantiate — b — A credit approximately equals one hour of weekly study, throughout an entire semester. Therefore, a subject with 13 credits implies an approximate dedication of 13 h per week. — Not a final answer
New hypothesis: c. {*Previous conclusion is repeated, removed for brevity*}

Concat — a,c — A credit is a measure of the dedication required for a subject. It is assumed that if a subject has more credits, it requires more hours of dedication. A credit approximately equals one hour of weekly study, throughout an entire semester. Therefore, a subject with 13 credits implies an approximate dedication of 13 h per week. — Final answer

References

1. Brown, T.B., Mann, B., Ryder, N., Subbiah, M., Kaplan, J., et al.: Language models are few-shot learners. In: Advances in Neural Information Processing Systems, vol. 33, pp. 1877–1901. Curran Associates, Inc. (2020). https://proceedings.neurips.cc/paper_files/paper/2020/file/1457c0d6bfcb4967418bfb8ac142f64a-Paper.pdf
2. Gao, Y., et al.: Retrieval-augmented generation for large language models: a survey (2024). https://arxiv.org/abs/2312.10997
3. Huang, J., Chang, K.C.C.: Towards reasoning in large language models: A survey. In: Rogers, A., Boyd-Graber, J., Okazaki, N. (eds.) Findings of the Association for Computational Linguistics: ACL 2023, pp. 1049–1065. Association for Computational Linguistics, Toronto (2023). https://doi.org/10.18653/v1/2023.findings-acl.67. https://aclanthology.org/2023.findings-acl.67
4. Izacard, G., et al.: Atlas: few-shot learning with retrieval augmented language models. J. Mach. Learn. Res. **24**(1) (2024)
5. Ji, Z., et al.: Survey of hallucination in natural language generation. ACM Comput. Surv. **55**(12) (2023). https://doi.org/10.1145/3571730
6. Kamalloo, E., Dziri, N., Clarke, C., Rafiei, D.: Evaluating open-domain question answering in the era of large language models. In: Rogers, A., Boyd-Graber, J., Okazaki, N. (eds.) Proceedings of the 61st Annual Meeting of the Association for Computational Linguistics, vol. 1: Long Papers, pp. 5591–5606. Association for Computational Linguistics, Toronto (2023). https://doi.org/10.18653/v1/2023.acl-long.307. https://aclanthology.org/2023.acl-long.307

7. Kim, S., et al.: Prometheus: inducing fine-grained evaluation capability in language models. In: The Twelfth International Conference on Learning Representations (2024). https://openreview.net/forum?id=8euJaTveKw

8. Lewis, P., et al.: Retrieval-augmented generation for knowledge-intensive NLP tasks. In: Advances in Neural Information Processing Systems, vol. 33, pp. 9459–9474. Curran Associates, Inc. (2020). https://proceedings.neurips.cc/paper/2020/hash/6b493230205f780e1bc26945df7481e5-Abstract.html

9. Nogueira, R.F., Cho, K.: Passage re-ranking with BERT. CoRR arxiv:1901.04085 (2019)

10. Reimers, N., Gurevych, I.: Sentence-BERT: sentence embeddings using siamese BERT-networks. In: Inui, K., Jiang, J., Ng, V., Wan, X. (eds.) Proceedings of the 2019 Conference on Empirical Methods in Natural Language Processing and the 9th International Joint Conference on Natural Language Processing (EMNLP-IJCNLP), pp. 3982–3992. Association for Computational Linguistics, Hong Kong (2019). https://doi.org/10.18653/v1/D19-1410. https://aclanthology.org/D19-1410

11. Shi, W., et al.: REPLUG: retrieval-augmented black-box language models. In: Duh, K., Gomez, H., Bethard, S. (eds.) Proceedings of the 2024 Conference of the North American Chapter of the Association for Computational Linguistics: Human Language Technologies, vol. 1: Long Papers, pp. 8371–8384. Association for Computational Linguistics, Mexico City (2024). https://aclanthology.org/2024.naacl-long.463

12. Touvron, H., Martin, L., Stone, K., Albert, P., Almahairi, A., et al.: Llama 2: open foundation and fine-tuned chat models (2023). https://doi.org/10.48550/arXiv.2307.09288. http://arxiv.org/abs/2307.09288

13. Valmeekam, K., Olmo, A., Sreedharan, S., Kambhampati, S.: Large language models still can't plan (a benchmark for LLMs on planning and reasoning about change). In: NeurIPS 2022 Foundation Models for Decision Making Workshop (2022). https://openreview.net/forum?id=wUU-7XTL5XO

14. Van Dalen, D.: Logic and Structure. Universitext, Springer, London (2013). https://doi.org/10.1007/978-1-4471-4558-5

15. Wei, J., et al.: Chain-of-thought prompting elicits reasoning in large language models. In: Proceedings of the 36th International Conference on Neural Information Processing Systems, NIPS '22. Curran Associates Inc., Red Hook (2024)

16. Xu, J., Fei, H., Pan, L., Liu, Q., Lee, M.L., Hsu, W.: Faithful logical reasoning via symbolic chain-of-thought (2024). https://arxiv.org/abs/2405.18357

17. Yao, S., et al.: Tree of thoughts: deliberate problem solving with large language models. In: Thirty-Seventh Conference on Neural Information Processing Systems (2023). https://openreview.net/forum?id=5Xc1ecxO1h

18. Zhao, X., et al.: Enhancing zero-shot chain-of-thought reasoning in large language models through logic. In: Calzolari, N., Kan, M.Y., Hoste, V., Lenci, A., Sakti, S., Xue, N. (eds.) Proceedings of the 2024 Joint International Conference on Computational Linguistics, Language Resources and Evaluation (LREC-COLING 2024), pp. 6144–6166. ELRA and ICCL, Torino (2024). https://aclanthology.org/2024.lrec-main.543

19. Zheng, L., et al.: Judging llm-as-a-judge with mt-bench and chatbot arena. In: Oh, A., Naumann, T., Globerson, A., Saenko, K., Hardt, M., Levine, S. (eds.) Advances in Neural Information Processing Systems, vol. 36, pp. 46595–46623. Curran Associates, Inc. (2023). https://proceedings.neurips.cc/paper_files/paper/2023/file/91f18a1287b398d378ef22505bf41832-Paper-Datasets_and_Benchmarks.pdf

Information Extraction from Electronic Health Records Written in Spanish for Epidemic Intelligence

Javier Petri[1(✉)], Pilar Barcena Barbeira[2], and Viviana Cotik[1,3]

[1] Departamento de Computación, FCEyN, Universidad de Buenos Aires, Buenos Aires, Argentina
javierpetri2012@gmail.com
[2] Programa de Innovación Tecnológica en Salud Pública, Departamento de Salud Pública, Facultad de Medicina, UBA, Buenos Aires, Argentina
[3] Instituto de Investigación en Ciencias de la Computación (ICC), CONICET-UBA, Buenos Aires, Argentina

Abstract. Automatic symptom detection from electronic health records is a valuable source for event-based surveillance systems. In this study, we develop tools to automatically detect symptoms associated with febrile illnesses in electronic health records written in Spanish. Therefore, we use a custom corpus, comprising 6228 expertly labeled and approximately 1 million unlabeled health reports. Our approach involved fine-tuning state-of-the-art named entity recognition models, including BiLSTM-CRF and transformer-based models like RoBERTa. We focused on domain-adaptive and task-adaptive models to enhance performance: the former were pretrained on biomedical corpora, while the latter were further pretrained on our unlabeled health reports. Despite computational constraints, our models demonstrated promising results, with RoBERTa-Clinico, a task-adaptive transformer model pretrained in our unlabeled corpus, showing the best micro recall performance (79.30), and 70.83 micro F1 score, which are comparable to results in similar studies. In this way, we contribute to the limited body of work in BioNLP in Spanish.

Keywords: Named entity recognition · BioNLP · Spanish electronic health records · automatic symptom detection · event-based surveillance

1 Introduction

In recent years, the amount of digital information within the medical domain has steadily increased, among other reasons, due to the growing use of electronic health record (EHR) systems. In order to identify trends, conduct epidemiological intelligence, and obtain useful information to streamline administrative tasks, the automatic extraction of information from EHRs is essential.

© The Author(s), under exclusive license to Springer Nature Switzerland AG 2025
L. Correia et al. (Eds.): IBERAMIA 2024, LNCS 15277, pp. 424–436, 2025.
https://doi.org/10.1007/978-3-031-80366-6_35

Epidemic intelligence, promoted by international organizations such as the Pan American Health Organization (PAHO) and the World Health Organization (WHO), is defined as the process of detecting, filtering, verifying, analyzing, evaluating, and investigating information about events or situations that may pose a threat to public health[1] [14]. It is based on "traditional" indicators and "non-traditional" events/signals, such as social media and instant messaging applications, news, rumors, and information from electronic health records [12, 22]. Both sources of information are important as they complement each other.

Indicator-based surveillance involves the collection, analysis, and interpretation of structured data from formal sources. In the event of suspected or confirmed events considered important for public health (e.g., dengue, COVID-19, measles, or exanthematous fever), healthcare personnel must notify health authorities.[2] On the other hand, event-based surveillance (EBS) serves as an early warning system, enabling a timely response to health events that may pose a risk to human health, thereby complementing notifiable disease reporting [21]. For example, by analyzing EHRs, an increase in consultations for fever or fever and rashes could be detected, which may indicate the presence of measles not yet detected by the health system. Additionally, EHRs can be used to perform retrospective searches to find the primary case.

In this work, which evolved from part of ARPHAI[3], a project funded by the International Development Research Centre (IDRC) and the Swedish International Development Cooperation Agency (SIDA), where the demand for this study was recognized[4], we conduct signal detection within the framework of EBS with the aim of automatically detecting symptoms associated with acute non-specific febrile syndrome, COVID-19, diarrhea, pneumonia, and influenza-like illness (ILI) in de-identified EHRs from the Argentine health system using natural language processing (NLP) techniques, particularly named entity recognition (NER). Implementing NER on EHRs is valuable as it can: aid in real-time automatic detection of symptoms and syndromes, facilitate mass searches to evaluate disease distribution or prevalence over time, and utilize the number of symptom occurrences to forecast disease progression, thereby enhancing decision-making and public health surveillance.

Some challenges of this task include the lack of publicly available annotated data in Spanish, particularly in the Rioplatense variant, to train the machine learning models for this task, the abundance of specific terminology of the medical domain, the presence of orthographical and grammatical errors, due to lack of available time to write them, and the linguistic variability of the Spanish language. These characteristics make NER in EHRs a more challenging task than in general domain corpora. Furthermore, a corpus had to be created, which can not

---

[1] https://www.paho.org/en/topics/epidemic-intelligence.

[2] In Argentina, it must be reported under National Law No. 15.465, and the WHO requires all member countries to have the capacity to detect events [20].

[3] https://www.ciecti.org.ar/arphai/.

[4] In ARPHAI, health records were annotated for NER, and an initial NER algorithm was devised.

be shared because medical information -even though anonymized- is sensitive. As far as we know there is only one previous work performed for the automatic detection of symptoms in Spanish (in Chilean variant) [1].

In order to automatically detect symptoms, we employed a variety of state-of-the-art models: BiLSTM-CRF along with Spanish pre-trained general-domain word embeddings, and embeddings trained by ourselves with a set of unlabeled EHRs; and transformer-based models from the BERT [11] and RoBERTa family [17], some pre-trained with general domain corpora in Spanish, others pre-trained with biomedical domain corpora, and others using our own unlabeled data. These methods are used to automatically detect symptoms, syndromes and hedges associated to them.

The rest of the paper is organized as follows. Section 2 reviews related work, Sect. 3 covers the data used, the manual labeling process, and the developed models. Section 4 presents the results, which are analyzed in Sect. 5. Finally, Sect. 6 outlines the conclusions and suggests directions for future work.

2   Previous Work

Chen et al. [6] studied approximately 150 research projects using NLP methods, as well as systems and datasets in English, addressing the COVID-19 pandemic to extract information from biomedical literature (scientific publications and abstracts) and detect symptoms in clinical notes (which describe, among others, diseases, symptoms, laboratory test results, and texts recounting patient visits to clinical institutions). Although this review covers a wide range of research, it is important to note that most do not use health records as a data source, and the implemented solutions are based on simple algorithms or pre-existing advanced systems (e.g., artificial neural networks and transformers), due to the need for quick solutions given the urgency of the circumstances. Furthermore, the authors explain the challenges posed by information extraction (IE) of EHRs, compared to IE in biomedical literature, and public health issues detection from news articles.

Currently, there is a shortage of publicly available annotated text (*corpora*) for the Spanish language. DisTEMIST (DISease TExt Mining Shared Task) [18] and CodiESP (eHealth CLEF 2020 - Multilingual Information Extraction) [19] challenges provided *corpora* in Spanish (Iberian variant) and NER tasks. In both cases, the datasets consist of clinical cases[5] instead of electronic health records (EHRs). SpRadIE (Spanish Radiological Information Extraction) [8] was a competition for performing NER on radiological reports written in Spanish (Rioplatense variant) [9,10]. In all three competitions, the best-performing systems utilized deep learning techniques. For example, the best-performing SpRadIE implementation was partially implemented with BERT [26].

---

[5] In medicine, a clinical case is the commented presentation of the health situation of a patient, or group of patients, which is exemplified as a "case" when it becomes the "individual realization of a more or less general phenomenon".

Finally, as far as we know, there is only one work in Spanish that tackles automatic symptom detection in clinical records similar to ours. [1]. In this work NER was performed using BiLSTM-CRF. It differs from ours in that while we work with EHRs written in Rioplatense Spanish (the variety spoken in Argentina and Uruguay), their study focuses on Chilean Spanish. Additionally, they use a single label for all symptoms and diseases (among other entities), whereas we categorize a wide range of symptoms with distinct labels for each.

3 Material and Methods

This section presents the data, dataset selection, annotation process, and the deep neural network models for automatic symptom detection.

3.1 Data

We worked with data from de-identified electronic health records from Argentina[6]. These records are written in Rioplatense Spanish and consist of unstructured text, which often includes descriptions of the reason for the visit, signs and symptoms of the patient, and the results of certain laboratory tests.

The original corpus contains 2,468,566 records spanning from October 4, 2016, to January 28, 2021, inclusive. Specific filters were applied to this dataset, primarily to exclude records from non-relevant specialties, such as Pharmacy, resulting in a refined dataset of 1,073,084 records.

3.2 Creation of an Annotated Dataset

An annotation schema and criteria were defined in the previously mentioned ARPHAI project, which was carried out, among others, by some of the authors of this work. To create it, we followed the MAMA (Model-Annotate-Model-Annotate) cycle [23]. A total of 33 different symptoms, associated with the five medical syndromes of interest -acute nonspecific febrile syndrome, COVID-19, diarrhea, pneumonia, and influenza-like illness- were identified. Each symptom corresponded to a named entity to be annotated. Besides, a single label was assigned for any of the syndromes. Additionally, labels for negation, speculation, temporal references, and conditional assertions were included. This resulted in a total of 46 labels.

The annotation criteria were written by a computational linguist, a Ph.D. in computer science with expertise in corpora creation, and two epidemiological residents. The annotation process was conducted by a team of three advanced medical students, and the previously mentioned epidemiology residents, who reviewed the quality of the annotations. All participants of the annotation are native Spanish speakers and are familiar with both reading (the first group) and writing (the second group) health records. Fig. 1, shows an EHR with its annotations.

---

[6] EHR anonymization was previously implemented in the ARPHAI project, and all confidentiality guidelines, including data access on a secured server, were followed.

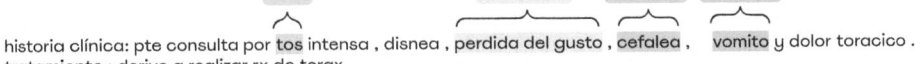

historia clínica: pte consulta por **tos** intensa , disnea , **perdida del gusto** , **cefalea** ,   **vomito** y dolor toracico .
tratamiento : derivo a realizar rx de torax.

Fig. 1. Example of an annotated EHR. This translates to English as: "clinical history: *pt* consults for intense cough[COUGH], dyspnea, loss of taste[DYSGEUSIA], headache[HEADACHE], vomiting[VOMITING] and chest pain. Treatment: referral for chest x-ray."

The annotation process consisted of a total of six annotation rounds, where annotators annotated EHRs with some overlap to calculate the inter-annotator agreement (IAA). Table 1 shows the number of annotated records, the percentage of overlap of the annotations, and the IAA obtained using the Cohen's Kappa coefficient (κ) [7] of each annotation round. As can be noticed, κ value's increasing trend changes to decline in the last annotation round. This is probably due to the fact that not all annotators were able to complete round 6.

Table 1. Number of annotated records per annotator per round, overlap percentage, and IAA (with Cohen's Kappa (κ)). * Not all annotators completed round 6

Rounds 1-3				Rounds 4-6			
Round	Rec. per annot.	Overlap	κ	Round	Rec. per annot.	Overlap	κ
1	100	40%	0.65	4	500	10%	0.65
2	200	10%	0.55	5	500	10%	0.68
3	200	30%	0.66	6*	2000	10%	0.50

3.3   Dataset Analysis

The annotated dataset consisted of 6228 EHRs and 46 labels. It was randomly divided into a development subset (4982 EHRs, 80% of the dataset), further divided into training and validation subsets (80%-20%), and a test subset (1246 EHRs). The development subset was used for model fine-tuning and training, while the test subset was used for a final evaluation of the best models.

4506 of the 6228 annotated EHRs had at least one annotation. The corpus has 414802 tokens, with words and punctuation counted as tokens, averaging 66 tokens and 323 characters per EHR.

Table 2 shows the total count of the most frequent named entities found by the annotators for each dataset, with a threshold of 400 occurrences.

In addition to these entities, the following were also included in the dataset: the symptoms: anorexia, anosmia, arthralgias, arthralgias myalgias, asthenia, chills, conjunctivitis, diarrhea, diarrhea with blood, dysgeusia, dysgeusia anosmia, general discomfort, hemorrhage, mild hemorrhages, myalgias, petechiae, pneumonia, retroocular pain, skin rashes, splenomegaly, tachypnea, thrombo-cytopenia, jaundice, leukopenia; the syndromes: diarrhea syndrome phenotype, diarrhea with blood syndrome phenotype, fever syndrome phenotype, pneumonia syndrome phenotype; and finally the other labels: conditional, distant temporal, negation, near temporal, speculation, and temporal conditional. Their distribution in the dataset can be viewed in the project's website.

Table 2. Types and quantity of the most frequent named entities annotated in the EHRs in each dataset. In parentheses, we show the number of distinct terms that have the same label (for example, the terms *fievre* and *fiebre* are different and both are labeled as *FEVER*). Tags were translated to English for better comprehension

Number of the most frequent labels and their distribution				
Label	#Training	#Validation	#Test	Total
PHENOTYPE_SYNDROME	703 (59)	159 (21)	219 (34)	1081 (76)
FEVER	1026 (97)	293 (45)	317 (47)	1636 (152)
ABDOMINAL_PAIN	515 (79)	127 (39)	136 (33)	778 (118)
COUGH	498 (51)	132 (17)	151 (18)	781 (66)
HEADACHE	472 (31)	115 (12)	156 (12)	743 (45)
VOMITING	420 (20)	122 (12)	163 (11)	705 (36)
ODYNOPHAGIA	352 (34)	96 (9)	108 (16)	556 (48)
NAUSEA	285 (14)	68 (7)	94 (9)	447 (19)
RHINITIS	254 (42)	65 (20)	87 (22)	406 (58)
Others	14603	3828	4372	22803
Total	19128	5005	5803	29936

3.4   Models

This section presents the type of models used: BiLSTM [25] and Transfomers [27].

BiLSTM. We performed three experiments, we trained: 1) a BiLSTM model with an initial layer of Spanish word embeddings, obtained from *fastText*[7], trained on general Spanish articles from Wikipedia; 2) a BiLSTM-CRF model [16]; and 3) in the previous model, we replaced the word embeddings by others, obtained from training *fastText* with a set of approximately 1M unlabeled clinical records -6M tokens- of our dataset.

---

[7] https://fasttext.cc/.

Transformers. Various models currently relevant to the field of the Spanish language were adapted, including some pre-trained models in English and two models that we created by further pretraining an existing Spanish model with our set of unlabeled EHRs. We worked with: 1) *BERT* and *RoBERTa* models, both trained in English, considering them as a baseline; 2) *BETO* [3], which is based on BERT and is trained on a generic corpus in Spanish [2] and an adaptation of BETO, named *BETO-clinical-wl-es*[8], that was created by continue pretraining BETO with a dataset of actual diagnoses from the Chilean Waiting List [1]; 3) two models based on the *RoBERTa* architecture: a) *roberta-base-biomedical-clinical-es* [4] (we will refer to it as RoBERTa-CLI-BIO), and b) *bsc-bio-ehr-es* [5] (from now on BSC-BIO-EHR-ES.). Both models were pre-trained exclusively with a Spanish corpus that contains medical data, the latter including EHRs similar to ours; and 4) we further pre-trained two models with our unannotated EHR dataset: a) *BETO-Clinico*, that has *BETO* as base model, and b) *RoBERTa-Clinico*, that has *roberta-base-biomedical-clinical-es* as base model.

3.5   Experiments

Due to the lack of a GPU (we used an Intelő Xeonő E5-2680 v2 with 20 CPUs and 65 GB of RAM), we were limited in hyperparameter testing and opted to use the ADAM (Adaptive Moment Estimation) [15] optimizer and its AdamW variant (in transformers) to enhance model performance, based on their proven effectiveness in deep learning [11,24,27].

During training, different combinations of hyperparameters were explored for all the architectures mentioned earlier (see Sect. 3.4), and the best configuration was selected by maximizing the micro F1 metric on the validation dataset. We also analyze the impact of training the models using two different versions of the training dataset: the complete set, which includes all 3985 records, mentioned in Sect. 3.3, and a filtered set that excludes EHRs, where no entities were found by the annotators, and that has 2913 records. Better and more consistent results were observed across all models when using the latter dataset.

BiLSTM. The BiLSTM based models were trained for a maximum of 40 epochs, using a patience parameter of 5 iterations to interrupt training if no improvements were observed in the loss function. Table 3 shows the hyperparameters that yielded the best results. Except for the number of epochs, all other hyperparameters were obtained through grid search as part of the ARPHAI project. For training our word embeddings with our unlabeled data and the *fastText library*, we used 20 epochs, a learning rate of 0.1, and a dimension of 300 for the representation vectors of each word.

Among the tested models, BiLSTM-CRF with our custom word embeddings layer, showed the best performance in the validation dataset (F1-score of 71.51% compared to 65.18% for the BiLSTM-CRF with generic Spanish Wikipedia embeddings, and 61.85% F1 for the BiLSTM).

---

[8] https://huggingface.co/plncmm/beto-clinical-wl-es.

Table 3. Hyperparameters used in the experiments

BILSTM parameters			
max epochs	40	Transformers parameters	
optimizer	ADAM	max epochs	10
batch size	32	optimizer	AdamW
initial learning rate	0.001	batch size	8
word dropout	0.25	initial learning rate	2e–5
BiLSTM layers	2	weight decay	0.01
BiLSTM hidden size	64		

Transformers. Experimentation with the different transformer models was performed with the HuggingFace *transformers library*. Fine-tuning for these models was carried out using the development dataset performing *five-fold cross-validation*. Table 3 shows the hyperparameters that yielded the best results.

Additionally, inspired by previous research [13], we implemented TAPT (Task-Adaptive Pretraining) and DAPT (Domain-Adaptive Pretraining) + TAPT. We used two models, *BETO* (general domain) and *roberta-base-biomedical-clinical-es* (biomedical domain), and further pretrained them with our unlabeled dataset using *Whole Word Masking*. Despite computational limitations restricting our training proccess, the results were promising, especially for the model resulting from performing TAPT on *roberta-base-biomedical-clinical-es*, which we named *RoBERTa-Clinico* model, which achieved the best results in the training phase.

3.6 Evaluation of Results

We measured our algorithms with exact match metrics (micro precision -P-, recall -R- and F1-score), calculated using the Python framework *seqeval*[9] and with a lenient (or approximate) match metric, based in the Jaccard Similarity[10]. Micro measures were used due to the dataset's imbalance. We also report macro P, R, and F1.

While F1 is the standard metric for NER, our focus on signal detection means that we prioritize recall over precision. Thus, with similar F1 scores, we prefer algorithms with higher recall.

4 Results

Although in the training phase we developed 11 different models, those who achieved the poorest performance in terms of F1 (or recall, when F1 was similar)

---

[9] https://github.com/chakki-works/seqeval.
[10] The Jaccard Similarity coefficient is calculated by dividing the size of the intersection of the sets by the size of the union of the sets.

were not evaluated in the test phase. In particular, we discarded all *BERT*-based models due to their lower performance compared to those based on *RoBERTa* and the first versions of BiLSTM based models. The models tested are called the *chosen models*.

Table 4 shows the results obtained with the test dataset for the *chosen models*, trained with the entire development dataset. Bootstrapping with 1000 resamples was used to estimate results variability. The table shows mean micro and macro precision, recall, and F1 (i.e. micro and macro metric means over the 1000 bootstrapping iterations). Standard errors are between 0.0005 and 0.0009 in all cases).

Table 4. Results on the test dataset for the *chosen models*, showing the mean precision, recall, and F1 scores. Both micro and macro means are provided. Additionally, the percentage of relative gain when using partial match for the micro metrics is shown

Model	Micro mean (%)			Macro mean (%)		
	Precision	Recall	F1-score	Precision	Recall	F1-score
BiLSTM-CRF+WE	**67.57** (6)	75.70 (4)	71.40 (5)	**54.88**	63.12	**56.99**
RoBERTa (English)	60.80 (8)	78.19 (5)	68.40 (7)	47.16	64.52	53.34
RoBERTa-CLI-BIO	66.71 (6)	78.45 (4)	**72.10** (5)	52.39	63.64	56.15
BSC-BIO-EHR-ES	64.61 (6)	78.58 (4)	70.91 (5)	49.76	63.29	54.29
RoBERTa-Clinico	64.03 (6)	**79.30** (4)	70.85 (5)	52.13	**66.15**	56.57

5   Analysis of Results

Table 4 shows that the model with the highest micro F1 is *RoBERTa-CLI-BIO* and the one with the highest micro recall is *RoBERTa-Clinico*, the fine-tuned model with our unannoted EHR corpus. On the other hand, *BILSTM-CRF+WE* has higher micro precision and lower micro recall than transformer-based models.

Recall outperforms precision in all the evaluated models. Given that analyzing the results we detected terms, that should have been labeled but were not, and also considering the low IAA (see Table 1), we suspected that the low precision is due to under-annotation. Therefore, we decided to evaluate the quality of the annotations. With this aim, we selected a random set of 100 EHRs belonging to the test set. One of the epidemiology residents reviewed them and corrected the previous annotations, when needed, following the annotation criteria originally established (described in Sect. 3.2). Subsequently, we calculated the mean micro metrics for these re-annotated data using the previously trained models *BiLSTM-CRF+WE* and *RoBERTa-Clinico* and observed considerable improvements: *BiLSTM-CRF+WE* achieved relative gains of 25%, 15%, and 19% for P, R, and F1 respectively. For *RoBERTa-Clinico*, the gains were 27%, 14%, and 21%.

As shown in Table 4, the mean macro metrics are worse than the mean micro metrics. The confusion matrix for the *RoBERTa-Clinico* model, which is not included in this paper, revealed several issues: Some terms, such as "diarrhea", "pneumonia" and "fever" can represent both symptoms and syndromes (with different labels, such as *diarrhea* and *diarrhea syndrome phenotype*), causing difficulties in both annotation and model labeling, often favoring the label with more support. This resulted in poor performance for the less-supported labels. Additionally, seven labels had fewer than 68 occurrences in the development dataset, leading to poor performance (F1 score of 0), which also affected the macro metrics. The confusion matrix and the entity-by-entity results table can be found on the project's website.

Our RoBERTa-Clinico model's metrics are comparable to those of Baez et al. [1], who achieved 75%, 71%, and 73% for P, R, and F1 scores, respectively in their multiclass model. They use a BiLSTM-CRF model with Spanish and clinical embeddings on their unannotated corpus and consider a single class for all diseases.

Finally, the results obtained using partial match for the micro metrics show an improvement in model performance across all metrics, as expected. Precision shows a relative improvement of 6% (except for *RoBERTa*, which increases by 8%), while recall improves by 4% and 5%. This translates into an increase in F1 score of approximately 5% (except for *RoBERTa*, which increases by 7%).

6   Conclusion and Future Work

In this work, we used a Spanish corpus created, among others, by some of the authors of this study, in the context of the ARPHAI project. The corpus contains 6228 EHRs labeled by experts with a total of 46 labels and approximately 1M EHRs with relevant unlabeled information (see Subsect. 3.1). Additionally, we implemented various NER models by fine-tuning models based on BiLSTM and pretrained transformers, some of general domain and others specifically pretrained for the biomedical domain in Spanish, which are currently considered state-of-the-art in NLP. We also performed further pretraining on two of these to generate models trained with our unlabeled data, and one of them, *RoBERTa-Clinico*, showed promising results despite the computational power limitations we had for training.

Our research fills a gap in Spanish BioNLP, in particular for Rioplatense Spanish, as we haven't found prior studies focusing on automatic symptom detection in EHRs of Argentina. Additionally, similar investigations in the Spanish-speaking realm are scarce. We fine-tune state-of-the-art models, achieving promising results despite our computing power constraints. While direct comparisons with existing research are challenging due to task variations, our results align well with reported outcomes in similar literature [1]. Furthermore, we analyzed two ways to evaluate our NER algorithms: the traditional *exact match* and a *partial match*, which in our opinion, is more suitable for this problem, since it retrieves relatively good approximations, obtaining significant improvements over the *exact match*.

Despite these strengths, the study is not without limitations that should be considered. The lack of other publicly available corpora in Spanish inhibits our ability to benchmark our work against others. Besides, given the limited size of our corpus and the limited computing power, we were constrained in our ability to continuing pre-training models with our unlabeled EHRs. An additional limitation to consider is that the low IAA may result in model performance worse than it actually is. In fact, based on the experiment of re-annotation conducted, explained in Sect. 5, we can conclude that with a correction of the annotations, the performance metrics would achieve higher results.

The potential creation of a real-time, automatic symptom and syndrome detection model is highly beneficial for event-based surveillance systems, providing crucial information for timely public health interventions. Automated EHR labeling enables efficient large-scale searches. This facilitates disease incidence evaluation across populations and regions over time by classifying large volumes of existing EHR. Additionally, the model can be integrated for real-time prospectively annotation, enabling event analysis and decision-making in public health.

As future work, it would be interesting to improve the annotation of the corpus, as the agreement between annotators turned out to be low, and certain underestimation in the annotation was detected. This is reinforced by the fact that when re-annotating a sample of the annotated EHRs (Sect. 5), the increase obtained for the precision metric almost doubled the improvement obtained for recall. Data augmentation could also be considered. Additionally, integrating dictionary-based solutions could enhance the performance of labels with insufficient occurrences, capturing specific and infrequent terms to improve model precision and coverage.

We believe our work represents a preliminary step towards the goal of developing a system that can be used in the real-world management of an EBS system. We acknowledge that there are additional challenges and aspects to consider in future research to achieve effective and functional deployment in operational environments.

Acknowledgement. This study was initiated and partially supported by ARPHAI, a project funded by IDRC and SIDA. It was later continued and further funded by the PICT-2021-GRF-TI- 0067 project, granted by Agencia I+D+i through FONCyT, Argentina.

Disclosure of Interests. The authors have no competing interests to declare that are relevant to the content of this article.

References

1. Báez, P., Villena, F., Rojas, M., Durán, M., Dunstan, J.: The Chilean Waiting List Corpus: a new resource for clinical named entity recognition in Spanish. In: Proceedings of the 3rd Clinical Natural Language Processing Workshop, pp. 291–300 (2020)

2. Cañete, J.: Compilation of large Spanish unannotated corpora. Zenodo, mayo de (2019)
3. Cañete, J., Chaperon, G., Fuentes, R., Ho, J.H., Kang, H., Pérez, J.: Spanish pre-trained bert model and evaluation data. arXiv preprint arXiv:2308.02976 (2023)
4. Carrino, C.P., et al.: Biomedical and clinical language models for Spanish: on the benefits of domain-specific pretraining in a mid-resource scenario (2021)
5. Carrino, C.P., et al.: Pretrained biomedical language models for clinical NLP in Spanish. In: Proceedings of the 21st Workshop on BioNLP, pp. 193–199. ACL (2022)
6. Chen, Q., et al.: Artificial intelligence in action: addressing the COVID-19 pandemic with natural language processing. Ann. Rev. Biomed. Data Sci. **4**, 313–339 (2021)
7. Cohen, J.: A coefficient of agreement for nominal scales. Educ. Psychol. Measur. **20**(1), 37–46 (1960)
8. Cotik, V., et al.: Overview of CLEF eHealth Task 1-SpRadIE: a challenge on information extraction from Spanish Radiology Reports. In: CLEF (Working Notes), pp. 732–750 (2021)
9. Cotik, V., Filippo, D., Roller, R., Uszkoreit, H., Xu, F.: Annotation of entities and relations in Spanish radiology reports. In: RANLP, pp. 177–184 (2017)
10. Cotik, V.: Extracción de información en informes radiológicos escritos en español. Ph.D. thesis, Universidad de Buenos Aires. FCEyN (2018)
11. Devlin, J., Chang, M.W., Lee, K., Toutanova, K.: BERT: pre-training of deep bidirectional transformers for language understanding. arXiv preprint arXiv:1810.04805 (2018)
12. Grein, T.W., et al.: Rumors of disease in the global village: outbreak verification. Emerg. Infect. Dis. **6**(2), 97 (2000)
13. Gururangan, S., Marasović, A., Swayamdipta, S., Lo, K., Beltagy, I., Downey, D., Smith, N.A.: Don't stop pretraining: adapt language models to domains and tasks. arXiv preprint arXiv:2004.10964 (2020)
14. Kaiser, R., Coulombier, D., Baldari, M., Morgan, D., Paquet, C.: What is epidemic intelligence, and how is it being improved in Europe? Weekly Releases (1997–2007) **11**(5), 2892 (2006)
15. Kingma, D.P., Ba, J.: Adam: a method for stochastic optimization. arXiv preprint arXiv:1412.6980 (2014)
16. Lample, G., Ballesteros, M., Subramanian, S., Kawakami, K., Dyer, C.: Neural architectures for named entity recognition. arXiv preprint arXiv:1603.01360 (2016)
17. Liu, Y., et al.: Roberta: a robustly optimized bert pretraining approach. arXiv preprint arXiv:1907.11692 (2019)
18. Miranda-Escalada, A., et al.: Overview of distemist at bioasq: automatic detection and normalization of diseases from clinical texts: results, methods, evaluation and multilingual resources. In: Working Notes of Conference and Labs of the Evaluation (CLEF) Forum. CEUR Workshop Proceedings (2022)
19. Miranda-Escalada, I.A., Gonzalez-Agirre, A., Armengol-Estapé, J., Krallinger, M.: Overview of automatic clinical coding: annotations, guidelines, and solutions for non-english clinical cases at CodiEsp Track of CLEF eHealth. In: CLEF Working Notes (2020)
20. Organization, W.H., et al.: Reglamento sanitario internacional (2005). Organizacion Mundial de la Salud (2008)
21. Organization, W.H., et al.: Early detection, assessment and response to acute public health events: implementation of early warning and response with a focus on

event-based surveillance: interim version. World Health Organization, Technical report (2014)

22. Paquet, C., Coulombier, D., Kaiser, R., Ciotti, M.: Epidemic intelligence: a new framework for strengthening disease surveillance in Europe. Eurosurveillance **11**(12), 5–6 (2006)

23. Pustejovsky, J., Bunt, H., Zaenen, A.: Designing annotation schemes: from theory to model. In: Handbook of Linguistic Annotation, pp. 21–72 (2017)

24. Ruder, S.: An overview of gradient descent optimization algorithms. arXiv preprint arXiv:1609.04747 (2016)

25. Schuster, M., Paliwal, K.K.: Bidirectional recurrent neural networks. IEEE Trans. Signal Process. **45**(11), 2673–2681 (1997)

26. Suárez-Paniagua, V., Dong, H., Casey, A.: A multi-BERT hybrid system for Named Entity Recognition in Spanish radiology reports. In: CEUR Workshop Proceedings, vol. 2936 (2021)

27. Vaswani, A., et al.: Attention is all you need. Adv. Neural Inf. Process. Syst. **30** (2017)

Social AI

From AI Act to Public-Private-People Partnerships: Building AI as a Global Public Good

Migle Laukyte[(⊠)]

Pompeu Fabra University, Law, Ramon Trias Fargas 25-27, 08005 Barcelona, Cataluña, Spain
migle.laukyte@upf.edu

Abstract. AI pervades all human activities, yet its benefits for the biggest part of the planet are still out of reach. This paper argues that the way to change is to involve civil society organizations in developing and deploying the AI. A timid attempt to do so is included in the first law on AI, that is European Union's Artificial Intelligence Act, which invites that deployers of AI systems to consult and collaborate with the civil society in making AI compliant with fundamental rights. As small advancement as it is, it opens the door to the possibilities to make AI much more human-centric and respondent to the real needs of communities where these systems would be deployed. This trend of involving civil society corresponds to the emerging trend of public-private-people partnerships as an alternative to public-private partnerships and could contribute to make AI a Global Public Good, that is, a good accessible to everyone everywhere and hence bring the benefits of the AI not only to the Global North, but also to the Global South.

Keywords: Artificial Intelligence · civil society · public-private-people partnership · Global Public Good

1 Introduction

Artificial Intelligence (AI) is advancing faster than our ability to grasp its ethical implications and work out legal rules to deal with them: usually we need more time to agree upon the regulatory measures that these advancements need to comply with to become ethically and legally acceptable. In fact, the advancements in AI ask us to address the problems AI gives rise to, such as discriminatory practices that stem from biased and incomplete data and lack of control and supervision of how AI is applied. We tried to address these problems through ethical guidelines, but once we have seen that ethics is not sufficient [1], we refer to (without abandoning ethical commitments) harder measures and introduced legal norms, which create obligations for the AI system developers and deployers to be more attentive to the outcomes these systems generate.

However, addressing the challenges of AI is not only a work for legislators, developers and deployers: people should have a role in this process too. This paper argues that to have a more ethically and legally compliant AI, we should give more space to the civil society, that is, all those organizations that are independent from public authorities and

L. Correia et al. (Eds.): IBERAMIA 2024, LNCS 15277, pp. 439–450, 2025.
https://doi.org/10.1007/978-3-031-80366-6_36

private interests and that represent the interests of individuals. Indeed, were these organizations involved in AI development from the very beginning of the AI hype, many of the problems—cases of bias and discrimination, denial of essential services and unequal treatment, among many—would not have taken place, because these problems would have been tackled much quicker by those who are accustomed to identify the threats and deal with the risks that expose the vulnerabilities of the individuals.

The involvement of civil society into the development and deployment of AI comes at a cost both for private and public sectors, who might not be willing to hear critical voices, that question their *modus operandi*. Yet, there is no other choice if we want AI to help us address global challenges, such as inequalities between Global South and Global North, climate change, poverty, hunger, modern slavery and many others. To be sure, we do not want AI to be another problem to add to this list. On the contrary, within the limits of possibilities and never losing the human rights perspective, we should use AI to make lives of everybody, but in particular, those of the most vulnerable ones, better: how to do that is not something that public and private sectors alone have achieved or could achieve. They need help, support and guidance: all of this can be provided by the civil society organizations, that work directly with the vulnerable communities and are much more knowledgeable of and sensitive to their everyday life experiences.

This article addresses the civil society involvement into AI deployment for a broader, globally relevant good by first of all tackling the small steps forward that the public sector is already making to include the civil society in this endeavor. Indeed, a few months ago the European Union (EU) has issued the world's first AI regulation (the so called AI Act).[1] Besides many issues it deals with, it establishes the fundamental rights impact assessment (FRIA) requirement, which not only aligns the AI with fundamental rights, but also opens the door to generate further positive impact in making civil society more involved not only in this impact assessment, but also by indirectly fostering the new trend that pervades co-construction and development of technologies and other advancements in different areas of social life, that is public-private-people partnership (4P). This is a new conception and advancement with respect to the already known model of public-private partnerships (3P): there are reasons why we must introduce and addition P of People in this equation and turn our attention to the needs of society, which is no longer represented by the P of Public.

---

[1] This work focuses on the EU AI Act because this is the only existing regulation on AI: other countries either rely on ethical guidelines (Australia), or specific administrative regulations on management of specific AI services (China), or are still working on a regulatory proposal (Japan) or drafting policies (Nigeria), or adopt sector specific approach to AI risks (UK, Singapore) or are the national AI laws are still pending of approval by national legislator (South Corea, USA). Bearing in mind the absence of legislation in other countries around the world, at least for the time being, this article can be considered Euro-centric in its approach. More information on AI legislative advancements worldwide, https://www.whitecase.com/insight-our-thinking/ai-watch-global-regulatory-tracker#articles.

In fact, advancements of 4Ps represent a progress toward a paradigm shift in our understanding of how AI could improve human lives and work towards re-conceptualization of AI not as private good but rather as Global Public Good, that is a good that is available for everyone everywhere and makes the lives of people better.

Global Public Good is a good that helps to improve human lives not only in Global North, but also in Global South where more often than not essential life needs are not met, healthcare is insufficient, and many other problems, that Global North countries would find incompatible with dignified human existence, persist. To be sure technology cannot solve all these problems, and this paper is not an example of techno-solutionism, according to which any existing problem of humanity has a technological solution or, as [2] puts it, the solution to world's problems lies in optimization of properties of machine-learning algorithm. Far from that: the world has many problems that are impossible to approach, not to say to solve, with the help of calculations or algorithms. And in addition to that some of these enormous planetary problems, such as climate change, are also caused by AI: for instance, in terms of energy consumption, AI is becoming unsustainable.

Bearing the above in mind, the paper is organized as follows: in the second part I introduce the AI Act and focus on fundamental rights impact assessment (FRIA). Although the article dedicated to FRIA does not explicitly refer to civil society organizations, I argue that it opens the door to the civil society participation in it by mentioning civil society organizations in Recital dedicated to FRIA.

Then I move on to discuss what this opening could mean to foster and push forward the collaborations between civil society organizations and the rest of the stakeholders, mainly public and private sectors, which so far have collaborated through 3P (public-private partnerships). I argue that there is a new possibility to include civil society into this model and turn it into 4P model by adding "people" ingredient or civil society ingredient: why this model is important and what benefits it could bring for the rest of the world in terms of AI is the topic for the third part.

The fourth part is dedicated to a broader discussion on the AI as something that cannot be seen as a nationally or regionally defined utility: in this part I explain the idea of AI as a Global Public Good and why and how the 4P model of building and using AI contributes in making AI a global technology that we could—and have a moral obligation to—share with the rest of the world, in particular, with the part of the world which is least technologically advanced and, at the same time, more vulnerable. This is how we could turn AI into a bridge between Global North and Global South.

The paper ends with conclusive remarks and suggestions for the future research.

2 Artificial Intelligence Act and Fundamental Rights Impact Assessment (FRIA): Opening the Door to Civil Society Organizations

Artificial Intelligence Act (AI Act) is the first law in the world that regulates the AI that is being sold or made available on the EU single market.[2] The law underwent many debates and has been subject to harsh critics and pressures from all the stakeholders, both private and public, who on the one hand either saw this Act as too demanding and burdensome for private sector, or too permissive and not sufficiently protective for vulnerable groups and society at large (among many, see [3] and [4]).[3]

The AI Act is based on the risk and classifies the AI systems into prohibited, high-risk, limited risk and minimal risk. Most of the systems deployed in the public sector are high-risk, because they are deployed to guarantee essential public services, to deal with critical infrastructures, be deployed in democratically vital environment such as judicial system or elections, or otherwise affect human lives. Bearing the importance of these systems in mind, the EU legislator has introduced many control and compliance instruments to make sure that any high-risk AI system is safe and secure in terms of fundamental rights and would not create new (or deepen old) inequalities, would not discriminate and would not negatively impact on vulnerable social groups.

Among many control and compliance mechanisms that this Act establishes and requires providers, deployers, importers, distributors and other stakeholders to comply with, what interests us most is the specific requirement for High-Risk AI systems, in particular if deployed within the public sector, namely, fundamental rights impact assessment (FRIA): this impact assessment requires the deployers of such systems to comply with specific instrument that should protect the fundamental rights from any AI-related risks.

2.1 Fundamental Rights Impact Assessment (FRIA)

FRIA is described in the art. 27 of the AI Act and states that before deploying certain high-risk AI systems for the first time, its deployers must perform an assessment of how this system could impact on fundamental rights.[4]

---

[2] The official full title is "Regulation (EU) 2024/1689 of the European Parliament and of the Council of 13 June 2024 laying down harmonised rules on artificial intelligence and amending Regulations (EC) No 300/2008, (EU) No 167/2013, (EU) No 168/2013, (EU) 2018/858, (EU) 2018/1139 and (EU) 2019/2144 and Directives 2014/90/EU, (EU) 2016/797 and (EU) 2020/1828 (Artificial Intelligence Act)Text with EEA relevance, https://eur-lex.europa.eu/legal-content/EN/TXT/?uri=CELEX:32024R1689, accessed 2024/05/19.

[3] Although geographically it is limited to the EU market, AI Act applies also to those who develop and deploy AI outside the EU, but want to sell it or use its output within the EU.

[4] What this assessment consists of is described in the art. 27.1: "(a) a description of the deployer's processes in which the high-risk AI system will be used in line with its intended purpose; (b) a description of the period of time within which, and the frequency with which, each high-risk AI system is intended to be used; (c) the categories of natural persons and groups likely to be affected by its use in the specific context; (d) the specific risks of harm likely to have an impact

As already briefly mentioned above, FRIA applies to limited set of high-risk AI systems, namely those systems that are (a) being deployed by the public sector, (b) deployed by private sector, providing public services, those which (c) evaluate people's credit trustworthiness or their credit score,[5] or those which (d) are used for risk assessment and pricing of life and health insurance. In all these cases FRIA is mandatory.

What interests us most in FRIA is not only its contents, but what we can read in the Recitals of the law: Recital 96 includes explanation that the deployers of these systems, when performing the FRIA, could involve not only people who might be affected by the system (for instance, citizens or special vulnerable groups such as immigrants, asylum seekers, people with disabilities, etc.), but also independent experts and civil society organizations. These organizations—defined by the EU as "organisational structure whose members serve the general interest through a democratic process, and which plays the role of mediator between public authorities and citizens"[6]—would not only help to assess the AI's real or hypothetical impact on fundamental rights but could also suggest how to act should these risks (impact) come into being.

Indeed, civil society organizations have been actively participating in drafting AI Act [5] and have always been active in keeping an eye on public and private initiatives of AI use that often lead to negative outcomes for the most vulnerable social groups, such as the famous case of SyRI in the Netherlands[7] or El Bosco system in Spain [6].[8]

Unfortunately, this suggestion is not part of the art. 27, but a part of Recital: Recitals are not legally binding and are used only when the problem of interpreting the norm arises. Yet this also means that the Recitals add to the meaning of the norm and help us understand what the legislator had in mind when drafting a particular norm. This is why we can argue that, in the case of AI Act, the legislator was suggesting that the deployers should rely on and involve the civil society organizations into its practices more, so much so when the AI system is about to be used for the public services or within a

---

on the categories of natural persons or groups of persons identified […]; (e) a description of the implementation of human oversight measures, according to the instructions for use; (f) the measures to be taken in the case of the materialization of those risks, including the arrangements for internal governance and complaint mechanisms.

[5] Systems to establish financial fraud are excluded from the FRIA.

[6] Full definition and its importance within the EU governance is explained here https://eur-lex.europa.eu/EN/legal-content/glossary/civil-society-organisation.html, accessed 2024/07/27.

[7] SyRI was a welfare fraud detection system used by Dutch government to elaborate profiles of possible tax evaders, which resulted in violation of fundamental right to privacy, established in the European Charter of Fundamental Rights, because the SyRI was not known to citizens nor they were informed about its application and functioning and the legislation on SyRI did not foresee the obligation to inform the data subjects (citizens) that they data was collected and that the reports on them have been generated. In 2020 The District Court of Hague prohibited its use. For more information, see https://www.uantwerpen.be/en/projects/aitax/publications/syri/, accessed 2024/05/18.

[8] El Bosco is a Spanish Government's system to determine whether a person is eligible for a energy bill subsidy: thousands of families have seen their applications denied and this called the attention of civil society organization CIVIO which started a battle to access the source code of the program which calculated eligibility. For the time being, this access is denied because on the Intellectual Property grounds.

public domain to guarantee essential support to those who need it most yet cannot afford to buy it on the private market.

We see this intention not only in the case of FRIA, but also in other norms and Recitals, which support the civil society involvement in drafting the codes of practices (art. 57), becoming part of advisory forum (art. 67), or participating in developing codes of conduct (art. 95). The role of civil society organizations is slowly becoming more legally acknowledged and hopefully will lead to granting it a more active role: to be sure the legislator would delegitimize itself should he acknowledge and grant a more substantial role to civil society in making AI fundamental rights-compliant. Indeed, ensuring fundamental rights is a role of a State and not of the civil society organizations. Yet, if it delegates the FRIA to the public or private deployers, the civil society organizations must be involved as well as a watchdog and as a control mechanism. Thus, this mention of civil society organizations in the Recitals could be read as a smart move to invite the civil society's participation without imposing it (and generating refusal from the other private and public stakeholders).

To be sure, there is no information or guidelines how to involve the civil society organizations: on the one hand, it is a positive move as otherwise the legislator might risk to involuntarily exclude certain organizations, yet on the other hand, lack of this information sends a message of uncertainty and vagueness. On of the possible ways to identify the civil society organizations is through a registry, for instance, similar to the one, called the Fundamental Rights Platform, used by the Fundamental Rights Agency of the EU.[9] This kind of registry—referred to as civil society cooperation channel—could be also used to identify the specific civil society organizations which best fit specific application at hand: for instance, if the FRIA needs to be prepared for an AI-based robotic assistance system to be deployed in the elderly residence, it should be mandatory to have involved in FRIA at least two civil society organizations working with elderly, for instance, organizations belonging to Age Platform Europe, the largest network in EU of organizations that represent elderly people's rights.[10]

But why this role is so important and what is the general trend in making civil society organizations more active in shaping the AI's deployment and use within the public domain? Why the collaboration with civil society organizations could be beneficial in the first place?

In what follows I explain why these collaborations are beneficial and that they represent a new pattern in collaborative effort between different sectors: indeed we are familiar with private-public partnerships (3P), but lately new ideas have been put on the table, and one of these ideas is public-private-people partnership (4P) as a new way to

---

[9] More about this platform, see https://fra.europa.eu/en/cooperation/civil-society, 2024/07/25.

[10] More about Age Platform Europe, see https://www.age-platform.eu/#, accessed 2024/07/27. There are many further ideas of how to operationalize the involvement of these organizations: for instance, by applying rotation principle, according to which the priority should be given to those organizations that have been least involved, so as to ensure as inclusive and participatory as possible involvement of these organizations and also to avoid a few organizations monopolizing the field of FRIA. Another idea could be to prioritize the involvement of small and local organizations thus offering them also a space to grow and learn to address and ass FRIA thus also contributing to create closer relationship between local population and private sector.

deal with contemporary problems of building co-existence between different sectors, social groups, interests and accountabilities.

3 Building a Partnership for AI

Public-private-people partnership (or 4P) scheme emerged as a global response to the difficulties to keep the citizens involved in (being part of) decision-making processes between public and private sectors. This scheme reflects an underlying broader and more structural critique of public-private partnership (3P) scheme: although the citizens are indirectly represented by the public sector in this partnership (public sector represents and sometimes is even directly elected by the citizens), this representation is not sufficient or there is general lack of trust in public sector's ability to truly represent and defend the interest of the citizens vis a vis the interests of the private sector (indeed, the cases of SyRI and El Bosco, mentioned above, are just two examples among many why this trust was lost). To be sure, public trust in general has a role to play in this equation: many researchers and institutions, among which the World Bank itself, have argued that there is a mutually reinforcing dynamic between trust in the public institutions and citizen engagement (for an overview, see [7]). In addition to that, a whole spectrum of nonmonetary determinants, such as end user inclusion, politics, citizens' sentiments, among other, have a role to play as well [8].

In the light of the above 4P model emerges as a response to the shortcomings of the 3P model, such as lack of transparency and limited input from the public [9]. Indeed, the fact that the society is not involved in the project (whatever that project might be), yet the project is dedicated to address social needs or its end-users are common citizens, is incoherent and can lead to negative consequences, that emerge when the providers and deployers of services make assumptions about reality, that have nothing to do with that reality. This is nothing new,[11] yet as much as we innovate and advance technologically, we continue repeating the same error, that is not putting the citizen at the center, but situating him or her in some hypothetical mute position of constant approval.

Furthermore, the dynamics of 3P is different with respect to 4P: in 3P the relationships are governed by clearly established formalisms (such as contractual arrangements), whereas in 4P these formalisms are not part of the ways in which the civil society organizations shape their participation, which by nature is much less formal and does not follow established dynamics of interaction [9].

Further difference between 4Ps and 3Ps is the dimension of the project in question: whereas 3P are enormous projects such as building of highways and airport management, 4Ps are still lagging behind in size and hence in the dimension of P of people, although

---

[11] See, for example, the OECD Recommendation of the Council on Digital Government Strategies (OECD/Legal/0406) (2014), which ten years ago called for "[…] engagement and participation of public, private and civil society stakeholders in policy making and public service design and delivery…", whole text is available at https://legalinstruments.oecd.org/en/instruments/OECD-LEGAL-0406, accessed at 2024/07/27.

some experiences are available to get the inspiration from, such as multi-disaster management in Fukushima [10], or urban regeneration and affordable housing provision [8], smart city initiatives [11], among others.[12]

In case of AI deployment, the 4P model promises much more than 3P: the cases of improper, discriminatory or rights-infringing uses of AI, that have been focus on many academic publications, reports and court cases, clearly show that the public sector not only failed to represent the citizens' interests, but also that the citizens cannot rely on the public sector to represent them and therefore need to be directly involved in making decisions on the deployment of these systems in public-interest-sensitive issues, such as social welfare or education.

In this sense we might see the 4P model as a model that fosters a more inclusive design and deployment of AI that would truly represent the interests of citizenry and push forward a normalization of such involvement: that is to say, to involve civil society in FRIA should lead to more collaborative initiatives between these sectors. More initiatives of this kind we have, less probable not only the infringement of fundamental rights will be, but also more aware the private (and public for that matter too) sector will become about the interests and problems that the end-users (people) must face. This awareness should lead to a more and better-informed public and private sector, which in the previous experiences has often lost the overall goal of their work in the public sector: that is to contribute to improve the lives of people and not to turn the technology (in our case, AI) against them.

The question is whether the 4P partnership could work to develop AI-based solutions: for the time being the academic literature does not advance any response to this question, so much so that the first steps of 3P for AI have just been made and their critical evaluation is still work in progress. However, there is an enormous potential in involving citizens more directly in the development and deployment of AI and this opportunity should not be lost.

One of the big problems is to convince the private sector that the involvement of civil society organizations is an added value to—and not a liability for—their business models. There are different possibilities in this regard: for instance, the collaboration with civil society organizations builds trust of citizens (consumers) and make a certain company recognizable to the local community or a certain social group (for example, if it is an AI tool deployed in education settings). Another suggestion is that the public sector itself would foster civil society involvement and, in a way, push the companies to collaborate with it by implicitly assuming that these organizations are inherent part of any project dedicated to address citizens' needs. Further suggestion would be to push the civil society organizations in becoming more proactive in constructively interacting with the private sector: probably hostile and negative attitudes are not functional in building collaborations and therefore a different approach is needed.

The assumption is that if we involve citizens in building the technologies—AI in this case—the citizens will make sure that this involvement would reflect on the way AI

---

[12] For instance, within the housing redevelopment, "people" dimension emerges because of the geographic reasons: the communities from adjacent areas to a particular redevelopment are concerned about its impact on the prices of their properties, increase or decrease of rents, and other effects.

functions and affects the society. This means that involving citizens is a form of doing FRIA in a continuous way: it is not just about the FRIA *before* the deployment, but a continuous FRIA throughout the life cycle of the AI system and doing FRIA even before an AI system takes shape.

But the citizens involvement in shaping AI does not stop at the boundaries of specific public services of individual communities, cities or states: overall objective of citizens involvement goes further and regards the change of paradigm of our understanding of AI not only as a product or business asset to provide the public service with, but also—and most importantly—as a Global Public Good that all the societies, in particular those in underdeveloped and poor countries, should have access to.

4   From 4P to AI as a Global Good

The debate on Global Public Goods—these goods whose benefits affects almost everyone everywhere [12] and which range from environment and culture to metric system [13]—moves around their accessibility and provision. Stiglitz [14] describes Global Public Good as a good that is not rivalrous—many people can use that good at the same time without any additional costs for doing that—and non-excludable, which means that nobody should be excluded from its use or that its use does not exclude other from doing the same. If applied to knowledge, my knowledge on how to write an article does not affect any other person's knowledge on how to write an article nor does it exclude anyone else from knowing that nor it de-valuates my knowledge.

In case of AI this is even more the case bearing in mind that a particular AI improves if used by many people (with due safeguards and guarantees for privacy and personal data protection, safety from manipulation, and other requirements). Therefore, from this perspective, AI should be considered as a Global Public Good which can and should be used by as many people as possible if this use improves their lives and benefits their wellbeing: it is about the knowledge that AI generates and about the solutions that people generate with the help of AI. In both cases we are talking about the AI-empowered knowledge as a Global Public Good.

The role of international cooperation and public-private (3P) and, in particular, public-private-people partnerships (4P) are essential in making this objective come true. Indeed, many times these Global Public Goods stem from national initiatives and partnerships, that can exercise political pressure in support of funding and delivery of these benefits. The need for such goods could push communities and private sector to consume those services and goods that could generate externalities that augment the intermediate public goods—such as knowledge and technology, but also norms and principles—and thus contribute to the provision of the Global Public Good [12]. Indeed, one of such externalities is AI Act and its timid yet clearly citizens-involvement-oriented approach to AI and to FRIA is a proof that we are moving in the right direction.

There is further aspect related to AI: besides understanding AI as knowledge good based on digital technologies (Global Public Good as such), it can also be seen as a means to build and share Global Public Goods: this means that the AI is not only a Global Public Good in itself, but also a tool to empower and push forward other Global Public Goods. This is especially relevant in addressing the knowledge- and digital technology divide between the Global North and Global South. For instance, such a Global Public Good as international regulations for civil aviation and telecommunications [15] relies on many procedures, stakeholders and technologies, yet the AI plays a particularly relevant role in making this Global Public Good of transport safety and communication reliability accessible almost all over the world. The same could be said about AI uses for diagnostics in medical domain or an AI that can monitors water quality and detect dangerous bacteria and harmful particles[13]: in both cases AI is functional to ensure the world with two Global Public Goods such as health and water respectively.

The way to access these Global Public Goods—which includes, but to be sure is not limited to, AI applications that add to the human wellbeing and help humans live a more dignified life—goes through strengthening commitments to (primary, secondary, tertiary, etc.) education [14]. Education is pivotal in this because it helps people access, understand, apply and, in general, use this knowledge.[14]

A further aspect that is worthwhile exploring is to look at AI through the international collaboration prism as a common good in itself (not to confuse with Global Public Good). Common good is not as an outcome of collective action which improves the world, but the collective action in itself which is valuable [17]. By this I mean that if the international community agrees that the benefits of AI should be distributed more equally and fairly among the developed and developing countries, agree that the benefits of AI should reach everyone around the globe and agree to work to achieve this goal to ensure the minimal standards of wellbeing to everyone, we would achieve an effective improvement of life quality in the (least) developed countries (through the Global Public Good prism), and also we would achieve a common good of working together as an international community for a common—and most importantly, good!—goal. In this sense the benefit comes not only from the result (Global Public Good), but also from the process itself (common good).

To be sure there are a few very serious critical aspects that hinder this paradigm of AI as a Global Public Good: without addressing them all here, enough is to mention the environmental problems related to the AI consumption and proprietary nature of AI. To be sure these are enormously problematic issues and they do not have a quick and easy solution: AI's environmental aspects are known and make part of the much broader map of environmentally devastating industries which are slow in finding alternatives to their usual practices that have altered the environmental equilibrium of the planet. The proprietary nature of AI weighs heavily on any AI initiative, yet there is part of AI community that do not fit into this dynamics and adopt open source and other models of technological development that privilege the social good of many over financial benefit of a few. We can only hope that such endeavors can become a common practice and push the private sector to rethink their duties in terms of corporate social responsibility and commitments with the rest of the planet.

Be it as it may, both the environment protection and unbalanced and abusive proprietary regimes are all issues on the agenda on civil society organizations: adding AI would not change, but rather justify, their work on these challenges and give more energy to engage in new ones.

---

[13] See for instance https://cleanwaterai.com/, accessed 2024/05/19.

[14] For an overview of AI applications for education from 1993 to 2020, see [16].

5   Conclusions

In this paper I have argued that the socially and ethically sound development of AI is based on the law, ethics, international politics among many, but also on civil society engagement: it is a common effort that should be functional to help us work on a common objective, that is an AI that would serve the humanity as a whole and not the part concentrated in rich countries.

I have argued that the first law on AI, that is the EU's AI Act, is making an attempt to involve the civil society organizations by timidly opening the possibility to involve them in the FRIA. FRIA is necessary to make sure that any AI system which is about to be deployed for the first time within the EU undergoes checks to make sure that it does not violate nor put to risk the fundamental rights of its citizens. Although the involvement of civil society organizations is not part of the specific article that deals with this assessment, we find the reference to the civil society organizations in Recitals, that are part of law used to explain the reasons and *raison d'être* of the norms.

Hence I argue that this is the step forward in making the AI more aligned with social values and needs of Global South vis-à-vis Global North: peoples' voices have not been heard and new emerging collaborative models to develop and deploy technologies should be the way to proceed. I argue that the people's (or civil society organizations') involvement in these collaborative models is a necessary condition to comply with socially sound and ethically compliant AI which would not only respect fundamental rights of EU citizens, but would exportable to other countries, regions and continents. Indeed, my last argument is that we cannot see AI as a national or regional matter: it has to reach the rest of the planet, in particular to contribute solving the problems that poor regions of the world need help in addressing. I am not advocating for techno-solutionism or new forms of colonialism: I argue that if we hold dear Universal Declaration of Human Rights, if we care about other human beings and future generations, we should make civil society organizations more actively involved in developing and deploying AI and help to make AI a Global Public Good, something that people could rely on and not doubt in its positive impact on their lives.

In terms of future research, there are numerous questions to be addressed: among many, these suggestions have to be operationalized and turned into widely adopted practices: otherwise, they remain naïve and empty exercises of wishful thinking. In this sense, sharing of good practices is paramount and collaborative and constructive attitude of all the stakeholders—and not just civil society organizations—is a game changer.

References

1. Munn, L.: The uselessness of AI ethics. AI Ethics **3**, 869–877 (2023)
2. De Cremer, D., et al.: The road to a human-centred digital society: opportunities, challenges and responsibilities for humans in the age of machines. AI Ethics **2**(4), 579–583 (2022)
3. Caroli, L.: Will the EU AI act work? lessons learned from past legislative initiatives, future challenges. IAPP (2024) https://iapp.org/news/a/will-the-eu-ai-act-work-lessons-learned-from-past-legislative-initiatives-future-challenges. Accessed 19 May 2024
4. Crenshaw, J.: Future of AI: EU ai act fails to strike sensible balance. us chamber of commerce (2023). https://www.uschamber.com/technology/future-of-ai-latest-updates. Accessed 19 May 2024

5. Edri, E., et al.: An EU artificial intelligence act for fundamental rights. a civil society statement (2022). https://edri.org/wp-content/uploads/2021/12/Political-statement-on-AI-Act.pdf. Accessed 18 May 2024

6. Cabo, D.: Fighting. blog of for algorithmic transparency in spain digital freedom fund (2020). https://digitalfreedomfund.org/fighting-for-algorithmic-transparency-in-spain/. Accessed 01 May 2024

7. Kumagai, S., Iorio, F.: Building trust in government through citizen engagement. working paper. Washington, D.C.: world bank group (2020). http://documents.worldbank.org/curated/en/440761581607070452/Building-Trust-in-Government-through-Citizen-Engagement. Accessed 01 May 2024

8. Miranda-Poggys, A.G., Morena, M.: A critique to public-private-people partnerships: from a definitional inconsistency to the partnering dilemma in today's housing conjunction. Sustainability **15** (2023), https://www.mdpi.com/2071-1050/15/6/4859. Accessed 01 May 2024

9. Perjo, L.: Public-private-people partnerships – a new concept to bring public and private actors and citizens together. Nordregio Mag. **4**(16), 3–4 (2016)

10. Kobashi, Y., et al.: Maturing of public-private-people partnership (4P): lessons from 4P for triple disaster and subsequently COVID-19 pandemic in Fukushima. J. Glob. Health **12** (2022). https://doi.org/10.7189/jogh.12.03028. Accessed 23 Apr 2024

11. Liu, T.S., et al.: Emerging themes of public-private partnership application in developing smart city projects: a conceptual framework. Built Environ. Proj. Asset Manage. **11**(1), 138–156 (2020)

12. Kaul, I.: Global public goods: explaining their underprovision. J. Int. Econ. Law **15**(3), 729–750 (2012)

13. Chin, M.: What are global public goods? international monetary fund (2021). https://www.imf.org/en/Publications/fandd/issues/2021/12/Global-Public-Goods-Chin-basics

14. Stiglitz, J.E.: Knowledge as global public good. In: Global Public Goods: International Cooperation in the 21st Century, pp. 308–325. Oxford University Press, Oxford (1999)

15. International task force on global public good: meeting global challenges: international cooperation in the national interest. report (2006). https://ycsg.yale.edu/sites/default/files/files/meeting_global_challenges_global_public_goods.pdf. Accessed 12 Apr 2024

16. Zhang, K., Aslan, A.B.: AI technologies for education: recent research & future directions. computers and education: Artif. Intell. **2** (2021). https://doi.org/10.1016/j.caeai.2021.100025. Accessed 10 May 2024

17. Deneulin, S., Townsend, N.: Public goods, global public goods and the common good. Int. J. Soc. Econ. **34**(1–2), 19–36 (2007)

Posters

Optimization of Generalized Assignment Problem for a Machinery-Aided Composting Process

Yael Andrade-Ibarra[1], Uriel Trejo-Ramirez[1], Oliver Cuate[1],
Adriana Lara[1], and Lourdes Uribe[2]

[1] Instituto Politécnico Nacional, ESFM, Edificio UPALM, C.P. 07730 Gustavo A.
Madero, Mexico City, Mexico
ocuatg@ipn.mx, alaral@ipn.mx
[2] Facultad de Ciencias Actuariales, Universidad Anáhuac México,
Huixquilucan, Mexico
lourdes.uribe@anahuac.mx
https://www.esfm.ipn.mx

Abstract. Currently, one of the leading human problems at a global
level is the management of organic waste. In this work, we model a composting process as a *Generalized Assignment Problem (GAP)* where we
aim to minimize the cost of composting a particular volume of specific
waste using specialized machinery. The proposed model is tested using
different approaches in different scenarios. Numerical experimentation
proved that B&B and the selected Evolutionary Algorithms are practical tools for addressing the composting and organic waste treatment
problem, finding optimal or near-optimal solutions for the defined objective function. This optimization directly resulted in a reduction in the
total cost of the composting process.

Keywords: Assignment · Composting · Evolutionary Algorithm

1 The Approach

Nowadays, one of the leading human problems at a global level is the management of organic waste. A considerable amount of this organic waste ends
up in landfills. Hence, composting is a transcendent topic that helps mitigate
the adverse effects of this worldwide problem. The aerobic composting method
consists of building waste piles created in parallel that are periodically rotated
to aerate the waste. For this type of composting process, there are specialized
machines for this process. These machines are pulled by a tractor, implying the
use of Diesel in each project. Details of each machine and technical information
about these machines can be consulted in [2,6]. Composting cycles refers to the
time needed to compost a specific type of waste. In recent years, different optimization approaches have been used to improve the composting process. These

L. Correia et al. (Eds.): IBERAMIA 2024, LNCS 15277, pp. 453–457, 2025.
https://doi.org/10.1007/978-3-031-80366-6

approaches involve both mathematical and artificial intelligence based methods [7–9]. In [3] a review of the optimization models proposed for this topic was carried out. In this work, the GAP was considered to tackle the problem. The objective of the GAP is to assign tasks to agents in such a way that all tasks are completed and the total cost is minimized. The GAP may be formulated as an *Integer Linear Programming (ILP)* model with binary variables. For the mathematical definition of the GAP please consult [5].

In this work, we revisit the application from [2] and propose a model for the composting problem as a GAP where we aim to minimize the cost of composting a particular volume of a specific waste using specialized machinery. The conditions and the characteristics of the composting process that must be taken into account are: (a) One composting cycle corresponds to 6 days, (b) each compost pile must be processed exactly 3 times per cycle, (c) once the compost pile is processed, it must be allowed to rest for one day before it can be processed again, (d) there are 8-hour workdays, with H hours being fully productive, (e) a pile cannot be processed simultaneously by more than one machine, (f) the volume processed for each cycle is constant and it is not uniform. Considering the data given in [2] and setting $H = 6$ as fully productive hours, the maximum processing capacity per day for SM $1800\,\mathrm{m}^3$, for MM $3000\,\mathrm{m}^3$, and for LM $4200\,\mathrm{m}^3$. Therefore, there is a maximum volume that can be processed per day considering different number of available machines. The model proposed below is designed for a single cycle. As each cycle will be repeated 18 times, thus, if we manage to optimize one cycle, we can replicate it for the others and thereby optimize the entire process.

Taking into account the assumptions stated, lets define c_{ij} as the cost of pile j being processed by machine i, t_{ij} as the time taken by machine i to process pile j and x_{ijk} is pile j is assigned to machine i on day k. Note that $x_{ijk} = 1$ if pile j is assigned to machine i on day k and $x_{ijk} = 0$ otherwise. Let $K = \{1, 2, \ldots, 6\}$ be the number of workdays, $N = \{1, 2, \ldots, n\}$ the number of piles and $M = \{1, 2, \ldots, m\}$ the number of available machines. The following linear model is proposed to minimize fuel costs in the composting process for n piles and m machines (see Eq. (1)).

$$\min \sum_{k=1}^{6} \sum_{i=1}^{m} \sum_{j=1}^{n} c_{ij} x_{ijk},$$

$$
\begin{aligned}
\text{s.t.} \quad & \sum_{j=1}^{n} t_{ij} x_{ijk} \leq 6 && \forall k \in K \text{ and } \forall i \in M, \\
& \sum_{i=1}^{m} x_{ijk} \leq 1 && \forall j \in N \text{ and } \forall k \in K, \\
& \sum_{k=1}^{6} \sum_{i=1}^{m} x_{ijk} = 3 && \forall j \in N, \\
& \sum_{i=1}^{m} x_{ij1} + 2 \sum_{i=1}^{m} x_{ij2} + \sum_{i=1}^{m} x_{ij3} = 2 && \forall j \in N, \\
& \sum_{i=1}^{m} x_{ij3} + 2 \sum_{i=1}^{m} x_{ij4} + \sum_{i=1}^{m} x_{ij5} = 2 && \forall j \in N, \\
& \sum_{i=1}^{m} x_{ij5} + \sum_{i=1}^{m} x_{ij6} = 1 && \forall j \in N.
\end{aligned}
\tag{1}
$$

Note that the first constraint describes the limitation that the machines must work for at most 6 h per day, which are the productive hours. While the second constraint indicates that a pile cannot be processed simultaneously by more than one machine. The third constraint states that each pile must be processed exactly 3 times in a week. Lastly, the remaining constraints indicate that the pile, once processed, must rest for exactly one day.

2 Numerical Results

In this section we present the experimental setting of the numerical experimentation. Two different scenarios were considered to test the proposed model. First we considered only three machines (one of each type) and different number of piles, then we considered a fixed number of piles and different number of machines. To solve both problems, two different approaches were taken into account. First, a branch and bound technique (B&B) was tested, we used the Matlab function available in the *MILP library* [4]. Then two evolutionary algorithms were employed. We used the genetic algorithm (GA) and Particle Swarm Optimization (PSO) implemented in *Pymoo Framework* [1]. As mentioned above,

Table 1. Numerical results for different scenarios

First scenario

Piles	GA				PSO				MILP	
	min f	*prom f*	*Max f*	*time*	*min f*	*prom f*	*max f*	*time*	*f(x)*	*time*
20	249.1252	**249.1252**	249.1252	22.5758	249.1252	249.1856	249.5537	30.9181	249.131	0.0208
(std.dev)	—	(2.84E-14)	—	(22.5758)	—	(0.1425)	—	(2.4275)		
22	298.8571	**298.8571**	298.8571	26.4881	298.8571	299.0574	299.8214	116.0662	300.1862	1.6874
(std.dev)	—	(5.68E-14)	—	(0.8615)	—	(0.1425)	—	(2.9582)		
24	336.3266	**336.4584**	336.6046	29.5761	336.2829	336.6067	336.9733	138.0508	337.6523	2.7028
(std.dev)	—	(0.1223)	—	(0.9080)	—	(0.1425)	—	(2.8653)		
26	373.7441	**373.8037**	373.9243	31.2217	373.7675	373.9965	374.5094	175.4071	375.0523	3.8522
(std.dev)	—	(0.04695)	—	(1.5039)	—	(0.1989)	—	(8.9852)		
28	391.2943	**391.4865**	391.9919	32.4525	391.3006	391.6647	391.9868	193.2557	392.6336	2.4102
(std.dev)	—	(0.2277)	—	(1.4998)	—	(0.1963)	—	(11.8871)		
30	422.9802	**423.1354**	423.6284	37.8623	422.9852	423.2916	423.8930	201.9562	424.5146	3.5487
(std.dev)	—	(0.1825)	—	(2.9190)	—	(0.2487)	—	(11.7070)		
32	456.5877	**456.6796**	456.8494	38.8693	456.6881	456.8748	457.0041	233.4533	458.063	2.6848
(std.dev)	—	(0.0761)	—	(2.5907)	—	(0.1117)	—	(13.3249)		
34	488.0852	**488.1712**	488.2512	48.1533	488.0494	488.6001	489.4876	270.0317	489.5162	4.3096
(std.dev)	—	(0.062648)	—	(2.9275)	—	(0.3571)	—	(21.7428)		
36	529.0803	**529.1799**	529.4520	56.2171	529.3900	533.9344	537.6745	318.6087	535.5007	5.6474
(std.dev)	—	(0.0957)	—	(5.7089)	—	(2.7348)	—	(16.9416)		
38	554.9801	**557.6354**	562.1027	59.4358	556.3207	561.4206	464.0978	459.3208	564.5548	4.4446
(std.dev)	—	(2.512)	—	(6.7145)	—	(2.2629)	—	(18.1620)		
40	573.8873	**574.2011**	574.7171	72.8439	575.1234	578.7676	583.0102	461.7576	585.1213	7.5728
(std.dev)	—	(0.4000)	—	(6.4000)	—	(2.0062)	—	(11.0079)		
42	624.2770	**626.1604**	628.4781	92.1710	625.0110	629.2156	633.8795	524.6892	640.3492	6.3962
(std.dev)	—	(1.4689)	—	(9.3143)	—	(1.9777)	—	(18.9794)		
44	660.4362	**664.0278**	667.1119	204.9671	661.5227	664.8941	669.3800	568.6045	681.4682	6.2068
(std.dev)	—	(2.0506)	—	(12.7037)	—	(1.9908)	—	(20.8917)		

Second scenario

Cases	GA				PSO				MILP	
	min f	*prom f*	*Max f*	*time*	*min f*	*prom f*	*max f*	*time*	*f(x)*	*time*
1	817.0549	**817.1389**	817.2558	158.4061	819.5569	821.9558	824.2744	992.5377	894.9596	9.9157
(std.dev)	—	(0.0613)	—	(10.8730)	—	(1.6537)	—	(16.5377)		
2	735.5334	**735.5391**	735.5455	104.1116	735.7135	736.3709	737.0621	878.9065	787.5351	4.3926
(std.dev)	—	(0.0035)	—	(7.0887)	—	(0.5301)	—	(13.6653)		
3	688.8684	**688.8684**	688.8684	103.4198	688.8684	**688.8684**	688.8684	847.4719	**688.8684**	0.0761
(std.dev)	—	(0)	—	(9.7561)	—	(0)	—	(12.4357)		

the composting process must satisfied specific restrictions. To enhance the efficiency of both the GA and PSO, we proposed a Repair operator. This operator determines if an individual meets all the constraints; if not, it identifies which constraints are violated and modifies the individual based on those constraints. For the first scenario we tested the selected algorithms in the proposed model considering from $n = 10$ piles up to $n = 44$ piles, increasing by 2 piles per experiment, and $m = 3$ machines (one of each type). For the second scenario we tested the proposed model considering $n = 60$ piles, and a total of $m = 5$ machines. In both cases, for the B&B algorithm, the GA and PSO we used as stopping criteria a maximum number of function evaluations ($feval_max = 50,000$).

Table 1 shows the obtained solutions starting with the first scenario and then the obtained results for the second scenario. It is important to recall that the entire process spans a total of 18 cycles. Analyzing the results from the first experiment, it is evident that the GA offers notable consistency and precision in solving the pile distribution problem (GA obtained better results in 24/24 cases). It is worth to notice that also PSO obtained promising results. In this case, one can see that beyond 35 piles, the savings increase significantly, reaching up to approximately 2.5% for 44 piles. For the second experiment, it can be observed that the GA provides significantly better solutions than MILP in cases 1 and 2; for case 1, there is an 8.7% saving in the entire process, while in case 2, there is a 4.3% saving. The provided graphs highlight the effectiveness of the GA compared to MILP when facing problems with different numbers of piles. Evolutionary algorithms show a clearer advantage, with MILP presenting higher costs. Thus, Evolutionary algorithms proved to be an effective tool for addressing the composting and organic waste treatment problem, finding optimal or near-optimal solutions for the defined objective function.

Acknowledgment.. Yael Andrade-Ibarra and Adriana Lara acknowledge support from project no. SIP20240568. Uriel Trejo-Ramirez and Lourdes Uribe acknowledges support from project no. SIP20241958. Oliver Cuate acknowledges support from project no. SIP20240570.

References

1. Blank, J., Deb, K.: pymoo: multi-objective optimization in python. IEEE Access **8**, 89497–89509 (2020)
2. De Jesús, N., Lara, A., Uribe, L.: Aplicación de un modelo flow shop con máquinas paralelas para optimización de ciclos de compostaje. Mem. RNAFM **28**, 445–453 (2023)
3. Malaysia, I.: Optimal process network for municipal solid waste management in. J. Cleaner Prod. **30**, 1e11 (2013)
4. MATLAB: intlinprog. https://la.mathworks.com/help/optim/ug/intlinprog.html. Accessed 20 May 2024
5. Nauss, R.M.: The generalized assignment problem. In: Integer Programming, pp. 55–72. CRC Press, Boca Raton (2005)
6. Román, P., Martínez, M.M., Pantoja, A.: Manual de compostaje del agricultor. Experiencias en América Latina. FAO, Organización de las Naciones Unidas para

la Alimentación y la Agricultura (FAO) (2015). http://www.fao.org/docrep/019/i3388s/i3388s.pdf. cobertura geográfica: América Latina y el Caribe

7. Sahu, A., Tapadar, R.: Solving the assignment problem using genetic algorithm and simulated annealing. In: IMECS, pp. 762–765 (2006)

8. Sethanan, K., Pitakaso, R.: Improved differential evolution algorithms for solving generalized assignment problem. Expert Syst. Appl. **45**, 450–459 (2016)

9. Sokac, T., et al.: Application of optimization and modeling for the composting process enhancement. Processes **10**, 229 (2022)

Configuring an LLM Chatbot as Practice Partner for Language Learning

Pablo Gervás[1]([⊠]) [iD], Carlos León[1] [iD], Mayuresh Kumar[2] [iD],
Gonzalo Méndez[1] [iD], and Susana Bautista[3] [iD]

[1] Facultad de Informática, Universidad Complutense de Madrid, 28040 Madrid, Spain
{pgervas,cleon,gmendez}@ucm.es
[2] Aligarh Muslim University, Aligarh, Uttar Pradesh, India
[3] Universidad Francisco de Vitoria, Madrid, Spain
susana.bautista@ufv.es

Abstract. Given appropriate prompts, chatbots based on Large Language Models (LLMs) can adapt both the content and the role in which they participate in the conversation. The present paper reports an experiment to configure one such chatbot to engage in conversation with a language learner so that it can propose engaging situations of the appropriate level of complexity, enact specific roles in the conversation and monitor learner responses. The proposed functionality is tested for a classroom of Hindi-speaking students of Spanish language at Aligarh Muslim University in India.

Keywords: AI in Education · Conversational AI · ChatGPT · Formal Language Learning · Usability Testing

1 Introduction

The present paper reports an experiment in the use of chatbots to help language students to practice their skills on a one-to-one basis with an accommodating partner. This is attempted by designing specific prompts to configure a chatbot to perform in this way.

This paper has been partially funded by the projects CANTOR: Automated Composition of Personal Narratives as an aid for Occupational Therapy based on Reminescence, Grant. No. PID2019-108927RB-I00 (Spanish Ministry of Science and Innovation), the ADARVE (Análisis de Datos de Realidad Virtual para Emergencias Radiológicas) Project funded by the Spanish Consejo de Seguridad Nuclear (CSN), project DARK NITE: Dialogue Agents Relying on Knowledge-Neural hybrids for Interactive Training Environments, Grant No. PID2023-146308OB-I00 (Spanish Ministry of Science and Innovation) and the SPARC project "Developing applications for learning Spanish in India using artificial intelligence and digital media", grant No. P2557 (Indian Ministry of Human Resource Development).

2   Previous Work

Our interest in using chatbots as tutoring systems [4] arose to face one of the challenges faced by second or foreign language learners is the fear of being evaluated negatively. Second language students are often afraid of being seen by other people making mistakes, a condition referred to as *foreign language anxiety* [2]. Chatbots [5] based on Large Language Models can engage in conversations that are linguistically correct and make sense from a pragmatic point of view. Given appropriate prompts, they can adapt both the content and the role in which they participate in the conversation [3]. In such interactions, the students can exercise their language skills dynamically, and they can benefit also from the ability of the chatbot to present students with virtual simulations of everyday life situations on which to train [1].

3   LLM Chatbot as Language Practice Partner

The experiment is carried out for a set of Hindi-speaking students of Spanish as a second language. Students were asked to provide the chatbot with a specific prompt:

> *I'm a student at university. I'm from India. I'm learning Spanish. I'm practising Spanish. I can speak native Hindi and English. I have A1 Spanish level. The conversation must be in A1 Spanish. Praise my advances. Use Spanish for the conversation. If I make any mistakes please explain them to me in English.*

Once the students are familiar with the idea of interacting with the system, they are asked to introduce the following prompt to drive the system to engage in role playing situations useful for language learning:

> *¿Puedes plantearme una situación en la que yo tengo que hablar español, y luego jugar tú el papel de alguno de los personajes, para que yo practique mi conversación en español?*[1]

4   Discussion

The results of the experiment were evaluated for linguistic correctness demonstrated by participants in the interactions, both human and non-human, and for usability of the chatbot as perceived by the students. A total of $N = 20$ students of Spanish participated in the session. There were 12 male participants (60%) and 8 female participants (40%). The conversations recorded during the session were reviewed by an experienced teacher of Spanish. Basic metrics computed over the contributions by the students and the chatbot are reported in Table 1.

---

[1] Can you set up a situation where I have to speak Spanish, and then you play the role of one of the characters, so that I can practice my Spanish conversation?

Table 1. Metrics on pedagogic strategies and linguistic correctness of recorded conversations. Unless specified otherwise, values given refer to number of instances detected of corresponding aspect.

Aspect	Average	Min	Max	SD
Interaction length (# turns)	24.95	5	56	14.57
Role playing span (# turns)	7.45	0	37	13.05
Exercises proposed	1.15	0	5	1.38
Clarifications requested	0.40	0	2	0.38
Errors	8.60	0	39	9.58
Praise	3.15	0	13	3.79
Encouragement	0.75	0	3	0.83

The metrics show a wide discrepancy in the values reported for the length of the interactions. The ratio between the average size of the spans of the interactions devoted to role playing conversation and the overall interaction length on average is low. The recorded interactions show instances of the chatbot providing both praise and encouragement to the student.

A summary of the statistics observed over linguistic correctness is provided in Table 2.

Table 2. Metrics on linguistic correctness.

Av. % student errors identified	47.48
Av. % appropriate corrections (over identified errors)	64.33
Av. % inappropriate corrections (over identified errors)	35.67

The responses for the part of the questionnaire that was specific for the task are reported in Table 3. In general responses are very positive (for all the questions the mode value was 5). Qualitative comments on these aspects were also gathered, and they all were positive or very positive (no negative comment was collected).

Table 3. Reported scores on relevant aspects.

	Mean	SD
Did you feel that the chatbot proposed conversation topics that allowed you to practice the targeted feature of the language?	4.60	0.68
Did you request clarifications?	4.20	1.15
Did you need to ask for explanation in your native language?	3.65	1.42
If you asked for clarification in English, did the chatbot remember to return to Spanish after explaining?	4.60	0.60
Did the chatbot remember to correct your mistakes?	4.30	0.92
Did you have to remind it to correct your mistakes?	4.05	1.32
Did you at any point receive praise from the chatbot?	4.25	1.12
Do you think you have learnt new languages skills?	4.65	0.59
Do you think the session helped you practice your Spanish?	4.80	0.52

5   Conclusions

The reported experiment shows evidence that, when used without any specific fine turning, LLM-based chatbots have the ability to propose everyday situations requiring conversations appropriate to exercise specific language topics (on average 7.7 dialog turns over around 25, with a student satisfaction of 4.60/5), and that students enjoy and engage with the system, even without previous training. The chatbots have also demonstrated the ability to explain the relevant concepts (4.20/5), and the ability to identify mistakes (47.48% errors identified, 4.30/5 perception by students) and correct them (64.33% valid corrections). The chatbot has also been seen to switch languages on demand (3.65/5) and to switch back to the target language after explanations (4.60/5). The observed interactions show instances of use of pedagogical strategies such as praise (on average 3.15 dialog turns over around 25, with 4.25/5 student perception) and encouragement (on average 0.75 dialog turns over around 25). In general terms, the chatbot responds to instructions by adapting the general trend of the conversation to particular topics requested by the students.

The chatbot shows a commendable flexibility to enter into conversations of specific interest to the student. In the students' conversation logs we have seen topics related to the Golden Age in Spain, the conquest of America, serial killers, shopping, or how to get a scholarship to study in Spain.

Other features desirable for language learning still need to be tested empirically such as the ability to switch languages on demand or the ability to adjust the complexity of explanations to different levels of expertise.

References

1. Al-Obaydi, L.H., Pikhart, M., Klimova, B.: ChatGPT and the general concepts of education: can artificial intelligence-driven chatbots support the process of language learning? Int. J. Emerg. Technol. Learn. (iJET) **18**(21), 39–50 (2023)
2. Horwitz, E.K., Horwitz, M.B., Cope, J.: Foreign language classroom anxiety. Mod. Lang. J. **70**(2), 125–132 (1986)
3. Liu, Z., Yin, S.X., Lee, C., Chen, N.F.: Scaffolding language learning via multi-modal tutoring systems with pedagogical instructions. arXiv preprint arXiv:2404.03429 (2024)
4. Nye, B., Mee, D., Core, M.G.: Generative large language models for dialog-based tutoring: an early consideration of opportunities and concerns. In: AIEDWorkshops (2023)
5. Wu, T., et al.: A brief overview of ChatGPT: the history, status quo and potential future development. IEEE/CAA J. Automatica Sinica **10**(5), 1122–1136 (2023)

Personalizing Learning with Intelligent Tutoring Systems: Leveraging GenAI and Predictive AI for Adaptive Education

Juan C. Zuluaga-Morillo[1] , Eduardo J. Tous-De la Ossa[1] ,
Álvaro J. Giraldo-Cadavid[1] , Carlos M. González-McMahon[1] ,
Gustavo A. Moreno-López[2] , Néstor D. Duque-Méndez[3] ,
Jaime A. Restrepo-Carmona[3] , and Jovani A. Jiménez-Builes[3(✉)]

[1] Contraloría General de La República, Bogotá CUN 111071, Colombia
[2] Politécnico Colombiano Jaime Isaza Cadavid, ANT 050022 Medellín, Colombia
[3] Universidad Nacional de Colombia, Bogotá CUN 16486, Colombia
jajimen1@unal.edu.co

Abstract. The goal of personalized education is to maximize each student's development and potential by tailoring the learning process to their individual needs. Intelligent Tutoring Systems (ITS) have emerged as powerful tools for achieving this goal. This article explores the potential of Generative Artificial Intelligence (GenAI) and Predictive Artificial Intelligence (PAI) to enhance learning personalization within ITS. Leveraging the capabilities of GenAI and PAI, ITS offers a highly effective method for customizing education and optimizing learning experiences. We review the fundamentals of ITS, GenAI, and PAI, examining their application in personalized education. A model is presented to demonstrate the successful integration of these technologies into educational environments. A case study highlights how ITS powered by GenAI and PAI delivers a tailored, adaptive learning experience that adjusts to each student's unique needs and learning pace. Observations reveal significant improvements in student satisfaction, motivation, and academic performance. The integration of GenAI and PAI into ITS represents a major advancement in the field of personalized education.

Keywords: Intelligent Tutoring Systems · Generative Artificial Intelligence · Predictive Artificial Intelligence · Personalized Learning

1 Introduction

Personalized learning refers to an educational approach that adapts to the individual needs, interests, learning styles and pace of each student, among other factors, to optimize their learning process [1]. This schema provides students with the opportunity to engage actively in their own learning process by obtaining training, resources, and academic activities that are tailored to their unique characteristics and needs [2].

Intelligent Tutoring Systems (ITS) represent a significant advancement towards personalized learning. ITSs use Artificial Intelligence (AI) technologies to tailor

L. Correia et al. (Eds.): IBERAMIA 2024, LNCS 15277, pp. 463–467, 2025.
https://doi.org/10.1007/978-3-031-80366-6

educational content and academic learning according to the needs of each student. In doing so, instructional technology systems (ITSs) can provide more effective and significant learning experiences by allowing students to advance at their own pace, receive personalized feedback, and concentrate on challenges [3]. Predictive Artificial Intelligence (PAI) and Generative Artificial Intelligence (GenAI) are pivotal components in enhancing personalized learning within ITSs. The GenAI is a tool that facilitates dynamic interaction with students by generating personalized educational content in real time and providing feedback that is specifically tailored to their needs [4]. In contrast, PAI employs advanced algorithms to analyze historical data and forecast the progress of students and their future needs, thereby facilitating the proactive adjustment of teaching and learning strategies [5].

This paper examines the potential of ITSs for personalizing the learning process by leveraging GenAI and PAI. It provides an overview of the status and applications of these technologies in the context of adaptive learning. Furthermore, this paper analyzes a model that serves as an illustration of the effectiveness of ITSs powered by GenAI and PAI within educational settings.

2   ITS Model Integrating AIGen & PAI

This section presents a model where an ITS architecture integrating GenAI and PAI is proposed, which can be implemented in different knowledge domains in an educational context (see Fig. 1). The subsections provide detailed explanations of the components comprising each of the three modules, aiming to clarify their operation. Functions are synonymous with the term components.

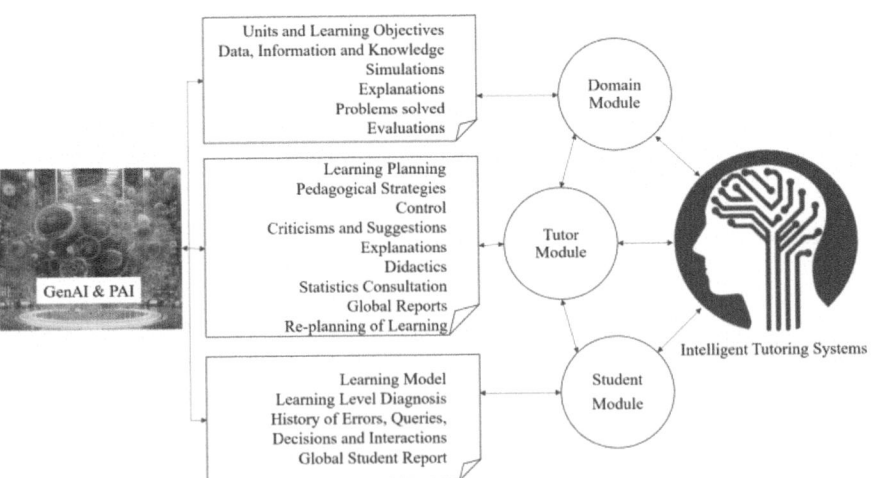

Fig. 1. Architecture of an ITS integrating GenAI and PAI technologies. Source: the authors.

AIGen & PAI play crucial roles in enhancing the functionality of ITS. IGen excels in creating dynamic and personalized learning content. This capability allows ITS to adapt instructional materials, questions, and exercises in real-time to meet the individual needs of each student. For example, AIGen can generate customized explanations or alternative problem-solving strategies based on a student's learning style or progress. This dynamic content generation makes learning more engaging and responsive, providing students with tailored pathways that adapt as they interact with the system. Additionally, AIGen can simulate complex scenarios and environments, particularly useful in fields requiring hands-on experience or abstract thinking, such as STEM education. On the other hand, PAI enhances ITS by analyzing student data to predict learning outcomes, performance gaps, and potential challenges. Using historical and real-time data, PAI can forecast when a student might struggle with a concept or need additional support. This predictive capability allows the system to intervene preemptively, adjusting the difficulty level, recommending specific exercises, or offering personalized feedback before the student falls behind. PAI can also identify trends in group behavior, helping instructors fine-tune their teaching strategies for larger cohorts.

2.1 Domain Module

These technologies facilitate the creation of personalized educational content that are specifically designed to meet the needs of each student. Additionally, they enable the anticipation and prediction of the challenges that each student might encounter throughout their learning process. GenAI can generate unique and relevant learning material, which are tailored to suit the preferences and learning style of every student. In contrast, PAI employs advanced analytics and historical data to predict areas of potential difficulty and provide early recommendations for optimizing teaching and learning.

2.2 Tutor Module

These technologies enable this module to offer personalized and adaptive tutoring, designed specifically for the unique needs of each student. GenAI facilitates the creation of innovative and contextualized pedagogical strategies, thereby enhancing the effectiveness of the module by providing materials and exercises tailored to individual learning styles. Conversely, PAI provides the module with the capability to anticipate areas of potential difficulty and adjust teaching strategies in real time.

2.3 Student Module

These technologies allow significantly personalizing and optimizing the educational experience. GenAI empowers the development of dynamic educational content tailored to the specific needs of every student, thereby promoting more profound and personalized learning. Conversely, PAI employs data analytics to forecast student progress, identify areas necessitating improvement, and provide accurate recommendations for adapting the educational approach. Implementing this proactive strategy not only enhances the efficacy of the learning process, but also fosters student autonomy and motivation through the provision of personalized resources and assistance that align with their academic trajectory and learning preferences.

3   Conclusions

The implementation of IA in education enriches the teaching and learning processes. Rather than prohibiting its use, educational institutions ought to encourage its implementation. Currently, there is a great deal of noise, misinformation, misconceptions, and false beliefs surrounding the use of IA in the classroom. The most prevalent fallacy is that IA will replace educators. On the contrary, IA enables them to simplify the execution of complex and repetitive tasks, freeing up time for other endeavors, such as staying updated on pedagogical issues and tool usage or providing better student service.

In summary, this research demonstrated that GenAI focuses on the creation of new content, whereas PAI concentrates on forecasting future results. In contrast to GenAI, which employs language models and generative neural networks, PAI applies techniques such as supervised and unsupervised machine learning. GenAI has applications in the generation of code, text, music, and images, whereas PAI is mainly utilized by students for data management. PAI can operate with smaller, less structured data sets than GenAI, which necessitates large, high-quality data sets. PAI models are frequently more interpretable than GenAI models, which can be challenging to interpret. Ethical concerns exist regarding the potential for GenAI to be misused just as they do regarding the discriminatory and biased nature of PAI.

GenAI and PAI, both disruptive technologies, offer enormous potential to revolutionize ITS, driving learning that is personalized, effective, and tailored to the individual needs of each student. The synergy of these technologies, despite their respective difficulties and the need for ethical and responsible approach to their implementation, has the potential to transform the educational domain by fostering engaging, meaningful, and successful learning experiences for all.

To maximize their positive impact on the future of education, it is vital to continue researching and developing these technologies while addressing their limitations and ensuring their responsible application. The integration of GenAI and PAI within the field of ITS presents a plethora of exciting possibilities, facilitating the transition towards a future of education that is more personalized, effective, equitable, and to some extent, massive, and affordable.

Acknowledgments. The authors extend their thank to DIARI (Dirección de Información Análisis y Reacción Inmediata) assigned to the Contraloría General de la República de Colombia and the Universidad Nacional de Colombia for the support received for carrying out the research: "Desarrollo y despliegue de un modelo de innovación para la integración de servicios que fortalezca la vigilancia y el control fiscal en la Contraloría General de la República que permita mejoras en la transparencia y rendimiento institucional", ID: 1550.

Disclosure of Interests. The authors have no competing interests to declare that are relevant to the content of this article.

References

1. Dumont, H., Istance, D., Benavides, F. (eds.): The Nature of Learning: Using Research to Inspire Practice. OECD Publishing, Paris (2018). https://doi.org/10.1787/9789264086487-en
2. Teo, T., Unwin, S., Scherer, R., Gardiner, V.: Initial teacher training for twenty-first century skills in the Fourth Industrial Revolution (IR 4.0): a scoping review. Comput. Educ. **170**, (2021). https://doi.org/10.1016/j.compedu.2021.104223
3. Baker, R.S., Inventado, P.S.: Educational data mining and learning analytics. In: Larusson, J., White, B. (eds.) Learning Analytics, pp. 61–75. Springer, New York (2014). https://doi.org/10.1007/978-1-4614-3305-7_4
4. Jie, G., Zhenan, S., Yonggang, W., Dacheng, T., Jieping, Y.: A review on generative adversarial networks: algorithms, theory, and applications. EEE Trans. Knowl. Data Eng. **35** (4), 3313–3332 (2023). https://doi.org/10.1109/TKDE.2021.3130191
5. Baker, R.S., Yacef, K.: The state of educational data mining in 2009: a review and future visions. J. Educ. Data Min. **1**(1), 3–17 (2009). https://doi.org/10.5281/zenodo.3554657

Improving Efficiency of QBF Planning with Mixed Linear Compact Tree Encodings

Frédéric Maris[(✉)] [ID]

IRIT, Université Toulouse 3,118 route de Narbonne, 31062 Toulouse, France
Frederic.Maris@irit.fr
https://www.irit.fr/ Frederic.Maris/

Abstract. The considerable improvements of Quantified Boolean Formula (QBF) solvers has made their use possible for the resolution of numerous problems in artificial intelligence and in particular automated planning. Even if QBF encodings for planning are still outperformed by those for SAT solvers, the gap is narrowing and it is interesting to use QBF language that allows one to produce more compact encodings. Several translations from STRIPS planning problems into QBF have been proposed, notably a compact tree encoding (CTE) [2]. In this article, we propose to explore a mixed approach with a new encoding based on both original linear SAT encoding and compact tree QBF encoding for planning. First experimental results show that the resolution time can then be significantly reduced with our encoding.

Keywords: Planning · Quantified Boolean Formula · Encodings

1 Introduction

In the SAT approach of planning, the problem is transformed into a propositional formula by encoding all possible solution plans of fixed length in a linear sequence of steps. A plan can be found using a SAT solver, possibly by incrementing the length of the sequence. Unfortunately, in the classical framework, the complexity of finding a solution to any problem is PSPACE-hard [1] and then the bound of the encoding could be exponential. To maintain a polynomial size, several planning encodings have been proposed which make use of the branching structure of a Quantified Boolean Formula (QBF) to reuse a single set of formulas that describes multiple distinct steps in the plan. Indeed, a binary tree is simulated using an alternation of universally quantified variables for branching and existentially quantified variables to describe the steps of the plan. The two possible assignments of a universal variable represent the first and second half of the plan split around that branch. The assignments of existential variables represent action choices and resulting states within a single step. The first approach which can be used for planning was introduced for general reachability [6]. In [2], authors showed that their Compact Tree Encodings (CTE) outperform this latter flat encoding because of the less important alternation of quantifiers. The

© The Author(s), under exclusive license to Springer Nature Switzerland AG 2025
L. Correia et al. (Eds.): IBERAMIA 2024, LNCS 15277, pp. 468–471, 2025.
https://doi.org/10.1007/978-3-031-80366-6

performances of CTE have been compared using several other encoding rules to describe the transitions in a plan [4]. In particular the rules adapted from the first SAT encoding for planning [5] based on explanatory frame-axioms outperform the other ones. In this paper, we introduce a QBF mixed CTE-based encoding in which each node of the tree will contain a bounded linear sequence of steps. By tuning the node width, we can improve the resolution performance.

2   Compact Tree Encodings with Node Width

The encoding that we introduce now is an adaptation of the CTE [2] with a node width (i.e. a bounded linear sequence of steps within each tree node). To describe the transitions in the plan, we will use the SAT encoding rules based on explanatory frame-axioms first introduced by [5] and implemented in the basic version of CTE by [4]. Let $\mathcal{F}$ be a finite set of *fluents* (atomic propositions). A STRIPS *planning problem* is a tuple $\langle I, \mathcal{A}, G \rangle$ where $I \subseteq \mathcal{F}$ is the set of initial fluents, $G \subseteq \mathcal{F}$ is the set of goal fluents and $\mathcal{A}$ is the set of actions. An action $a \in \mathcal{A}$ is a tuple $\langle Cond(a), Add(a), Del(a) \rangle$ where $Cond(a) \subseteq \mathcal{F}$ is the set of fluents required to be true in order to execute a, $Add(a) \subseteq \mathcal{F}$ and $Del(a) \subseteq \mathcal{F}$ are the sets of fluents respectively added and removed by the action a. We denote by d the fixed depth of the CTE and by w the node width. For each $i \in \{1...d\}$ and $l \in \{1...w\}$, we define a propositional variable b_i and a set of propositional variables $X_{i,l} = \mathcal{A}_{i,l} \cup \mathcal{F}_{i,l}$ where $\mathcal{A}_{i,l} = \{a_{i,l} : a \in \mathcal{A}\}$ and $\mathcal{F}_{i,l} = \{f_{i,l} : f \in \mathcal{F}\}$. At a given depth i, the values of variables in $X_{i,l}$ depend on the node selected by the values of universal branching variables $b_{i+1} \ldots b_d$. Figure 1 shows the three possibilities for selecting two consecutive steps to define transitions in the CTE formula. We use the following quantifiers as a prefix to our formula to construct the branching structure.

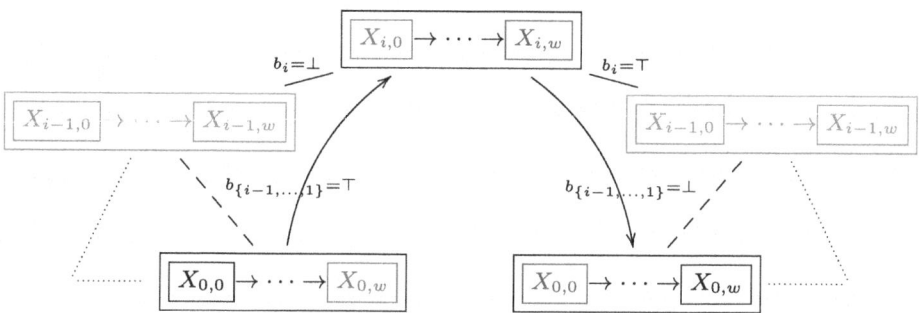

Fig. 1. Transitions in a mixed Linear CTE following the branching structure of a QBF: $X_{i,l} \rightarrow X_{i,l+1}$ (in a linear sequence within a tree node), $X_{0,w} \rightarrow X_{i,0}$ (from leaf to node on the left), and $X_{i,w} \rightarrow X_{0,0}$ (from node to leaf on the right).

On the one hand, any action which do not have all conditions in the initial state cannot be executed in the first transition in the plan. Then, in the first linear step within the leftmost leaf of the tree the propositional variables corresponding to such action must be false. On the other hand, in the state after the last transition in the plan, all goal fluents must occur. Hence, the propositional variables corresponding to such fluent must be true in the last linear step within the rightmost leaf of the tree.

If an action a is executed in a transition of the plan, then each effect of a occurs in resulting state and each condition of a is required in previous state.

If the value of a propositional variable corresponding to a fluent changes between two consecutive states from false to true, then an action which produces this change is executed in the plan transition between these states.

Similar extra encoding rules that we do not describe here are also required to describe explanatory frame axioms for negative changes from true to false and for the first transition in the plan from the initial state (i.e. for the first linear step within the leftmost leaf of the tree). Finally, a last encoding rule is required to prevent negative interactions. Contradictory effects are already disallowed by previous rules (positive and negative effects of actions within a same step). We further states: if an action removes a fluent which is needed by another action, then these two actions cannot be both executed in a same plan transition.

$$\bigwedge_{i=0}^{d} \bigwedge_{l=0}^{w} \bigwedge_{\substack{(a,a') \in \mathcal{A}^2 \\ a \neq a' \\ Cond(a) \cap Del(a') \neq \emptyset}} \left(\neg a_{i,l} \vee \neg a'_{i,l} \right)$$

3   Experiments and Perspectives

The first experimental results show that resolution performance is improved when we know how to choose the right node width for a given type of problem. We give plan generation times in Table 1 for several classical planning benchmarks calculated using TouIST planning module [3]. For SAT encodings based on explanatory frame-axioms [5], the plan length starts from 1 and is incremented, until a solution is found, by a given number of new steps each time the solver runs (+1 or ×2). For CTE with different node widths, the tree depth is incremented until a solution is found, then QBF solver is run recursively assigning values of branching variables b_i to focus on each node to extract the plan.

It would now be interesting to explore how a machine learning approach, particularly based on a classifier, would make it possible to automatically find the right node width value to obtain the best performance.

Table 1. Plan generation time (in seconds) for SAT with incremental plan length, and CTE with different node widths and incremental tree depth. For CTE, the value given in parentheses is the depth for which a plan is found; * represents timeout over 10 mn.

Planning Problem	SAT +1	SAT ×2	CTE $w = 1$	CTE $w = 2$	CTE $w = 3$	CTE $w = 4$	CTE $w = 5$	CTE $w = 6$
BlocksWorld-6	17.65	**13.07**	(2) 24.85	(1) **14.47**	(1) 24.05	(1) 38.89	(1) 49.95	(1) 64.84
Gripper-6	2.76	1.86	(3) 3.31	(2) 2.14	(2) 3.23	(1) **1.21**	(1) 1.41	(1) 1.60
Gripper-7	253	49.12	(3) 4.66	(3) 22.96	(2) 4.67	(2) 14.59	(1) 2.57	(1) **2.35**
Ferry-8	*	39.76	(4) 70.09	(4) *	(3) 14.46	(3) 508	(2) **7.40**	(2) 15.16
Hanoi-5	*	131	(4) 111	(4) 196	(3) 125	(3) 221	(2) **51.81**	(2) 103

References

1. Bylander, T.: The computational complexity of propositional STRIPS planning. Artif. Intell. **69**(1–2), 165–204 (1994)
2. Cashmore, M., Fox, M., Giunchiglia, E.: Planning as quantified Boolean formula. In: Proceedings of the 20th European Conference on Artificial Intelligence (ECAI 2012)
3. Fernandez, J., et al.: TouIST: a friendly language for propositional logic and more. In: Proceedings of the 29th International Joint Conference on Artificial Intelligence (IJCAI 2020)
4. Gasquet, O., Longin, D., Maris, F., Régnier, P., Valais, M.: Compact tree encodings for planning as QBF. Intel. Artif. **21**(62), 103–113 (2018)
5. Kautz, H.A., Selman, B.: Planning as satisfiability. In: Proceedings of the 10th European Conference on Artificial Intelligence (ECAI 1992)
6. Rintanen, J.: Partial implicit unfolding in the Davis-Putnam procedure for quantified Boolean formulae. In: Proceedings of the 8th International Conference Logic for Programming, Artificial Intelligence, and Reasoning (LPAR 2001)

Adversarial Attacks in a Shallow Neural Network to Classify Android Malware

Leidy Marcela Aldana$^{(\boxtimes)}$ (ID) and Jorge E. Camargo (ID)

Universidad Nacional de Colombia, Bogotá, Colombia
{laldanab,jecamargom}@unal.edu.co

Abstract. This article presents the implementation of adversarial attacks *(FGSM, ZOO, Deepfool, PGD)* on a shallow neural network, it classifies whether an Android application is malware or not. To evaluate the efficacy of the various adversarial attacks, the loss and accuracy obtained in each attack are compared, it allows the identification of the optimal attack. This article makes a significant contribution to the field of information security by comparing different adversarial attacks in the context of a malware dataset.

Keywords: Malware · Static analysis · Adversarial attack

1 Introduction

Malicious software, or malware, represents a significant threat to system security. Once infected, a device can become a conduit for malware to spread to other devices in close proximity. Malware is designed for a range of devices and systems, including Android. Android is an open-source system with an architecture that enables the execution of applications in *Android Application Package* format (APK). According to the reference [1], Android is installed on 70.68% of mobile devices. Malware detection on Android has three types of analysis [2]:

1. *Static Analysis:* The examination of the code is conducted without the APK being executed.
2. *Dynamic Analysis:* APK is executed to scan for malware, but this process is time-consuming and requires significant computational resources.
3. *Hybrid Analysis:* It synthesises the preceding two analyses, combining their respective strengths.

While neural networks have demonstrated excellent performance in classification tasks, they are susceptible to adversarial attacks [3]. An adversarial attack is a method of making subtle modifications to an input [4], which causes the classification function to fail in the face of such alterations, this paper compares adversarial attacks in the context of a malware dataset.

L. Correia et al. (Eds.): IBERAMIA 2024, LNCS 15277, pp. 472–475, 2025.
https://doi.org/10.1007/978-3-031-80366-6

2   Background

A systematic literature review of adversarial attacks in malware scenarios (realized by the author [5]) indicates that decision trees, naive bayes, and linear support vector machine (SVM) are the machine learning models most susceptible to adversarial attacks. In contrast, *neural networks* are observed to exhibit greater resilience. Moreover, the authors of [2] posit that *neural networks* and deep learning achieve high accuracy and strong fault tolerance. Three additional models (deep neural network, convolutional neural network, and gradient boosting) are analyzed by the author of [6], employing five adversarial attacks; this author notes that successful results have been obtained in the image domain but not in the malware domain. In reference [7], the generation of adversarial examples in malware is subject to three principal challenges: preservation of the application format, preservation of executability and preserve malignant features.

3   Methodology

Android Malware Dataset for Machine Learning dataset available in Kaggle (reference [9]) was utilized, it has a total value of $15,036$ rows. It comprises one row for each APK executable and 215 columns, each column represents the results of the feature extraction process carried out by the author [9], who performed the static analysis of each application. A shallow neural network was chosen according to the best accuracy, it receives the input of 215 features, a hidden layer with the rectified linear unit (ReLU) activation function and an output layer with sigmoid activation function for classification between two classes, malignant and benign. The neural network receives 80% of the data for training and the remaining 20% for validation, learning process has 7 epochs and *Adam* optimizer. Four adversarial attacks are implemented to generate entries that can look like benign apks, despite the fact that they are, in fact, malware entries, with the intention of maximizing the loss [8] and to affect the classification.

1. **Fast Gradient Sign Method (FGSM):** It is gradient-based in order to create adversarial examples; in this instance, the value of $\epsilon = 0.1$ was used as the smallest possible value in order to prevent any potential impact on the malicious or benign functionality of the distinct applications.
2. **Deepfool:** It uses more computational resources than the FGSM attack since it proposes an iterative process [11], a small value of $1e^{-10}$ is added to avoid divisions by zero and $\epsilon = 0.1$ was used.
3. **Zeroth Order Optimization (ZOO):** It generates adversarial examples with the proposed values $\epsilon = 0, 1$ and $size = 0, 1$.
4. **Projected Gradient Descent (PGD):** It is an iterative version of FGSM [10], which generates the adversarial examples with $\epsilon = 0.1$ and $\alpha = 0.01$.

4   Evaluation

Accuracy and loss were calculated for the 7 training epochs, accuracy obtained on the training and validation dataset without any attack is between 0.96 and 0.99, while loss has low values, because its values are between 0.12 and 0.01 for different epochs, those performance metrics were obtained without any adversarial attack. Figure 1 shows the performance metrics from Table 1, accuracy and loss in the neural network, first two columns in training and validation and others are the adversarial attacks implemented in this dataset about malware.

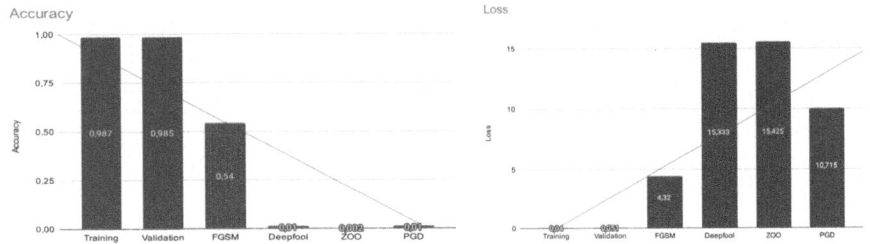

Fig. 1. Accuracy and loss before and after adversarial attacks implementation

Table 1. Performance metrics

	Training	Validation	FGSM	Deepfool	ZOO	PGD
Accuracy	0,987	0,985	0,54	0,01	0,02	0,01
Loss	0,04	0,05	4,32	15,33	15,42	10,71

5   Discussion

Adversarial attacks were implemented as a gray box, where the attacker knows the number of features that the model receives; he knows that the model receives 215 columns for each application, this is presented as a limitation because in the real context a black box attack may occur, where the attacker is completely unaware of the data and the system architecture. To solve this, it would be necessary to implement the research in a larger architecture, where users upload the APK and obtain the result in a graphical interface. Implementing dynamic and hybrid analysis of Malware characteristics can possibly obtain more accurate results in the implementation of adversarial attacks; it can also be complemented by creating a controlled environment to run the malware application, as originally found in the data set, and build the generated adversarial examples, to verify its malicious functionality from an attack perspective.

6   Conclusion

Neural networks to malware analysis requires a dataset composed by multiple features, such as permissions, attempts, command strings, API calls, among others. The FGSM adversarial attack shows good results when computing capacity and time are limited, this attack allows creating adversarial examples quickly, other attacks present good results although they take more time. According to the analysis carried out, it is observed that using a variety of attacks is viable, combining the advantages of different adversarial attacks. Deepfool and ZOO attack shows the best results, taking more computing resources and time processing, it presents a better loss providing more robustness, being viable for future research.

Acknowledgments.. This research was supported by the Research Group UNSecure-Lab and *Universidad Nacional de Colombia*, Bogota, Faculty of Engineering.

References

1. Muzaffar, A., Ragab, H.W., Lones, M., Zantout, H.: An in-depth review of machine learning based Android malware detection. 3rd edn. Computers & Security. https://doi.org/10.1016/j.cose.2022.102833. ISSN 0167-4048
2. Liu, K., Xu, S., et al.: A review of Android malware detection approaches based on machine learning, IEEE. CCBY (2020). https://doi.org/10.1109/ACCESS.2020.3006143
3. OWASP, Machine Learning Security Top Ten. Open source initiative license: CC BY-SA 4.0 (2023). https://owasp.org/www-project-machine-learning-security-top-10/
4. Bishop, C., Bishop, H.: Deep learning foundations and concepts (2024). https://doi.org/10.1007/978-3-031-45468-4. eBook. ISBN: 978-3-031-45467-7
5. Martins, N., et al.: Adversarial machine learning applied to intrusion and malware scenarios: a systematic review. Article. IEEE (2020). https://doi.org/10.1109/ACCESS.2020.2974752
6. Qi, X., et al.: Adversarial example attacks against intelligent malware detection: a survey (2022). https://doi.org/10.1109/ICAML57167.2022.00068
7. Ling, X., et al.: Adversarial attacks against windows PE malware detection: a survey of the state-of-the-art, article. Comput. Secur. (2023). arXiv:2112.12310v5
8. Yan, S., Ren, J., Wang, W., et al.: Survey of adversarial attacks on Imagttack and defense methods for malware classification in cyber security (2023). Digit. Object Identifier. https://doi.org/10.1109/COMST.2022.3225137
9. Tiwari, S.: Android malware dataset for machine learning, dataset. Kaggle: CC BY 4.0 (2021). https://www.kaggle.com/datasets/shashwatwork/android-malware-dataset-for-machine-learning
10. Fanyou, W., Gazo, R., Haviarova, E., Benes, B.: Efficient project gradient descent for ensemble adversarial attack (2019). arXiv:1906.03333v1
11. Moosavi-Dezfooli, S., Fawzi, A., Frossard, P.: DeepFool: a simple and accurate method to fool deep neural networks. IEEE Xplore (2016). https://www.computer.org/csdl/proceedings-article/cvpr/2016/8851c574/12OmNAle6j6

Harnessing Deep Learning for Detection of Violence and Vandalism

Tiago Ribeiro[1], Juan Pavón[2], José Machado[1], Paulo Novais[1],
and Manuel Rodrigues[1(✉)]

[1] LASI/ALGORITMI Centre, University of Minho, Guimarães, Portugal
pg50779@alunos.uminho.pt, {jmac,pjon,mfsr}@di.uminho.pt
[2] Institute of Knowledge Technology, Universidad Complutense Madrid, Madrid, Spain
jpavon@fdi.ucm.es

Abstract. This research investigates various artificial intelligence techniques to detect violence and vandalism in surveillance videos, assessing four models: Conv3D, ConvLSTM, LRCN, and CNN-BiLSTM. These models use supervised learning with convolutional and recurrent neural networks to capture spatial and temporal features. The dataset used is 'Real Life Violence Situations Dataset.' Future work will extend training, test diverse datasets, and explore advanced architectures to improve model performance for real-world surveillance.

Keywords: Smart Surveillance · Artificial Intelligence · Deep Learning

1 Introduction

In today's complex societal landscape, crime prevention is of paramount importance. Detecting and preventing violence, vandalism and stressful situations is crucial for ensuring public safety [1, 2]. Traditional reactive approaches are being superseded by proactive strategies facilitated by Deep Learning (DL) models, a subset of Artificial Intelligence (AI) [3]. By harnessing DL, security agencies can detect instances of violence and vandalism with greater accuracy and foresight [3]. This proactive shift aligns with the concept of Smart Surveillance, integrating intelligent data analysis with surveillance technologies to identify potential threats before they materialize [4].

2 Context and Challenges

Detection and prevention of violence and vandalism are paramount, demanding innovative and adaptable strategies, shifting from traditional reactive approaches, responding to incidents after they occur, that have limitations in anticipating and

This work has been supported by FCT - Fundação para a Ciência e Tecnologia within the R&D Units Project Scope: UIDB/00319/2020.

preventing such crimes effectively [1, 2]. The emergence of Smart Surveillance (SS), powered by advances in DL [4], signifies a pivotal shift towards proactive and preventative crime prevention [4]. However, while promising, its implementation encounters significant challenges. Ensuring data quality and integrity is a foremost challenge, as DL models rely on meticulously collected and unbiased datasets [5, 6]. Ethical and privacy concerns also loom large, with SS raising questions about mass monitoring and individual privacy violations [7, 8]. Proactive policies are essential to safeguard citizens' rights [7, 8]. Despite these hurdles, SS holds immense potential to enhance public safety operations and protect communities more effectively. By tackling these challenges head-on, public security agencies can fully leverage DL and SS technologies, fostering a safer environment for all.

3   Data Collection and Preprocessing

The chosen dataset, 'Real Life Violence Situations Dataset,' offers diverse real violence scenarios [9]. Its accessibility, balanced distribution of violent and non-violent videos, and acceptable video quality make it ideal for training and evaluation. The dataset consists of 1000 videos each of violence and non-violence sourced. Violence videos depict real-life street fights, while non-violence videos include various human actions. Violence videos have an average duration of 5.4 s, with 159.8 frames, and an average resolution of 599×481 pixels, with a frame rate of 29.6. Non-violence videos have an average duration of 5.1 s, with 127.6 frames, and an average resolution of 418×318 pixels, with a frame rate of 24.9. Data processing involves extracting 32 frames from each video, sized $128 \times 128 \times 3$, at equal time intervals throughout the video. This process includes determining total frames, calculating intervals and extracting frames. Frames are manually labeled based on video category and converted into one-hot encoded vectors for simplified class probability calculations.

4   Model Selection and Architectures

Various DL methods and techniques were explored, each tailored to different algorithms and architectures, employing diverse supervised learning techniques. Criteria were set to select model characteristics and parameters, considering the data nature and detection objectives. Implementation used well-established machine learning libraries and tools. Multiple model versions were developed with specific parameter adjustments, enabling exploration of different configurations and refinements. These iterations were pivotal in identifying the most effective combination of elements for detecting events of interest. The selected models wore: Conv3D [10], ConvLSTM [11], LRCN [12], CNN-BiLSTM [13].

5 Training Process

Each set of frames, that represents a video, consists of 32 frames, sized $128 \times 128 \times 3$. This creates a large input that demands substantial computing resources. To manage this, the dataset was divided into batches for computational efficiency. With 2000 videos, each batch contains 32 videos, resulting in 62 complete batches and one unused partial batch, with each batch having a balanced number of violent and non-violent videos. To evaluate the model's performance, 7 test batches were used. These batches were consistently used throughout the epochs to ensure an accurate and consistent evaluation. With a total of 224 videos, the test set provided a solid basis for measuring the model's generalization capacity and identifying possible overfitting problems. A custom cycle mimicking the fit method was implemented, allowing progressive learning. The train-on-batch approach was used, feeding each batch individually to the model and adjusting its parameters accordingly. To prevent the model from learning specific data patterns, the order of videos within each batch and the order of the batches themselves were shuffled. This promotes generalized learning and reduces overfitting. The same set of test batches was used throughout the epochs for consistent performance evaluation. Data augmentation techniques, such as random rotations, translations, zooming, inversions, and shearing, were applied to combat overfitting and improve generalization. These techniques increased the diversity of the training set, generating a wide range of training examples from the original data.

6 Results and Discussion

The CNN-BiLSTM model stands out with an accuracy of 93.750%, leveraging both convolutional and LSTM layers for spatial and temporal analysis, respectively. ConvLSTM and LRCN models also perform exceptionally well, achieving an accuracy of 92.857% and 92.243%. These models effectively capture spatiotemporal patterns, contributing to accurate event detection. Although the Conv3D model achieves 91.071%, its performance exhibits more pronounced fluctuations, indicating potential instability in generalization. Comparison in Table 1 and Fig. 1. Variations in test accuracy reveal differences in model stability. Conv3D shows significant fluctuations, indicating generalization issues, while other models display stable accuracy trends. Although all models perform well, Conv3D needs adjustments for better stability. Regarding loss, ConvLSTM excels with closely aligned test and training loss curves, suggesting effective generalization and minimal overfitting. Other models show larger discrepancies, with Conv3D potentially overfitting. Conv3D and ConvLSTM use three-dimensional convolutions for spatial and temporal features, while CNN-BiLSTM and LRCN use spatial convolutions followed by LSTM layers for temporal analysis. Both approaches are effective, with no significant advantage of one over the other, underscoring their validity.

Table 1. Comparison of the results from the different models

Model	Ver	Conv. Filters	Conv. Kernel Sizes	Pooling Layers	Dropout Layers	LSTM units	Accuracy
CNN-BiLSTM	#9	32, 64, 128, 128	All (3, 3)	All (2, 2)	All 0.1	64	93.750%
ConvLSTM	#3	16, 20, 24, 28	All (3, 3)	All (1, 2, 2)	All 0.1	None	92.857%
LRCN	#6	32, 64, 128, 128	All (3, 3)	All (2, 2)	All 0.1	128	92.243%
Conv3D	#8	32, 64, 64, 128	All (3, 3, 3)	(1, 2, 2), (1, 2, 2), (1, 4, 4) and (1, 4,4)	All 0.1	None	91.071%

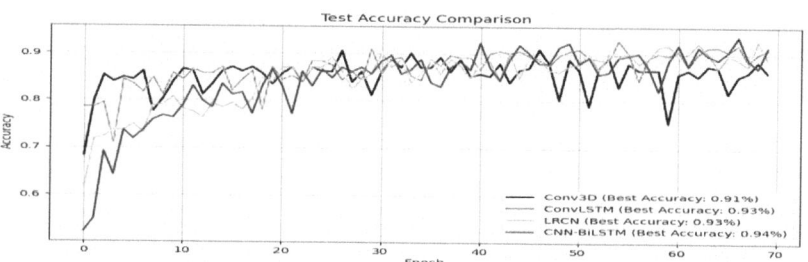

Fig. 1. Comparison between the test accuracy of the four models over the epochs

7 Conclusions and Future Work

Among the models tested, the CNN-BiLSTM emerged as the most effective, achieving an accuracy of 93.75%. It effectively combines CNNs for spatial feature extraction with BiLSTM networks for temporal feature extraction. The ConvLSTM and LRCN models also demonstrated strong performance, showcasing robust generalization and integration of Conv2D and LSTM layers. However, the Conv3D model, despite using 3-dimensional convolutions, was less effective and showed a tendency to overfit. The CNN-BiLSTM model effectiveness highlights its potential usability in real-world applications where accurate and reliable detection is crucial. Future work should focus on refining the model, optimizing its hyperparameters and architecture design, and training the model with more extensive and diverse datasets to improve robustness and generalization. Additionally, advanced data augmentation techniques and testing in real-time scenarios will further ensure their usability and effectiveness in practical settings. Furthermore, exploring new architectures like Transformers for video analysis could enhance both accuracy and efficiency. In conclusion, while the CNN-BiLSTM model demonstrated superior performance, continuous research and refinement are essential for advancing intelligent surveillance systems. The ultimate goal is to develop a practical, reliable, and highly accurate system for real-world detection of violence and vandalism, ensuring both usability and effectiveness.

References

1. Feldstein, S.: The Global Expansion of AI Surveillance, vol. 17. Carnegie Endowment for International Peace, Washington, DC (2019)
2. Brayne, S.: Big data surveillance: the case of policing. Am. Sociol. Rev. **82**(5), 977–1008 (2017)
3. Tsakanikas, V., Dagiuklas, T.: Video surveillance systems-current status and future trends. Comput. Electr. Eng. **70**, 736–753 (2018)
4. Zhu, S., Chen, C., Sultani, W.: Video anomaly detection for smart surveillance. In: Computer Vision, pp. 1–8. Springer, Cham (2020). https://doi.org/10.1007/978-3-030-03243-2_845-1
5. Sreenu, G., Durai, S.: Intelligent video surveillance: a review through deep learning techniques for crowd analysis. J. Big Data **6**(1), 1–27 (2019)
6. Shidik, G.F., et al.: A systematic review of intelligence video surveillance: trends, techniques, frameworks, and datasets. IEEE Access **7**, 170457–170473 (2019)
7. Almeida, D., Shmarko, K., Lomas, E.: The ethics of facial recognition technologies, surveillance, and accountability in an age of artificial intelligence: a comparative analysis of US, EU, and UK regulatory frameworks. AI Ethics **2**(3), 377–387 (2022)
8. Adams, A.A., Ferryman, J.M.: The future of video analytics for surveillance and its ethical implications. Secur. J. **28**, 272–289 (2015)
9. Soliman, M.M., et al.: Violence recognition from videos using deep learning techniques. In: 2019 Ninth International Conference on Intelligent Computing and Information Systems (ICICIS), pp. 80–85 (2019)
10. Park, J.H., Mahmoud, M., Kang, H.S.: Conv3D-based video violence detection network using optical flow and RGB data. Sensors **24**(2), 317 (2024)
11. Lin, Z., Li, M., Zheng, Z., Cheng, Y., Yuan, C.: Self-attention ConvLSTM for spatiotemporal prediction. In: Proceedings of the AAAI Conference on Artificial Intelligence, vol. 34, issue number 07, pp. 11531–11538, April 2020
12. Massa, L., Barbosa, A., Oliveira, K., Vieira, T.: LRCN-RetailNet: a recurrent neural network architecture for accurate people counting. Multimedia Tools Appl. **80**, 5517–5537 (2021)
13. Halder, R., Chatterjee, R.: CNN-BiLSTM model for violence detection in smart surveillance. SN Comput. Sci. **1**(4), 201 (2020). https://doi.org/10.1007/s42979-020-00207-x

Magic Matching: A Virtual Assistant for Financial Trading Using Machine Learning

Melisa Arena, Belén Olivera, Santiago Pérez, Paola Romay,
and Victor Sabbia[✉]

Brokerware, Montevideo, Uruguay
info@brokerware.com.uy, vsabbia@brokerware.com.uy
http://www.brokerware.ai/

Abstract. In the dynamic field of financial trading, decision-making is critical for success. This paper presents "Magic Matching" a virtual assistant designed to support financial advisors by leveraging machine learning techniques to enhance trading strategies. Our system mimics experienced traders using a cascade model with XGBoost, predicting buy/sell decisions and asset selection from historical data since 2018, achieving a 60% accuracy, this translates in an alignment with the trading decisions in this dataset. We detail our methodology, including data preprocessing, model selection, hyperparameter tuning, and performance evaluation. Our analysis shows the systemâĂŹs effectiveness, making it a valuable tool in trading by facilitating decision-making for advisors, reducing their time investment, and enabling them to manage more portfolios with better recommendations.

Keywords: Financial advisor · Behavioral trading modeling · Recommendation systems

1 Introduction

Financial Advisors manage investor portfolios in Broker Dealers, each exhibiting unique preferences and behaviors. This variability, influenced by personal biases and investor preferences, is reflected in portfolio asset choices. While other works often focuses on profit maximization [2, 4], we aim to explore how it can improve decision-making and trading strategies by analyzing behavioral patterns and biases rather than solely focusing on profit.

2 Related Works

At present, as far as we know, there are no works with this approach in the field. The recommendation systems classical approaches are Content Based Filtering and Collaborative filtering [3], while the former identify relevant products according to the characteristics of the items previously consumed or acquired by

© The Author(s), under exclusive license to Springer Nature Switzerland AG 2025
L. Correia et al. (Eds.): IBERAMIA 2024, LNCS 15277, pp. 481–485, 2025.
https://doi.org/10.1007/978-3-031-80366-6

the user, the latter makes connections between users that have common interests, with the idea that *you might like item 2 because other people who liked item 1, also liked item 2.*

On tabular data, [1] concludes that the state of the art solutions are done with ensemble models like XGBoost.

3   Data Design

The dataset includes obfuscated data from several companies since 2018, a normalization was necessary to reconciled discrepancies between sources. Data is centered around accounts, assets, trades, and asset quotes (Fig. 1).[1] Data exploration showed minimal impact of demographics on trading decisions. The correlation matrix (Fig. 2) indicated strong correlations between transaction prices and asset quotes from the same or previous days. Key variables included financial product risk and country risk, aiding in feature identification for model training. Temporal effects were considered by including previous days' asset quotes and transaction details.

To address class imbalance (Table 1), we focused on frequently traded sub-assets. Sub-assets with fewer than 10 transactions were excluded due to insufficient data for effective model training. Techniques such as SMOTE were not applied in this study but could be considered in future work.

Table 1. Comparative Data Distribution by Date

SubAssetClass	2023-09-01 to 2023-11-01 Percentage	2023-11-01 to 2024-01-27 Percentage
Corporate Bonds	0.50	0.17
Funds	0.29	0.76
Otc fixed income	0.21	0.05

---

[1] This information cannot be made public due to privacy concerns of third-party users. However, interested parties may contact us to request access to the obfuscated dataset. Access may be granted upon signing a Non-Disclosure Agreement.

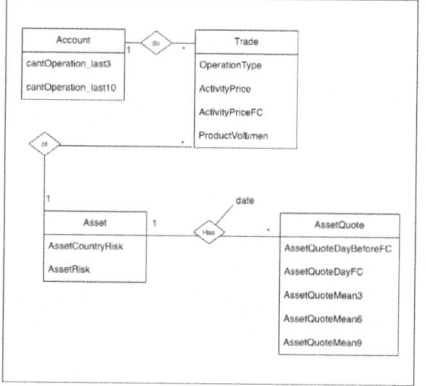

Fig. 1. Entities and their relationships.

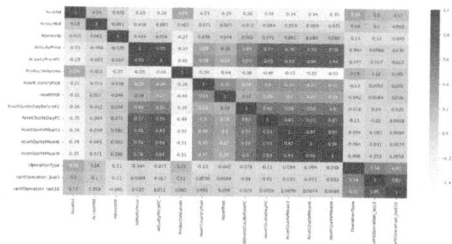

Fig. 2. Correlation matrix.

4   Implementation

We evaluated a reduced dataset, scaling to full historical data.

4.1   Approaches Used

Collaborative and Content-Based Filtering. Grouping assets by account IDs using **AgglomerativeClustering** followed by KNN faced challenges with high-dimensional vectors and complexity in distinguishing buy/sell operations, so this solution was discarded.

Multitarget Classification. Three algorithms were tested: **XGBoost**, a neural network, and SVM. **XGBoost** and the neural network handled tabular data effectively, with SVM providing robust classification for SubAssetClass. **Cascade Model.** The implementation of models with the three mentioned algorithms (Fig. 5) yielded metrics for accuracy and processing time. These metrics demonstrated that XGBoost was the most effective model for predicting both side decisions and SubAssetClass. Confirming the state-of-the-art solution for tabular data according to [1], XGBoost consistently outperformed other methods in this specific context. It is also discussed why we finally opted for a staged model, where the first decides the side of the operation, then the SubAssetClass, and finally the Asset for the recommendation. Figure 3 shows the results using a single multitarget model to predict buy/sell and SubAssetClass. As we know that the choice between buy and sell affects the results, we decided to test a cascade model predicting buys and sells with another recommending sub asset classes shown in tables of Fig 4.

Algorithm	Accuracy	Comments
XGBoost MT 16	0.5810	Multitarget
XGBoost MT 16	0.6689	Only buy data
XGBoost MT 16	0.6980	Only sell data
NN 16	0.5505	
NN 16	0.6917	Only buy data
NN 16	0.7492	Only sell data

Model	Binary Accuracy
NN binary	0.9268
SVM binary	0.8906
XGBoost binary	0.9126
Model	Cascade Multitarget Accuracy
NN binary → XGBoost	0.7106
XGBoost binary → XGBoost	0.7268
SVM binary → XGBoost	0.7216
XGBoost binary → NN	0.6925
NN binary → NN	0.682
SVM binary → NN	0.6883

Fig. 3. Accuracy results for XGBoost **Fig. 4.** Results for binary and multitarget models and Neural Network data

March 2024 results validated the cascade model with a 59.74% accuracy in predicting Operation Type, SubAssetClass, and Asset.

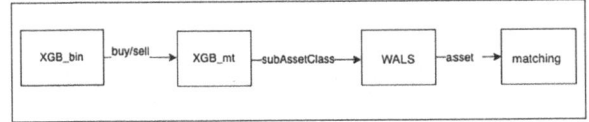

Fig. 5. Process diagram.

Hyperparameter Tuning. XGBoost hyperparameters were tuned using 5-fold GridSearch with ROC-AUC evaluation. For SVM, training with linear and polynomial kernels were too slow.

4.2   Selection of Final Recommendations

For the final recommendation of specific assets, we employed Weighted Alternating Least Squares (WALS), an effective matrix factorization technique for this context that captures the underlying relationships and preferences. This method is well-suited for tabular data, offering personalized recommendations based on latent factors. The dot product of account and asset embeddings within the predicted SubAssetClass yields a score, from which top recommendations are selected. The final step involves the matching process, which uses a FIFO system to store the predictions in the database.

5   Conclusions

We demonstrated the emulation of financial advisors' trading behavior, achieving a 60% alignment with actual decisions. Boosting methods, particularly XGBoost, proved being the most effective for handling tabular data. The cascade model provided accurate recommendations, making the system a valuable tool for financial advisors. This promising accuracy indicates that the recommendations are relevant, suggesting they hold significant potential for effectively guiding asset choices.

References

1. Hwang, G.: Machine learning: foundations and principles (2021)
4. Li, J., Zhang, Y., Yang, X., Chen, L.: Online portfolio management via deep reinforcement learning with high-frequency data. Inf. Process. Manage. **60**, 103247 (2023)
3. Pazzani, M.J., Billsus, D.: Content-based recommendation systems. In: Brusilovsky, P., Kobsa, A., Nejdl, W. (eds.) The Adaptive Web. LNCS, vol. 4321, pp. 325–341. Springer, Heidelberg (2007). https://doi.org/10.1007/978-3-540-72079-9_10
4. Taghian, M., Asadi, A., Safabakhsh, R.: A reinforcement learning-based encoder-decoder framework for learning stock trading rules. Expert Syst. Appl. **244**, 114279 (2023)

Design of a Machine Learning Algorithm for Precipitation Prediction in Riobamba Using Neural Networks

Gladys Urquizo[1,2](✉) ⓘ, Cristina Ramos[1,2] ⓘ, Juan Villacr[3] ⓘ,
and Alex Pozo[2] ⓘ

[1] Universidad Nacional de Chimborazo, Riobamba, Ecuador
gladys.urquizo@unach.edu.ec
[2] Escuela Superior Politécnica de Chimborazo, Riobamba, Ecuador
[3] Universidad Técnica Equinoccial, Riobamba, Ecuador
http://unach.edu.ec/

Abstract. This study aims to predict daily precipitation from meteorological data of the city of Riobamba using neural networks. The extreme values present in this phenomenon influenced the generation of weights in the network, representing a learning difficulty for the models. To overcome these challenges, 2 LSTM models and a CNN were implemented. The LTSM network provided an acceptable fit, matching the predicted trends with the observed ones, with an RMSE of 2.7648, indicating that the model performance was reliable and consistent.

Keywords: Artificial Neural Networks · Precipitation Prediction · Meteorological Data · Model Performance Evaluation

1 Introduction

The Total Water Storage in a river basin comprises surface water, soil moisture, groundwater, local snow, and ice. It has been affected worldwide and shows a trend below the values measured in previous years [1], the overexploitation of water resources, drought, and climate change have caused a decrease in freshwater storage in many regions [2]. This leads to unfavorable precipitation scenarios owing to a change in the rainfall cycle, which directly affects the management and reserves of water sources, groundwater, river flows, and the agricultural [3].

The study and application of machine learning as a predictive tool has gained significant momentum across various domains [4], with the development of neural networks contributing to this objective [5]. In the case of predicting climatological variables, advancements in new algorithms have optimized computational resources while enhancing predictive accuracy [4]. Neural networks offer substantial capabilities and functionality compared to traditional methods, facilitating function approximation and pattern recognition [6].

L. Correia et al. (Eds.): IBERAMIA 2024, LNCS 15277, pp. 486–489, 2025.
https://doi.org/10.1007/978-3-031-80366-6

This investigation compared recurrent and convolutional neural networks for estimating daily precipitation in Riobamba. The principal objective of this analysis is to determine which type of artificial neural network (ANN) model is most suitable for understanding precipitation patterns.

2   Methodology

2.1   Study Area and Data

Study Area. Riobamba City is the capital of the province of Chimborazo and has an area of 2,812.59 hectares. It is located 2754 m above sea level, at 1° 41' 46" South latitude and 0° 3' 36" West longitude from the meridian of Quito, which corresponds to the Ecuador Central Highlands [7]. For this study, a conventional meteorological station located at the Escuela Superior Politécnica de Chimborazo (ESPOCH) at an altitude of 2471 m.s.n.m with coordinates 758398 N and 9816965 S zone 17 was used.

Data. Daily maximum, average, and minimum ambient temperature (°C), relative humidity (%), wind speed (m/s), and precipitation (mm) data recorded during synoptic hours over 32 years (1990–2022) were used as variables predictors. One of the challenges when working with meteorological data is that there is normally a high number of missing values, and the imputation is not simple because of the characteristics of the measurements. To address this, data completion was performed using the cubic spline interpolation method [8], which is particularly effective for smooth functions, provides a good fit for observed data patterns and can produce smooth curves that closely match actual variations in the data, thereby maintaining the integrity of the original measurements [10]. The observations did not follow a normal distribution according to the Kolmogorov-Smirnov test.

2.2   Structure and Development of the Models

Input Data. The models were based on the program developed in Python by Greg Hogg for the prediction of climatological data using the open code module TensorFlow and Keras, which allows the construction, training, evaluation, and validation of deep automatic learning models [9]. The data frame has a total of 12053 observations for each variable, and the instances or samples of information that enter the models are structured in batches of six rows and six columns and allow the specification of the time interval in which the prediction is made. The database was segmented into 70% for training, 15% for validation, and 15% for validation, and the results were evaluated using the root mean square error (RMSE) metric between the observations and model predictions.

Model 1 LSTM. This model is of the sequential type, the input layer works with six neurons corresponding to six predictor variables. The model included hidden LSTM layers with a memory capacity of 64 neurons in each layer, utilizing a rectified linear unit activation function. The output layer is configured as a single neuron with a linear activation function, because the prediction variable is numerical. The model was trained over 10 epochs, and the loss function and mean square error were analyzed to ensure proper learning and validation of the model.

Model 2 CNN. The creation of this model involved the development of a CNN.

The aim of the extraction of structured information is for the algorithm to work with convolution in only one dimension for each row, as opposed to the two dimensions with which these algorithms typically work. The parameter values used were a convolution filter set to 64, kernel size of 2, with a ReLU activation function, the dimension of the resulting sequence is 5, and it has a filter of 64. The result of flattening the remaining dimensions of the input is 320. The dense layers 3 and 4 have six neurons and an activation function ReLu, respectively, whereas the dense layer 5 has one neuron and a linear activation function. The learning rate of the network is 0.0001, it has a regularization with a coefficient of 0.02, and has 10 training epochs.

Model 3 LSTM1. The neural network structure of this model is similar to

Model 1. It works with standardized data and aims to better manage data variability, thereby improving predictive performance.

3   Results and Discussion

The three models presented an acceptable descent of the batch gradient, a decrease in the loss and mean square error indicators and are pioneers in the area of prediction of climatological variables using Artificial Intelligence for Riobamba. The loss and mean square error indicators indicate that the model with the lowest RMSE for precipitation prediction is the LSTM model with an RMSE of 2.7923, followed by the CNN model with an RMSE of 2.8085, which generated less favorable results, while the LSTM1 model generated an RMSE value of 0.8843. The higher RMSE of the implemented CNN model compared to model 1 suggests the presence of limitations in the CNN model, such as the inability to capture complex patterns, a possible loss of valuable information and restricted learning of the model due to regularization. In contrast, the reduction in the RMS of the LSTM1 model can be attributed to the impact of standardization. This reduction in RMSE suggests greater accuracy of the predictions and optimizes the performance of the model. However, when the data was denormalized, a larger error was observed. Results that agree with recent studies on precipitation prediction, in [3], three LSTM models and one SARIMA model with monthly precipitation data showed a higher accuracy of the LSTM over the SARIMA in precipitation prediction, with RMSE values of 8.15 for Hue, 12.72 for Alouoi and a slightly lower accuracy in Namdong with an RMSE of 12.81.

4   Conclusions

The developed models exhibited a slightly increasing trend, consistent with observed patterns at the ESPOCH Meteorological Station. However, the significant variability in station data complicates the accurate prediction of extreme weather events. Notably, the LSTM model excelled in capturing the temporal dynamics of precipitation, achieving an RMSE of 2.7923, outperforming the CNN model in accuracy. Daily observed precipitation ranged from 0 to 18.90 mm, with RMSE values of 2.7648 and 2.8055 for the LSTM model in the training and validation sets, respectively, indicating robust and consistent performance. Additionally, normalizing data for the LSTM1 model resulted in negligible differences compared to models trained without normalization. This work is pioneering in applying artificial neural networks to predict climatological variables in Riobamba. By comparing recurrent and convolutional neural networks, this study identified the LSTM model as the most suitable for understanding and predicting precipitation patterns in this region. By comparing our results with other studies and acknowledging the limitations in capturing extreme values, we aim to provide a more comprehensive understanding of precipitation prediction using artificial neural networks.

Acknowledgments. Thanks to the "Predict Climate Behavior Patterns in Chimborazo" project and GEEA-ESPOCH for their collaboration and data provision.

References

1. Rodell, M., et al.: Emerging trends in global freshwater availability. Nature **557**(7707), 651–659 (2018)
2. Boser, A., Caylor, K., Larsen, A., Pascolini-Campbell, M., Reager, J.T., Carleton, T.: Field-scale crop water consumption estimates reveal potential water savings in California agriculture. Nat. Commun. **15**(1), 2366 (2024)
3. Giang, N.H., Wang, Y., Hieu, T.D., Thinh, N.T.: Monthly precipitation prediction using neural network algorithms in the Thua Thien Hue Province. J. Water Clim. Change **13**(5), 2011 2033 (2022)
4. Li, H., Li, M.: Modeling of precipitation prediction based on causal analysis and machine learning. Atmosphere **14**(9), 1396 (2023)
5. Barzola-Monteses, J., Gomez-Romero, J., Fajardo, W.: Hydropower production prediction using artificial neural networks: an Ecuadorian application case. Neural Comput. Appl. **34**(16), 13253–13266 (2022)
6. Luk, K.C., Ball, J.E., Sharma, A.: A study of optimal model lag and spatial inputs to artificial neural network for rainfall forecasting. J. Hydrol. **227**(1–4), 56–65 (2000)
7. Alcaldía de Riobamba. Plan de desarrollo cantonal
8. Schoenberg, I.J.: Contributions to the problem of approximation of equidistant data by analytic functions. Q. Appl. Math. **4**(2), 112–141 (1946)
9. Hogg, G.: Multivariate time series forecasting using LSTM, GRU & 1d CNNs'
10. Baltazar, J.-C., Claridge, D. E.: Study of cubic splines and Fourier series as interpolation techniques for filling in short periods of data. Solar Energy (2002)

Comparative Evaluation of Algorithms for Adaptive Learning Analytics: A Systematic Review of Higher Education Applications

Pablo Andres Quijano-Cabezas[(✉)] [ID], Néstor Duque-Méndez[ID],
and Jovani Alberto Jiménez-Builes[ID]

National University of Colombia, Bogotá, Colombia
{paquijanoc, ndduqueme, jajimenl}@unal.edu.co

Abstract. Adaptive Learning Analytics (ALA)—a discipline at the intersection of data science and education—emerges as a promising approach to enhance learning effectiveness and adaptability. This systematic review aims to identify the most frequently used algorithms in ALA within higher education, based on literature from 2018 to 2024. It seeks to provide insights into the general performance and specific uses of each algorithm in ALA applications, in order to obtain a point of comparison that contributes to informed decision-making.

Keywords: Learning Analytics · Adaptive Learning Analytics · Personalized Learning · Algorithm Performance · Data Science in Education

1 Introduction

Learning analytics personalizes education by tailoring insights to student needs. A subset, Adaptive Learning Analytics (ALA), focuses on delivering personalized content to optimize learning outcomes by adjusting to individual learner characteristics such as learning styles and prior knowledge levels [1]. While there are many methods in learning analytics, the choice of algorithm depends on the specific use case, as no single algorithm is best for all tasks, courses or groups [2]. Consequently, a problem lies in identifying the most suitable ALA algorithms in each educational context, given the lack of a universally optimal solution. This systematic review aims to identify the most frequently used algorithms in ALA applications in higher education based on literature, to obtain a point of comparison, comparing their performance and uses. The article is structured with sections on materials and methods, results, discussion, and conclusions.

2 Materials and Methods

This review considers guidelines and recommendations from PRISMA 2020 statement [3], and the review process proposed by [4] with a three-stage methodology called planning, implementation and reporting. One of the fundamental parts in the planning stage is the creation of the research questions in line with the proposed objectives.

L. Correia et al. (Eds.): IBERAMIA 2024, LNCS 15277, pp. 490–495, 2025.
https://doi.org/10.1007/978-3-031-80366-6

- *Q1: What are the most frequently used algorithms in ALA applications?*
- *Q2: What are the specific uses of the algorithms in ALA applications?*
- *Q3: Which algorithms have performed best in the specific uses of ALA?*

To define the research, the following related words were considered: *Adaptive Learning Analytics (ALA), adaptive, personalization, educational data, higher education, learning and teaching.* Additionally, inclusion criteria (covering the topic and aligning with the research questions), exclusion criteria (other areas, length, structure), and temporal criteria (2018–2024) were contemplated. The search gathered 590 documents from Scopus (385), Science Direct (135), and Web of Science (70) databases, selected for their extensive range of peer-reviewed publications related to the topic. After a close examination these were reduced to 17 papers.

3   Results

Table 1 synthesizes which algorithms had the best performance for the uses found.

Table 1. Best performance ALA algorithms comparison per task.

Source: authors

Uses	Algorithm	Performance
Performance categorization (Due to the large number of algorithms, only the top two for each metric are presented)	Random Forest - RF	Accuracy: 0,76 [5]; 0,788 [6] F-score: 0,7702 [6]; 0,896 (Bagging) [7]
	Decision Tree - DT	Accuracy: 0,644 [8]; 0,947 [9]; 0,8083 (Early stopping with XGBoost) [10]; 0,755 (Bagging) [5]
	Neuronal Network - NN	Accuracy: 0,7778 [9]
	NN (Multi-Layer perceptron - MLP)	Accuracy: 0,99 [7]; 0,99 [11] F-Score: 0,978 [11]
	Logistic Regression - LOG	Accuracy: 0,9925 [11]; 0,8725 (Traditional) [9]; 0,7292 (Blended) [9] F-score: 0,983 [11]
	Stacking (Naïve Bayes - NB, DT, MLP, LOG meta)	Accuracy: 0,989 [7] F-score: 0,904 [7]
	XGBoost	Accuracy: 0,783 [6] F-score: 0,764 [6]
	Naïve Bayes	Accuracy: 0,633 [8]
	Deep feed forward neural network (merging grades)	Accuracy: 0,79 [10]

(continued)

Table 1. (*continued*)

Uses	Algorithm	Performance
Early Warning Systems - at risk student (Due to the large number of algorithms, only the top two for each metric are presented)	Supply Vectors Machine SVM	Accuracy: 0,9508 [12]; 0,8964 (50% semester) [13] Precision: 0,8964 (SMOTE oversampling) [12] F-Score: 0,7402 (50% semester) [13]
	DT	F-Score: 0,79 (6 W) [14]; 0,7467 (50% sem) [13]
	NB	Accuracy: 0,8929 (50% semester) [13] F-Score: 0,7723 (50% semester) [13]
	KNN	F-Score: 0,8 (6 W/weeks) [14]; 0,7025 [15]
	RF	MAE: 7,1 [16]
	BART	MAE: 6,5 [16]
	Bayesian Network Classifier- BNC	F-score: 0,7725 (4 weeks) [15]
Prediction of Dilatory Behavior/procrastination	RF	Accuracy: 0,9508 (Combined predictors) [17] MAE: 7,27 (Combined predictors) [17]
	Gradient Boosting Machine - GBM	Accuracy: 0,6689 (Subjective predictors) [17] MAE: 9,11 (Subjective predictors) [17]
	Bayesian model with random slopes - BARS	Accuracy: 0,6901 (Objective predictors) [17] MAE: 7,57 (Objective predictors) [17]
Performance prediction	RF	RMSE:1,775 [5] MAPE:16,4% [18]
	DT	RMSE:1,65 [5]
	SVM	RMSE:1,55 [5]
Enrolment decision prediction	DT	Accuracy: 0,7136 [19]
	LOG	Accuracy: 0,7075 [19]
	NB	Accuracy: 0,7004 [19]
Predicting student dropout and retention	MLP	Accuracy: 0,9486 [20]
	LOG	Accuracy: 0,8182 [20]
	SVM	Accuracy: 0,8121 [20]
Personality modelling	NB	Accuracy: 0,492 [21] Precision: 0,523 [21]

4   Discussion

This review examined frequently used algorithms in ALA, addressing next RQ.

Q1: What are the most frequently used algorithms in ALA applications?: Several identified algorithms within machine learning, deep learning, and ensemble methods. *ML:* These include, ordered by frequency, SVM, RF, KNN, NB, DT, LOG, LDA (Linear Discriminant Analysis), and BNC. *DL:* NN and its variations such as MLP and DFNN. *EN:* DT and RF bagging, boosting techniques like AdaBoost and XGBoost, stacking (NB, DT, MLP, and LOG), and voting methods (NB, DT, and MLP).

Q2: What are the specific uses of the algorithms in ALA applications?: Performance categorization, early warning systems for at-risk students, enrollment decision prediction, performance prediction, personality modeling, and procrastination prediction. The most application cases were the first two categories. Therefore, beyond performance, future research should explore a comprehensive view of student needs that includes engagement, mental health, and personalized learning paths, integrating social, emotional, and behavioral data. Researchers must address data volume, quality, type, and computational resource challenges, especially for deep learning, and consider integration with current educational systems.

Q3: Which algorithms have performed best in the specific uses of ALA?: For performance categorization, the best-performing algorithms were MLP, LOG, DT, and Stacking (NB, DT, MLP, LOG meta). For early warning systems for at-risk students, the top performers were SVM, KNN, NB, and BART. Predicting procrastination was best handled by RF, GBM, and BARS. For performance prediction, RF, DT, and SVM were used, while DT, LOG, and NB were effective for enrollment decision prediction. MLP, LOG, and SVM were used for predicting student dropout and retention, and NB was used for personality modeling. Each application has unique characteristics, so it's crucial to analyze the context for more appropriate comparisons. Most applications used ML algorithms. Ensemble and DL algorithms are probably less widespread due to high computational resource requirements, complexity in implementation and interpretation, and the need for high-quality large-scale datasets. These challenges can be barriers for educational institutions. Despite this, the potential benefits of ensemble and DL algorithms in ALA justify further exploration and investment to overcome these obstacles.

5   Conclusions

Selecting suitable algorithms for ALA applications could consider the described criteria as a useful reference point, however it is essential to also consider other variables related to the problem and context. Since there is no single best algorithm for all tasks, comparing similar methods is crucial. For future work, it is proposed to conduct additional studies that deepen the comparative analysis of specific algorithms in different educational contexts and for various applications, addressing current research limitation such as research questions and information sources. Furthermore, it would be beneficial to investigate the integration of new algorithmic approaches, such as deep learning and ensemble methods, in the field of ALA to improve the efficiency and adaptability of adaptive learning systems.

Disclosure of Interests. The authors have no competing interests to declare that are relevant to the content of this article.

References

1. Sarıyalçınkaya, A.D., Karal, H., Altinay, F., Altinay, Z.: Reflections on adaptive learning analytics. In: Presented at the (2021). https://doi.org/10.4018/978-1-7998-7103-3.ch003
2. Namoun, A., Alshanqiti, A.: Predicting student performance using data mining and learning analytics techniques: a systematic literature review. Appl. Sci. **11**, 237 (2020). https://doi.org/10.3390/app11010237
3. Page, M.J., et al.: The PRISMA 2020 statement: an updated guideline for reporting systematic reviews. BMJ **71** (2021). https://doi.org/10.1136/bmj.n71
4. Kitchenham, B., Charters, S.: Guidelines for performing SLR in software engineering (2007)
5. Zhang, Y., et al.: Educational data mining techniques for student performance prediction: method review and comparison analysis. Front Psychol. **12**, (2021). https://doi.org/10.3389/fpsyg.2021.698490
6. Md, S., Krishnamoorthy, S.: Student performance prediction, risk analysis, and feedback based on context-bound cognitive skill scores. Educ. Inf. Technol. (Dordr) **27**, (2022). https://doi.org/10.1007/s10639-021-10738-2
7. Butt, N.A., et al.: Performance prediction of students in higher education using multi-model ensemble approach IEEE Access. **11**, 136091–136108 (2023). https://doi.org/10.1109/ACCESS.2023.3336987
8. Yildirim, D., Gülbahar, Y.: Implementation of learning analytics indicators for increasing learners' final performance. Technol. Knowl. Learn. **27**, (2022). https://doi.org/10.1007/s10758-021-09583-6
9. Hamim, T., Benabbou, F., Sael, N.: Survey of machine learning techniques for student profile modelling Int. J. Emerg. Technol. Learn. **16**, 136–151 (2021). https://doi.org/10.3991/ijet.v16i04.18643
10. Adnan, M., Alarood, A.A.S., Uddin, M.I., Rehman, I.U.: Utilizing grid search cross-validation with adaptive boosting for augmenting performance of machine learning models. PeerJ Comput. Sci. **8**, (2022). https://doi.org/10.7717/peerj-cs.803
11. Abdelhafez, H.A., Elmannai, H.: Developing and comparing data mining algorithms that work best for predicting student performance Int. J Inf. Commun. Technol. Educ. **18**, 1–14 (2022). https://doi.org/10.4018/ijicte.293235
12. Lu, O.H.T., Huang, A.Y.Q., Yang, S.J.H.: Impact of teachers' grading policy on the identification of at-risk students in learning analytics. Comput. Educ. **163**, (2021). https://doi.org/10.1016/j.compedu.2020.104109
13. Baneres, D., Elena Rodriguez-Gonzalez, M., Serra, M.: An early feedback prediction system for learners at-risk within a first-year higher education course IEEE Trans. Learn. Technol. **12**, 249–263 (2019). https://doi.org/10.1109/TLT.2019.2912167
14. Azcona, D., Hsiao, I.-H., Smeaton, A.F.: Detecting students-at-risk in computer programming classes with learning analytics from students' digital footprints User Model. User-Adap. Inter. **29**, 759–788 (2019). https://doi.org/10.1007/s11257-019-09234-7

15. Kustitskaya, T.,A., Kytmanov, A.A., Noskov, M.V.: Early student-at-risk detection by current learning performance and learning behavior indicators Cybern. Inf. Technol. **22**, 117–133 (2022). https://doi.org/10.2478/cait-2022-0008

16. Howard, E., Meehan, M., Parnell, A.: Contrasting prediction methods for early warning systems at undergraduate level Internet High. Educ. **37**, 66–75 (2018). https://doi.org/10.1016/j.iheduc.2018.02.001

17. Imhof, C., et al.: Prediction of dilatory behavior in eLearning: a comparison of multiple machine learning models IEEE Trans. Learn. Technol. **16**, 648–663 (2023), https://doi.org/10.1109/TLT.2022.3221495

18. Rincon-Flores, et al.: Predicting academic performance with Artificial Intelligence (AI), a new tool for teachers and students. In: IEEE Global Engineering Education Conference, pp. 1049–1054 (2020). https://doi.org/10.1109/EDUCON45650.2020.9125141

19. Ab Ghani, N.L., Che Cob, Z., Mohd Drus, S., Sulaiman, H.: Student enrolment prediction model in higher education institution: a data mining approach. In: Othman, M., Abd Aziz, M., Md Saat, M., Misran, M. (eds.) SYMINTECH 2018. LNEE, vol. 565, pp. 43–52. Springer, Cham (2019). https://doi.org/10.1007/978-3-030-20717-5_6

20. Nithya, S., Umarani, S.: An identification of the prominent learner behavioral features to predict MOOC dropouts using hybrid algorithm J. Theor Appl. Inf. Technol. **101**, 1261 1274 (2023)

21. Tlili, A.: Automatic modeling learner's personality using learning analytics approach in an intelligent moodle learning platform Interact. Learn. Environ. **31**, 2529–2543 (2019). https://doi.org/10.1080/10494820.2019.1636084

Audio-Based Violence Detection Using Spectrograms and Deep Learning

Bruno Campos[1], Carlos Rodríguez-Domínguez[2], Miguel J. Hornos[2],
and Manuel Rodrigues[1(✉)]

[1] 1LASI/ALGORITMI Centre, University of Minho, Braga, Portugal
pg50275@alunos.uminho.pt, manuel.rodrigues@algoritmi.uminho.pt
[2] Software Engineering Dept, Research Centre for Information and
Communication Technologies (CITIC-UGR), University of Granada, Granada,
Spain
{carlosrodriguez,mhornos}@ugr.es

Abstract. Violence detection in audio is crucial for public safety and preventing harmful behaviors. With the increasing availability of audiovisual content and growing security concerns, there is a need for automatic systems capable of identifying signs of violence in sound media. This study explores recent approaches to violence detection in audio that incorporate visual representations and deep neural networks, especially Convolutional Neural Networks (CNNs). The efficiency of lightweight models is evaluated, highlighting the importance of efficiency in real-world scenarios with limited computational resources. The results indicate a shift towards visual representations and efficient deep neural networks, promising improvements in audio violence detection with practical applicability.

Keywords: Violence · Audio detection · Artificial intelligence · Deep neural networks

1 Introduction

Violence is a broad and complex term that encompasses various forms of physical, psychological, or emotional harm [7]. It can manifest in several ways, including interpersonal violence, that can have severe consequences on physical and mental health, social cohesion, and overall well-being [9]. According to [5], 1 in 10 individuals reported experiencing physical violence in the five years preceding the survey. In Portugal, between 2020 and 2022, there was a significant increase of 25.5% in the number of services provided by APAV's Services [2]. These figures underscore the interest in researching a real-time violence detection system to help identify and address these critical issues more efficiently and effectively. Traditionally, violence recognition methods primarily relied on video detection, demanding high performance

This work has been supported by FCT - Fundação para a Ciência e Tecnologia within the R&D Units Project Scope: UIDB/00319/2020.

L. Correia et al. (Eds.): IBERAMIA 2024, LNCS 15277, pp. 496–500, 2025.
https://doi.org/10.1007/978-3-031-80366-6

hardware and software for recording. However, the alternative technique of using audio for violence detection has been gaining increasingly more prominence in the field [4]. Deep learning algorithms have proven effective in recognizing and classifying audio signals, providing a novel dimension to violence detection into the surveillance ecosystem.

2 Related Works

Recent approaches to violence detection in audio have shifted towards using visual representations of audio signals and leveraging deep neural networks, especially CNNs. This paradigm shift involves converting audio signals into visual spectrogram representations and allowing deep learning models to learn hierarchical features directly from these representations. The work of [1] showcased the effectiveness of combining CNNs and ensemble classifiers for acoustic scene classification. The late fusion model, which merged the advantages of both models, demonstrated higher accuracy compared to individual models. In [6, 10], the authors tested various deep learning models, including CNN, ResNet, VGG and MobileNet (LDLM - lightweight deep learning model) for violence detection in audio. The study indicated that CNNs yielded the best results, emphasizing the power of deep learning models in capturing intricate patterns in audio data, and the shift towards LDLM for real-world applications with constrained computational resources, that was also demonstrated in [11]. The work by [8] introduced TreePat 23, a model using wavelet bands and a raw audio signal for violence detection. Despite incorporating visual representations, the model used shallow classifiers (KNN and SVM) and achieved high classification performance. [3] proposed deep learning approaches for classifying speech data containing violent behavior, using Mel-spectrogram images as input for both traditional deep neural networks and LDLM. These lightweight models demonstrated exceptional performance, highlighting the efficiency and effectiveness of simpler architectures. In [12], the authors compared CNNs and Shallow Networks, emphasizing the significance of data augmentation and appropriate training configurations.

Concluding, recent related works underscore the advancement of violence detection in audio towards data-driven strategies and visual representation using deep learning and LDLM. These approaches prioritize efficiency, generalization, and real- world applicability.

3 Audio-Based Violence Detection

3.1 Dataset – Preprocessing

Three publicly available datasets were used in this work, encompassing a variety of violence audios, (Violence Detection Dataset, Audio-based Violence Detection Dataset, Real Life Violence Situations Dataset). The raw audio data underwent several preprocessing steps to ensure data quality and consistency: a) conversion to (.wav) files to ensure compatibility with subsequent processing steps; b) resampling to a minimum frequency of 44.1 kHz to ensure a common sampling rate across all audio files, which is crucial for tasks like feature extraction and spectrogram generation; c) data cleaning procedures were implemented to address corrupted or missing audio files (ex. Manually correcting labels from violent to non-violent where violence was only present in video and not in the accompanying audio); d) segmentation into fixed duration clips of 5 s, enabling the generation of Mel spectrograms.

3.2   Dataset – Final

The final dataset consisted of 7288 spectrograms generated from audio files, 79.3% non-violent and 20.7% violent. A balanced Dataset was achieved reducing non-violent videos via random sampling. To further enhance the size and robustness of the dataset, data augmentation techniques were implemented (horizontal flipping) as well as normalization techniques (scaling pixel intensity), Fig. 1. Finally, the dataset was split into training, validation, and testing sets using a 70/10/20 split ratio, respectively. This final dataset, prepared through meticulous preprocessing, balancing, augmentation, and splitting steps, provides a robust foundation for training and evaluating the violence detection models.

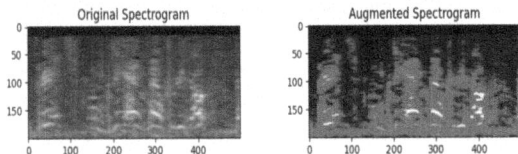

Fig. 1. Example of an augmented spectrogram.

3.3   Classification

Models The following models were used: ResNet, MobileNet, EfficientNet, and VGG. All models were initialized with weights pre-trained on the ImageNet 1K V2 dataset, and a final fully-connected layer with ReLU activation and dropout was added. Additionally, the initial layers' parameters were frozen to leverage the pretrained features for efficient training and overfitting prevention.**Training**: Hyperparameter sweeps were conducted using Weights & Biases (Wandb) to optimize the model training process: Epochs: 100; Batch Size: Quantized log-uniform distribution; Learning Rate: Uniform distribution between 0 and 0.0002; Optimizer: Adam, Adagrad, AdamW, RMSprop; Loss Function: CrossEntropyLoss; Fully-connected Layer Size: 256, 512, 1024, 2048; Dropout Rate: 0.1, 0.3, 0.4, 0.5. To prevent overfitting, early stopping with a patience of 5 epochs was implemented. Additionally, a learning rate scheduler (ReduceLROnPlateau) was employed to reduce the learning rate by a factor of 0.1 if the validation accuracy plateaued for 2 consecutive epochs.

A mobile application (Fig. 2) was developed and used, following a structured process at 5-s intervals: a) capturing a segment of audio data; b) creating the spectrogram c) using a lightweight PyTorch model for provisional classification within the

Fig. 2. Mobile App

app; d) spectrogram and classification label are sent back to the server, for more advanced and accurate classification that is sent back to the user, enabling real-time audio classification with progressive refinement, leveraging the strengths of both lightweight and advanced models.

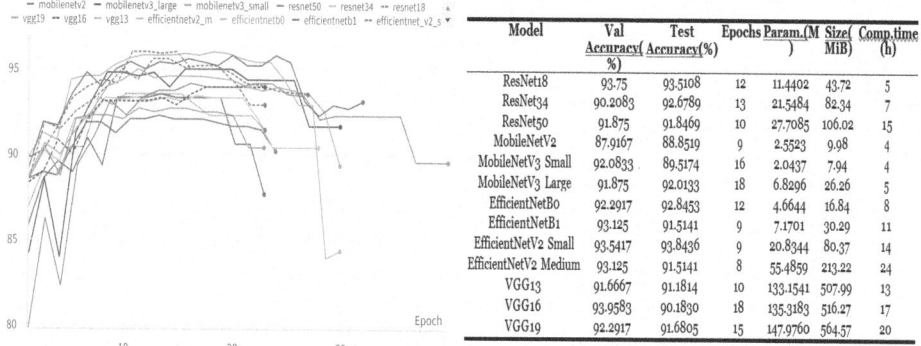

Model	Val Accuracy(%)	Test Accuracy(%)	Epochs	Param.(M)	Size(MiB)	Comp.time (h)
ResNet18	93.75	93.5108	12	11.4402	43.72	5
ResNet34	90.2083	92.6789	13	21.5484	82.34	7
ResNet50	91.875	91.8469	10	27.7085	106.02	15
MobileNetV2	87.9167	88.8519	9	2.5523	9.98	4
MobileNetV3 Small	92.0833	89.5174	16	2.0437	7.94	4
MobileNetV3 Large	91.875	92.0133	18	6.8296	26.26	5
EfficientNetB0	92.2917	92.8453	12	4.6644	16.84	8
EfficientNetB1	93.125	91.5141	9	7.1701	30.29	11
EfficientNetV2 Small	93.5417	93.8436	9	20.8344	80.37	14
EfficientNetV2 Medium	93.125	91.5141	8	55.4859	213.22	24
VGG13	91.6667	91.1814	10	133.1541	507.99	13
VGG16	93.9583	90.1830	18	135.3183	516.27	17
VGG19	92.2917	91.6805	15	147.9760	564.57	20

Fig. 3. Maximum Validation Accuracy per Epoch and Comparison of Models

4 Conclusions and Future Work

Figure 3 provides an overview of the model's training performance over each epoch tested on the validation set. Efficient- NetV2 Small achieves the highest test accuracy, however its complexity might limit its deployment on resource-constrained devices. EfficientNetB0 presents a compelling alternative in this case. Overall, the accuracy of all models increases as the number of epochs increases, reaching its maximum around 10 epochs and then declining around 20 epochs. This quick learning process can be attributed to the effectiveness of both the CNN architectures and transfer learning. The trade-off between accuracy, model complexity and training time needs to be carefully considered when selecting a CNN architecture for real-world applications. The mobile application developed for real-time violence classification show-cases the practical application of this research, and stands as a valuable contribution.

One area to explore is the effectiveness of ensemble methods that combine multiple CNN architectures to achieve better classification performance compared to individual models. Additionally, it would be beneficial to investigate the potential of transferring knowledge from pre-trained models on related audio classification tasks, such as speech emotion recognition or audio event detection, to enhance performance with less training data specific to violence detection.

References

1. Alamir, M.A.: A novel acoustic scene classification model using the late fusion of convolutional neural networks and different ensemble classifiers. Appl. Acoust. **175**, 107829 (2021). https://doi.org/10.1016/j.apacoust.2020.107829
2. APAV: Estatisticas apav relatorio anual 2022 (2022). https://apav.pt/apav_v3/index.php/pt/estatisticas-apav. Accessed 24 October 2023
3. Bakhshi, A., Garcia-Gomez, J., Gil-Pita, R., Chalup, S.: Violence detection in real-life audio signals using lightweight deep neural networks. Procedia Comput. Sci. **222**, 244–251 (2023). https://doi.org/10.1016/j.procs.2023.08.162
4. Duraes, D., Veloso, B., Novais, P.: Violence detection in audio: evaluating the effectiveness of deep learning models and data augmentation. Int. J. Interact. Multimedia Artif. Intell. **8**(3), 72 (2023). https://doi.org/10.9781/ijimai.2023.08.007
5. FRA: Crime, safety and victims' rights – fundamental rights survey (2021). https://fra.europa.euen/publication/2021/fundamental-rights-survey-crime. Accessed 24 October 2023
6. Gomes, E.D.: Deep learning-based algorithm for violence detection in audio data, January 2022. https://hdl.handle.net/1822/83542
7. Jacquin, Kristine M.: "violence". Encyclopedia Britannica (2023). https://www.britannica.com/topic/violence. Accessed 24 October 2023
8. Metehan, A., et al.: A novel tree pattern-based violence detection model using audio signals. Expert Syst. Appl. **224**, 120031 (2023). https://doi.org/10.1016/j.eswa.2023.120031
9. Rivara, F., et al.: The effects of violence on health. Health Aff. **38**(10), 1622–1629 (2019). https://doi.org/10.1377/hlthaff.2019.00480
10. Veloso, B., Durães, D., Novais, P.: Analysis of machine learning algorithms for violence detection in audio. In: González-Briones, A., et al.: PAAMS 2022. CCIS, vol. 1678, pp. 210–221 (2022). Springer, Cham. https://doi.org/10.1007/978-3-031-18697-4_17
11. Veloso, B.C.: Análise de algoritmos de machine learning para deteção de violência em áudio, January 2023. https://hdl.handle.net/1822/86577
12. Zhu-Zhou, F., Tejera-Berengue, D., Gil-Pita, R., Utrilla-Manso, M., Rosa-Zurera, M.: Computationally constrained audio-based violence detection through transfer learning and data augmentation techniques (2023). https://doi.org/10.2139/ssrn.4476618
13. Andrade, G., Rodrigues, M., Novais, P.: A survey on the semi supervised learning paradigm in the context of speech emotion recognition. In: Arai, K. (eds.) IntelliSys 2021. LNCS, vol. 295, pp. 771–792. Springer, Cham (2022). https://doi.org/10.1007/978-3-030-82196-8_57

Capturing Collective Memories
of the Disappeared with Artificial Intelligence

Tomas Laurenzo[(✉)] [ID]

University of Colorado Boulder, Boulder, CO 80309, USA
tomas@laurenzo.net

Abstract. This paper presents a project that aims to prototype a system that utilizes LLM-based conversational interactions to capture and store elements of Uruguayans' collective memory of the disappeared during its latest dictatorship. Memories that would otherwise be inaccessible, forgotten, or that would leave out the oft-untold personal narratives that form a significant part of collective memories.

Keywords: Machine Learning · Conversational Interfaces · Collective Memory

1 Introduction

While South America is far from constituting a univocal block, its countries, among their many similarities, share similarly painful periods in their recent history which is marked by a series of military dictatorships that took control of many South American countries during the 1960s–1980s.

In the Southern Cone, military dictatorships took control in Brazil (1964), Chile (1973), Uruguay (1973), Argentina (1976), establishing regimes marked by ruthless repression of all opposition, and the capture and execution of an unprecedented number of people.

With a stated objective of cleansing their society of what they considered subversive elements polluted by socialist and communist ideas, these regimes also deepened their countries' geopolitical roles, solidifying their peripheral ('third world') status [1].

2 Collective Memory

Collective memory, a concept pertaining to the social sciences, differs notably from the typical memory categories studied in neuroscience; instead, it encompasses the aggregation of individual brains, minds, and memories that are part of a specific social group, thereby constituting a crucial element of its cultural identity.

Although much of collective memory is housed in the human brain, its creation and maintenance are regarded as distinctive aspects of a social group's collective behavior and cultural outputs, becoming foundational to the group's own identity.

© The Author(s), under exclusive license to Springer Nature Switzerland AG 2025
L. Correia et al. (Eds.): IBERAMIA 2024, LNCS 15277, pp. 501–504, 2025.
https://doi.org/10.1007/978-3-031-80366-6

Collective memories are propagated through cultural, rather than genetic, mechanisms throughout the lives of the group's members [2].

In this context, social regulations and cultural artefacts play a determining role, affecting which memories prevail, and how they are stored, transformed, and communicated within and across social groups and sub-groups.

Post-dictatorship (i.e. democratic) governments across the Southern Cone have systematically established 'politics of oblivion', effectively working against the construction and dissemination of collective memories pertaining to the dictatorships.

These politics, always executed under a pretense of "moving forward" and "leaving behind' a painful-yet-shared period –about which, in reality, there is no consensus– has precipitated a social "eternal returning" to these issues [3].

These eternal returns have grown in frequency, size, and effectiveness, with organizations, individuals, and sometimes governmental agencies, working towards the identification of remains, the prosecution of criminals, and the uncovering of the truth about individuals who have been forcibly disappeared.

Artists, especially filmmakers, have contributed greatly to the creation of new collective memories through the representation of specific personal histories [4].

This is particularly important for artworks (among other cultural artefacts) often act as "places of memory", i.e. places where "memory crystallizes and secretes itself"; the places where the exhausted capital of collective memory condenses and is expressed [3].

3   Uruguay

Uruguay's most recent military dictatorship (1973–1985) exhibited serious violations of human rights and fundamental freedoms, including more than 15.000 political prisoners, systematic torture in detention centers, summary executions, and enforced disappearance of persons.

Forced disappearance is a permanent and continuous violation of multiple rights, such as the right to personal liberty, to humane treatment, to life, and to recognition as a person before the law, which subsist so long as the disappeared person's whereabouts have not been established or their remains identified. It is a continuous crime against the missing persons, their families, and humanity.

Within Latin America historical heterogeneities, Uruguay offers a favorable context to investigate (and to act) in topics related to its recent past, due to the following factors:

- A small population of ∼3.5 million.
- A highly articulated society that has struggled to find ways to cope and reconcile with its recent past.
- An effective implementation of the 'politics of oblivion', during 1985–2000, signed by its Law 15.848, which granted amnesty to military responsible for crimes committed during the dictatorship.
- An effective process of "memory waves", that is, social moments working towards the establishing of new, alternative collective memories, marked by the establishing

of institutions (mostly NGOs) working towards this. In particular, two NGOs, SERPAJ and HIJOS, have been instrumental in the prosecution of 'crimes against humanity', as well as in the clarification of the outcome of forced disappearances (for example, in the identification of children of disappeared who have been appropriated by military families).

4 LLMs and Conversational Interfaces

Machine Learning (ML) is a field of Artificial Intelligence (AI) that investigate how algorithms can learn from observations and data, as opposed to having explicitly programmed behavior.

Over the past two decades ML has become one of the most important techniques in information technology, providing solutions to problems that seemed unsolvable, and outperforming humans in areas until now considered impossible to approach computationally [6].

Large Language Models (LLM) such as OpenAI's ChatGPT, are ML models trained on extremely large corpora of natural language (i.e. 'the web'), utilizing techniques such as tokenization and embedding, infer semantic relationships from statistical correlations in the training data. This statistical inference allows LLMs to capture multiple-level semantic information present in the training data, and to apply it in the processing of new information.

Among many outstanding results (such as automatic musical composition, writing, and software coding), LLMs are extraordinarily effective in supporting conversational interfaces. These interfaces often succeed in allowing users of channel all their interaction needs through natural language (such as English or Spanish).

These interaction designs have been shown to be extremely effective in making users feel that they are interacting with an intelligent system, and in leveraging users' existing conversational capabilities [5].

Even more, conversational interfaces are also effective when used by persons that would otherwise struggle, such as elderly or disabled users. They are also effective in enabling interaction to persons who would normally struggle due to lack of education (from computational literacy to analphabetism), or trust in automated systems.

5 Project Objectives

The project has two main goals, 1) to prototype a system that utilizes a custom LLM to aid in capturing and storing elements of Uruguayans' collective memory of the disappeared, that would otherwise be inaccessible, storing the oft-untold personal narratives that form a significant part of collective memories; and 2) to develop an on-line, interactive data-visualization system and artwork that permits to explore and navigate the collected information.

The project hopes to contribute to the understanding of the significance and importance of forced disappearances. In addition to the software, its documentation, and the properly anonymized data collected -all of which will be accessible online as open-source nature– we expect the project to be appropriated by organizations and persons, and so, it will create new spaces of political reflection and artistic collaboration world-wide.

6 Art as Research

Rhetorics conceived in geopolitically powerful contexts very often ignore the different relationships between art and politics that systematically appear in the geopolitical periphery, very specially in Latin America. Consequently, there is a need for a richer artistic and sociopolitical vocabulary that not only explores these practices, but that also avoids the false dichotomy inherent to the modernizing discourse, always focusing on the arduous task of developing yet another replacement of the stories surrounding "the other". In this context, explicitly political art is able to contest prevalent forms of political and disciplinary subjectivities, allowing for original forms of activist art and, more importantly significantly collaborating towards disarticulation of some of the processes which systematically normalize conditions of structural oppression, via a militant reclamation of the sensible.

Research that incorporates artistic creation as a core component has proven successful in creating design and technological alternatives, with artists not only contributing culturally relevant artefacts, but also generating novel approaches that expand the scope of existing analysis. This project intends to continue this trajectory, specifically focusing on potential synergies between artists, machine learning, and the corpora the project hopes to produce.

Acknowledgments. This project is supported by a RIO Arts & Humanities Grant award from University Colorado Boulder.

Disclosure of Interests. The author has no competing interests to declare.

References

1. Achugar, M.: What we remember: The construction of memory in military discourse (2008)
2. Kvasnička, V., Pospíchal, J.: Artificial intelligence and collective memory. In: Sinčák, P., Hartono, P., Virčíková, M., Vaščák, J., Jakša, R. (eds.) Emergent Trends in Robotics and Intelligent Systems. AISC, vol. 316, pp 283–291. Springer, Cham (2015). https://doi.org/10.1007/978-3-319-10783-7_31
3. Montaño, E.A.: Places of memory: Is the concept applicable to the analysis of memorial struggles? The case of Uruguay and its recent past. Cuadernos del CLAEH (2008)
4. Sipprelle, D.S.: Cinema Remembers: Forming a Collective Memory of Military Dictatorship. digitalrepository.trincoll.edu (2014)
5. Wang, B., Li, G., Li, Y.: Enabling conversational interaction with mobile ui using large language models. In: Proceedings from Proceedings of the 2023 CHI Conference on Human Factors in Computing Systems, New York, NY, USA (2023)
6. LeCun, Y., Bengio, Y., Hinton, G.: Deep learning. Nature **521**(7553), 436 (2015)

Using Deep Learning Models in Clinical Histories in Order to Find Precursors of Diseases

Juli Climent Querol[1]([⊠]), Gonzalo Hernández Ortega[1],
and Ernest Valveny Llobet[2][iD]

[1] Department of Artificial Intelligence, ASHO, 08018 Barcelona, Spain
juli@asho.net
[2] Department of Computer Science, Universitat Autònoma de Barcelona,
08193 Bellaterra, Spain

Abstract. This study demonstrates the use of advanced natural language processing (NLP) and deep learning techniques to analyze complex clinical histories in Spanish and Catalan. The focus is on detecting atrial fibrillation and prior stroke incidents, key precursors to severe cardiovascular issues. The approach employs a combination of Convolutional Neural Network (CNN), Bidirectional Long Short-Term Memory (BiLSTM), and Conditional Random Field (CRF) models. The CNN extracts local text patterns, the BiLSTM processes contextual information bidirectionally, and the CRF optimizes sequence labeling, achieving 99.4% precision. This method provides clinicians with valuable insights for early risk assessment and proactive healthcare.

Keywords: Atrial fibrillation · Clinical histories · Disease precursors · Natural Language Processing · Predictive analytics

1 Introduction

Hospitals face challenges with unstructured historical patient data, making manual extraction costly and labor-intensive. Given the large amount of untapped data, it is necessary to find a way to automate the information extraction process in order to analyze more reports in less time and even improve comprehensiveness.

The utilization of advanced NLP techniques, particularly employing deep learning architectures, has shown promise in uncovering critical information embedded within complex clinical histories [1]. Atrial fibrillation and stroke, in particular, have been identified as crucial precursors to severe cardiovascular complications [2]. For instance, Elkin et al. used a mixture of free text and a structured database with ICD-9 codes. In this study, they conclude that in just one year, this detection could lead in the USA to the prevention of 176,537 strokes and 10,575 deaths [3].

This work focuses on using Named Entity Recognition (NER) models to automate data extraction for the neurology services of CSAPG (Consorci Sanitari

L. Correia et al. (Eds.): IBERAMIA 2024, LNCS 15277, pp. 505–509, 2025.
https://doi.org/10.1007/978-3-031-80366-6

Alt Penedès-Garraf). The goal is to identify patients who experienced a stroke and had prior atrial fibrillation, determining the first occurrence date of atrial fibrillation before each stroke. An example is shown in Fig. 1.

Initially, a dictionary approach using regular expressions structures the information. Then, a CNN-BiLSTM-CRF architecture disambiguates entities and relates precursor episodes to detected disease episodes. This architecture has shown promise in capturing complex patterns and relationships present in textual clinical data [4, 5].

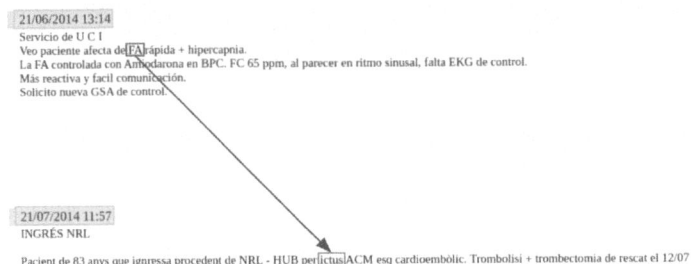

Fig. 1. Information extracted from clinical courses. After detecting atrial fibrillation (FA) prior to a stroke, it is associated with the subsequent stroke immediately (ictus) while retaining the date of onset of both.

The developed pipeline achieved a 0.994 accuracy in identifying atrial fibrillations preceding strokes in a cohort of 300 reports, proving to be an effective alternative to manual extraction.

2   Research Methodology

The methodology for this work follows different phases: Entity detection, NER disambiguation and Relations extraction:

Entity Detection: The goal was to identify terms indicating atrial fibrillation episodes, specifically those coded as I48 in ICD-10. Initial dictionaries included 26 relevant acronyms and variations of the disease's names in Catalan and Spanish. Preannotation of 100 clinical courses with 193 regex expressions and Levenshtein distance addressed misspellings, although some false positives and ambiguities persisted. A NER model was then used for disambiguation. Additionally, some detected terms were negated or referred to as antecedents.

NER Disambiguation: To address the issue of detecting entities that did not refer to atrial fibrillation, a disambiguation process was applied. This involved selecting a context window of five words before and after each detected entity. The task was approached as a NER problem, tagging each acronym by its specific meaning. The architecture used for this task combines CNN, BiLSTM and CRF layers. The CNN extracts local patterns from the text, while the BiLSTM captures temporal dynamics and dependencies. The CRF layer then models transitions between labels to enhance the accuracy of entity recognition [5]. In Fig. 2, a schematic of the CNN-BiLSTM-CRF architecture can be observed.

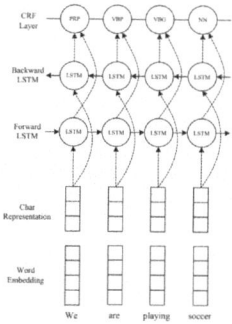

Fig. 2. CNN-BiLSTM-CRF architecture [6].

Relations Extraction: The structured and refined clinical course information

documented episodes of atrial fibrillation and their dates, correlating them with subsequent stroke episodes. The model linked atrial fibrillation episodes to the next recorded stroke, ensuring no overlap.

Human in the loop: Annotators labeled part of the dataset, training an initial

model. Subsequent predictions were manually validated and corrected, improving accuracy iteratively. Documents were split into chunks, filtered into positive and negative detections, and processed through multiple iterations, achieving 99% accuracy. A custom negation algorithm and regex searches for background detection refined the final dataset. The final results were sent to the hospital and the report was validated by experts that analyzed the predictions.

3   Results and Discussion

3.1   Dataset

The dataset used for this study consisted of clinical course documents, which include all daily annotations for patients' hospital stays. These documents were written in both Catalan and Spanish, often mixing languages within the same document or paragraph. To streamline the process, a single model was trained to handle both languages due to their linguistic similarities and the computational cost of maintaining separate models.

A list of all stroke episodes for each patient was provided by the hospital and the occurrences of "fa" were labeled by specialized annotators according to their meanings. The dataset included 3,200 clinical courses, with 1,788 predominantly in Spanish and 1,412 in Catalan. These documents were divided into chunks, each associated with a date. Later, the dataset was split into a training group and a test group. The training group consisted of 2,900 progressively annotated clinical courses, while the test group included 300 clinical courses (150 in Spanish and 150 in Catalan) containing information about stroke patients, evaluated to achieve the results.

3.2   Results

The developed algorithm effectively extracted entities from clinical texts in both Catalan and Spanish, accurately identifying atrial fibrillation episodes and tracking subsequent strokes. The BiLSTM-CNN-CRF neural network achieved a 97% accuracy in disambiguating ambiguous terms.

Results were evaluated based on stroke episodes, marked as 1 if atrial fibrillation occurred before a stroke, and 0 if not. Among 358 stroke episodes in 300 reports, 354 were correctly analyzed in relation to atrial fibrillation. To evaluate these results, the clinical annotators manually reviewed each predicted report of the test group, and the detailed results can be consulted in Table 1.

Table 1. Metrics. In the 'Disambiguation' column, metrics regarding disambiguation are represented. In the 'AF before ictus no disambiguation' column, metrics regarding the final detection of atrial fibrillation episode before stroke without the application of the disambiguation model can be seen. On the other hand, in the 'AF before ictus disambiguation (CNN+BiLSTM+CRF)' column, metrics regarding the final detection of atrial fibrillation episode before stroke with the application of the disambiguation model can be seen.

Metrics	Disambiguation	AF before ictus no disambiguation	AF before ictus disambiguation (CNN+BiLSTM+CRF)
Precision	0.971	0.681	0.994
Recall	0.970	0.679	0.982
F1-Score	0.971	0.680	0.988

4   Conclusion and Future Work

This study involved annotating a clinical dataset to detect atrial fibrillation in extensive patient histories. Using a human-in-the-loop approach, the dataset was manually labeled after an initial regex preannotation, refining the process until the CNN-BiLSTM-CRF model achieved 99.4% accuracy after clinical validation.

The success of this approach suggests its potential for safe deployment in production. It could revolutionize hospital data mining practices, enabling the exploitation of historical clinical records and the tracking of diseases and their precursors.

References

1. Sandoval, A.M., et al.: Biomedical term extraction: NLP techniques in computational medicine. Int. J. Interact. Multimedia Artif. Intell. (2018, InPress)
2. Benjamin, E., et al.: Independent risk factors for atrial fibrillation in a population-based cohort the Framingham heart study. J. Am. Med. Assoc. **271**, 840–844 (1994)
3. Ashburner, J., et al.: Natural language processing to improve prediction of incident atrial fibrillation using electronic health records. J. Am. Heart Assoc. **11** (2022)
4. Chiu, J.P.C., et al.: Named entity recognition with bidirectional LSTM-CNNs. Trans. Assoc. Comput. Linguist. **4**, 357–370 (2015)
5. Kocaman, V., Talby, D.: Biomedical named entity recognition at scale. In: Del Bimbo, A., et al.: ICPR 2021. LNCS, vol. 12661, pp 635–646. Springer, Cham (2021). https://doi.org/10.1007/978-3-030-68763-2_48
6. Ma, X., et al.: End-to-end sequence labeling via bi-directional LSTM-CNNs-CRF. CoRR, abs/1603.01354 (2016)

Evaluating State-of-the-Art Extractive Summarization Methods for Brazilian Portuguese

Germano Antonio Zani Jorge[1]([✉]), Davi Alves Bezerra[1], Clarissa Castell[2],
and Thiago Alexandre Salgueiro Pardo[1]

[1] Instituto de Ciências Matemáticas e de Computação, Universidade de São Paulo,
Avenida Trabalhador São-carlense, 400, São Carlos, SP 13566-590, Brazil
[2] SiDi, Rua Aguaçu, 171 - Ed., Jacarandá - Campinas, SP, Brazil

Abstract. Automatic Text Summarization is an important task for managing large volumes of information. Summarization methods include extractive, which select text passages directly from the source(s), and abstractive ones, which generate new sentences from the original content [10]. Some current state-of-the-art extractive single-document summarization methods like PreSumm [6], HiStruct+ [12], and MemSum [4] have shown success in English corpora but are under-explored for other languages. In this paper, we test such methods for Brazilian Portuguese, considered a low-resource language. Besides the traditional ROUGE informativeness evaluation metric [5], we also use the more recent BLANC metric [13], which assesses summary fluency and coherence. We show that the methods also perform well for Portuguese.

Keywords: Extractive Summarization · Natural Language Processing · Brazilian Portuguese

1 Introduction

Text Summarization is a fundamental task in Natural Language Processing (NLP) and is essential for handling large amounts of information. The task aims to condense information from one or more source documents, producing summaries that preserve the main information of the original content [7].

Overall, summarization can be classified into two main types: extractive and abstractive. Extractive summarization selects sentences or phrases directly from the original text(s), while abstractive summarization generates new sentences that capture the original meaning of the text(s) [10]. Additionally, summarization can be categorized as single or multi-document, depending on how many input documents are processed [3].

Despite the recent popularity of Large Language Models (LLMs) [1] and their use for abstractive summarization and other NLP tasks, this paper focuses on extractive single-document summarization. The reasons are that LLM-based methods present an explainability deficit, require expensive computational infrastructure, and are subject to hallucination, which may hinder the proposal

L. Correia et al. (Eds.): IBERAMIA 2024, LNCS 15277, pp. 510–514, 2025.
https://doi.org/10.1007/978-3-031-80366-6

of preserving the original meaning of the texts. Moreover, some more recent extractive summarization methods have shown good performance (mainly for the English language), without the above LLM issues. Some of the state-of-the-art methods include PreSumm [6], Histruct+ [12] and MemSum [4].

To evaluate such methods, the area has traditionally used ROUGE (Recall-Oriented Understudy for Gisting Evaluation) informativeness metric [5], based on the overlap of n-grams among automatic and reference (human-produced) summaries. However, ROUGE is limited as it only computes superficial text similarity, without considering the semantic quality and coherence of the summaries. To address these limitations, this work also employs BLANC [13], a more recent metric that evaluates the fluency and coherence of the generated summaries by using pre-trained language models to perform the token masking task alongside the summaries generated, trying to predict the next token with the help of the summaries and then calculating their helpfulness.

Our contributions include testing the state-of-the-art extractive summarization methods for Brazilian Portuguese[1] as well as some classical baseline methods. We show that the methods also perform well for Portuguese.

This paper is organized as follows: Sect. 2 describes the methodology, Sect. 3 presents the main results, and Sect. 4 concludes this paper.

2   Methodology

For running the experiments, we selected the well-known **CSTNews** [2] dataset. **CSTNews** is a Brazilian Portuguese news corpus with 150 texts (collected from various Brazilian media sources) and their respective human-produced summaries.

The experiments consisted of model training for PreSumm, MemSum, and Histruct+ methods and running the algorithms for the classical TextRank [8] and Centroid-embedding [11] methods.

We evaluated the models using **ROUGE** [5] and **BLANC** [13] metrics. We used one of the most commonly used ROUGE variants, ROUGE-1, which computes unigram (individual words) overlapping between automatic and reference summaries.

3   Results

Table 1 presents the evaluation results of the summarization methods on the CSTNews corpus. One may notice that the Centroid-embedding model exhibits the highest BLANC score (57.74), significantly outperforming other models, which indicates its superior capability in generating coherent summaries in the Portuguese language, but it does not perform well according to ROUGE, which indicates PreSumm as the best method (55.55). Table 2 shows a summary generated by the Centroid-embedding method.

---

[1] Brazilian Portuguese is the native language of the authors.

Table 1. Summarization results

Model	BLANC	ROUGE-1
PreSumm	25.66	55.55
MemSum	41.31	51.48
HiStruct+	26.01	54.58
TextRank	26.57	46.47
Centroid-embedding	57.74	44.79

Table 2. Comparison between an original text and a centroid-embedding generated summary

Original Text
A seleção brasileira masculina de vôlei conseguiu, nesta sexta-feira, a sétima vitória consecutiva na Liga Mundial ao derrotar a Finlândia por 3 sets a 0 - parciais de 25/17, 25/22 e 25/21 -, em jogo realizado na cidade de Tampere, na Finlândia. Invicto na competição, o Brasil está tranqüilo na liderança do Grupo B. Os finlandeses estão na terceira colocação, com três vitórias e quatro derrotas. Portugal e Argentina - que duelam duas vezes neste final de semana, em Portugal - completam a chave. O time comandado pelo treinador Bernardinho só encontrou um pouco mais de dificuldades no segundo set. No terceiro, mesmo com vários reservas como o levantador Marcelinho e Samuel, os brasileiros conseguiram fechar a partida com tranqüilidade. Brasil e Finlândia se enfrentarão novamente neste sábado, às 12h30 (horário de Brasília), com transmissão ao vivo do canal de TV a cabo SporTV. Nas duas últimas rodadas da fase de classificação da Liga Mundial, a seleção brasileira receberá portugueses e finlandeses. A fase final da competição deste ano acontecerá na Rússia.
Centroid-embedding Generated Summary
A seleção brasileira masculina de vôlei conseguiu, nesta sexta-feira, a sétima vitória consecutiva na Liga Mundial ao derrotar a Finlândia por 3 sets a 0 - parciais de 25/17, 25/22 e 25/21 -, em jogo realizado na cidade de Tampere, na Finlândia.Portugal e Argentina - que duelam duas vezes neste final de semana, em Portugal - completam a chave.

It is possible to conclude that some summarization methods perform well for Portuguese, but more investigation is necessary to identify the best one, which also requires a deeper look at the evaluation metrics. In the end, human evaluation may be necessary to select the best method (at a higher cost).

4   Final Remarks

We have evaluated some state-of-the-art extractive methods for Brazilian Portuguese. In the future, we intend to evaluate more methods, including other extractive ones, as well as to consider LLM-based methods in order to determine the impact that they may bring to the task (despite their bottlenecks that we cited before). Other corpora may also be used. For instance, we are aware of a recent corpus, RecognaSumm [9], which contains over 130.000 journalistic texts in Brazilian Portuguese, and we plan to perform experiments with it.

Another issue for future research is how such methods perform for other text genres, which, as it is widely known, may significantly influence the results.

Acknowledgments. We express our gratitude to SiDi for funding this research project. The author Thiago Alexandre Salgueiro Pardo also thanks the Center for Artificial Intelligence of the University of São Paulo (C4AI - http://c4ai.inova.usp.br/, funded by the São Paulo Research Foundation – FAPESP grant 2019/07665-4 – and by the IBM Corporation) and the support of the Ministry of Science, Technology and Innovation (Law N. 8,248, of October 23, 1991, within the scope of PPI-SOFTEX, coordinated by Softex and published as Residence in TIC 13, DOU 01245.010222/2022-44).

Disclosure of Interests.. The authors have no competing interests to declare.

References

1. Bommasani, R., et al.: On the opportunities and risks of foundation models. arXiv preprint arXiv:2108.07258 (2021)
2. Cardoso, P.C., et al.: CSTNews-a discourse-annotated corpus for single and multi-document summarization of news texts in Brazilian Portuguese. In: Proceedings of the 3rd RST Brazilian Meeting, pp. 88–105 (2011)
3. El-Kassas, W.S., Salama, C.R., Rafea, A.A., Mohamed, H.K.: Automatic text summarization: s comprehensive survey. Expert Syst. Appl. **165**, 113679 (2021)
4. Gu, N., Ash, E., Hahnloser, R.H.: MemSum: Extractive summarization of long documents using multi-step episodic markov decision processes. arXiv preprint arXiv:2107.08929 (2021)
5. Lin, C.Y.: ROUGE: A Package for Automatic Evaluation of Summaries (2004)
6. Liu, Y., Lapata, M.: Text summarization with pretrained encoders. CoRR abs/1908.08345 (2019)
7. Mani, I.: Automatic Summarization. Natural language processing (2001)
8. Mihalcea, R., Tarau, P.: TextRank: bringing order into text. In: Proceedings of the 2004 Conference on Empirical Methods in Natural Language Processing, pp. 404–411 (2004)
9. Paiola, P.H., et al.: RecognaSumm: a novel Brazilian summarization dataset. In: Proceedings of the 16th International Conference on Computational Processing of Portuguese, pp. 575–579 (2024)
10. Radev, D.R., McKeown, K., Hovy, E.: Introduction to the Special Issue on Summarization (2002)
11. Rossiello, G., Basile, P., Semeraro, G.: Centroid-based text summarization through compositionality of word embeddings. In: Giannakopoulos, G., et al. (eds.) Proceedings of the MultiLing 2017 Workshop on Summarization and Summary Evaluation Across Source Types and Genres, pp. 12–21 (2017)

12. Ruan, Q., Ostendorff, M., Rehm, G.: HiStruct+: Improving extractive text summarization with hierarchical structure information. arXiv preprint arXiv:2203.09629 (2022)
13. Vasilyev, O., Dharnidharka, V., Bohannon, J.: Fill in the BLANC: Human-free quality estimation of document summaries. arXiv preprint arXiv:2002.09836 (2020)

Assessing AI's Persuasive Power: Can an AI Agent Influence Belief in News?

Jean Gabriel Nguema Ngomo[✉] and Ana Cristina Bicharra Garcia

Federal University of State of Rio de Janeiro, Rio de Janeiro - Urca, Brazil
mvojgnn@edu.unirio.br, cristina.bicharra@uniriotec.br

Abstract. The objective of this study is to investigate the ability of AI to influence people in recognizing fake news. We first explored some existing theories from the literature to better understand and characterize how AI might exert such influence. We then conducted an online experiment with fake news related to politics in the Brazilian context. The study showed apparent evidence of AI's impact on people, helping them recognize fake news, even overcoming beliefs of some participants.

Keywords: Artificial Intelligence · Fake news detection · AI influence · AI warnings · AI human interaction

1 Introduction

Today the dissemination of false information on the Internet has been a major issue for society, resulting in negative impacts in several fields, such as politics and health. It has become urgent to face and investigate ways to combat them.

Artificial Intelligence (AI) has played an increasingly important role in society, with countless successful use cases, from intelligent systems for the general public such as chatbots, specific intelligent systems such as those that detect diseases, predict share prices on stock exchanges, to name a few examples. In this sense, several approaches seek to automatically recognize Fake News through AI models. However the effectiveness of these intelligent systems will strongly depend on their use by human beings, who, faced with an AI prognosis regarding a given piece of information, will decide trusting or not in the fruit of an AI's reasoning. In this context our research question is: How can AI influence peoples perception of Fake News? Our research hypothesis is that Intelligent systems can influence the way people perceive the veracity of news.

2 Background

In this study we explored the principles of persuasion proposed by Cialdini [1], which considers the persuasion to be a form of social influence, defined as a change in one's beliefs, behavior, or attitudes due to external pressure that may be real or imagined [1]. Our main focus is on the authority principle, which

© The Author(s), under exclusive license to Springer Nature Switzerland AG 2025
L. Correia et al. (Eds.): IBERAMIA 2024, LNCS 15277, pp. 515–519, 2025.
https://doi.org/10.1007/978-3-031-80366-6

stipulates people are more likely to be persuaded by information that comes from an authority, such as an expert on a particular subject, especially in uncertainty cases. AI has all the characteristic of an authority because it is known as an expert, and as so it can therefore exercise its social influence, convincing people to change their attitudes or opinions. On the other hand, beliefs and attitudes towards important societal issues, such as political issues, can be examined from the social judgement theory (SJT) [3]. According to SJT, any individual has a certain stand regarding the message topic, conceptually known as his anchor [3], consisting in one of the three latitude zones [3]: latitude of acceptance, where a message is received with positive regard, accepting it; latitude of rejection, where a message is received with skepticism and negative regard, rejecting it; and latitude of non-commitment, where the message is received with neutrality or lack of commitment [2, 3].

3   Methodology

To investigate our hypothesis, we conducted an online experiment, where people gave their opinions on the veracity of a sample of Brazilian political-related news, considering veracity prediction on these news providing by a AI model. For this purpose, we built a web system, fed with news snippets extracted from two of the most known Brazilian fact-checking agencies: "G1:Fato ou Fake" and "Lupa"[1]. Moreover, we implemented a simple AI classifier building a two-layer neural network using the Keras framework. Thus, each news fed in our experiment system had not only its label (false or true) but also its AI verdict (false or true). This web system was made available for two weeks in our research institution web site so that people can participate anonymously.

Participants had to assess two different news, randomly selected from the system sample. They were presented with the news and asked to determine their content's veracity. Then after presenting the AI agent's verdict on the same information, we asked participants if they had changed their first judgment. By analyzing their willingness to change their opinion in light of the AI's input, we explore whether personal beliefs - specifically on polarizing topics such as abortion decriminalization, gun control, and privatization of public companies-affect susceptibility to persuasion by AI agents. Finally, participants were also exposed to crowds opinion on the same news, where they could change their opinion again. Here we explore a well-known evidence from literature that the crowd breeds wisdom [7], which usually leads people to conforms to opinion of majority, verifying if participants are more persuade by AI than crowd.

Before news assessment, each participant had to provide some demographics information (gender range, age range, education level, income range), as well as his belief profiling by informing his position on two polarizing topics ("Agree", "disagree", "no opinion"). This information defined his anchor. We make the

---

[1] https://g1.globo.com/fato-ou-fake/ and https://lupa.uol.com.br/.

tools built in this study available on github: experiment's web system and AI-based veracity news classifier[2].

4   Results and Discussion

In total, 78 people participated in our experiment, 53% female (F) and 47% male (M). From the participants' response, we define accuracy as the number of times news items were correctly classified by all participants among the total number of expected correct classification. This accuracy increased from 66% (the case of initial opinion on news without AI) to 70%, participants opinions under AI predictions. This improvement in accuracy in recognizing the veracity of the news in the presence of AI suggests that the AI was perceived by participants as an authority, an expert who can be trusted to properly classify news as false or real. These evidences are in line with reports in the literature as in [4, 6].

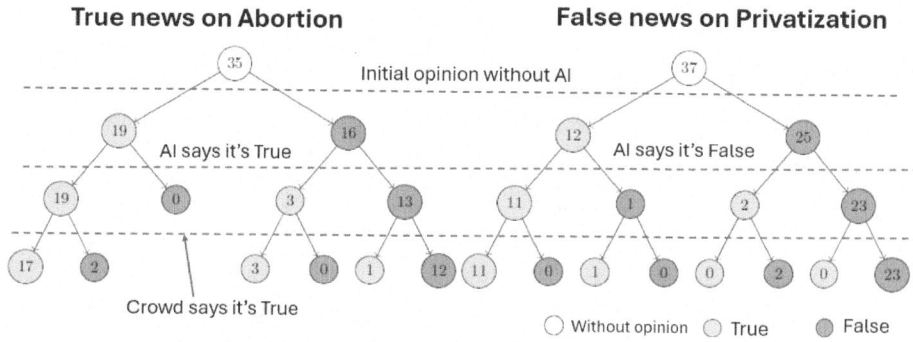

Fig. 1. Analysis of the participant's attitude in two news

In Fig. 1, we illustrate the impact of AI and crowd on participants' anchors, tracking their answers in binaries trees for two news randomly assigned to some participants. The root contains the number of tasks that used a given news, while intermediate nodes capture the number of true (yellow) or false (orange) answers in each step. More specifically, a true news on abortion legalization (T1) was answered 35 times. Since This news brings negative information about the legalization of abortion, a large majority of those whose agreed with T1 initially classified this news wrongly as false (16 times), what was according to their anchors. Next, faced with the opposite diagnosis of AI, 3 participants aligned themselves with AI, but a 13 remained within their latitude of rejection zone. Finally, under crowd opinion, 11 of those latter did not change their opinion, showing loyalty to their anchors. In the same Fig. 1, we observe similar behaviors, but with a false news about the privatization of public services (T2).

---

[2] https://github.com/JeanGabrielNguemaN/AIBeliefsInfluence.

This tracking of participants' behavior highlighted AIs influence on participants, suggesting that, when playing this role of authority, AI has the potential to overcome people's beliefs (their anchors), to align them with the veracity of the news, or to align beliefs of many with truth about the news.

With respect to demographics, our results suggest older people are more resistant to align themselves with AI than younger people. This can be understood by the fact that new generations are more immersed in technology than older people. The results obtained do not allow us to say that more educated people are more influenced by AI, nor are people with higher incomes. However, as our sample was small, other studies are needed to verify these findings.

5 Conclusion

In our study AI acted as a social actor, an authority capable of persuading people to recognize fake news. The presence of AI prediction along with news helped participants to have more accurate opinions on news veracity. A reasonable number of participants changed their opinion, aligning themselves with the AI's prognosis, even renouncing their beliefs. For others, AI served to confirm their beliefs, but also correctly classify the news. The AI also outperformed slightly the crowd with respect to opinion changing, already known for generating intelligent consensus, thus confirming the AI influence on the participants. We found no significant differences in the impact of AI on recognizing fake news by gender, education level or income.

The novelty of our approach lies in the combination of belief profiling and AI interaction to assess its trust. While much of the literature focuses on AI's technical accuracy or ethical implications, we delve into the psychological and social dimensions, questioning how deeply entrenched beliefs interact with AI's authority in truth verification. Moreover, our research sets the stage for future experiments to introduce explainability features to AI agents, examining whether explaining AI decisions strengthens user trust and further influences belief revision [5]. By addressing these questions, we aim to uncover key drivers of AI trustworthiness and inform the development of more effective AI systems in the fight against misinformation.

Acknowledgments.. This study was financed in part by the Coordenaço de Aperfeiçoamento de Pessoal de Nível Superior - Brasil (CAPES) - Finance Code 001.

References

1. Cialdini, R.B., Goldstein, N.J.: The science and practice of persuasion. Cornell Hotel Restaur. Adm. Q. **43**(2), 40–50 (2002)
2. Darity, W.: Social Judgment Theory. Macmillan, Detroit (2008)
3. Hovland, C., Sherif, M.: Social Judgment: Assimilation and Contrast Effects in Communication and Attitude Change. Greenwood, Westport (1980)

4. Lu, Z., Li, P., Wang, W., Yin, M.: The effects of AI-based credibility indicators on the detection and spread of misinformation under social influence. Proc. ACM Hum. Comput. Interact. **6**, 1–27 (2022)
5. Reis, J.C.S., et al.: Explainable machine learning for fake news detection. In: Proceedings of the 10th ACM Conference on Web Science (2019)
6. Shabani, S., Sokhn, M.: Hybrid machine-crowd approach for fake news detection. In: IEEE 4th International Conference on CIC, pp. 299–306 (2018)
7. Zhi, D., Xing, H.: Harnessing the wisdom of crowds. Manage. Sci. **66**(5), 1847–1867 (2020)

Author Index

A

Acevedo, Daniel 132
Adamatti, Diana 360
Aguilera, Nicolás 147
Andrade-Ibarra, Yael 453
Archila, John 110
Arena, Melisa 481

B

Backes, André Ricardo 3
Baqueta, Jeferson José 372
Barbeira, Pilar Barcena 424
Barone, Dante Augusto Couto 323
Batard Lorenzo, David 161
Bautista, Susana 458
Becerra, Israel 197
Belo, Orlando 335
Belzarena, Diego 122
Benavides, Facundo 221
Berry, Andrew 74
Betancourt, Nicolas 50
Bezerra, Davi Alves 510
Billa, Cleo 360
Bisso, Indalecio Carboni 132
Bof, Patricia 323
Botti, Vicente 26
Bottrighi, Alessio 237
Burgos, Valeria 132

C

Callejas, Sofía 74
Camargo, Jorge E. 472
Campos, Bruno 496
Canul-Reich, Juana 62
Castell, Clarissa 510
Castro, Graciana 122
Ceballos, Ignacio Fernández 132
Celis, Luis Fernando 99
Chiappone, Francesco 14
Cifuentes, Eiker J. 14

Cotik, Viviana 132, 424
Courtois, María Florencia 132
Couto, Javier 387
Cruz-Mendoza, Mariana-Carolyn 62
Cuate, Oliver 453
Cutrona, Nicolas 310

D

Dantas, Thiago 360
de Oliveira, Alessandro Bof 323
de Wolff, Taco 173
Di Maro, Vicenzo 14
Di Somma, Andrea 14
Díaz, Agustina 147
Dominguez, Enrique 287
Duarte, Ana 335
Duque-Méndez, Néstor D. 463, 490

E

Ernst, Erik 400
Espinosa-Bedoya, Albeiro 262
Estradé, María José 147

F

Farias, Giovani 360
Fariello, María Inés 122
Federico, Bliman 209
Fernández-Rodríguez, Jose David 287
Fuentes, Yhon 38

G

Gálvez, Ramiro H. 400
Garcia, Ana Cristina Bicharra 515
García-Ojeda, Juan C. 347
Gervás, Pablo 458
Giraldo-Cadavid, Álvaro J. 463
Giret, Adriana 26
Gonçalves, Eder 360
González-McMahon, Carlos M. 463
Gravano, Agustín 400

Guillot, Dominique   310
Gulín González, Jorge   161

H
Heras, Marcos Las   132
Hernández, Mateo R.   14
Hoffman, Romina   122
Hornos, Miguel J.   496

J
Jiménez-Builes, Jovani Alberto   262, 463,
    490
Jorge, Germano Antonio Zani   510
Juarez Juarez, Maria Gabriela   26

K
Kawulok, Michal   185
Kuban, Patryk   185
Kumar, Mayuresh   458

L
Lara, Adriana   453
Laukyte, Migle   439
Laurenzo, Tomas   501
Lecumberry, Federico   122, 147
León, Carlos   458
Lira, Hernan   74, 173
Llobet, Ernest Valveny   505
Lockhart, Carolina   132
Loera-Ponce, Jesús Alejandro   197
Lopes, Victor Hugo Schneider   323

M
Machado, José   298, 476
Maconi, Antonio   237
Malvaez-Hernández, Carolina   62
Manzanera, Antoine   110
Marcela Aldana, Leidy   472
Marcondes, Francisco S.   298
Maris, Frédéric   468
Martí, Luis   74, 173
Martin, Llofriu   209
Martínez Pazos, Jorge Felix   161
Martínez, Fabio   99, 110
Martínez-Santos, Juan Carlos   14
Mastropasqua, Nicolas   132
Medeiros, Gonçalo   298
Meléndez-Armenta, Roberto Ángel   62
Méndez, Gonzalo   458

Mercado-Ravell, Diego A.   197
Moncecchi, Guillermo   387, 412
Monzón, Pablo   221
Mordecki, Gabriel   387
Moreno-López, Gustavo A.   463
Morveli-Espinoza, Mariela   250
Musé, Pablo   147
Musitelli, Mateo   122
Muvdi, David   14

N
Nera, Stefano   237
Ngomo, Jean Gabriel Nguema   515
Nieves, Juan Carlos   250
Novais, Paulo   298, 476

O
Oliveira, Pedro   298
Olivera, Belén   481
Olmos, Juan   99
Orellana García, Arturo   161
Ortega, Gonzalo Hernández   505

P
Pablo, Monzon   209
Pardo, Thiago Alexandre Salgueiro   510
Pavón, Juan   476
Pérez, Santiago   481
Petri, Javier   424
Piovesan, Luca   237
Pozo, Alex   486
Puertas, Edwin   14

Q
Querol, Juli Climent   505
Quijano-Cabezas, Pablo Andres   490

R
Raina, Erica   237
Ramos, Cristina   486
Restrepo-Carmona, Jaime A.   463
Ribeiro, João Batista   3
Ribeiro, Tiago   476
Risk, Marcelo   132
Rocha, Anderson   38
Rodrigues, Manuel   476, 496
Rodríguez-Domínguez, Carlos   496
Romay, Paola   481

Rosá, Aiala   412
Rován, Ernesto   221

S

Sabbia, Victor   481
Sanchez-Pi, Nayat   74, 173
Sastre, Ignacio   412
Serrano-Pérez, Jonathan   275
Soriano-Vargas, Aurea   38
Sucar, L. Enrique   275
Surribas-Sayago, Gonzalo   287

T

Tacla, Cesar A.   372
Tarsia, Luciano   132
Terenziani, Paolo   237
Tous-De la Ossa, Eduardo J.   463
Trejo-Ramirez, Uriel   453

U

Uribe, Lourdes   453
Urquizo, Gladys   486

V

Valentin-Coronado, Luis Manuel   197
Velasco, Mauricio   50
Villacr, Juan   486
Villacrés, Juan   87

Y

Yari, Yessenia   38

Z

Zapata-Medina, Daniel   262
Zimmer, Sofía   147
Zuluaga-Morillo, Juan C.   463

The manufacturer's authorised representative in the EU is Springer
Nature Customer Service Centre GmbH, Europaplatz 3, 69115 Heidelberg,
Germany. If you have any concerns regarding our products, please
contact ProductSafety@springernature.com

Printed and bound by CPI Group (UK) Ltd, Croydon, CR0 4YY
29/04/2026
02099546-0006